Composite Construction

Composite Construction

Edited by David A. Nethercot

CRC Press
Taylor & Francis Group
Boca Raton London New York

CRC Press is an imprint of the
Taylor & Francis Group, an **informa** business

A CHAPMAN & HALL BOOK

CRC Press
Taylor & Francis Group
6000 Broken Sound Parkway NW, Suite 300
Boca Raton, FL 33487-2742

First issued in paperback 2019

© 2003 by Taylor & Francis Group, LLC
CRC Press is an imprint of Taylor & Francis Group, an Informa business

Publisher's Note
Part of this book has been prepared from camera-ready-copy supplied by the authors

ISBN-13: 978-0-415-24662-0 (hbk)
ISBN-13: 978-0-367-44669-7 (pbk)

This book contains information obtained from authentic and highly regarded sources. Reasonable efforts have been made to publish reliable data and information, but the author and publisher cannot assume responsibility for the validity of all materials or the consequences of their use. The authors and publishers have attempted to trace the copyright holders of all material reproduced in this publication and apologize to copyright holders if permission to publish in this form has not been obtained. If any copyright material has not been acknowledged please write and let us know so we may rectify in any future reprint.

Except as permitted under U.S. Copyright Law, no part of this book may be reprinted, reproduced, transmitted, or utilized in any form by any electronic, mechanical, or other means, now known or hereafter invented, including photocopying, microfilming, and recording, or in any information storage or retrieval system, without written permission from the publishers.

For permission to photocopy or use material electronically from this work, please access www.copyright.com (http://www.copyright.com/) or contact the Copyright Clearance Center, Inc. (CCC), 222 Rosewood Drive, Danvers, MA 01923, 978-750-8400. CCC is a not-for-profit organization that provides licenses and registration for a variety of users. For organizations that have been granted a photocopy license by the CCC, a separate system of payment has been arranged.

Trademark Notice: Product or corporate names may be trademarks or registered trademarks, and are used only for identification and explanation without intent to infringe.

British Library Cataloguing in Publication Data
A catalogue record for this book is available
from the British Library

Library of Congress Cataloging in Publication Data
Nethercot, D.A.
Composite construction / David A. Nethercot.
p.   cm.
Includes bibliographical references and index.
ISBN 0-415-24662-8
1. Composite construction.   2. Composite materials.   I. Title.
TA664 .N48 2003
620.1′18—dc21
                                                    2002042805

Visit the Taylor & Francis Web site at
http://www.taylorandfrancis.com

and the CRC Press Web site at
http://www.crcpress.com

Contents

Contributors   viii
Foreword   ix
Acknowledgments   xi

1 Fundamentals   1
DAVID A. NETHERCOT

1.1 Introduction 1
1.2 History 1
1.3 Basic concepts 3
1.4 Material properties 5
1.5 Shear connectors 6
1.6 Design for ULS 16
1.7 Design for SLS 17
1.8 Composite systems 18
1.9 Current usage 18
1.10 Concluding remarks 20
1.11 References 20

2 Composite Beams   23
HOWARD D. WRIGHT

2.1 Introduction 23
2.2 Types of beam 24
2.3 Basic behaviour 28
2.4 Ultimate strength design 31
2.5 Calculating the deflection 34
2.6 Shear connector behaviour 35
2.7 Continuous beams 40
2.8 Beams with composite slabs 43
2.9 Current design and future development 44
2.10 References 46

3 Composite Columns   49
YONG C. WANG

3.1 Introduction 49
3.2 Composite columns under axial load in cold condition 51
3.3 Composite column under combined axial load and bending moments at ambient temperature 58
3.4 Effect of shear 68
3.5 Load introduction 68
3.6 Composite columns in fire conditions 68
3.7 Summary 76
3.8 Acknowledgement 77
3.9 References 77
3.10 Notations 78

4 Instability and Ductility   81
ALAN R. KEMP

4.1 Introduction and elastic buckling theory 81
4.2 Ultimate resistance of composite columns 90
4.3 Continuous composite beams 91
4.4 Ductility considerations for compact beams 101
4.5 References 116

5 Composite Floors   121
J. BUICK DAVISON

5.1 Introduction 121
5.2 Current practice 122
5.3 Behaviour as formwork 124
5.4 Composite behaviour 130
5.5 Dynamic behaviour 146
5.6 Concentrated loads and slab openings 147
5.7 Fire resistance 149
5.8 Diaphragm action 153
5.9 Slim floor decking 154
5.10 References 157

6 Composite Connections   161
DAVID B. MOORE

6.1 Introduction 161
6.2 Types of composite connections 163
6.3 Design principles 165

    6.4   *Classification of composite connections 167*
    6.5   *Capacity of composite connections 169*
    6.6   *Ductility of composite connections 185*
    6.7   *Stiffness of composite connections 194*
    6.8   *Summary 199*
    6.9   *References 200*

7 Composite Frames   203
GRAHAM H. COUCHMAN

    7.1   *Introduction 203*
    7.2   *Principles of frame behaviour 207*
    7.3   *Frame analysis and design 221*
    7.4   *Design using software 229*
    7.5   *Conclusions 232*
    7.6   *References 233*

Index   235

Contributors

Graham H. Couchman
Steel Construction Institute
Silwood Park
Buckhurst road
Ascot
Berks
SL5 7QN

J. Buick Davison
The University of Sheffield
Department of Civil & Structural
  Engineering
Sir Frederick Mappin
  Building
Mappin Street
Sheffield
S1 3JD

Alan R. Kemp
Faculty of Engineering
University of Witwatersrand
1 Jan Smuts Avenue
Johannesburg
2001 South Africa

David B. Moore
BRE
PO Box 202
Watford
Herts
WD2 7QG

David A. Nethercot
Imperial College
Department of Civil & Environmental
  Engineering
London SW7 2AZ

Yong C. Wang
School of Civil Engineering
University of Manchester
Oxford Road
Manchester
M13 9PL

Howard D. Wright
University of Strathclyde
James Weir Building
75 Montrose Building
Glasgow
G1 1XJ

Foreword

Composite Construction has developed significantly since its origins approximately 100 years ago when the idea that the concrete fire protection around columns might be able to serve some structural purpose or that the concrete bridge deck might, with advantage, be made to act in conjunction with the supporting steel beams was first proposed. Take-up in practice and began in earnest shortly after the end of the Second World War and progress has been particularly rapid during the past 20 years. Indeed, it is now common to ask, "Why is this not acting compositely?" when looking to improve the efficiency of a structural steelwork design. In those countries where steelwork enjoys a particularly high market share e.g. for high-rise buildings in the UK and Sweden, the extensive use of composite construction is a major factor.

Early approaches to the design of composite structures generally amounted to little more than the application of basic mechanics to this new system. However, it was soon realised that this particular medium possessed features and subtleties of its own and that effective usage required that these be properly understood and allowed for. Composite construction is now generally regarded as a structural type in its own right, with the attendant set of design codes and guidance documents. The most comprehensive and up to date of these is the set of Eurocodes—specifically EC4 that deals exclusively with composite construction. It is not the purpose of this textbook to serve as a commentary on the Eurocodes. Rather, it is an explanatory and educational document, presenting the technical basis for many of the newer concepts, design procedures and applications of composite construction in buildings. Inevitably, it makes some reference to the Eurocodes but only in the sense that their procedures often represent formal statements of the most appropriate simplified implementation of our current understanding. For convenience and consistency it adopts their notation.

The authors—each an acknowledged expert in the topic on which they have written—have selected their own way of presenting the subject matter. In all cases the intent has been to share the technical basis and background to design so that extrapolation and intelligent use beyond the obvious is possible. The book is not claimed to be comprehensive or to represent a full state of the art. It should be regarded as helpful background reading for all those wishing to acquire a better appreciation and understanding of the major developments in the use of composite construction for building structures.

The first Chapter of this book traces the key historical steps in the development and understanding of Composite Construction and introduces the main fundamental features. The next two deal with basic elements—horizontal beams and vertical columns—showing how the combined action of the concrete and the steel member may be synthesised to give a more efficient load resisting arrangement. A relatively new development is the deliberate use of composite action in beam to column connections, thereby requiring them to be treated as partial strength and semi-rigid for design purposes as explained in Chapter 4. Because buckling is a key item when dealing with the response of steel members, its importance for composite elements—especially beams—is then considered in some detail. Building floor systems now often comprise arrangements with two-way spanning composite action and several such arrangements are discussed in Chapter 6. The final Chapter deals with the interaction of beams, columns and joints in presenting a complete treatment for the design of non-sway composite frames that recognises the actual behaviour more closely than does conventional treatments based on consideration of individual components.

This book is collaborative effort, with all the Chapter authors having made an equal contribution. Its preparation has inevitably involved delivery against deadlines and the required instructions. My thanks to Howard, Yong, David, Buick, Alan and Graham for their patience and cooperation. Production has benefited from the firm but sympathetic guidance of the publishers—particularly Alice Hudson. The coordination and final preparation of the manuscript was just one of the tasks handled so efficiently by my PA Alice Kwesu.

David A. Nethercot

Acknowledgments

Considerable effort has been made to trace and contact copyright holders and secure replies prior to publication. The authors apologise for any errors or omissions.

Extracts from Eurocode 4, Eurocode 3 and BS 5950 Part 3: 1990 are reproduced with the permission of BSI under licence number 2002SK/0204. Eurocodes and British Standards can be obtained from BSI Customer Services, 389 Chiswick High Road, London W4 4AL. (Tel+44 (0) 20 8996 9001).

Figures from Steel Construction Institute publications are reproduced with kind permission from the Steel Construction Institute.

Acknowledgments are also required for the following:

Chapter One—Fundamentals
David A. Nethercot

Figure 1.1 reproduced with kind permission from the ASCE from: Moore, W.P., Keynote Address: An Overview of Composite Construction in the United States, Composite Construction in Steel & Concrete, ed. C.D. Buckner & I.M. Viest, Engineering Foundation, 1988, pp. 1–17.

Figures 1.2 and 1.3 reproduced from: David A. Nethercot, Limit States Design of Structural Steelwork, Spon Press.

Figures 1.4, 1.5, 1.6 and 1.8 reproduced from: Johnson, R.P., Composite Structures of Steel & Concrete Volume 1 Beams, Slabs, Column & Frames for Buildings, 2nd edition, Blackwell Scientific Publications.

Figure 1.16 reproduced from: Lam, D., Elliott, K.S & Nethercot, D.A., Structures and Buildings, ICE Proceedings

Chapter Two—Composite Beams
Howard D. Wright

Figure 2.4 reproduced from: Mullett, D.L., Composite Floor Systems, Blackwell Science Ltd.

Chapter Three—Composite Columns
Yong C. Wang

Tables 3.4, 3.5 and 3.6 are reprinted from Journal of Constructional Steel Research, 51, Kodur, V.K.R., Performance-based fire resistance design of concrete-filled columns, pp. 21–36, 1999, with permission from Elsevier Science.

Chapter Five—Composite Floors
J. Buick Davison

Figure 5.20 reproduced from: Composite Slab Behaviour and Strength Analysis. Part 1 calculation Procedure, Daniels, Byron J., Crisinel, Michael, Journal of Structural Engineering, Vol. 119, 1993—ASCE.

Figure 5.26 reproduced courtesy of Corus plc.

Chapter Six—Composite Connections
David B. Moore

Figures 6.1, 6.13, 6.16, 6.17 and 6.18 reproduced with kind permission of Building Research Establishment Ltd.

CHAPTER ONE

Fundamentals

David A. Nethercot

1.1 INTRODUCTION

The term "composite construction" is normally understood within the context of buildings and other civil engineering structures to imply the use of steel and concrete formed together into a component in such a way that the resulting arrangement functions as a single item. The aim is to achieve a higher level of performance than would have been the case had the two materials functioned separately. Thus the design must recognise inherent differences in properties and ensure that the structural system properly accommodates these. Some form of interconnection is clearly necessary.

Since its introduction, the utilisation of composite action has been recognised as an effective way of enhancing structural performance. In several parts of the world a high proportion of steel structures are therefore designed compositely. Design codes, textbooks, specialist design guides, descriptions of projects and research papers directed to the topic exist in abundance; many of these are referred to in the present text.

This opening chapter covers the general background to the subject necessary for a proper appreciation of the following six chapters, each of which concentrates on a specific topic. Its coverage is thus much broader and its treatment of particular aspects of composite construction rather more elementary and less detailed than will be found elsewhere in the book. Readers already possessing some knowledge of the subject may well prefer to go directly to the chapter(s) of interest.

Because the book focuses on the use of composite construction in building structures it does not attempt to cover those structural phenomena that are unimportant in this context. Thus items such as fatigue, temperature effects, corrosion, impact response etc., which need to be properly addressed when designing bridges, offshore structures, tunnels, military installations etc., are not included.

1.2 HISTORY

In both the opening Address (1) to the first in the series of Engineering Foundation Conferences on Composite Construction (2–5) and a Keynote paper to a US–Japan Symposium on the subject (6), authors from the same US consulting firm have traced the very early use of composite construction in North America. The year 1894 is stated as the period in which concrete encased beams were first used in a bridge in Iowa and a building in Pittsburgh. The earliest laboratory tests on encased columns took place at Columbia University in 1908, whilst composite beams were first tested at the Dominion Bridge Works in Canada in

1922. By 1930 the New York City building code recognised some benefit of concrete encasement to steelwork by permitting higher extreme fibre stresses in the steel parts of the encased members. Welded shear studs were first tested at the University of Illinois in 1954, leading to publication of a design formula in 1956 and first use for both bridge and building projects that same year. In 1951 a partial interaction theory was proposed, also by the team from Illinois. Metal decks first appeared in the 1950s, with the first recorded use of through deck stud welding on the Federal Court House in Brooklyn in 1960. It was not until 1978, however, that this arrangement was recognised in the AISC specification.

Early usage in Japan has been recorded by Wakabayashi (7), who refers to the use of concrete encasement to improve both fire and earthquake resistance, dating from about 1910. Termed "steel reinforced concrete" or SRC, this form of construction quickly became popular for buildings of more than 6 storeys. Its integrity was demonstrated by the good performance of structures of this type in the great Kanto earthquake of 1923. Research on the topic in Japan did not start until the 1930s; codes came much later, with the first SRC code produced by the Architectural Institute of Japan (AIJ) appearing in 1958. Work on a design guide for composite bridges started in 1952 but the document was not published until 1959. Beginning with the earliest beam tests in 1929 and column tests in the same year, research studies flourished in the 1950s and 1960s, leading to a heavy concentration since then on understanding the behaviour of composite construction when subjected to seismic action.

One of the earliest substantial documents devoted to composite construction was the book by Viest, Fountain & Singleton (8). Published in 1958, it referred to early usage of the technique in the United States by 1935 and a patent by Kahn dated 1926 (Figure 1.1). The book was written to complement the American Association of State Highway Officials (AASHO) 1957 Specification

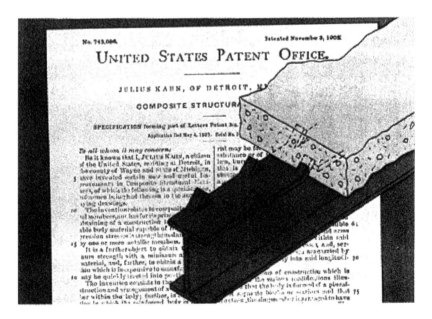

Figure 1.1 Kahn Patent of 1926

Fundamentals 3

that covered the use of composite beams for bridges. In addition to referring to several North American examples of composite bridges constructed in the 1940s and 1950s, it includes an Appendix outlining the use of composite beams in building construction. A full set of Rules covering the design of composite beams was provided in the 1961 American Institute of Steel Construction (AISC) Buildings Specification.

Parallel developments had been taking place in Europe—especially as part of the post-war reconstruction in Germany. Reporting on this in 1957, Godfrey (9) refers to "research in Germany, Switzerland and elsewhere" providing the basis for their "Provisional Regulations for the Design of Girders in Composite Construction" published in July 1950. Four years later the topic was addressed more formally in DIN 1078. In a follow-up paper Sattler (10) reported on numerous examples of the use of composite construction for both bridges and buildings in Germany. This extended to the use of concrete filled tubes as columns but for beams relied on far more complex forms of shear connection than the welded studs which were at that time already being introduced into the United States by the Nelson company. In the discussion to these papers, British researchers Chapman & Johnson referred to both research in progress and buildings under construction that had been designed compositely at Imperial College and Cambridge University. Reports on this work appeared a few years later (11–15), papers on cased stanchions (16), early UK composite bridge applications (17) and background studies for buildings (18) having appeared in the late 1950s.

Thus by the mid 1960s the structural engineering community in the UK was appreciative of the merits of composite construction. It had for example been employed for a number of Government designed buildings (19), essentially in the form of composite beams but with the novel feature that these utilised precast lightweight aggregate concrete panels and planks (20). Early ideas for composite columns that saw the 1948 edition of BS 449 simply permit the radius of gyration to be taken as something greater than that for the bare steel member had, as a result of the 1956 tests (21), been extended to using the full concrete area. The later BRS column tests (22), including considerations of beam-column behaviour, saw the development of soundly based interaction formulae approaches to design. Ref. 22 includes an interesting reference to several very early sets of tests on concrete encased columns conducted during the period 1912–1936. Much of this British work was then brought together into the first comprehensive composite code, CP 117, published in 3 parts (23), covering: simply supported beams in buildings, beams for bridges and composite columns. For building structures this was later replaced by the Part 3 of the limit states based code BS 5950, although only the Part 3.1 dealing with simply supported beams and Part 4 covering composite slabs were ever completed (24).

Fascinating accounts of research into various aspects of composite construction conducted during different parts of the period 1940–1990 in the USA, UK, Japan, Australia and Germany are available in the third Engineering Foundation Volume (4).

1.3 BASIC CONCEPTS

The essence of composite construction is most readily appreciated by considering its most commonly used application, the composite beam. To begin with a very simple illustration, consider the beam consisting of two identical parts shown in

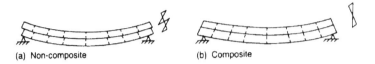

Figure 1.2 Basic mechanics of composite action

Figure 1.2. In the case of Figure 1.2a both parts behave separately and move freely relative to each other at the interface, whilst in the case of Figure 1.2b both parts are constrained to act together. For case a longitudinal slip occurs as indicated by the movement at the ends, whereas in case b plane sections remain plane. It is readily demonstrated using elastic bending theory that case b is twice as strong and four times as stiff as case a. Now consider the steel/concrete arrangement of Figure 1.3a. The two parts are now of different sizes and possess

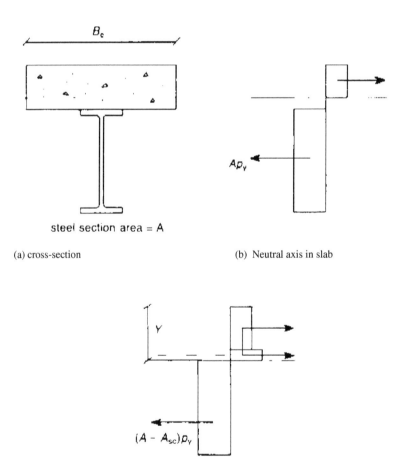

Figure 1.3 Stress block representations

Fundamentals

different stress–strain characteristics. Assuming for the purpose of the illustration that the neutral axis of the composite section is located at the concrete/steel interface and that full interaction is ensured so that no slip occurs, the distributions of strain and a corresponding stress block representation of stresses at the assumed ultimate condition will be as shown in Figures 1.3b and 1.3c respectively. Using the latter, considerations of cross-sectional equilibrium permit the moment of resistance to be readily calculated. Although the member's neutral axis will clearly not always fall at the interface, good design will attempt to locate it close to this position as representing the most efficient use of the strengths of the two different materials (concrete acting in compression and steel acting in tension).

For such cases the resulting equilibrium calculations to determine the moment of resistance are only slightly modified.

The use of plastic methods to determine strength, as employed in the above illustration, are now commonplace when dealing with composite construction. Although extensive elastic treatments exist, it has been found that, providing certain rules are observed e.g. relating to potential instability in parts of the steel section, the ability of the shear connection to resist the interface slip etc., then a relatively simple plastic approach is both easier to use and leads to higher resistances.

1.4 MATERIAL PROPERTIES

When designing composite elements it is usual to adopt the same properties for steel and for concrete as would be the case when designing structural steelwork or reinforced concrete. Thus codes of practice covering composite construction, such as EC4, normally simply summarise the relevant sections from the steelwork and concrete documents—EC3 and EC2 in the case of EC4.

Within the limit states framework of design, it is customary to work with characteristic values of material strengths. Whilst these are normally defined with reference to a suitable fractile in the assumed statistical distribution of the material strengths, for structural steel nominal values are usually taken as characteristic values due to the quality control processes used in its manufacture and thus the basis on which it is supplied. The design values to be used in structural calculations are derived directly by dividing the characteristic by the appropriate partial factor. In the case of shear connectors it is customary to work with component strengths—defined on the basis of testing—rather than material strengths.

As an illustration, the relevant sections of EC4 covering: concrete, steel reinforcement, structural steel, metal decking and shear connections, together with their key recommendations, are summarised in Table 1.1.

Concrete is specified in terms of its compressive strength, as measured in a cylinder test, f_{ck}. Grades between 20/25 and 50/60 are permitted. Characteristic tensile strengths are also provided; for lightweight concretes tensile values should be modified by the correction factor:

$$\eta = 0.30 + 0.70 \, (\rho/2000) \tag{1.1}$$

in which ρ is the oven-dry unit mass in kg/m^3.

Table 1.1 Material properties specified in EC4

Concrete	f_{ck}, f_{ctm}, $f_{ctk0.05}$, $f_{ctk0.95}$, total long-term free shrinkage strain, E_{cm} modular ratio
Reinforcing steel	f_{sk}, f_t, minimum ratio of f_t/f_{sk}, ε_u as given in EN 10 080
Structural steel	f_y, f_u or refer to EN 10 080
Profiled steel sheeting	f_{yb}
Shear connectors	Refers to Chapter 10 of EC4 to obtain P_{Rd}, δ_v & P_{Rk} $f_u/f_y < 1.2$, ε_u on a gauge length of $5.65\sqrt{A_o} < 12\%$

Nominal values for the total long-term free shrinkage strain due to setting are provided for typical environments. Creep may normally be covered by using an effective modulus equal to $E_{cm}/3$ when dealing with deflections due to long-term loads, with E_{cm} for short-term loads being the tabulated secant modulus for the particular grade. Some cautions on the extent to which shrinkage and creep effects are adequately treated by present simplified approaches have been provided by Leon (25).

Characteristic yield strengths of plain and ribbed bars f_{sk} are provided in EN 10 080. For structural steel, as previously mentioned, nominal values of yield strength f_y and ultimate strength fu are given, with grades Fe 360, 430 and 510 being covered. These should be used as characteristic values. Nominal yield strengths for profiled steel sheeting for use as characteristic values when dealing with composite slabs are also provided.

In the case of shear connectors the characteristic resistance P_{Rk} is to be based on a 5% fractile of test results using a homogenous population of specimens. The design resistance P_{Rd} is then equal to P_{Rk}/δ_v. Further discussion of the behaviour of shear connectors, including details of the test methods used to obtain P_{Rk}, is provided in Section 1.5 of this chapter.

1.5 SHEAR CONNECTORS

Several early forms of shear connector, used principally for bridges—especially in the United States and Germany—are illustrated in Figure 1.4. By comparison with today's near universal use of welded, headed shear studs of the type shown in Figure 1.5, they are cumbersome and expensive but provide significantly higher strengths. Studs range typically between 13 & 25 mm in diameter, although since the welding process becomes significantly more difficult and therefore expensive for diameters exceeding about 20 mm, 19 mm studs are by far the most commonly used. Since the resistance developed by a stud depends (among other things) on the thickness t of the flange to which it is welded, a limit of d/t of 2.5 is specified in EC4. The steel used to manufacture studs typically has an ultimate tensile strength of at least 450 N/mm² and an elongation of at least 15%. Stud resistances, depending on size and other factors, of up to about 150 kN are achievable using simple welding procedures. Studs have equal strengths in all directions and provide little interference to the positioning of reinforcement.

Stud strengths are normally obtained from "push-off" tests, in which a load–slip curve is determined using a standard test arrangement. Several variants of these have been recommended, designed to both replicate more closely the

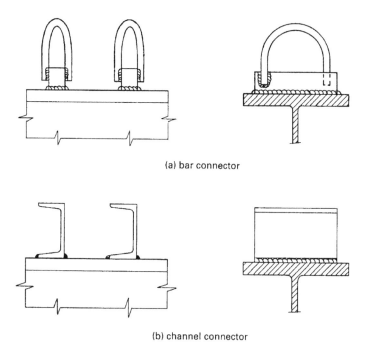

Figure 1.4 Early forms of shear connector

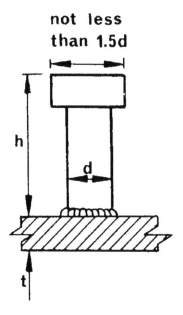

Figure 1.5 Headed shear stud

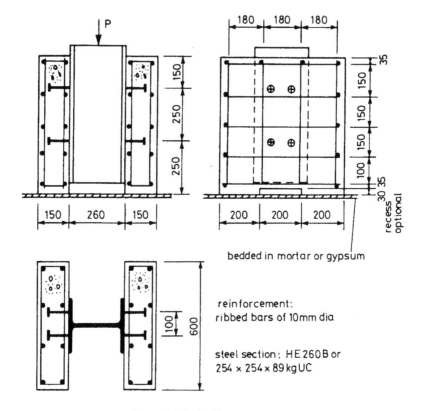

Figure 1.6 Push-off test arrangement

conditions experienced by studs in the compression region of a composite beam and to ensure greater consistency of results from notionally identical arrangements and procedures. Figure 1.6 illustrates the standard EC4 test arrangement. Johnson (26) lists those factors which influence the load–slip relationship obtained from a push-off test as:

1. number of connectors in the test specimen,
2. mean longitudinal stress in the concrete slab surrounding the connectors,
3. size, arrangement, and strength of slab reinforcement in the vicinity of the connectors,
4. thickness of concrete surrounding the connectors,
5. freedom of the base of each slab to move laterally, and so to impose uplift forces on the connectors,
6. bond at the steel-concrete interface,
7. strength of the concrete slab, and
8. degree of compaction of the concrete surrounding the base of each connector.

A typical load–slip relationship is given as Figure 1.7. This exhibits significant ductility or deformation capacity, a property that is necessary for the redistribution of forces between shear connectors that is an essential requirement of the

Shear Connection

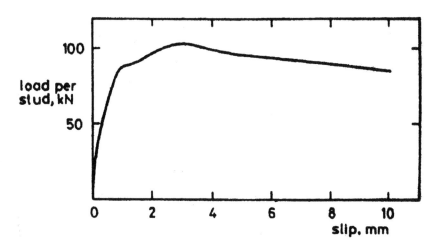

Figure 1.7 Typical load–slip relationship

usual plastic method of designing the shear connection in composite beams. For studs with $h/d > 4$ EC4 requires designers to use the lower of the values for stud strength P_{Rd} of:

$$P_{Rd} = \frac{0.8 f_u (\pi d^2 /4)}{\gamma_v} \tag{1.2}$$

$$P_{Rd} = \frac{0.29 d^2 (f_{ck} E_{cm})^{1/2}}{\gamma_v} \tag{1.3}$$

These cover the two possible modes of failure: reaching the maximum load when the concrete fails or shearing of the stud. Using $\gamma_v = 1.25$ and $f_u = 450 \text{ N/mm}^2$, the first equation will normally govern for concrete grades in excess of C30/37. An explanation of the basis and significance of equations 1.2 and 1.3 may be found in ref. 26.

In addition to resisting horizontal shear at the steel/concrete interface, headed shear studs also prevent any tendency for the slab to lift off the steel resulting from the deformations occurring in the system. EC4 covers this design requirement generally by means of an uplift check using a nominal tensile force of 10% of the design shear resistance of the connectors. Headed studs automatically satisfy this.

For forms of shear connector other than headed studs some design expressions are given in Section 6 of EC4. However, for proprietary components e.g. mechanically fastened connectors, including the recently introduced Perfobond arrangement, it is customary to refer to the design data supplied by the manufacturers themselves. This will have been obtained from the results of

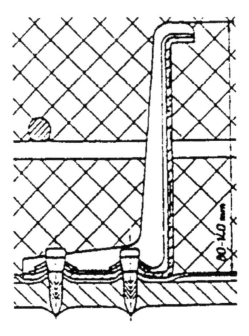

Figure 1.8 Mechanically-fixed shear connector

push-off tests; Section 10.2 of EC4 covers the design of a suitable test programme, the conduct of the tests and the assessment of the results so as to derive characteristic strengths.

In situations where site welding of shear studs is unattractive, an alternative is to use mechanically fastened shear connectors. One such product is illustrated in Figure 1.8. The advantages associated with the use of shear connectors attached using the shot-firing process as compared with welded studs have been stated as (27):

1. the fixing process is not sensitive to protective coatings on either the beams or profiled-steel sheeting, and is unaffected by moisture or inclement weather,
2. the fastening can be performed by semi-skilled labour after a short training period,
3. the fastening pins are easily checked visually,
4. the connectors are secured by powder-activated fastening which eliminates the need for an electricity supply,
5. and the whole system is compact.

Tests indicate that such components typically only develop around 40% of the resistance of the standard 19 mm welded stud. They also tend to be more expensive.

A further development is the use of a strip of connector material of the form shown in Figure 1.9. First proposed in Germany (28), this has been given the name "Perfobond". Interaction is developed by concrete engaging with the perforations, the strip being cut and attached by welding as required to the

Fundamentals 11

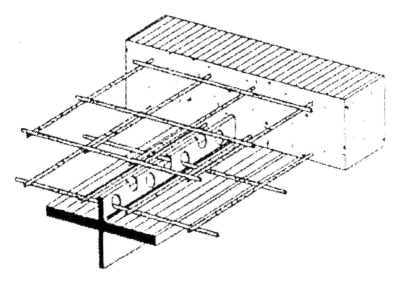

Figure 1.9 "Perfobond" shear connector

beam flange. Push off tests (28) have been used to devise a design equation for the case where concrete strength (rather than yielding of the Perfobond strips) governs:

$$P_{Rd} = 1.6 l d^2 f_{ck}/\gamma_v \qquad (1.4)$$

Further studies (29, 30) have suggested that the influence of other variables e.g. hole size and spacing, should be included in the strength formula. They have also raised some concerns that with certain combinations of variables the Perfobond arrangement may not deliver the level of ductility required by EC 4 to permit shear connector design to be based on the use of a plastic approach. A refinement of Perfobond replaces welding with the use of powder actuated fasteners (31), also modifying the profile by using an array of slots. In a further development the same form of fastening has been used with pieces of metal deck profile (31), a modification of the earlier idea (32) illustrated in Figure 1.10. Further developments in the form of curved

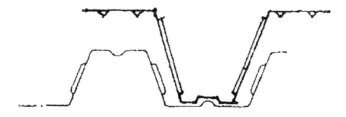

Figure 1.10 Shear connection using sections of decking

strips have also been proposed and a limited number of push-off tests performed (33).

1.5.1 Influence of Slab Type

Equations 1.2 and 1.3 were devised from test data for solid slabs. Other arrangements are, of course, possible. Of particular importance in building construction is the type of composite slab illustrated in Figure 1.11, in which the concrete is cast directly on top of metal decking (more properly referred to as profiled steel sheeting). This provides permanent formwork during the curing operation and then acts as bottom reinforcement to the slab spanning transversely between the beams. Full details of its use are presented in Chapter 5 covering floor systems.

The presence of the sheeting means that the system of forces to which a shear connector, attached by through deck welding to the beam's top flange in the trough region, is subjected differs from that experienced by a stud in a solid slab. Figure 1.12 illustrates this. The most important feature is that the large fraction of load carried by the weld collar is now very significantly reduced. As a result stud strengths in composite slabs should not be obtained directly for equations 1.2 and 1.3.

EC 4 addresses this through the use of a pair of reduction factors, k_1 and k_t one each for slabs with ribs parallel to or perpendicular to the beam:

Figure 1.11 Use of metal decking

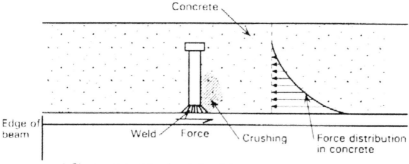

a) Shear connector in plain slab

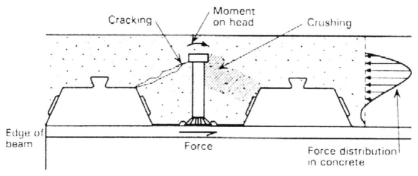

b) Shear connector fixed through profiled sheeting

Figure 1.12 Forces on shear studs

$$k_1 = 0.6 \frac{b_o}{h_p} \left(\frac{h}{h_p} - 1\right) \leq 1.0 \tag{1.5}$$

$$k_t = \frac{0.7}{\sqrt{N_r}} \frac{b_o}{h_p} \left(\frac{h}{h_p} - 1\right) \leq 1.0 \tag{1.6}$$

Most of the terms in equations 1.5 and 1.6 are defined in Figure 1.13; h should be limited to $h_p + 75$ mm, N_r is the number of studs in one rib where it crosses the beam and must not exceed 2.

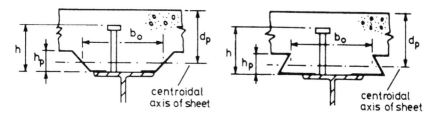

Figure 1.13 Definition of terms used in equations 1.5 and 1.6

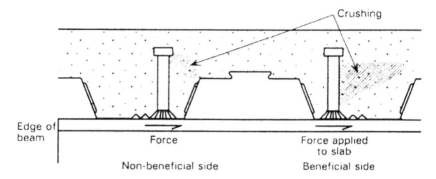

Figure 1.14 Effect of position of stud

The basis for these formulae comes largely from test programmes conducted in North America (32). Their suitability for all design situations is of some concern and in both the United States (33) and the U.K. (34) attempts are being made to improve upon them. Since these involve the giving of explicit recognition in the design expressions to a larger number of influencing factors, the resulting equations are, inevitably, more complex. Particular concerns centre around the practical point that in order to provide additional stiffness to the decking when acting to support the wet concrete a longitudinal stiffener is often formed in the trough. This prevents central positioning of the studs, thereby creating the "strong" and "weak" orientations of Figure 1.14. Other influencing factors include sheeting thickness, sheeting material strength and the use of studs arranged in pairs. A point that would appear to have received little attention is the popularity of the re-entrant deck profiles of Figure 1.15 for which very little test data are available.

Another slab variant, based on removing the need to support the slab whilst it hardens, is the use of precast concrete slabs (hollow core units or hcu) with a small amount of *in situ* concrete placed over the beam's flange as illustrated in Figure 1.16. For this arrangement studs may be welded off-site as part of the steel fabrication process. Push off tests (36) specifically configured so as to provide data on this arrangement have suggested that when P_{Rd} is governed by stud strength (equation 1.2) no change is necessary and that when concrete failure controls equation 1.3 should be modified to:

$$P_{Rd} = \frac{0.29 \alpha \beta \varepsilon d^2 (\omega f E_{ck})^{1/2}}{\gamma_v} \tag{1.7}$$

In which $\alpha = 0.2\ (h/d+1),\ 1.0$

f_{ck} = average concrete cylinder strength, taken as $0.8 \times$ average cube strength of the *in situ* and precast concrete

E_{ck} = average value of the elastic modulus of the *in situ* and precast concrete

$\beta = 0.5\ (g/70+1) < 1.00$ is a factor to account for the gap width g and $g > 30$ mm

Fundamentals

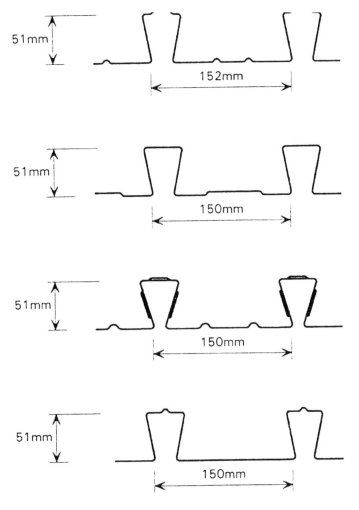

Figure 1.15 Re-entrant decking profiles

$\varepsilon = 0.5 \, (\phi/20 + 1) < 1.0$ is a factor to account for the diameter of the transverse high tensile tie steel reinforcement and $\phi > 8$ mm
$\omega = 0.5 \, (w/600 \, \text{I})$ is a factor to account for transverse joints
W = width of hollow core unit.

Since equation 1.7 is based on the use of 125 mm × 19 mm TRW-Nelson studs in a 150 mm deep slab, caution is necessary before extending its application to other arrangements. A valid criticism of all existing methods for predicting stud strength is their reliance on test data—whether directly from push-off tests or, in some cases, by working back from actual beam tests. So far attempts to devise a theory based on the application of mechanics to the topic have not been successful, although recent application of F.E. techniques to model the

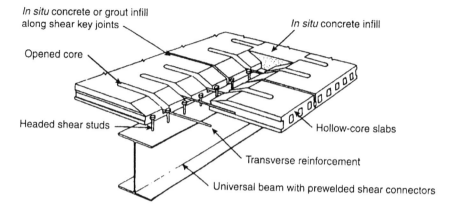

Figure 1.16 Composite action using precast slabs

detailed behaviour of studs (38) show promise. Although written from the perspective of criticism of the current treatment of the topic in United States Codes, the recent review by Leon (25) discusses several concerns on the appropriateness of current empirically based methods of shear stud design. In particular, he questions the advisability of extrapolation to cover several of today's practical arrangements using data obtained from tests specimens having different basic properties. Comfortingly, the EC4 treatment is regarded as more measured and less open to criticism.

1.6 DESIGN FOR ULS

Whilst early approaches to the design of composite members and systems were based on elastic or permissible stress concepts, it was soon appreciated that the nature of the materials involved and their forms of interaction made an ultimate strength format not only more logical but, for the majority of situations, easier to operate and more competitive in outcome. Thus the 1956 AASHO and 1961 AISC (39) documents followed the working stress approach, as did the UK's CP 117 (23). However, many actual designs of that period (13,16) were based on more fundamental approaches that anticipated much of the ultimate strength methodology.

1. Loss of equilibrium of the structure or any part of it, considered as a rigid body.
2. Failure by excessive deformation, rupture, or loss of stability of the structure or any part of it, including shear connection, supports and foundations.

The first of these requires comparison of the design effects of destabilising and stabilising actions and is actually a general requirement for all forms of construction. The second involves determination of the design value of internal force, moment, combination etc. for comparison with the corresponding design resistance. Guidance on dealing with the action side of these checks using appropriately practical combinations of permanent, variable and accidental

Fundamentals

actions is provided in Chapter 1 of ref. 26; the present text is principally concerned with the resistance side.

1.7 DESIGN FOR SLS

Adoption of the limit states design philosophy has highlighted the need to give proper attention to ensuring that structures perform adequately under in-service conditions—something that tended to be rather more taken for granted when using the working stress approach. In particular, explicit consideration of each condition that might render the structure unfit for use is now required. EC4 lists 5 such conditions:

1. Deformations or deflections which adversely affect the appearance or effective use of the structure or cause damage to finishes or non-structural elements.
2. Vibration which causes discomfort to people, damage to the building or its contents or which limits its functional effectiveness.
3. Cracking of the concrete which is likely to affect appearance, durability or water-tightness adversely.
4. Damage to concrete because of excessive compression, which is likely to lead to loss of durability.
5. Slip at the steel–concrete interface when it becomes large enough to invalidate design checks for other serviceability limit states in which the effects of slip are neglected.

The load arrangements (combinations and factors) appropriate for SLS verification are specified in cl. 2.3.4 and explained in ref. 26. For the two most common requirements of deflection control and crack control, Chapter 5 of EC4 provides detailed guidance. This includes procedures for calculating stresses and deflections based on the use of elastic theory, suitably modified to allow (as appropriate) for:

- Shear lag.
- Incomplete interaction due to slip and/or uplift, cracking and tension stiffening of concrete in hogging moment regions.
- Creep and shrinkage of concrete.
- Yielding of steel—especially if unpropped construction is employed.
- Yielding of reinforcement in hogging moment regions.

In common with normal practice in reinforced concrete construction, cracking is to be controlled through the provision of an appropriate arrangement of the reinforcement. This may be achieved either through the actual calculation of crack widths and ensuring that they do not exceed the limits agreed with the client or by following bar-spacing rules. Johnson (26) suggests that cracking is only likely to become an issue for composite structures using encased beams or where the top surfaces of continuous beams are exposed to corrosion. This latter situation is, of course, very unlikely for buildings.

One particular SLS is the condition during concreting, for which the popularity of unproppped construction means that an intermediate form of the structure e.g. Bare steel beam, metal decking, must support the construction loads. Worldwide experience shows that risk of a structural failure is greatest during the construction phase due to the combination of: absence of helpful "non-structural" components that will be present in the final condition, greater inability to accurately determine the applied loading and a tendency to give less attention to the structural checking. An interesting discussion of precisely how these issues might be addressed in a way that balances the requirements of safety and efficiency alongside the practicalities of the construction stage has been provided by Leon (25).

1.8 COMPOSITE SYSTEMS

It is becoming increasingly clear that the design of composite structures needs to be viewed as the provision of a complete system. Thus whilst many of the present design checks operate at the detailed level e.g. strength of shear studs, or in terms of a defined component e.g. composite beam or column, interaction between the various parts of the structure so that recognition of load resisting properties of the complete system are explicitly allowed for is receiving increasing attention. Some of this stems directly from the severe demands made in structures in seismically active zones and the need to mobilise all potentially helpful structural contributions to provide adequate safety in a cost effective way. Some of the benefits of this line of approach are, however, being utilised in producing more efficient solutions to arguably, less severe challenges. Some particular illustrations of this are provided in Chapters 5, 6 and 7—covering joints, floors and frames. Readers wishing to venture further into this topic will need to consult recent, specialised papers dealing with framing systems, frame-core interaction and overall structural response (40).

1.9 CURRENT USAGE

Composite action is presently most often utilised between beams and slabs—in the form of building floors or bridge decks—in certain types of column—particularly in very tall buildings where extremely high compressive loads must be resisted. A proper appreciation of the structural performance of these systems frequently requires an understanding of the various forms of instability that have the potential to influence behaviour i.e. overall buckling for columns and either local buckling of component plate elements or distorsional buckling of partially restrained cross-sections for beams. Recently, the potential benefits of utilising various degrees of continuity in frame systems through the use of beam to column connections that can transmit appropriate levels of moment and possess some degree of rotational stiffness have been investigated and design procedures for certain well defined arrangements proposed. The six topics of: beams, columns, buckling, floors, connections and frames that make up the main body of this text therefore represent a combination of the principal topics of current and likely future interest. Clearly not every aspect of every topic can be covered in a text of manageable size. The aim herein has therefore been to balance a description of fundamental behaviour with presentation of design applications in such a way that the reader can see how to apply the principles in practice, appreciate sufficient of the scientific basis that safe extrapolation to previously unforeseen situations is possible and be guided to more specialised sources when a more detailed explanation of any particular narrow specialty is required.

Fundamentals   *19*

1.9.1 Beams

Almost certainly, the most frequent use of composite construction is for beams, in which a part of the slab acts with the steel section to provide a structural member with greater strength and stiffness than the bare steel section. Chapter 2 explains ways in which this is achieved, including the design checks required and their basis in the mechanisms of load transfer that develop within composite beams.

1.9.2 Columns

Composite columns tend to be used either when the bare steel section is unable to develop sufficient resistance to cope with the design loading or, in certain more specialised applications, where the clever combination of the two materials permits very economic solutions to be devised. An important feature of their use is ensuring that the concrete takes its share of the load, a matter that often requires careful attention to detailed aspects of load introduction. Members are commonly either filled tubes or encased open sections; although similar in concept, certain subtle differences in detailed aspects of behaviour need to be recognised. The chapter demonstrates how a clever distillation of the full understanding of the behaviour of composite columns that is now available can lead to relatively simple design approaches.

1.9.3 Buckling

Although the obvious way to use steel and concrete in combination is to restrict compression to the concrete, using the steel member to withstand tension, it is, of course, not always possible to arrange for such a straightforward solution. It is therefore necessary to consider how various forms of instability might influence the behaviour of composite members. For example, the walls of filled tubes are supported on only one face, continuous composite beams will have regions of the steel member exposed to compression, even arrangements at connections may well involve the transfer of some compressive load through parts of the steel members etc. This chapter therefore deals principally with the local and/or overall buckling modes possible in composite construction; in the case of the latter, the importance of distortion of the cross-section that arises from the restraining influence of the concrete is an important feature. Since the economic design of composite systems often requires that a degree of moment redistribution take place, the associated issue of ductility or available rotation capacity as it is influenced by buckling is also addressed.

1.9.4 Floors

In buildings, composite beams will normally comprise longitudinal steel members acting with part of the floor slab. There are a number of ways in which this might be arranged and this chapter describes several of the more popular systems and identifies the particular structural features associated with their behaviour. It therefore develops the more fundamental concepts presented in the chapters on beams (and buckling) and applies them to a number of different arrangements.

1.9.5 Connections

Although it is still the usual practice to design beam to column and beam to beam joints as if they were bare steel, there is an increasing realisation that significant benefits are available by deliberately providing some degree of load transfer between members through the use of composite joints. The chapter presents the principles, notes that it is still reasonable to treat certain types of joints in composite frames where no special provision has been made in detailing as "simple connections", identifies the key behavioral features in deliberately configured composite joints and indicates how relatively simple design procedures may be derived from the recently obtained clear understanding of the structural behaviour of composite connections.

1.9.6 Frames

When some degree of continuity is introduced into the framing through the use of properly designed composite joints, it is possible to achieve enhanced structural performance as compared with the more usual "simple construction" arrangements. This chapter explains how these benefits may be achieved in a relatively simple way, providing certain restrictions on the governing parameters are observed.

1.10 CONCLUDING REMARKS

It is the purpose of this, introductory, chapter to set the scene for the following, much more specialised, chapters. Thus it has described the basic concept of steel/concrete composite construction as employed in buildings, has traced some of the history of the early development and introduction into practice of the material and has discussed certain aspects of behaviour and design that are necessary for a proper appreciation of what follows. Readers requiring more information on any of the specific topics mentioned should consult the appropriate references.

1.11 REFERENCES

1. Moore, W.P., "Keynote Address: An Overview of Composite Construction in the United States", Composite Construction in Steel & Concrete, ed. C.D. Buckner & I.M. Viest, Engineering Foundation, 1988, pp. 1–17.
2. Buckner, C.D. & Viest, I.M. eds., "Composite Construction in Steel & Concrete", Engineering Foundation, 1988.
3. Eastering, W.S. *et al.* eds., "Composite Construction in Steel & Concrete II", Engineering Foundation, 1992.
4. Buckner, C.D. & Shahraz, B.M. eds., "Composite Construction in Steel & Composite III", Engineering Foundation, 1996.

5. Kennedy, D.J.L. *et al.* eds., "Composite Connection IV", Engineering Foundation, 2000.
6. Griffis, L., "State of the Art Report: Composite Frame Construction in the United States; Composite & Hybrid Structures", ed. S. Goel & H. Yamanouchi, University of Michigan, 1992.
7. Wakabayashi, M., "Japanese Standards for the Design of Composite Buildings; Composite Construction in Steel & Concrete", ed. C.D. Buckner & I.M. Viest, Engineering Foundation, 1988, pp. 53–70.
8. Viest, I.M., Fountain, R.S. & Singleton, R.C., "Composite Construction in Steel & Concrete", McGraw-Hill, New York, 1958.
9. Godfrey, G.B., "Post-War Developments in German Steel Bridges & Structures", Structural Engineer, Feb. 1957, pp. 53–68.
10. Sattler, K., "Composite Construction in Theory & Practice", Structural Engineer, Vol. 4, No. 2, Apr. 1964, pp. 115–125.
11. Chapman, J.C., "Composite Construction in Steel & Concrete: Behaviour of Composite Beams", Structural Engineer, Vol. 4, No. 2, Apr. 1964, pp. 115–125.
12. Chapman, J.C. & Balakrishanan, S., "Experiments on Composite Beams", Structural Engineer, Vol. 42, No. 11, Nov. 1964, pp. 369–383.
13. Cassell, A.C., Chapman, J.C. & Sparkes, S.R., "Observed Behaviour of a Building of Composite Steel & Concrete Construction", Proceedings ICE, Vol. 33, April 1966, pp. 637–658.
14. Barnard, P.R. & Johnson, R.P., "Ultimate Strength of Composite Beams", Proceedings ICE, Vol. 32, Oct. 1965, pp. 161–179.
15. Barnard, P.R. & Johnson, R.P., "Plastic Behaviour of Continuous Composite Beams", Proceedings ICE, Vol. 32, Oct. 1965, pp. 180–197.
16. Johnson, R.P., Finlinson, J.C. & Heyman, J., "A Plastic Composite Design", Proceedings ICE, Vol. 32, Oct. 1965, pp. 198–209.
17. Faber, O. "More Rational Design of Cased Stanchions", Structural Engineer, Vol. 34, March 1956, pp. 88–109.
18. Reiner, S.M., Wright, K.M. & Bolton, D., "The Design & Construction of Pelham Bridge, London" Structural Engineer, Dec. 1958, pp. 399–407.
19. Wood, R.H., "Composite Construction", The Structural Engineer Jubilee Issue, July 1958, pp. 135–139.
20. Creasy, L.R., "Composite Construction", The Structural Engineer, Vol. 42, No. 12, Dec. 1964, pp. 411–422.
21. Silhan, S.A.G. & Westbrook, R.C., "Composite Construction in Government Buildings", Conference on Structural Steelwork, BCSA, London, Sept. 1966, pp. 191–201.
22. Stevens, R.F., "Encased Stanchions", The Structural Engineer, Vol. 43, No. 2, Feb. 1965, pp. 59–66.
23. CP 117, "Composite Construction in Structural Steel & Concrete. Part 1: Simply Supported Beams in Building, 1965, Part 2, Beams for Bridges", British Standards Institution, London 1967.
24. BS 5950, "The Structural Use of Steel in Building, Part 3:1, Code of Practice for Design of Simple & Continuous Composite Beams", British Standards Institution, London, 1990.
25. Leon, R., "A Critical Review of Current LRFD provisions for Composite Members", Proceedings SSRC 2001, Fort Lauderdale, pp. 189–208.
26. Johnson, R.P., "Composite Structures of Steel & Concrete Volume 1 Beams, Slabs, Column & Frames for Buildings", 2nd edition, Blackwell Scientific Publications, Oxford, 1994.

27. Thomas, D.A.B. & O'Leary, D.C., "Composite Beams with Profiled—Steel Sheeting & Non-Welded Shear Connectors", Steel Construction Today, Vol. 2, No. 4, August 1988, pp. 117–121.
28. Leonhardt, F., Andra, V. & Harre, W., "Neues Vorteilhaftes Verbundmittel für Stahlverbund – Tragwerke mit hoher Dauerfestigkeit", Beton und Stahlbetonbau, No. 12, 1987, pp. 325–351.
29. Nishida, T. & Fugi, K., "Slip Behaviour of Perfobond Rib Shear Connectors & its Treatment in FEM", Engineering Foundation, 2000.
30. Machacek, J. & Studnicka, J., "Perforated Shear Connectors", Steel & Composite Structures, Vol. 2, No. 1, Feb. 2002, pp. 51–66.
31. Fontana, M. & Beck, H., "Novel Shear Rib Connector with Power Activated Fasteners", Engineering Foundation, 2000.
32. Shanit, G., Chryssanthopoulos, M.K. & Dowling, P.J., "New Profiled Unwelded Shear Connectors in Composite Construction", Steel Construction Today, Vol. 4, No. 5, Oct. 1990, pp. 141–146.
33. Galjaard, H.J.C. & Walraven, J.C., "Behaviour of Different Types of Shear Connecting for Steel-Concrete Structures", Structural Engineering, Mechanics and Computation, Elsevier Science Ltd., Cape Town, 2001, pp. 385–392.
34. Grant, J.A., Fisher, J.W. & Slutter, R.G., "Composite Beams with Formed Metal Deck", Engineering Journal American Institute of Steel Construction, Vol. 14, No. 1, 1977, pp. 24–42.
35. Rambo-Roddenbury, M. *et al.*, "Performance and Strength of Welded Shear Studs", Engineering Foundation, 2000.
36. Johnson, R.P., "Shear Connection—Three Recent Studies" Engineering Foundation, 2000.
37. Lam, D., Elliott, K.S. & Nethercot, D.A., "Push off Tests on Shear Studs with Hollow Covered Floor Stabs" Structural Engineer, Vol. 76, No. 9, May 1998, pp. 167–174.
38. Lam, D. & El-Labody, E., "Finite-Element Modelling of Headed Stud Shear Connections in Steel – Concrete Composite Beam", Structural Engineering, Mechanics and Computation, ed. A. Zingoni, Elsevier, Amsterdam 2001, pp. 401–410.
39. Viest, I.M., "Studies of Composite Construction at Illinois and Lehigh, 1940–1978", Composite Construction in Steel and Concrete III, ed. C.D. Buckner & B.M. Shahrooz, Engineering Foundation, 1996, pp. 1–14.
40. Hajjar, J., "Composite Construction for Wind and Seismic Engineering", ibid., pp. 209–228.

CHAPTER TWO

Composite Beams

Howard D. Wright

2.1 INTRODUCTION

Composite beams may well be the most common form of composite element in steel frame building construction and have been the major form for mid range steel bridges. As the main criterion for design is flexure, the study of composite beams provides an explanation to the associated behaviour of many other composite elements such as composite slabs, composite columns subject to lateral loads and composite joints between beams and columns.

As a basis for the work in this chapter, composite beams are defined as; "elements resisting only flexure and shear that comprise two longitudinal components connected together either continuously or by a series of discrete connectors". Furthermore it is assumed that the two components are positioned directly one above each other with their respective centroids vertically above each other.

Composite beams vary in behaviour from the situation when the connection between the two layers is non-existent to the situation where the bond between the layers approaches infinite stiffness and strength. There is also the influence of the contrast in material properties of the two layers. A weak and flexible layer will have little influence in the overall strength and stiffness of the beam. Consequently the influence of the difference in strength and stiffness of the components and in particular the strength and stiffness of the connection between them plays a vital role. In addition, the stress states in each component that exist prior to connection may also affect behaviour. The overall analysis and design of composite beams is, therefore, significantly more complex than for single material beams.

In providing a staged approach to learning about this behaviour, the chapter will first provide a general description of a range of common composite beams, their components and the possible sequence of construction that will cause variable stress states throughout the beam depth.

Most beams are designed initially for ultimate strength using load factor methods and the chapter will begin with the analysis of simple span composite elements with zero initial stress in all components. The ultimate capacity of the composite element will be considered for the condition when the connection between the components is able to fully resist the forces applied to it (full connection). This is possibly the most common situation; however, over the last two decades the use of beams in building construction has led to many instances when the connection cannot resist all of the forces applied (partial connection).

This pattern will be repeated when considering the serviceability limit-state of composite beams. In this case the condition when the connection between the components is considered infinitely stiff the beam is said to have full interaction. Whilst this is often assumed in design it is theoretically impossible and cases where the connection has more limited stiffness (partial interaction) often needs to be considered. This is a complex case as the beam is statically indeterminate and it will be shown that for design purposes an approximation is generally assumed.

Critical to the overall behaviour of composite beams is the specific behaviour of the connection. By far the most common connector is now the welded stud that has evolved as the cheapest product and the easiest device to weld on site. The connector and the welding process will be described along with its behaviour under the predominant shear load. In particular the requirements for ductility and the variation in strength and stiffness when it is used with proprietary slab systems will be explained.

In recent years the use of continuous composite beams has become more common. The benefits of continuity come with a design price in that composite beams behave very differently in this situation and can almost never provide the same strength or stiffness in the negative moment regions. Consequently semi-continuous design is normal with redistribution of the support moments in a similar way to that used in RC beams design.

One of the most common forms of slab used with composite beams is the composite slab. This in itself is a composite system (see chapter 7) and its shape and behaviour influences the beam design. A more traditional slab is the pre-cast concrete plank. With recent developments this too is being used as a part of a composite beam again with very specific influences on beam design (Lam *et al.* 1998).

Finally in this chapter current practice in composite beam design will be described. Given the fact that virtually all steel frame buildings now use some form of composite system current practice is quite varied and new developments are constantly being introduced. These inevitably also stretch the application of current design codes and may in some cases conflict with them.

2.2 TYPES OF BEAM

The generic form of composite beam comprises the combination of a solid concrete slab attached to a rolled steel section (normally of I shape). The slab will be designed to carry the floor load spanning between parallel beams but may also take compression perpendicular and along the beam line if it is connected to the steel section. This arrangement is shown in Figure 2.1. The connection between steel and concrete must be sufficient to control longitudinal shear and any uplift forces. The longitudinal forces generated by this connection must transfer fully from the steel section into the wider slab (normally by resisting the transverse shear developed between slab and beam). Shear stud connectors and typical transverse shear reinforcement is shown in Figure 2.2.

As already mentioned the concrete slab may be a composite deck (see chapter 7). In this case the connection is more difficult. The predominant form of connection is the "Through Deck" welded stud. This construction uses high output arc and forge weld equipment to burn through the steel sheet and weld the stud to the beam below. Figure 2.3 shows a typical composite beam formed with a composite slab. The profiled shape of the deck reduces the amount of concrete available to carry compression in the beam and the situation of the stud in a relatively narrow flute of concrete may also change its behaviour

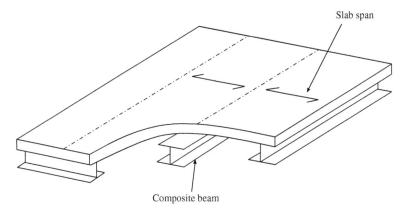

Figure 2.1 Typical composite beam arrangement

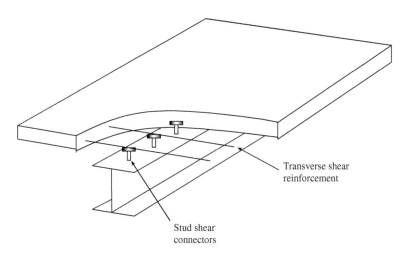

Figure 2.2 Composite beam showing shear connection and transverse shear reinforcement

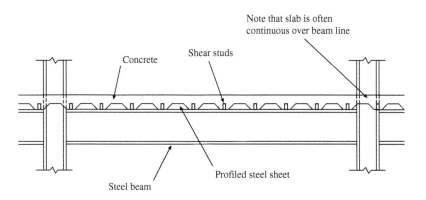

Figure 2.3 Composite beam formed with a composite floor slab

under load and may reduce its strength and stiffness. The presence of the steel deck does, however, mean that additional transverse shear reinforcement may often be omitted.

A variation on this theme is the "Slimflor" system (Lawson *et al.* 1997) that uses a deep deck laid upon bottom flange extensions to the steel section. The connection between concrete and steel provided simply by chemical and friction bond is sufficient to obtain composite action. The steel section is often relatively stocky and often of column rather than classical beam proportions. This allows the overall beam to be relatively shallow and the system forms an obvious rival to the RC concrete flat slab. Figure 2.4 shows a diagram of this system.

For both systems there is a possible further advantage associated with construction process. The steel deck provides a platform for concrete operations and the steel beam alone may be designed to take the load of wet concrete, workmen and tools (i.e. the dead load). Once hardened the composite action between steel and concrete is often more than sufficient to carry the additional service load. This allows a simple and speedy construction process that avoids temporary propping.

Whilst unpropped construction was considered a novelty in the UK in the 1980s (Wright *et al.* 1987) it is, of course, often the only option open to bridge constructors. Composite bridge construction has been popular since the middle of the last century. The shear connection in bridge beams is normally attached to the steel in the fabrication shop and steel decks are rarely used, although other

Figure 2.4 Slimflor® beam (SCI 1997)

forms of sacrificial formwork, such as glass fibre panels and pre-cast slab forms are common.

The basic concepts developed for the generic concrete steel beams described above have also been applied to several other forms of composite beam.

Composite slabs, described in more detail in Chapter 7, are, perhaps a special case as the failure load is more often dictated by shear and the predominant design method is semi-empirical. A natural derivative of composite slabs is the profiled composite beam system (Oehlers et al. 1994) shown in Figure 2.5.

Slabs may also be formed from dry panels attached to the steel decking with screws. This form of construction has been named profiled steel sheet dry board (PSSDB) construction (Wright et al. 1989).

Double skin composite construction (DSC) (Wright et al. 1991) and its derivative "Bisteel" (Corus 1999) comprise two skins of steel plate with a layer of concrete cast between. Long stud shear connectors provide both longitudinal and vertical shear resistance in DSC whilst friction welded bars link the two skins in the "Bisteel" case. Figure 2.6 shows the two systems. Initially the system was devised for submerged tube tunnel construction where the steel skin provided a form and protection to the concrete from seawater (Narayanan et al. 1987).

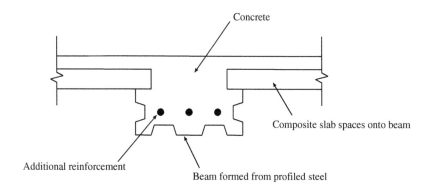

Figure 2.5 Section through a profiled composite beam

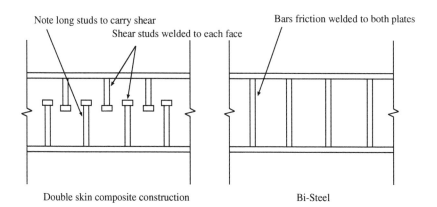

Figure 2.6 Double skin composite and Bi-Steel®

Many other possible uses have been suggested many associated with the, as yet unproved, blast and impact resistance.

A similar form of construction is the use of plates to strengthen or repair concrete structures. Plates may be glued or bolted to the soffit or sides of the beam, providing tension and shear reinforcement. These plates are often made of fibre composite plates (Concrete Society 2000) although steel plates may also be used (Nguyen et al. 1997).

Composite beam behaviour also occurs in situations where concrete is used to encase a steel section, possibly for fire protection. Full encasement has been common for external beams in building frames for many years, however rarely has the benefit of composite behaviour been used.

In all these cases the basic behaviour through the elastic and plastic phases are relatively similar. By far the most common forms of construction are the generic and composite slab/composite beam systems. Consequently these systems will be used to describe the behaviour and its analytical modelling in more detail in the following sections.

2.3 BASIC BEHAVIOUR

To describe the development of stresses within the section during increasing load we will first assume propped construction. Figure 2.7 shows a possible composite beam section along with the strain profile and subsequent stress profile assuming pure bending and elastic material properties. All of the assumptions used in classic simple beam theory apply. It may be noted that the elastic neutral axis (ENA) is shown in the concrete slab in this figure but this may well be in the steel depending upon the relative geometry and material properties of the two components.

The figure shows that the linear strain profile creates a bilinear stress profile due to the different material stiffness of concrete and steel. The forces present in the section will depend upon geometry and have been conveniently shown as applying to the concrete, steel flanges and steel web.

An ENA high in the section is normal, as the amount of concrete available to carry compression is large and is a direct function of the width of slab assumed to act with the beam. This width is often termed the effective width of slab. Experiments on composite beams (Harding 1991) show that shear lag in the plane of the slab leads to a variation of longitudinal stress, the pattern of which is shown in Figure 2.8a for a beam subject to a four point load. The fall-off in stress shown at

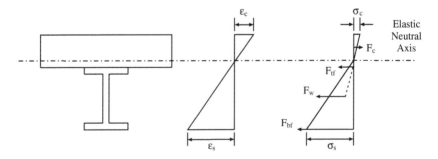

Figure 2.7 Composite beam section; strains and stresses

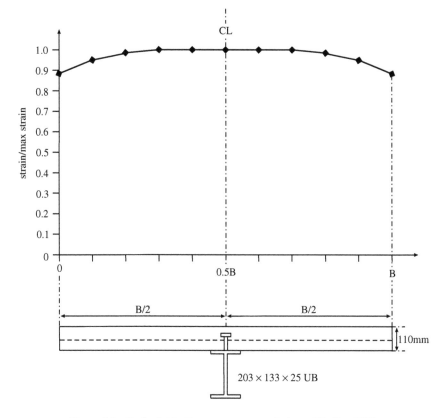

Figure 2.8a Strain distribution across a composite beam (Harding 1991)

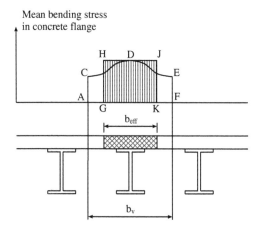

Figure 2.8b Effective breadth concept

the edges of the slab is indicative of a reduction in the capacity of the slab to carry the longitudinal force. Figure 2.8b shows the transformation of the stress distribution ACDEF into the equivalent rectangular distribution GHJK of effective width b_{eff}. The extent of shear lag has been found to be dependent upon the span and load pattern on the beam (Ansourian 1975). For design an effective breadth of slab shown of between $1/3$ and $1/4$ the beam span is often assumed (Eurocode 4, 1994).

An intrinsic assumption of simple beam theory is that plane sections remain plane and hence that shear deformation through the section is negligible. However if we consider the likely deformation of the composite section with and without connection (Figure 2.9) we can see that horizontal deformation may occur at the concrete steel boundary. For this deformation (or slip) to be restrained the connection between the two layers must have a finite stiffness. However for simple beam theory to hold this connection must be infinitely stiff, an unlikely situation as we will discover later in the chapter. In most loading cases on simple beams the distribution of shear along the section is continuous leading to the conclusion that any connection between the two layers should also provide a continuous or "smeared" resistance. However in most beams the connection is by discrete shear studs although the spacing of these is normally such that a smeared connection may be safely assumed.

There are two effects of providing connection that has different properties to either main component. Firstly the connection may fail in shear before either of the other components reach their own failure stress. This is known as partial connection and for some beams is the main design criterion. Secondly the connection may deform giving rise to relative movement along the boundary and the effect of increased shear deformation in the beam as a whole. This is termed partial interaction and occurs to some extent in all beams whether fully connected or not. Figure 2.10 shows the effect of partial interaction on the strain profile and describes end slip, the resulting phenomenon associated with it.

As noted in the previous section a major advantage for building construction and the normal situation for bridge construction is the fact that the beam may be unpropped during construction. This gives rise to a stress history that affects the final stress distribution. This is described diagrammatically in Figure 2.11. It can be seen that in this example the combination of stresses in the steel beam

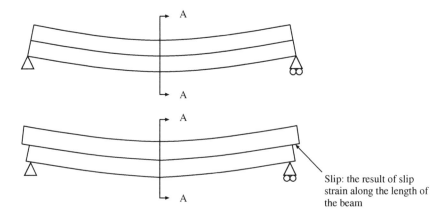

Figure 2.9 Behaviour of fully connected and unconnected beams

Composite Beams

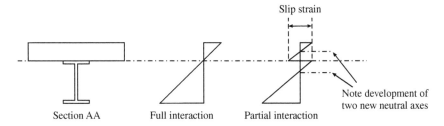

Figure 2.10 Effect of partial connection on strain

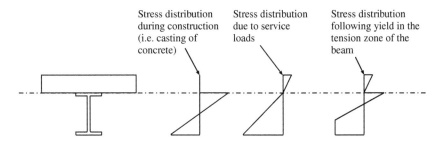

Figure 2.11 Stress history of unpropped beam during construction

due to construction loads and those due to service loads give rise to yield in the tension zone of the beam. This situation led to the inclusion of a design check in BS5950 Part 3 (British Standards Institute 1990). It can be shown that for the determination of ultimate strength this is not important, however it may affect the final deflection of the beam (Nethercot *et al.* 1998).

2.4 ULTIMATE STRENGTH DESIGN

The design of composite beams is generally based upon limit state principles with the ultimate limit state being established using the load factor method. As already introduced in chapter 1 this assumes that the entire depth of concrete is subject to its maximum design stress and the whole depth of steel is subject to yield stress (plastic section analysis). Figure 2.12 shows a typical section and the stress pattern at this plastic state.

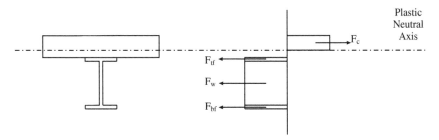

Figure 2.12 Section, stress and force distributions at plastic moment

In this example the plastic neutral axis (PNA) is assumed in the concrete although as seen later it could well be in the steel flange or web. The concrete below the PNA is in tension and is assumed cracked and therefore carries no load. The full depth of concrete down to the neutral axis is assumed to take the design stress ($0.45 \times f_{cu} \times A_c$). The 0.45 allows for a partial safety factor and for the non-linear (parabolic) stress distribution. In this case the entire steel section is subject to its maximum tensile stress ($p_y \times A_s$). These stresses give rise to the forces in the concrete, steel flanges and web as shown.

To establish the position of the PNA it is first required to calculate the force resulting from the maximum stress down the full depth of concrete and then use the principle of equilibrium of horizontal forces.

The ultimate moment of resistance is easily calculated by taking moments of these forces about any point. In this example moments of forces are taken about the neutral axis and the final moment of resistance is

$$M_r = (0.45 \times f_{cu} \times A_c \times l_c) + (p_y \times A_f \times l_{ft}) + (p_y \times A_w \times l_w) + (p_y \times A_f \times l_{fb}) \quad (2.1)$$

So far it has been assumed that the connection between the concrete and steel is complete or full and the forces shown in the diagram may be fully developed. Figure 2.13 shows a simplification of the effect of these forces on the whole beam. The force in the concrete at the plastic hinge is distributed between those connectors between this point and the end of the beam. The connectors are usually discrete and an assumption is often made that the force is distributed equally to each connector. This places an onerous ductility requirement on the connectors as will be demonstrated later.

We have already mentioned that in buildings the steel section is often designed to take the construction load (dead load) and that the final composite beam is required only to take the additional service load. By providing sufficient connectors to carry the full force in the concrete the moment of resistance of the composite beam may be much higher than required. There is, therefore, the opportunity to reduce the number of connectors so that a better match of applied and resisting moment may be achieved. This is termed partial connection and provides some economy in the provision of connectors.

The saving in connector cost is however marginal and the main reason for the use of partial connection is often based more on the inability to provide sufficient connectors to carry the full concrete force. This is particularly the

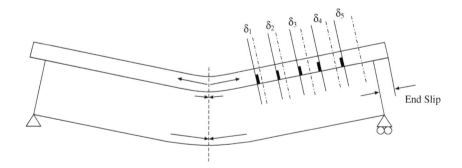

Figure 2.13 Forces and resulting deformation of connectors

Composite Beams

case when composite slabs are used where the spacing of the lower flanges of the steel sheeting dictates that only one stud may be welded in each trough. The reduced efficiency of the connector when welded through deck will be discussed later.

Figure 2.14 shows the forces resulting from a partial connection design of a typical composite beam. The force in the concrete (shown in Figure 2.12) is changed to a maximum connection resistance. The remaining calculation is identical to that previously described. In this figure a composite slab spanning perpendicular to the beam has been assumed. It should be noted that because of the profiled nature of the slab the concrete below the top flange of the sheeting is assumed not to act. The centreline of force acting through the connectors is assumed to be the mid-height of concrete above this top flange. For a similar beam with full connection only this area of concrete would be assumed to act.

The degree of connection may be plotted against the moment of resistance achieved for any beam. A typical graph is shown in Figure 2.15. The curve so produced is often approximated by a straight line and gives rise to the idea of a linear variation in connection strength. When the ratio of service load to dead load is very low the degree of connection may be reduced substantially. In practice connection ratios below about 40% are regarded with suspicion as they rely very heavily on every connector in the beam working to its maximum capacity (Eurocode 4, 1994).

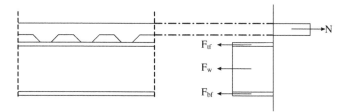

Figure 2.14 Forces due to partial connection

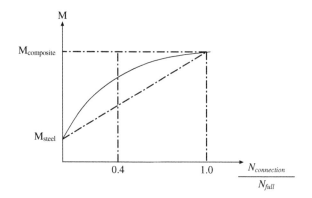

Figure 2.15 Moment capacity and degree of connection

2.5 CALCULATING THE DEFLECTION

The deflection of the steel section under the action of construction loads may be obtained using standard steel design methods. Once the concrete has hardened the unfactored service load is normally used to evaluate the deflection of the composite beam for the serviceability limit state.

Figure 2.16 shows a typical composite beam section along with the transformed section that is used to calculate a section modulus (I value) that may be used along with the steel Young's modulus (E value) to obtain a suitable bending stiffness for the composite beam. This technique involves transforming the effective breadth of concrete slab to an equivalent breadth of steel slab, hence the common name for the process of the "transformed section method". The relationship between the actual concrete breadth and equivalent steel breadth is in the inverse ratio of their respective Young's modulus.

$$\frac{b_{eff}}{b_{act}} = \frac{E_{steel}}{E_{concrete}} \qquad (2.2)$$

Once a transformed section has been established the elastic neutral axis is calculated by taking moments of the cross section area about any point. For the section shown in Figure 2.16 and taking moments about the top surface

$$\text{depth of neutral axis} = \frac{[(A_c \times d_c) + (A_s \times d_s)]}{A_c + A_s} \qquad (2.3)$$

The I value may then be determined using parallel axis theory.

$$I_{total} = I_c + I_s + (A_c \times r_a^2) + (A_s \times r_b^2) \qquad (2.4)$$

The determination of deflection described so far is only for the bending stresses within the beam and assumes full interaction between the two layers. If the stiffness of the connection is thought to be low, possibly as a result of

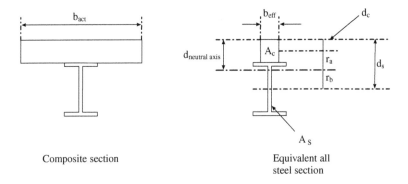

Figure 2.16 Transformed section used to calculate elastic stiffness

partial connection being adopted, then this may have to be taken into account. In theory infinite connection stiffness is impossible, however any flexibility in the connection may be ignored for beams designed for full connection.

Figure 2.10 has described the possible strain profile through a composite beam where the connection stiffness is not infinite. It may be seen that there is a strain discontinuity at the boundary between the two materials. This strain is dependent upon the stiffness of the two materials and on the connection stiffness and consequently renders any partially interactive beam as statically indeterminate. Solutions to this problem have been established and reproduced in texts (Johnson 1994). In general these solutions are for relatively simple load cases and beam support conditions. For more general solutions finite element and finite difference methods may be adopted (Ansourian 1975, Roberts 1984).

For general design a more simple method is needed. The most common of these assumes a linear variation between deflection and the degree of connection in a similar way to that used for degree of connection. The formulaic representation of this is

$$d_c = d_{fc}\left[1 + k\left(1 - \frac{N}{N_f}\right)\left(\frac{d_a}{d_c} - 1\right)\right] \qquad (2.5)$$

2.6 SHEAR CONNECTOR BEHAVIOUR

It may be appreciated that a critical part of a composite beam is the connection. Various types of connection and the required behaviour have been described in Chapter 1.5. For the design processes described in the previous sections each connector is assumed to carry its full design load and for full connection is expected to remain virtually infinitely stiff.

Figure 2.17 shows a shear stud connector exposed following a test on a full-scale composite beam. The connector is approximately 100 mm long, 19 mm in diameter and has a 10 mm thick head. In this case the stud is a "through deck" connector and probably constitutes the most common size and type of connector used in building construction. Larger 22 mm shear studs are often used in bridge beams but these are rarely welded through a deck.

The head of the connector prevents the concrete slab from lifting away from the steel and this leads to some tension in the shank of the connector. Despite this the predominant action on the connector is shear.

The behaviour of an individual connector may be examined using a model test known as the push test where small sections of concrete slab are connected to a length of beam. The steel is pushed through the two slabs putting a pure shear load on the connection. Figure 2.18 shows a typical graph of the load deformation response of a shear stud connector in a solid concrete slab established using a push test. It can immediately be seen that the response is non-linear and that there is a substantial post yield part to the graph. The failure of the connection in this case is associated with concrete crushing in the slab, however the connector may fail in shear with not dissimilar results.

The fact that either concrete crushing or steel yield may occur means that for design the connector must be checked for both failure modes. For connectors

Figure 2.17 Shear connection exposed following test to failure of composite beam (Harding 1991)

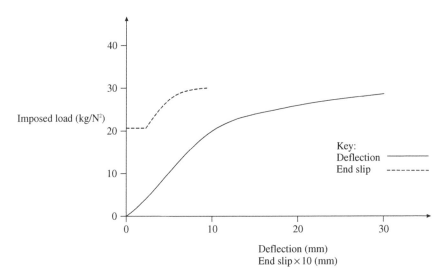

Figure 2.18 Typical load slip plot from push test

in a solid slab most codes use derivatives of the following semi-empirical formulae (Eurocode 4, 1994)

$$P_{Rd} = 0.8 f_u ((\pi d^2/4)/\gamma_v) \tag{2.6}$$

$$P_{Rd} = 0.29 \alpha d^2 \sqrt{(f_{ck} E_{cm})/\gamma_v} \tag{2.7}$$

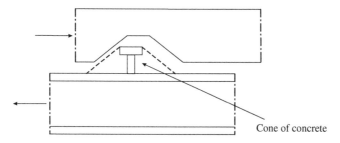

Figure 2.19 Possible cone failure

There have been attempts to model a different concrete failure mechanism (Jayas and Hossain 1988) which assumes that the concrete slab shear around the stud leaving a failure cone as shown in Figure 2.19, however these have not been adopted in design codes.

If we consider Figure 2.13 again it may be noted that in order to carry a design load the connector must deform. For a plastic hinge to develop (as assumed in the ultimate load analysis), the total compressive force at the hinge must be carried by all the connectors between the point of maximum moment and the support. In order to share this force each connector must deform by a progressively increasing amount. This leads to the situation of those connectors close to the maximum moment deforming by a small amount but those near the support deforming by a much larger amount but all still needing to carry the same design load. Consequently the connectors must have exceptional ductility.

This ductility may be achieved through the careful specification of the steel used in the stud and by ensuring that the beam is not over long. (The longer the beam the more the connectors close to the supports must deform). The design load of the connector must also be chosen to ensure that this ductility requirement is satisfied as described in the next two paragraphs.

It may be noted that the equations used for connector strength include an additional reduction factor (e.g. 0.8 in equation 2.7). This is necessary for two reasons, as a safety factor against the possible failure of the connection due to loss of ductility and to account for the discrete nature of the connection.

In Figure 2.20 the relationship between moment capacity and degree of shear connection is shown. This is a more detailed description of the effects of connection to that given in Figure 2.15. Between A and B the connection is very low, the concrete slab has little effect and the strength of the beam is that of the steel beam alone. As the connection ratio increases more of the concrete slab is effective (B to C in the figure). At about 50% connection the ductility of the connectors becomes important and, in practice, the fully plastic moment can only be achieved by the provision of $\frac{1}{2}$ as many again shear connectors. This is shown in the line C to D and E to F to G when strain hardening of the steel beam is taken into account. The dotted line in the figure provides a connection ratio of about 1.25 that virtually guarantees the full plastic moment capacity of the beam. Hence the 0.8 value in the determination of stud strength.

In Figure 2.21 the applied bending moment for a uniformly distributed load on a simple span composite beam is compared to the resisting moment provided by a number of discrete connectors. The applied moment follows a continuous and parabolic curve. The resisting moment depends upon the number of connectors

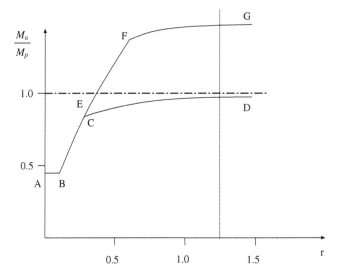

Figure 2.20 Relationship between moment capacity and degree of connection

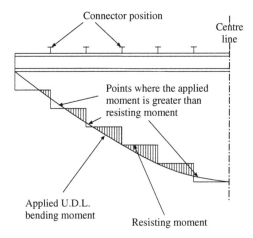

Figure 2.21 Applied bending moment diagram compared with resistance moment

between any point and the support as discussed previously in the chapter. Consequently it follows a stepped form. The shaded portions in the figure denote parts of the beam where the moment of resistance is less than the applied moment. Clearly this is unsafe. In order to overcome this the 0.8 reduction factor artificially ensures that the actual resisting moment is 25% higher than the design resistance moment as shown in the figure.

Readers should note that whether the increase is associated with satisfying ductility requirements or whether it satisfies the discrete nature of the resistance it involves an approximate 25% increase in the number of connectors provided.

Composite Beams

A further design consideration is the detailing and spacing of the connectors. Clearly in order to develop their full strength connectors need to be surrounded by sufficient concrete. Consequently they should be sufficiently long to anchor in the slab but not so long as to provide too little cover of concrete over their head. It is also good practice to ensure that reinforcement in the slab lies below the head of the connector. Unfortunately this is often not the case in beams formed with a composite slab where the light shrinkage mesh used to reduce hogging cracks over the beam line is often positioned on the stud head. The use of mesh or light bars to reduce hogging cracking and shrinkage is important as a possible failure mode of the connection is due to longitudinal splitting of the slab along the line of stud connectors. Such reinforcement constitutes a form of transverse shear resistance covered in the next but one paragraph.

For full connection, particularly in beams formed with a composite slab satisfying the criterion for stud spacing is often difficult. It is quite common to see two connectors welded side by side. In this case the design resistance of the connection must be reduced leading to the requirement for more stud connectors. In this case partial connection design is useful.

A further design and detailing requirement is for transverse shear resistance. Figure 2.22 shows a cross section of a composite beam and possible planes along which the shear developed in the connection may have difficulty in transferring.

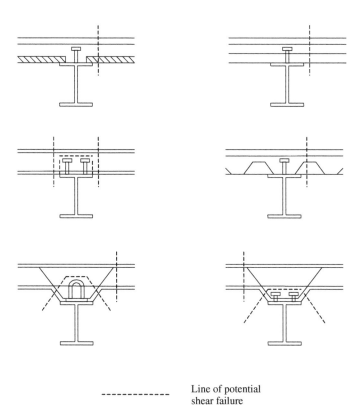

 Line of potential shear failure

Figure 2.22 Potential transverse shear failure planes

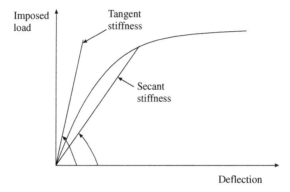

Figure 2.23 Tangent and secant connector stiffnesses

For solid concrete slabs it is necessary to provide additional reinforcement across these planes. The design of this steel is again semi-empirical but follows the form given in EC4

$$\gamma_{Rd} = 2.5 A_{cv} \eta \tau_{Rd} + A_e \frac{f_{sk}}{\gamma_s} + \nu_{pd} \qquad (2.8)$$

For composite beams formed with composite slabs the steel deck may provide sufficient reinforcement to satisfy this requirement.

It has already been noted that for general design the stiffness of the connection is often approximated as the degree of connection in the linear interaction equation (2.5). However if a more detailed analysis of beam stiffness is needed then the connection stiffness must be determined. For most situations a secant stiffness to the design load as shown in Figure 2.23 is acceptable. However it should be noted that for beam deflections approaching the ultimate load the connectors close to the beam support will be more likely to have passed their ultimate load and will have very low stiffness (Wright 1989).

2.7 CONTINUOUS BEAMS

So far we have assumed that the beam section is predominantly subjected to sagging bending with the concrete in compression and steel in tension. However the concrete slab is often cast continuously over supports and points to the potential for continuous beam behaviour to be mobilised. This is dependent upon the connection between beam and column and this is explored more fully in Chapter 4.

Figure 2.24 shows a typical composite beam showing the continuous nature of the slab and the forces that may be developed at the beam column connection. (Note that there is potential for confusion regarding the connection provided by shear studs at the concrete steel interface in the beam and the joint between the beam and column, hence the qualification "beam column"). The tensile force in the slab area may be carried by additional reinforcement and compressive force across the column may be carried by additional web stiffeners. However there is also the transfer of longitudinal shear between the concrete and steel to consider and the fact that the bottom part of the steel section is in compression and therefore both flange and web are prone to buckling.

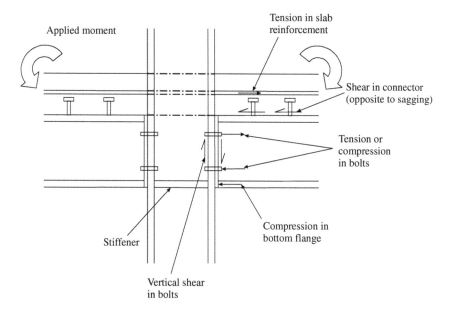

Figure 2.24 Forces on the beam column connection in a continuous composite beam

Let us first consider the resistance of tension forces in the concrete slab. In order to develop the moment capacity of a fully continuous beam the overall section must have a similar capacity in the hogging and sagging regions. This leads to the concept of a fully rigid beam column connection. Given the ability of the concrete slab to take high compression forces in the sagging region the composite moment at mid-span is normally very high. At the support the concrete is in tension, is likely to be cracked and all the tension forces are carried in reinforcing bars placed in the concrete. To generate sufficient tension resistance the reinforcement needs to be very large and there are complications regarding the available cover to the bars and consequential cracking in the concrete. For most practical situations developing the full support moment is unreasonable and the beam column connection carries a proportion of the moment. This is known as partial strength beam column connection.

In carrying only a proportion of the support moment there is an intrinsic assumption that the beam column connection is able to rotate to a certain degree whilst carrying the design resisting moment. The beam column connection requires certain ductility. This is often difficult to reconcile with the requirements for serviceability as the concrete will crack as the tension reinforcement within it strains beyond yield. To minimise these cracks the reinforcement must be well distributed and preferably relatively small diameter bars. The bars must also extend beyond the area likely to be subjected to hogging moments. Detailing rules, for the reinforcing bars in a continuous beam, are therefore important (British Standards Institute 1994).

The requirements for ductility also impinge upon the design of any stiffeners across the compression zone in the column web. For fully rigid beam column behaviour the connection is assumed not to deform however in semi-rigid beam column connections this is permitted as long as the beam to column connection has adequate rotation capacity. Ideally for semi-rigid beam column connections web stiffeners should be avoided as they place additional ductility requirements on the bolts and end plates and add to the overall cost of the connection.

Unlike in the case of sagging moment at the support the concrete is in tension and most probably cracked. Consequently the shear connectors are placed in an area of cracked concrete that offers less triaxial support. Early design guides suggested increasing the number of connectors in the hogging region (CIRIA 1983). Current recommendations require the hogging region to be designed for full connection (British Standards Institution 1994). It is also the case that the effective breadth of the slab section is less in this region and is disturbed by the presence of the column continuing between floors. The effective breadth of the slab over the support is taken as

$$b_{eff} = l_0/8 \tag{2.9}$$

Finally the lower flange of the steel section at the support is subject to compression. The moments carried by the beam at this point are high. The potential for local buckling of the constituent plates and lateral torsional (and distortional) global mode is high (the upper flange may be held in position by the concrete slab). This may be particularly acute for local buckling as the lower part of the beam will have reached and possibly exceeded yield. It is important to ensure that the lower part of the web is classified correctly which for slender webs may entail an elastic analysis of the section (Johnson 1994). Again Chapter 4 covers this in more detail.

Fortunately the actual beam column connection normally constitutes a restraint for the more global buckling mode and the maximum possible buckle length is generally relatively short as shown in Figure 2.25. Oehlers and Bradford

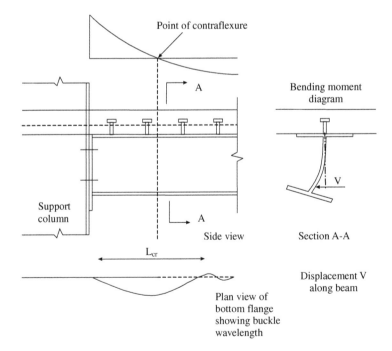

Figure 2.25 Maximum buckle length in continuous beam

2.8 BEAMS WITH COMPOSITE SLABS

As stated earlier in the chapter the use of composite slabs in building construction has been very popular over the last thirty years. In order to establish composite beam behaviour the composite slab must act with the steel section and there are a number of aspects of this that deserve special comment.

Figure 2.26 shows the cross section geometry of a typical composite slab/composite beam arrangement. In this case the flutes of the steel deck run perpendicular to the steel section although situations where the flutes run parallel are also common. The concrete in the deck flutes is discontinuous and is unlikely to resist any longitudinal forces. Hence the effective area of concrete in the composite section is above the upper flange of the deck. In calculating the force in the concrete or the transformed area of concrete this has the effect of reducing the total force in the concrete and lowering the neutral axis. However in most common situations the neutral axis lies within the area between the top and bottom flanges of the steel sheeting.

It can also be noted in Figure 2.26 that the stud shear connectors are located in the flutes of the deck. The predominant force on the connector is through the concrete above the deck flange and is therefore more eccentric to the stud base than is the case in a solid slab. In addition the stud is situated in an area of unstressed concrete is less likely to be able to resist the same force as a stud situated in a solid slab. In fact the failure mode of the connector may be very different. Lloyd and Wright (1990) suggest that the predominant failure is one of shear across a plane of concrete as shown in Figure 2.27. Johnson and Yuan (1999) however suggests that this is just one of a number of failure modes dependant upon the position of the connector. Current codes of practice base the resistance of through deck welded connectors on the full capacity reduced by a factor that empirically takes into account the deck and stud geometry as noted in equations 1.4 and 1.5.

Figure 2.26 Forces in composite beam utilising a composite slab

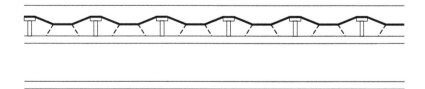

Figure 2.27 Possible concrete failure plane in composite beam formed with a composite slab

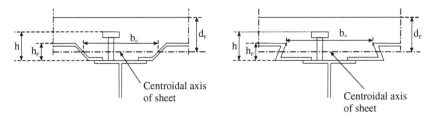

Figure 2.28 Geometry of deck for calculation of reduction formula

$$k_l = 0.6(b_o/h_p)[(h/h_p)-1] \le 1.0 \qquad (2.10)$$

Figure 2.28 describes the geometry associated with this formula.

A further influence requiring consideration for composite slab/composite beams is that of the effective transfer of longitudinal shear between the steel section and slab. The presence of the steel deck means that there is a lower requirement for reinforcement. Decking lying transverse to the steel section provides most resistance to shear forces and it is often the case that no additional reinforcement is required. For decking lying longitudinal to the beam the likely shear surface (see Figure 2.22) is reduced and additional reinforcement is more common. Design rules for both cases are provided in most codes (British Standards Institute 1990 and Eurocode 4, 1994).

Most composite beams are relatively slender in comparison to conventional cases where the steel section carries the whole load. However composite slab/composite beams tend to be even more so and are often used in office situations where their flexibility may be perceived as being "lively". It is generally recognised and described in the work of Chen and Richie (1984), that if the natural frequency of a floor is less than 4 Hz the response of the floor to foot traffic may cause disquiet to occupants. In practice the natural frequency of a floor is a complex combination of factors which include the effect of adjacent spans, floor finishes and partitions. However for general design purposes the frequency of a beam may be established using the formula given by Wyatt (1989)

$$f \approx 18/\sqrt{\delta_{sw}} \qquad (2.11)$$

2.9 CURRENT DESIGN AND FUTURE DEVELOPMENT

This chapter has described the various engineering aspects that influence the concept, design and performance of composite beams. It may be noted that the composite nature of the element adds to the complexity of the design process especially if partial connection and interaction are used. However design using composite beams can provide significant economy through reduced materials more slender floor depths and quicker construction.

These have tended to be taken to the limit and it is reported (Leon 2000) that current practice in America is leading to almost 20 m spans with shallow steel section sizes and degrees of partial connection as low as 20%. The code (AISC 1993) used in America is less restrictive than British and European codes where such spans and connection levels are not permitted. Leon argues that the reason that failures have not been reported lies in the use of high overall load factors and the use of relatively simple and conservative plastic design methods. In the case of serviceability requirements the American code fails to address the possibility of yielding in the un-propped steel during construction, shrinkage, creep, vibration limits and the extent of cracking over the supports. Despite this long slender beams are being constructed without major problems and it will be hard to introduce code rules that suggest that these are now unsafe or unserviceable.

A further aspect that has assisted in the adoption of composite beam systems is the development of fire engineering methods. Initially fire regulations have required that the steel section be protected using casing in fire resistant boards, however the use of sprayed fire coatings, normally vermiculate-based cements meant that protection could be applied quickly and therefore cheaply. From 1988 the Steel Construction Institute has promoted fire engineering (SCI 1988). This method allows engineers to calculate the fire load and the temperature rise in unprotected steel concrete composite sections. With appropriate reinforcement in the slab and for moderate periods of fire resistance this allows the steel to be unprotected. The benefits of this are obvious in both reducing wet trades and reducing construction times.

As composite beams are normally subject to uniform loading they are required to resist high moments and relatively low vertical shear forces. This has led to the development of two structural forms that have become relatively common in current construction.

The first is the use of a castellated beam rather than solid section. This allows the depth of the steel section to be high without incurring a weight penalty. Westok (Westok 2002) manufacture a cellular beam in which the castellations are cut as circular holes. This is particularly useful as the resulting holes may be used to route ventilation ducting in the floor depth rather than below the beam.

Another form of beam developed to assist in the integration of services is the Fabsec® beam (Fabsec® 2002). Rather than use rolled steel the sections are fabricated using plate. This allows the shape of the beam to be changed along its length. This profiling may be applied to flange or web. For simple span beams this may result in a deep section at mid-span and a shallow section at the support allowing services to be routed close to the column lines where the overall floor depth is reduced. Fabsec® can also incorporate cellular webs allowing considerable optimisation of the overall section.

Recent developments in Britain and Europe include the use of the Slimflor® beams mentioned earlier in the chapter. Initially these beams were fabricated using steel plates in the form of a top hat box of from a column section with a flat plate welded to the lower flange to carry the deep decking (Mullet 1998). The system works in conjunction with a deep cassette form of deck described more fully in Chapter 7.

Corus have recently developed a special asymmetric section with a wider lower flange and a ribbed surface to the upper flange to assist in composite connection. The geometry of the section was developed not only with strength properties in mind but also to withstand fire, using principles of fire engineering. This has led to a relatively thick flange capable of taking some moment when fire has reduced the capacity of the lower flange. The Slimflor® system is a direct competitor to concrete flat slab construction and is expected to span between six and nine metres on an approximately square column grid. It may also be possible to extend this by the use of continuity and to this end a novel "stem-girder" system has been proposed by Kim *et al.* (2001).

The "stem-girder" concept in this reference is a development of semi-continuous construction described earlier in the chapter and in Chapters 4 and 5. It is an attempt to make the design more easy to understand by avoiding the need to establish the moment rotation behaviour of the connection. The author's experience in dealing with local industry is that design charts such as those produced by Corus (2000) and software tools (CADS 2001) are used extensively without a real appreciation of the complex behaviour of the beam or beam to column joint.

2.10 REFERENCES

AISC (1993) *Load and Resistance Factor Design Specification for Structural Buildings*. American institute of Steel Construction Chicago Illinois.

Ansourian P (1975) *"An application of the method of finite elements to the analysis of composite floor systems"*. Proc Inst Civil Engineers, London Part 2, Vol 59, 699–726.

British Standards Institution (1990) *Code of Practice for Design of Simple and Continuous Beams*, London.

Eurocode 4 (1994) *Design of Composite Steel and Concrete Structures Part 1.1 General Rules and Rules for Buildings (with UK National Applications Document)*. London.

CADS (2001) *SMART Engineer*, Computer software produced by Computer and Design Services Ltd.

Chien EYL and Ritchie JK (1984) *Composite Floor Systems*. Canadian Institute of Steel Construction, Willowdale, Ontario.

CIRIA (1983) *Composite Beams and Slabs with Profiled Steel Sheeting*. Report 99 CIRIA, London.

Concrete Society (2000) *Design Guidance for Strengthening Concerete Structures using Fibre Composite Materials*. The Concrete Society Technical report No 55.

Corus (1999) *Bi-Steel™ Design and Construction Guide*, Vol 1, Corus.

Corus (2000) *Corus Construction Centre*. CD produced by Corus Scunthorpe.

Fabsec (2002) *Product Brochure*, Fabsec Ltd, Brooklands Court, Tunstall Rd, Leeds, LS11 5HL.

Harding PW (1992) *Composite Floors with Profiled Steel Sheeting*, PhD Thesis, University of Wales Cardiff.

Jayas BS and Hosain MU (1988) *Behaviour of Headed Studs in Composite Beams: Push-out Tests*. Canadian Journal of Civil Engineering, Vol 15, pp. 240–253.

Johnson RP (1994) *Composite Structures of Steel and Concrete Vol 1 Beams, Slabs, Columns and Frames for Buildings*, Blackwell Scientific Publications, London.

Johnson RP and Buckby RJ (1986) *Composite Structures of Steel and Concrete Vol 2 Bridges*, 2nd edition, Collins Professional and Technical Books, London.

Johnson RP and Yuan H (1999a) *Existing Rules and New Tests for Stud Shear Connectors in Troughs of Profiled Sheeting*. Proc Inst Civil Engineers Structures and Buildings, Vol SB128, No 3.

Johnson RP and Yuan H (1999b) *Models and Design Rules for Stud Shear Connectors in Troughs of Profiled Sheeting*. Proc Inst Civil Engineers Structures and Buildings, Vol SB128, No 3.

Kim B, Wright HD and Cairns R (2001) *A Study of Continuous Stem Girder Systems*. Structural Engineering and Mechanics, Vol 11, No 5, 469–484.

Lam D, Elliot KS and Nethercot DA (1998) *"Push-off tests on shear studs with hollow-cored floor slabs"*. The Structural Engineer, Vol 76, No 9, pp. 167–174.

Lawson RM, Mullet DL and Rackham JW (1997) *Design of Asymmetric Slimflor® using Deep Composite Decking*. The Steel Construction institute, Ascot Berks. UK.

Leon R (2001) *A Critical Review of Current LRFD Provisions for Composite Members*. Proc Structural Research Stability Council 2001, Fort Lauderdale, pp. 189–208.

Lloyd RM and Wright HD (1990) *Shear Connection between Composite Slabs and Steel Beams*. Jour of Constructional Steel research, No 15, pp. 255–285.

Mullet DL (1998) *Composite Floor Systems*. Blackwell Science/The Steel Construction Institute Oxford.

Narayanan R, Wright HD, Evans HR and Francis RW (1987) *Double-skin Composite Construction for Submerged Tube Tunnels*. Steel Construction Today. Vol 1, pp. 185–189.

Nethercot DA, Li TQ and Ahmed B (1998) *Plasticity of Composite Beams at Serviceability Limit State*, The Structural Engineer, Vol 76, No 15, pp. 289–293.

Nguyen NT and Oehlers DJ (1997) *Simply Supported Bolted Side-plated Beams with Transverse and Longitudinal Partial Interaction*. Research Report No R144. Department of Civil and Environmental Engineering. University of Adelaide.

Oehlers DJ and Bradford MA (1995) *Composite Steel and Concrete Structural Members Fundamental Behaviour*. Pergamon.

Oehlers D, Wright HD and Burnet M (1994) *Flexural Strength of Profiled Beams*. Jour American Society of Civil Engineers, Vol 120, No 2, February, pp. 378–393.

Roberts TM (1984) *Finite Difference Analysis of Composite Beams with Partial Interaction*. Computers and Structures, Vol 21, No 3.

Westok (2002), *Product Brochure*, Westok Structural Steel, Horbury Junction Industrial Estate Wakefield, England WF4 5ER.

Wright HD (1989) *The Deformation of Composite Beams with Discrete Flexible Connection*. Jour Constructional Steel Research, No 15, pp. 49–64.

Wright HD, Evans HR and Burt CA (1989) *Profiled Steel Sheet/Dry Board Composite Floors*. The Structural Engineer, Vol 67, No 7, pp. 114–120, 129.

Wright HD, Evans HR and Harding PW (1987) *The Use of Profiled Steel Sheeting in Floor Construction*. Jour Constructional Steel Research, 7, pp. 279–295.

Wright HD, Oduyemi TOS and Evans HR (1991) *The Design of Double Skin Composite Elements*. Jour Constructional Steel Research, 19, pp. 97–110.

Wright HD, Oduyemi TOS and Evans HR (1991) *The Experimental Behaviour of Double Skin Composite Elements*. Jour Constructional Steel Research, 19, pp. 111–132.

Wyatt TA (1989) *Design Guide on the Vibration of Floors*. The Steel Construction Institute (SCI)/CIRIA.

CHAPTER THREE

Composite Columns

Yong C. Wang

3.1 INTRODUCTION

3.1.1 Types of Composite Cross-section

Composite columns offer many advantages over bare steel or reinforced columns and are becoming increasingly used in multi-storey and tall buildings. A particular advantage of using a composite column is the reduction in column cross-sectional area, this being especially desirable in tall buildings where load is high and space is usually at a premium. Another important consideration is fire resistance.

Composite columns are available in many different types of cross-section a number of which are shown in Figure 3.1. Among these, the steel section encased in concrete (Figure 3.1a) perhaps represents the earliest type of composite cross-section. Initially, due to low grade, concrete was merely used as insulation to provide the steel section's fire resistance. But later research studies showed that by using better quality concrete, significant enhancements in the column strength were possible, enabling smaller steel sections to be used.

Nowadays, owing to unattractive appearance and the need for temporary formwork for concrete casting, composite columns made of steel sections encased in concrete are less often used than concrete filled hollow sections (Figure 3.1b). Moreover, by using steel tubes as permanent formwork, construction speed is increased. Because of the inherently high fire resistance of concrete filled columns,

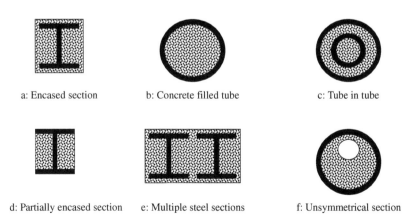

Figure 3.1 Types of composite cross-section

fire protection of steel is in many cases not necessary and the steel can be exposed to achieve attractive appearance. Since construction speed is an important advantage, reinforcement is usually not used. But when required, a convenient reinforcement method is to insert a second tube inside the main one (Figure 3.1c).

To eliminate temporary formwork while still using universal sections in composite columns, partial encasement may be used (Figure 3.1d). In this type of column, concrete is cast between the flanges of the steel section. Since the steel web is protected from fire attack, fire resistance of this type of composite column is reasonably high.

If the applied load is particularly heavy, e.g. in the bottom storey of a tall building, composite columns can be made by encasing two or more universal steel sections together into concrete (Figure 3.1e) or by concrete filling tubes made of large welded plates.

The above mentioned composite cross-sections are all symmetrical. Sometimes, unsymmetrical cross-sections may become unavoidable. For example, building services ducts may be arranged within concrete of a composite cross-section (Figure 3.1f) or due to difficult access the steel section may have to be aligned towards one side in an encased composite cross-section.

3.1.2 Objectives of This Chapter

This chapter is divided into two main parts: dealing with both the cold and fire conditions. Many research studies have been carried out on both aspects of the composite column behaviour. For detailed information on previous research studies, the reader is referred to the chapter on composite columns in the latest SSRC guide (SSRC 1998) and a review article by Uy (1998), both have provided an extensive list of references on this topic. This chapter will provide a brief description of those aspects of composite column behaviour that should be considered in design calculations. This is followed by more detailed discussions on the design provisions for composite columns included in Eurocode 4 Parts 1.1 and 1.2 (CEN 1992b, 1994). Since Eurocode 4 is limited to composite columns with concrete cylinder strength less than 50 N/mm^2 and thick steel plates where there is no local buckling, results of some recent research studies will be introduced to extend the range of applicability of this design method.

Usually, there is no need for special consideration of composite column behaviour under construction loading. However, if a composite column is made by concrete filling a steel tube of welded thin-walled plates, consideration should be given to lateral deflection of the steel plate under combined axial loading and lateral loads from wet concrete (Uy and Das 1999). To limit this deflection to an acceptable level, bracing may be necessary. One provision of bracing may be based on the design of temporary formwork for reinforced concrete columns. However, this is beyond the scope of this chapter and only composite behaviour will be considered here.

Design calculations for composite columns in each country follow their national standards. The UK design method is provided in the bridge standard BS 5400 Part 5 (BSI 1979). The British building code BS 5950 Part 1 (BSI 2000) also has a set of clauses for encased steel columns, however, this method is based on experimental information of some time ago and is very rarely used in practice. The applicable European standard is Eurocode 4 Part 1.1 (CEN 1992b, hereafter to be referred to as EC4). Available international design guides include the CIDECT design guide (Bergmann, Matsui, Meinsma and Dutta 1995) for concrete filled columns and that by Roik and Bergmann (1992), both having very a similar approach to EC4. Other composite design methods include those of the American Institute of Steel Construction (AISC 1996), the American Concrete Institute (ACI 1995) and the Architectural Institute of Japan (Matsui, Mitani, Kawano and Tsuda 1997). Whilst the Japanese standard is based on the allowable stress approach, all other standards adopt the limit state approach. Both ACI and AIJ standards are based on the concrete design approach and treat the steel section in a composite cross-section as reinforcement. The AISC treatment is similar to EC4 and adopts the steel column design approach.

For fire safety design, Eurocode 4 Part 1.2 (CEN 1994, hereafter to be referred to as Part 1.2) has a detailed design method for composite columns and the British Standard BS 5950 Part 8 (BSI 1990) also deals with this subject. However, discussion is currently taking place to introduce the Part 1.2 method into BS 5950 Part 8 and since Eurocodes are set to replace different national standards in Europe and are basis of the aforementioned international design guides, the presentation of design methods in this chapter will follow EC4 for cold design and Part 1.2 for fire design.

Eurocodes adopt the limit state design philosophy and partial safety factors are used to deal with variability and uncertainty in loading and material properties. Since the basis of design calculations is not affected by different values of partial safety factors, these partial safety factors are set to unity to simplify presentation in this chapter.

For the cold conditions, presentation in this chapter starts from the relatively simple calculations of columns under pure compression and progresses up to composite columns under combined axial compression and biaxial bending. For fire resistance calculations, since Part 1.2 is based on EC4, this chapter will only present modifications to the design method that are appropriate to the fire situation.

3.2 COMPOSITE COLUMNS UNDER AXIAL LOAD IN COLD CONDITION

3.2.1 Strength of Concrete

EC4 is applicable only to normal strength concrete having a cylinder strength of up to 50N/mm^2. As columns are mainly designed to carry compressive load, it is sometimes more economical to use high strength concrete whose cylinder strength may be up to 100N/mm^2. Although high strength concrete may be used to encase steel sections, high strength concrete failure is brittle so it is more likely that such concrete will be used to infill hollow steel sections to form composite columns. Some recent research studies suggest that EC4 may be easily modified to design high strength concrete filled columns. These modifications will be introduced in this chapter.

3.2.2 Local Buckling of Steel

To fully utilize steel strength, local buckling of a steel section should be prevented so that it does not occur before the steel reaches its yield stress. Thus, steel column sections should conform to Class 3 (semi-compact) or better. In composite columns, concrete strain at peak stress is about 0.0035 while the yield strain of grade S355 steel is only 0.00175. Therefore, when a composite cross-section is under axial compression and both steel and concrete attain the same strain, the steel will yield before the concrete reaches its peak compressive stress. For the concrete to reach its peak stress, the steel plates will therefore have to undergo further strain without local buckling to allow the concrete to take up additional load until it has also reached its peak stress. Thus, according to the definition of section classification, the steel plates should be Class 2 (Compact) to prevent local buckling in the composite cross-section. In this sense, requirements to control local buckling are more stringent in a composite column than in a bare steel column. However, since concrete provides substantial restraint to prevent local buckling of steel in a composite cross-section, the allowable steel width to thickness ratio is often significantly higher than in a bare steel cross-section.

EC4 is only applicable to composite columns where local buckling of the steel plates is prevented.

For steel sections encased in concrete, provided the code requirements of minimum thickness for concrete cover are met, local buckling will not occur and full steel strength can be achieved.

For other types of composite cross-section, Table 3.1 lists the maximum allowable steel plate width to thickness ratio to prevent local buckling. These are contrasted with the requirements for bare steel cross-sections from Eurocode 3 Part 1.1 (CEN 1992a) for steel structures.

It can be seen from Table 3.1 that concrete filling rectangular hollow sections can significantly improve the local buckling strength of steel. This is because the buckling pattern of an unfilled rectangular tube consists of both inwards and outwards buckles along the tube length. With concrete filling, the inwards buckles are unable to form forcing a higher buckling mode. In contrast, the buckling half wavelength for an unfilled circular steel tube is small and consists predominantly of a single circumferential outwards buckle. Consequently, concrete filling does not enhance the buckling strength of steel.

Using the limits in Table 3.1 for composite cross-sections, it may be checked that for Corus (formerly British Steel) produced steel sections, local buckling will not occur in concrete filled circular hollow sections or in partially encased I or H sections. For concrete filled rectangular hollow sections, only a few sections will not meet the requirements of Table 3.1. These are large sections with the least steel thickness and are identified in Table 3.2 for grade S355 steel.

Table 3.1 Allowable limits of steel width to thickness ratio for local buckling

Section type	Eurocode 4 Part 1.1: Composite requirement	Eurocode 3 Part 1.1: Steel requirement
Rectangular hollow section	52ε	42ε
Circular hollow section	$90\varepsilon^2$	$90\varepsilon^2$
Partially encased I-section	44ε	30ε

Table 3.2 Sections not satisfying requirements of Table 3.1

Section type	Grade S275 steel	Grade S355 steel
SHS	None	300×300×6.3
RHS	None	250×150×5.0
		300×200×6.3
		400×200×8.0
		450×250×8.0
		500×300×10.0

3.2.3 Concrete Filled Thin-Walled Columns

It is sometimes more economical to use thin-walled steel tubes in concrete filled columns. Uy (1999) carried out experimental and theoretical studies of the local buckling of concrete filled thin-walled rectangular sections. The width to thickness ratio of the steel plates was up to 100. He compared the test results with predictions of EC4 and found that EC4 could still be used provided the effective section of the steel plates was used in the EC4 calculations. O'Shea and Bridge (1999) tested concrete filled circular hollow sections with steel diameter to thickness ratio in excess of 200. They concluded that the EC4 method may be used without modification for local buckling.

3.2.4 Squash Load of a Composite Cross-Section

If local buckling of steel is prevented before the concrete reaches peak stress, the maximum resistance to axial compression of a composite cross-section is given by:

$$N_{pl,Rd} = A_a f_y + 0.85 A_c f_{ck} + A_s f_{sk} \tag{3.1}$$

where A is area and f design strength of material. Subscripts a, c and s refer to steel, concrete and reinforcement respectively. For concrete, the cylinder strength is used and the constant 0.85 is a factor to allow for concrete deterioration due to environmental exposure and splitting. For concrete filled steel sections where concrete is protected, its cylinder strength can be fully developed and the constant 0.85 may be omitted from equation 3.1.

Equation 3.1 is generally applicable to composite cross-sections. For concrete filled circular hollow sections, concrete confinement should be included. For as concrete approaches plastic failure, its Poisson's ratio increases drastically. But this rapid lateral dilation is restrained by high stiffness of the steel hollow section whose Poisson's ratio is much lower. This gives rise to confinement, leading to an increase in failure strength of the concrete. Both rectangular and circular steel sections can confine concrete, but the confinement effect in rectangular sections is non-uniform and small and may be safely ignored. In design calculations, the advantage of concrete confinement is only considered for concrete filled circular steel sections. When steel is restraining concrete, a tensile stress is also produced in the circumferential direction of the steel tube, leading to a reduction in steel strength in the longitudinal direction. Despite this, the net effect is always an increase in the axial resistance of the composite cross-section.

For long columns or columns under large bending moments, concrete failure strain is small and its Poisson's ratio is similar to that of steel. Therefore, both steel and concrete undergo similar lateral expansion under longitudinal compression and steel does not confine concrete. In general, the squash load of a concrete filled circular steel section including confinement effect may be calculated from:

$$N_{pl,Rd} = \eta_2 A_a f_y + A_c f_{ck}\left(1 + \eta_1 \frac{t}{D}\frac{f_y}{f_{ck}}\right) + A_s f_s \tag{3.2}$$

where

$\eta_1 = \eta_{10}\left(1 - \frac{10e}{D}\right)$ but >0.0 with $\eta_{10} = 4.9 - 18.5\bar{\lambda} + 17\bar{\lambda}^2$ and

$\eta_2 = \left(\eta_{20} + (1-\eta_{20})\frac{10e}{D}\right)$ but <1.0 with $\eta_{20} = 0.25(3+2\bar{\lambda})$

where e is eccentricity, D outer diameter of the steel section and $\bar{\lambda}$ column slenderness (to be defined in equation 3.8).

In realistic applications, it is worthwhile taking advantage of the improved squash load of a concrete filled CHS section only when the column slenderness is very low and eccentricity is small. This usually occurs in the bottom storey columns of tall buildings. Even for these cases, when the column slenderness is higher than 0.5, concrete confinement is negligible and the enhanced strength minimal.

For high strength concrete (cylinder strength > 80 N/mm²) filled columns, the results of tests by O'Shea and Bridge (1999) and Wang (1999b) indicate that the effect of confinement is small. EC4 overestimates the confinement effect and should be ignored.

As can be seen later in this chapter (equation 3.8), the column slenderness is dependent on the squash load $N_{pl,Rd}$ of the composite cross-section. In order to avoid iteration, concrete confinement should be ignored when calculating the column squash load to be used in equation 3.8.

3.2.5 Euler Load

For slender columns, instability will occur before the composite cross-section reaches its squash load. For a perfect column with ideal pin supports at ends, the column buckling load is given by the Euler load:

$$N_{cr} = \frac{\pi^2 (EI)_e}{L_e^2} \tag{3.3}$$

while $(EI)_e$ is the effective flexural rigidity of the composite cross-section and L_e the column effective length.

The effective flexural rigidity of a composite cross-section is given by:

$$(EI)_e = E_a I_a + 0.8 E_{cd} I_c + E_s I_s \tag{3.4}$$

where E is the modulus of elasticity of a material and I the second moment of area of the component material about the relevant principle axis of the composite cross-section. For convenience, equation 3.4 is used with uncracked composite cross-section, and crack and long-term concrete behaviour is accounted for by using a reduced concrete modulus of elasticity.

The reduced modulus of elasticity of concrete E_{cd} is obtained from:

$$E_{cd} = E_{cm}/\gamma_m = E_{cm}/1.35 \qquad (3.5)$$

where E_{cm} is the concrete secant modulus. γ_m (=1.35) in equation 3.5 appears as a material safety factor. However, this factor in combination with a factor of 0.8 in equation 3.4 is used to account for the effect of cracking in concrete so that design calculations may be carried out on an uncracked section. Therefore, this factor should not be changed when carrying out evaluation of test results where partial safety factors are usually set at unity.

If a composite column were perfect and under pure axial compression, the column would shorten uniformly, without causing any additional bending moment. However, in realistic columns, due to initial imperfections, second order bending moments are generated. These bending moments give rise to strain gradients (curvature) in composite cross-sections. Under the influence of creep and shrinkage, these strain gradients are magnified, leading to increased bending moment and earlier failure of a composite column.

To account for the effects of creep and shrinkage, the effective design stiffness of concrete is further reduced in Eurocode 4 and E_{cd} in equation 3.4 is replaced by E_c:

$$E_c = E_{cd}\left(1 - 0.5\frac{N_{G,Sd}}{N_{Sd}}\right) \qquad (3.6)$$

where $N_{G,Sd}$ is the permanent (long-term) design load and N_{Sd} the total design load.

Equation 3.6 is obtained by assuming that under permanent design load only ($N_{G,Sd}/N_{Sd}=1$), concrete stiffness is reduced to half its short-term value.

Creep and shrinkage induce secondary bending moments in a composite column by magnifying the lateral deflection due to initial imperfection. But depending on the column slenderness and eccentricity, the effects of creep and shrinkage may be small enough to be neglected. Therefore, when using equation 3.6, the following ranges of application should be observed:

- The development of second order bending moments depends mainly on the column slenderness, the more slender a column, the higher the second order bending moment. Thus, the effect of creep and shrinkage is more pronounced in more slender columns. In short columns, the effect of creep and shrinkage is small and may be neglected. EC4 has set some limits on column slenderness below which the effect of concrete creep and shrinkage does not have to be considered. In most cases, composite columns have a slenderness lower than these limits.
- The importance of second order bending moments depends on the magnitude of the primary bending moment. With small primary bending moments, secondary bending moments are relatively important and creep induced P–δ effect should be considered. In contrast, under large primary bending moments, the creep induced bending moment is

relatively small and may be ignored. Eurocode 4 sets the upper limit of eccentricity at twice the relevant dimension of the composite cross-section above which the secondary bending moment may be ignored. This is a very high limit and it is not very often that columns have to sustain such large bending moments.

3.2.6 Column Design Strength

The composite cross-section squash load and Euler buckling load are upper bounds on column strength. Under realistic conditions, various imperfections will generate second order bending moment and the column design strength will be lower. For composite column design, either the steel or concrete based approach may be used. EC4 takes the steel column design approach and the strength of a composite column is calculated using:

$$N_{Rd} = \chi N_{pl,Rd} \tag{3.7}$$

where χ is the column strength reduction factor and is a function of the column slenderness $\bar{\lambda}$.

In EC4, the slenderness of a composite column is defined as:

$$\bar{\lambda} = \sqrt{\frac{N_{pl,Rd}}{N_{cr}}} \tag{3.8}$$

It should be pointed out that to simplify presentation, material partial safety factors are not included in this chapter. In actual design calculations, when material safety factors are included in design calculations of the column squash load $N_{pl,Rd}$, a smaller value will be obtained than given by equation 3.1. This would give a lower column slenderness from equation 3.8, leading to a higher column strength reduction coefficient χ. This represents an artificial increase in the column design strength that should not be used. Hence when calculating values to be used in equation 3.8, the material partial safety factors should not be used. In addition, for concrete filled circular columns, the column slenderness should be calculated without including the effect of concrete confinement.

The relationship between the column strength reduction factor χ and the column slenderness is given by a column buckling curve. Figure 3.2 shows three column buckling curves used in EC4. The selection of a column buckling curve depends on the column cross-section type and its axis of buckling. Based on calibration against test results, column buckling curve "a" may be used for concrete filled steel sections and column buckling curves "b" and "c" used for concrete encased steel sections bending about the major and minor axis of the steel section.

3.2.7 Unsymmetrical Sections

The aforementioned calculation method is only suitable to composite columns with symmetrical cross-sections. For unsymmetrical composite cross-sections, this

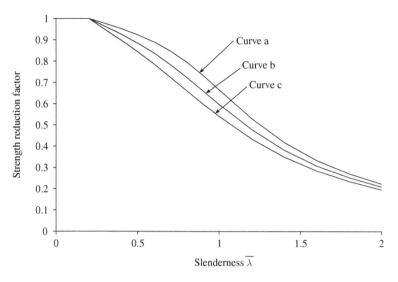

Figure 3.2 Column buckling curves

method can still be used. However, in this case, even under pure compression, a column with an unsymmetrical cross-section should be designed for combined axial compression and bending. This is because the resultant compression force acts at the elastic centroid of the composite cross-section according to the distribution of axial stiffness, but the column axial resistance acts at its plastic centroid according to the distribution of axial resistance. For an unsymmetrical cross-section, these two centroids do not coincide.

In lieu of carrying out full design calculations for combined compression and bending, the simple method of Roik and Bergmann (1990) may be used. In this method, the column design resistance for axial load with regard to the plastic centroid ($N_{Rd,pl}$) is given by:

$$N_{Rd,pl} = \chi_{pl} N_{pl,Rd} \tag{3.9}$$

where the column strength reduction factor with regard to the plastic centroid (χ_{pl}) is related to that with regard to the elastic centroid (χ_{el}) according to:

$$\chi_{pl} = \left(\frac{\alpha}{2} - \sqrt{\frac{\alpha^2}{4} - \frac{\chi_{el}}{\bar{\lambda}^2}} \right) \tag{3.10}$$

where α is given by:

$$\alpha = \left(\chi_{el} + \frac{1.1}{\bar{\lambda}^2} \right) \tag{3.11}$$

The column slenderness $\bar{\lambda}$ and strength reduction factor with regard to the elastic centroid (χ_{el}) are calculated in the same way as for a symmetrical cross-section.

3.3 COMPOSITE COLUMN UNDER COMBINED AXIAL LOAD AND BENDING MOMENTS AT AMBIENT TEMPERATURE

3.3.1 Axial Load–Bending Moment (N–M) Interaction Diagram of a Composite Cross-Section

To design a composite column under axial compression combined with bending moments, it is necessary to evaluate the axial load–bending moment (N–M) interaction diagram of the composite cross-section. This diagram gives the failure condition of the composite cross-section under combined axial load and bending moment and forms the basis of design for the composite column.

Refer to Figure 3.3, the general procedure for evaluating the N–M interaction diagram of a composite cross-section is as follows:

1. Set the concrete strain at the furthest compression fibre to its crushing strain.
2. Assume an arbitrary position for the neutral axis (NA). Assuming the strain distribution in the composite cross-section is linear, strains in the composite cross-section are now determined (Figure 3.3b).
3. Evaluate the stress distribution of the composite cross-section according to its strain distribution and stress–strain relationships of the constituent materials. It is assumed that concrete does not have any tension resistance. The stress distributions are now obtained as in Figures 3.3c and 3.3d.
4. The axial load is obtained by integration of stress over the whole composite cross-section. The bending moment is obtained by taking moments about the plastic centroid of the cross-section. This determines one point in the N–M interaction diagram.
5. Changing positions of the neutral axis, other points in the N–M diagram are obtained.

Due to the assumption that concrete has no tensile strength, the N–M interaction curve is convex as shown in Figure 3.4.

The above described general approach is time consuming and is best carried out by computers. To simplify calculations, the concrete stress distribution may be approximated by a uniform stress block with a reduced depth of compression as shown in Figure 3.3c. For a composite cross-section with encased steelwork, this reduction factor γ is (Oehlers and Bradford 1995):

$$\gamma = 0.85 - 0.007(f_{ck} - 28) \leq 0.85 \tag{3.12}$$

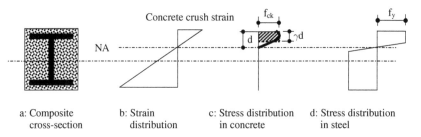

a: Composite cross-section   b: Strain distribution   c: Stress distribution in concrete   d: Stress distribution in steel

Figure 3.3 Determination of the N–M interaction diagram of a composite cross-section

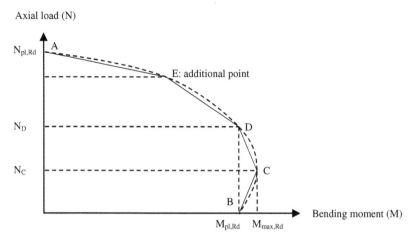

Figure 3.4 Axial force (N)–bending moment (M) diagram of a composite cross-section

3.3.2 Simplified Method for Symmetrical Cross-sections

For a composite cross-section symmetrical about the axis of bending, Roik and Bergamann (1992) have proposed a simple method to evaluate its N–M interaction diagram. This method is adopted in EC4 and is described in the following section. As shown in Figure 3.4, in this method, instead of determining the continuous N–M interaction curve, only a few key points in the curve are determined. The N–M curve is then constructed by joining these key points by straight lines.

When evaluating these key points, rigid-plastic material behaviour is assumed. Thus, steel is assumed to have reached yield in either tension or compression. Concrete is assumed to have reached its peak stress in compression and its tensile strength is zero.

However, results of a recent study by O'Shea and Bridge (1999) indicate that for high strength concrete ($f_{ck} \geq 100$ N/mm^2) filled columns, the assumption of rigid-plastic behaviour in the simplified approach is unconservative and the general procedure should be used. Similarly, if steel of very high strength is used, steel may not fully yield at concrete crushing and the general approach should be used (Uy 2001).

The key points in Figure 3.4 to be evaluated are:

A Squash load point $(N_{pl,Rd}, 0)$,
B Pure flexural bending point $(0, M_{pl,Rd})$,
C The maximum bending moment point $(N_C, M_{max,Rd})$,
D Point $(N_D, M_{pl,Rd})$ with bending moment equal to the pure bending moment capacity,
E A point between A and D to refine the N–M interaction diagram.

The squash load $N_{pl,Rd}$ has been determined in equation 3.1. In the following paragraphs, values of N_D, N_C, $M_{pl,Rd}$ and $M_{max,Rd}$ are determined.

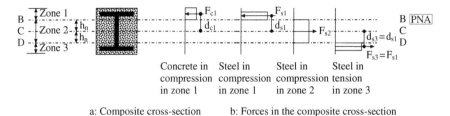

Concrete in   Steel in   Steel in   Steel in
compression   compression   compression   tension
in zone 1   in zone 1   in zone 2   in zone 3

a: Composite cross-section   b: Forces in the composite cross-section

Figure 3.5 Forces in a composite cross-section

N_D

This is the axial compression force at which the composite cross-sectional bending moment resistance is the same as that under pure bending. Refer to Figure 3.5 and consider the position of the plastic neutral axis (PNA) of the composite cross-section under pure bending. Assume that the part below the PNA is in tension and the part above in compression. Since concrete is assumed to have no tensile resistance, the plastic neutral axis for pure bending must be above the axis of symmetry C–C. Assume it is at a distance h_n from the axis of symmetry at B–B.

If the composite cross-section is divided into three zones as shown in Figure 3.5, the forces in the composite cross-section may be divided into four parts: concrete in compression in zone 1 above the plastic neutral axis (F_{c1}), steel in compression in zone 1 above the plastic neutral axis (F_{s1}), steel in tension in zone 2 within a distance h_n on either side of the axis of symmetry (F_{s2}) and steel in tension in zone 3 (F_{s3}). From symmetry:

$$F_{s1} = F_{s3} \tag{3.13}$$

Since the resultant axial force in the composite cross-section is zero under pure bending, it follows:

$$F_{c1} + F_{s1} = F_{s2} + F_{s3} \tag{3.14}$$

Making use of equation 3.13, it follows:

$$F_{c1} = F_{s2} \tag{3.15}$$

The plastic bending moment capacity of the composite cross-section under pure bending is that about the axis of symmetry (C–C) and is then given by:

$$M_{pl,Rd} = F_{c1} * d_{c1} + 2 * F_{s1} * d_{s1} \tag{3.16}$$

Now consider the case where the plastic neutral axis is moved from B–B to D–D, forces in both concrete and steel in zone 2 changing from tension to compression. But since the centroid of these forces is at the axis of symmetry, their bending moment contribution is zero. Thus the bending moment in the composite cross-section is unchanged by moving the plastic neutral axis from B–B to D–D, but there is now a net compressive force in the composite cross-section. This force is N_D in Figure 3.4 and it is given by:

$$N_D = F_{c1} + F_{c2} + F_{s1} + F_{s2} - F_{s3} = F_{c1} + F_{c2} + F_{s2} \quad (3.17)$$

Recognizing $F_{c1} = F_{c3}$ and substituting into equation 3.15, it follows $F_{s2} = F_{c3}$ hence

$$N_D = F_{c1} + F_{c2} + F_{c3} = N_{c,Rd} \quad (3.18)$$

where $N_{c,Rd}$ is the compressive resistance of the entire concrete part of the composite cross-section.

$M_{max,Rd}$

The bending moment of a composite cross-section is taken about the axis of symmetry. Therefore, the maximum bending moment is obtained by placing the plastic neutral axis at the axis of symmetry of the composite cross-section. This conclusion can be obtained by examining the change in the bending moment of the composite cross-section by making a small change in the position of the plastic neutral axis.

Refer to Figure 3.6. Assume positive bending moment is anti-clockwise and causes tension below and compression above the plastic neutral axis. Consider the plastic neutral axis is moved up from the axis of symmetry. Compared to the case where the plastic neutral axis is at the axis of symmetry, this movement in the plastic neutral axis gives a net increase in the tension force. This net tension force acts at an eccentricity above the axis of symmetry, causing a clockwise bending moment about the axis of symmetry, which reduces the bending moment of the composite cross-section. Similarly, if the plastic neutral axis is moved down from the axis of symmetry, a net compressive force is obtained acting below the axis of symmetry, again causing a clockwise bending moment about the axis of symmetry and a reduction in the bending moment of the composite cross-section. Thus, moving the plastic neutral axis of the composite cross-section either above or below the axis of symmetry, the

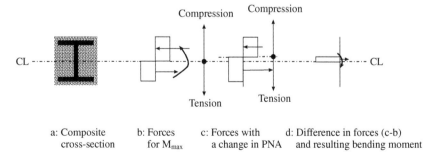

Figure 3.6 Determination of the plastic neutral axis (PNA) for the maximum bending moment

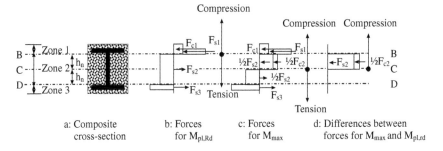

Figure 3.7 Forces for determining $M_{pl,Rd}$

bending moment of the composite cross-section is reduced, implying that the bending moment is the maximum if the plastic neutral axis is placed at the axis of symmetry. The maximum bending moment of the composite cross-section is then given by:

$$M_{max,Rd} = \left(W_{pa}f_y + \frac{1}{2}W_{pc}f_{ck} + W_{ps}f_{sk}\right) \qquad (3.19)$$

where W_{pa}, W_{pc} and W_{ps} are plastic modulus of steel, overall concrete and reinforcement about the axis of symmetry of the composite cross-section. The steel plastic modulus can be obtained from steel section book. Expressions for concrete and reinforcement plastic modulus can be easily derived. The coefficient "$1/2$" in equation 3.19 is a result of the assumption that concrete has no tensile strength and only the compressive strength contributes to the bending moment capacity.

$M_{pl,Rd}$

Comparing the stress diagram of the composite cross-section for the maximum bending moment in Figure 3.7c with the stress diagram for pure bending (zero axial force) in Figure 3.7b, the differential stress diagram (Figure 3.7d) has stress resultants $1/2 F_{c2}$ for concrete and F_{s2} for steel within a distance h_n above the axis of symmetry. Thus, the plastic bending moment capacity of the composite cross-section under pure bending may be evaluated using:

$$M_{pl,Rd} = M_{max,Rd} - (W_{pan}f_y + \frac{1}{2}W_{pcn}f_{ck} + W_{psn}f_{sk}) \qquad (3.20)$$

where W_{pan}, W_{pcn} and W_{psn} are plastic modulus of steel, concrete and reinforcement within zone 2. To obtain their values, the value of h_n should be found.

In EC4, detailed analytical equations have been provided for calculating h_n for various types of composite cross-section. Due to restriction on the length of this chapter, these equations are not repeated here.

N_C

The corresponding axial compression to the maximum bending moment is:

$$N_C = F_{s2} + \frac{1}{2}F_{c2} = F_{c1} + \frac{1}{2}F_{c2} = \frac{1}{2}(F_{c1} + F_{c2} + F_{c3}) = \frac{1}{2}N_{c,Rd} \qquad (3.21)$$

Additional Point E

Due to the way in which the N–M interaction curve is used to calculate the column strength (to be explained in the next section), using a polygonal curve will result in overestimation of the column strength compared with using the exact N–M curve. In order to reduce this overestimation, sometimes an additional point (point E) in the N–M curve is evaluated. In EC4, this point corresponds to the PNA being at half way between the compression edge of the composite cross-section and (D–D) line in Figure 3.5. Detailed analytical equations are provided in EC4 for the determination of point E in the N–M curve.

3.3.3 Strength of Composite Column with Axial Force and Bending Moment About One Axis

Determination of the composite column strength is based on the N–M interaction diagram of the composite cross-section that has been obtained in the previous section.

This N–M interaction diagram is the failure surface of a composite cross-section under compression and bending about one axis. The failure load of a composite column is always less than the capacity of the composite cross-section and this comes about because of a secondary moment associated with the column's straightness imperfection. Refer to Figure 3.8, which shows the non-dimensional N–M interaction curve of a composite cross-section, normalized with respect to $N_{pl,Rd}$ and $M_{pl,Rd}$. For a composite column under pure compression, its failure load is calculated according to equation 3.7 and is indicated by χ. This implies that at the instant of column failure, the column straightness imperfection causes an equivalent bending moment μ_k to cause failure at the most highly stressed cross-section of the composite column.

If an additional external column bending moment is applied, the column compressive resistance will be even smaller than the column resistance under

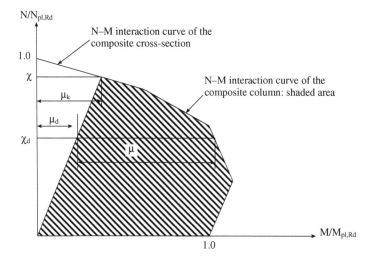

Figure 3.8 Design N–M diagram for a composite column under compression and uniaxial bending

pure compression, thus the second order bending moment arising from imperfection is also reduced. Assume that the imperfection-induced second order bending moment varies linearly with column axial load. Under the applied load χ_d, the second order bending moment is μ_d and the usable bending moment resistance is μ. Thus, the interaction curve for the composite column (as opposed to the composite cross-section) is the shaded part in Figure 3.8.

For design checks, the following equation should be satisfied:

$$M_{Sd} \leq 0.9 \mu M_{pl,Rd} \qquad (3.22)$$

The constant 0.9 in equation 3.22 is used to account for approximations in the determination of the N–M interaction curve of the composite cross-section.

In this calculation, it is assumed that there is no reduction in the column bending moment capacity due to lateral torsional buckling when it is subjected to pure bending. Since the lateral torsional stiffness of a composite column is high, this assumption is reasonable.

3.3.4 Column Strength Under Biaxial Bending

To check the composite column capacity against combined compression and biaxial bending, the N–M interaction curves of the composite cross-section about both principal axes should be evaluated. From these two N–M interaction curves, the bending moment capacities of the composite column under axial compression are separately obtained as in Section 3.3.3. When calculating these values, it is assumed that the second order bending moment arising from axial compression acting on imperfection is only effective in the expected plane of column buckling. In the other plane, column deflection and second order bending moment are assumed small. Thus, if the composite column is expected to buckle about the z–z plane, the column bending moment capacities μ_z and μ_y are obtained as shown in Figure 3.9.

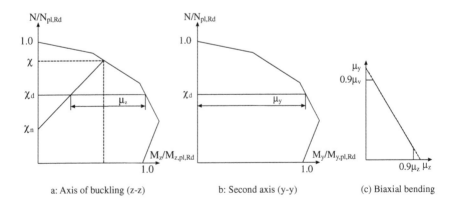

Figure 3.9 Design N–M interaction diagrams for biaxial buckling

Composite Columns

Having obtained μ_z and μ_y, a linear interaction equation is used for biaxial bending:

$$\frac{M_{y,Sd}}{\mu_y M_{pl,y,Rd}} + \frac{M_{z,Sd}}{\mu_z M_{pl,z,Rd}} \leq 1 \qquad (3.23)$$

In addition, as described in Section 3.3.3 for combined compression and uniaxial bending, equation 3.22 should also be satisfied for bending about each axis separately.

Thus, the interaction curve for a composite column under biaxial bending is as shown in Figure 3.9c.

Sometimes, the axis of buckling of a composite column may not be obvious. Under this circumstance, the above design check should be carried out for both axes.

3.3.5 Column Bending Moments

When a column is under combined compression and bending, the bending moment may be regarded as comprising of three parts as shown in Figure 3.10: the primary bending moment (Figure 3.10b), the secondary bending moment arising from initial imperfections (Figure 3.10c) and the secondary bending moment from P–δ effect (Figure 3.10d). All these bending moments should be considered in design calculations to give the maximum overall bending moment in the column.

Effect of Imperfections

When calculating the second order bending moments, a column is often assumed to have an equivalent half-sine form of initial imperfection. When acting with equal end bending moments, the maximum second order bending moment then directly adds to the primary bending moment. With other applied

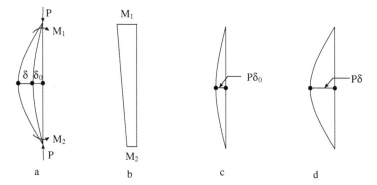

a: Column loads and deflections
b: Primary bending moment distribution
c: Bending moment distribution from initial deflections
d: Bending moment distribution from displacements

Figure 3.10 Bending moments in a column

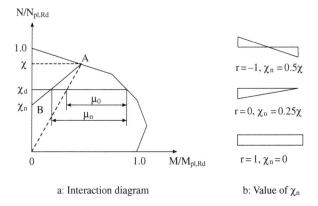

a: Interaction diagram   b: Value of χ_n

Figure 3.11 Effect of bending moment distribution on N–M interaction diagram

end bending moments, because the maximum second order bending moment is not at the same position as the maximum applied bending moment, the effect of the second order bending moment is less severe. However, it is difficult to evaluate precisely the secondary bending moment induced by initial imperfections. In EC4, this is considered by increasing the axial load at which the column N–M interaction diagram is affected.

Refer to Figure 3.11. If the primary bending moment distribution is uniform, the effect of the secondary bending moment is immediate and the column bending moment capacity is reduced from the beginning of loading as indicated by line OA. If the primary bending moment distribution is non-uniform, the effect of the secondary bending moment is delayed until a much higher load so that the available bending moment capacity is higher. This is shown in Figure 3.11 by line BA. The design bending moment capacity is μ_n, instead of μ_0. The position B, at axial compression χ_n, depends on the ratio of the primary bending moments at the column ends. In EC4, χ_n is given by:

$$\chi_n = \chi \frac{1-r}{4} \quad \text{with} \quad -1 \leq r \leq 1 \qquad (3.24)$$

where r is the ratio of the numerically smaller to the larger end bending moment. Figure 3.11b shows values of χ_n for the three common cases of column bending moment diagram. Obviously, χ_n is higher with a higher gradient in the column bending moment distribution, giving a higher value of μ_n.

P–δ Effect

Due to the second order (P–δ) effect, the column bending moment will be larger than that obtained from the first order analysis. The increase in the column bending moment is high for slender and more heavily loaded columns. Conversely, it is negligible for columns of low slenderness or for columns on which the applied compressive force is low. Under the latter conditions, the P–δ effect does not need be considered. EC4 defines columns of low slenderness as:

$$\bar{\lambda} \leq 0.2(2-r) \qquad (3.25)$$

In the case of a column under transverse loading, $r = 1$.

Table 3.3 Value of β

Moment distribution	β
Column with transverse loading	1.0
End moments	$\beta = 0.66 + 0.44r$, but $\beta \geq 0.44$

The definition of low compressive force is when the design compression load is less than 10% of the column Euler buckling load, i.e.

$$\frac{N_{Sd}}{N_{cr}} \leq 0.1 \qquad (3.26)$$

In other cases, second order analysis has to be carried out to obtain the increased column bending moment. In the absence of such a refined analysis, the increased column bending moment may be approximately obtained by magnifying the first order bending moment, i.e.

M (design column bending moment) $= k\, M$ (first order bending moment) (3.27)

where the magnification factor k depends the column bending moment distribution, obtained using:

$$k = \frac{\beta}{1 - \dfrac{N_{Sd}}{N_{cr}}} \geq 1.0 \qquad (3.28)$$

in this equation, values of β are given in Table 3.3.

3.3.6 Unsymmetrical Sections

In this case, depending on the direction of the column bending moment, the N–M interaction curve of an unsymmetrical composite cross-section about the same axis of bending will be different. Thus, when determining the N–M interaction curve of a unsymmetrical cross-section, the direction of bending moment should be observed. Here, the bending moment should be calculated with regard to the plastic centroid of the uncracked composite cross-section. Also the N–M interaction curve cannot be obtained using the simplified method in Section 3.3.2, instead, the N–M interaction curve can only be obtained following the general procedure outlined in Section 3.3.1.

Other design considerations, such as local buckling of steel (Section 3.2.2) and shear resistance (Section 3.4) are evaluated in the same way as for symmetrical sections.

3.4 EFFECT OF SHEAR

Since a column mainly resists compression, the effect of shear force will be small and can often be neglected. In rare occasions where the effect of shear force has to be included, shear force may be assumed to be resisted by the shear area of the steel component of the composite cross-section. The shear area of a composite cross-section is the same as that of the steel section.

3.5 LOAD INTRODUCTION

So far, it has been implicitly assumed that the applied load is properly transferred to steel and concrete to obtain composite action. This is achieved by ensuring that in the region of load introduction, the shear resistance at the interface of the steel section and the concrete is not exceeded. However, available design guidance on this issue is not clear and the design approach is usually to use a construction detailing to eliminate this problem (Bergmann *et al.* 1995; Roik and Bergmann 1992).

3.6 COMPOSITE COLUMNS IN FIRE CONDITIONS

3.6.1 General Introduction

Columns are critical parts of a building structure so that when subjected to fire attack, their stability should be maintained to prevent structural collapse that can lead to fire spread. The traditional approach to assess fire performance of a composite column is similar to other forms of structural element, i.e. the performance of a composite column in fire is assessed by performing standard fire resistance tests (BSI 1987). However, since fire testing is expensive and standard fire test conditions do not usually reflect realistic situations, there is a widespread trend to use the so called "fire engineering method" or "performance based approach" to evaluate fire performance of structural elements.

The essence of the performance based approach is to achieve the specified fire performance criteria by finding solutions that are based on engineering principles. This is in contrast to the traditional prescriptive approach in which a specific solution is prescribed to a general problem without considering its many design variables. For example, in the prescriptive approach, the fire resistant design of a steel member simply means the specification of a fire protection thickness depending on the required fire resistance rating and its section factor, which gives a measure of the surface area exposed to fire attack relative to the volume of material being heated. This approach does not address other issues such as the importance of this structural element in an entire structure, its specific loading and support conditions or the specific characteristics of the fire attack.

In contrast, in the performance based approach, these factors are taken into account explicitly. To evaluate the load bearing capacity of a structural member under fire conditions, the following three general steps are followed in the performance based approach:

- The realistic temperature–time relationship of the fire attack is evaluated. For an enclosure fire, this relationship depends on the amount of

combustible materials, the ventilation condition and the construction materials of the fire enclosure.
- Under the realistic fire exposure, the temperature field in the structural member is calculated.
- Under elevated temperatures, both steel and concrete will experience reductions in their strength and stiffness, thus resulting in a loss in the load carrying capacity of the structural element. To ensure structural stability, the reduced load carrying capacity of the structural element should not be less than the externally applied forces prevailing at the time of the fire.

Many fire tests have been performed on composite columns and they have shown that composite columns have inherently high fire resistance. The results of these tests have been used to calibrate numerical methods and to develop fire safety design methods for composite columns.

The Eurocode design method for fire resistance of composite columns is given in section Part 1.2.

3.6.2 Introduction to Part 1.2 Design Methods

In Part 1.2, the fire resistance of a composite column may be determined using one of the following three methods:

1. Tabulated method
2. Simple calculation method
3. Advanced calculation method.

The tabulated method directly gives values of desired column parameters (e.g. column dimensions, amount of reinforcement and concrete cover to reinforcement) to achieve the required standard fire resistance time according to the level of the applied load. This method has been developed from results of the standard fire resistance tests on composite columns. Although this method is easy to use, the designer's choice is limited to the standard fire resistance only and to composite columns that are similar to the test specimens.

On the other hand, the simple calculation method is based on the column design method at ambient temperature that has been described in previous sections, being modified to take into account the effect of elevated temperatures. This method is only suitable for composite columns under axial compression.

In the advanced calculation method, numerical methods such as the finite element technique are employed to obtain the column fire behaviour, either as an individual member, or as part of a framework. The advanced calculation method employs basic principles of heat transfer and structural mechanics and knowledge of the constitutive relationships of steel and concrete at elevated temperatures. It is flexible and can be used to deal with any composite cross-section, combined axial load and bending moments and interaction of composite columns with other structural members. However, the advanced calculation method is usually developed as part of research activity and seldom available in a suitable form to engineers. More importantly, the execution and interpretation of numerical results require the user to have specialist knowledge about modelling behaviour of structures under fire conditions.

Thus, the simple calculation method will most likely be used by practising engineers and so it will be described in this section.

3.6.3 Loading Condition

As seen in previous discussions, the design method for composite columns under combined axial compression and bending is complex. With the introduction of temperature effects, it becomes very difficult to develop a simple design method to deal with the general case of composite columns under combined compression and bending under fire attack. Thus, in Part 1.2, columns are treated as predominately under compression, with bending moments playing only a minor role. This is justified for the following reasons:

1. For columns, it is often assumed that they are fully exposed to fire attacks on all sides. Thus, for symmetric sections under pure compression, temperature distribution in the column cross-section is symmetric and there is no thermal bowing to introduce bending moments.
2. In simple construction, any bending is due to connection eccentricity and a simple method (to be explained later) may be used.
3. In continuous construction using composite columns, as the column approaches its limit state under fire attack, it undergoes significant loss of stiffness relative to adjacent structural members so that very little bending moment is transmitted to the composite column. This has been demonstrated theoretically (Wang 1999a) and experimentally (Kimura *et al.* 1990).

For a column with an unsymmetrical composite cross-section, bending moments are unavoidable. Under this circumstance, the advanced method may have to be employed to calculate the column's fire resistance.

In this chapter, fire design methods are concerned with composite columns under compression only.

3.6.4 Simple Calculation Method of Eurocode 4 Part 1.2

The simple calculation method is based on the calculation method for ambient temperature design, given in Section 3.2. In this method, the squash load and rigidity of the composite cross-section at elevated temperatures should be determined. However, the confinement effect should be excluded.

Since the thermal conductivity of concrete is low, the temperature distribution in a composite cross-section is highly non-uniform. This means when calculating the squash load and rigidity of the composite cross-section, the composite cross-section has to be divided into a number of layers of similar temperature. The squash load and rigidity of the composite cross-section are obtained by summing up contributions from all layers. For the rigidity, Part 1.2 introduces modification factors such that:

$$(EI)_{fi} = \Sigma \varphi_a E_{a,T} I_a + \Sigma \varphi_c E_{c,T} I_c + \Sigma \varphi_s E_{s,T} I_s \qquad (3.29)$$

where symbol E is the secant modulus of elasticity of material at elevated temperatures and I the second moment of area.

Composite Columns

In equation 3.29, φ is an empirical factor to allow for material non-linear stress–strain relationships at elevated temperatures, its value being dependent on the final stress of the constituent material. Eurocode 4 Part 1.2 has given values of these factors for different types of column under different standard fire ratings. For concrete, this factor is 0.8, being the same as for ambient temperature design (equation 3.4). For steel and reinforcement, this factor is 1.0 in most cases, being close to 1.0 in other cases. These factors have been empirically obtained and apply only to the standard fire resistance. Their introduction complicates the calculation procedure. Moreover, due to inaccuracy in other important variables (e.g. degradation of strength and elasticity of materials at elevated temperatures), such complication appears unnecessary. Thus, it is recommended that equation 3.29 be replaced by:

$$(EI)_{fi} = \Sigma E_{a,T} I_a + 0.8 \Sigma E_{c,T} I_c + \Sigma E_{s,T} I_s \tag{3.30}$$

This equation is now the same as equation 3.4 for ambient temperature design, except that the ambient temperature elasticity of material is replaced by the elevated temperature value.

The reduced squash load and rigidity of the composite cross-section at elevated temperatures are substituted into equations 3.8 and 3.7 to determine the column slenderness and strength. However, instead of using different column buckling curves for different types of composite cross-section, for fire design, only column buckling curve "c" is used (see Figure 3.2). Using column buckling curve "c" will give a lower column strength than using other curves. This is to allow for some additional detrimental effects of fire such as slight non-uniform temperature distribution in the column.

3.6.5 Effects of Structural Continuity on Column Effective Length in Fire

When considered as part of a frame, the fire performance of a column will be affected by structural continuity afforded by the adjacent members. Consider only columns under axial loads only, the effects of the adjacent members on a column are to provide axial restraint to its thermal expansion and additional rotational restraint as the column stiffness reduces at high temperatures. The axial restraint is mainly from adjacent beams to the column and is likely to produce additional compressive force to reduce the column fire resistance. The additional rotational restraint to the column comes from the adjacent columns outside the fire compartment that are cold. The effect of the additional rotational restraint to the column will be beneficial.

Part 1.2 considers only the beneficial effect of structural continuity by reducing the column effective length. It states that:

> in the case of a steel frame in which each storey comprises a separate fire compartment with sufficient fire resistance, in an intermediate storey the buckling length of a column $L_{e,fi} = 0.5L$ and in the top storey the buckling length $L_{e,fi} = 0.7L$, where L is the system length in the relevant storey.

This statement is illustrated in Figure 3.12.

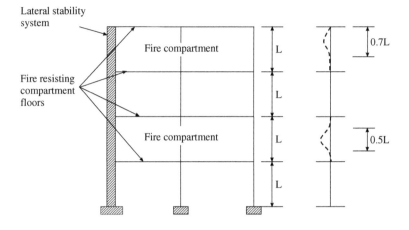

a: Locations of fire exposed columns   b: Deformed shape and buckling length

Figure 3.12 Column buckling length for fire design according to Eurocode 4 Part 1.2

The effect of axial restraint to the column thermal expansion should be considered. Otherwise, there is a danger that the column may be under-designed for fire safety.

It is difficult to calculate the exact increase in the column axial compression force due to restrained thermal expansion. To overcome this problem and to compensate for ignoring the effect of restrained thermal expansion, the column effective length may instead be increased from those recommended by Part 1.2 so as to reduce the column resistance. The UK practice is to use 0.7L and 0.85L respectively, replacing 0.5L and 0.7L in Figure 3.12. This brings fire design in alignment with the ambient temperature values for the effective length of a column with fixed ends.

It has already been mentioned that the improved rotational restraint to a column is a result of the adjacent cold columns' bending stiffness providing increasing rotational restraint, while the additional compression force is due to restraint on the column thermal expansion provided by the adjacent floors and beams. As long as the building is compartmented, cold columns continuous outside the fire compartment will always impose rotational restraint to the hot column under consideration. Thus, the beneficial effect is always present in the structure. On the other hand, the restraint to the column thermal expansion occurs only where there is both restraint and differential thermal expansion at the same floor level. Therefore, the detrimental effect will disappear if either of these two factors does not exist.

Figure 3.13 shows two cases where the detrimental effect of the restrained column thermal expansion does not occur so that recommendations in Part 1.2 may be used. The first case may occur in a frame with ideally pinned beam-column connections. Here the restraint stiffness is zero. In the second case, columns on the same floor level are heated in such a way that there is no difference in their thermal movements. This is possible only when all columns have the same strain and are heated uniformly.

Composite Columns

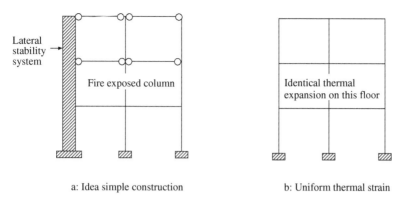

Figure 3.13 Two cases of columns without restraint to thermal expansion

Since even nominally simple construction can have substantial connection stiffness, it would be difficult to design for the first case. On the other hand, it becomes complicated to identify a design case where there is no differential thermal movement. Thus, the simplest and safest way is to always allow for restrained column thermal expansion in fire design.

3.6.6 Simplified Methods for Concrete Filled Columns in Fire

Advantages of concrete filled columns have already been stated in the introduction to cold design. Combined with good fire performance, they are becoming a very attractive column solution. However, as can be seen in previous descriptions, although the general design calculation method is not difficult to understand, its implementation in practice may be tedious. Not only does the user have to carry out numerical analysis to determine the temperature distribution, but also when calculating the squash load and rigidity of the composite cross-section, calculations become tedious by the need to divide the composite cross-section into many layers. In this section, some simplified calculation methods are described for evaluating the fire resistance of concrete filled composite columns. Ease of implementation is particularly important during the initial design stage when decisions are made on whether and how composite columns should be constructed. These simple methods will enable the engineer to quickly establish whether a particular concrete filled composite column is sufficient to provide the required fire resistance before detailed calculations become necessary. At this stage, these methods are only suitable to deal with the standard fire resistance while simplified design methods for realistic fire exposure are still under development.

Unprotected and protected columns are dealt with separately.

Unprotected Columns

The National Research Council of Canada (NRCC) carried out many standard fire resistance tests on concrete filled composite columns. Fire exposure was according to ASTM E119-88 (ASTM 1990) which is very similar to the standard

Table 3.4 Values of f_1 for plain and fibre reinforced concrete filled columns

Type of aggregate	Plain concrete	Fibre reinforced concrete
CHS		
Siliceous	0.07	0.075
Carbonate	0.08	0.085
SHS		
Siliceous	0.06	0.065
Carbonate	0.07	0.075

Table 3.5 Values of f_1 for bar reinforced concrete filled columns

Aggregate type	% steel reinforcement	Concrete cover thickness (mm)	f_1 for CHS	f_1 for SHS
Siliceous	<3%	<25	0.075	0.065
		≥25	0.08	0.07
	≥3%	<25	0.08	0.07
		≥25	0.085	0.075
Carbonate	<3%	<25	0.085	0.075
		≥25	0.09	0.08
	≥3%	<25	0.09	0.08
		≥25	0.095	0.075

fire exposure in the UK (BSI 1987). Test parameters included size of steel tube, steel wall thickness, load level, concrete strength and type of aggregate. From these fire test results, regression analyses were carried out and the following equation has been obtained (Kodur 1999):

$$FR = f_1 \frac{(f_{ck} + 20)}{(l_{e,fi} - 1000)} D^2 \sqrt{\frac{D}{N_{sd}}} \qquad (3.31)$$

where f_1 is a constant.

Values of f_1 depend on the type of aggregate, the shape of the steel section and whether concrete is plain, bar reinforced or fibre reinforced. Values of f_1 may be taken from Tables 3.4 and 3.5.

It will be noticed that equation 3.31 does not contain any reference to the steel section. This is a result of diminishing resistance from the unprotected steel section at high temperatures. Nevertheless, this is a conservative assumption and the contribution of the steel section may be included in detailed design calculations. The difference between siliceous and carbonate aggregates is because carbonate aggregates have substantially higher heat capacity than siliceous aggregates, leading to much lower concrete temperatures and higher fire resistance. Since this method has been developed based on the results of the standard fire resistance tests, the limits of application in Table 3.6 should be observed:

Table 3.6 Limits of application for equation 3.31

Variable	Plain concrete	Bar-reinforced concrete	Steel fibre-reinforced concrete
Fire resistance time	≤120 min	≤180 min	≤180 min
Axial load	≤1.0 times factored concrete core resistance	≤1.7 times	1.1 times
Effective length	2–4 m	2–4.5 m	2–4.5 m
Concrete cylinder strength	20–40 N/mm^2	20–55 N/mm^2	20–55 N/mm^2
Steel tube size			
Circular	140–410 mm	165–410 mm	140–410 mm
Square	140–305 mm	175–305 mm	100–305 mm
Steel reinforcement	N/A	1.5–5%	N/A
Concrete cover to rebar	N/A	20–50 mm	N/A
Local buckling	No	No	No

Protected Columns with Unreinforced Concrete

For protected composite columns, an additional design parameter is the steel temperature. For a given steel temperature, Wang (2000) has found that the squash load and rigidity of a composite cross-section may be linearly related to the standard fire exposure time by joining two points. At zero fire exposure time, it may be assumed that the steel is at its maximum temperature and the concrete remains cold. If the fire exposure is sufficiently long, the concrete would be heated to the same temperature as that of the steel. The minimum time (in minutes) taken to reach uniform temperature in the composite cross-section is numerically equal to the outside dimension (in mm) of the composite cross-section. This procedure is illustrated in Figure 3.14.

3.6.7 Effect of Eccentricity

Design for combined axial load and bending moment is already complicated for composite columns at ambient temperature. With the introduction of non-uniform temperature distribution due to fire exposure, accurate treatment can only be obtained by sophisticated numerical methods. Bending moment in a column comes from two sources: eccentricity and bending moment transferred from the adjacent beams in continuous construction. In continuous construction, although the column bending moment at ambient temperature may be large, results from experiments (Kimura *et al.* 1990) and numerical simulations (Wang 1999) indicate that this bending moment reduces to almost zero as the column approaches failure in fire. Thus, at the fire limit state, it is only necessary to deal with bending moment from eccentricity.

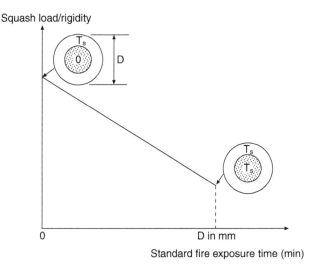

Figure 3.14 Squash load/rigidity of a protected concrete filled column

In Part 1.2, the reduced column axial compressive strength under bending moment from eccentricity is calculated using:

$$N_{Rd,e,fi} = N_{Rd,fi} \frac{N_{Rd,e}}{N_{Rd}}$$

where N_{Rd} is the strength of composite column at ambient temperature and $N_{Rd,e}$ is the reduced cold column strength due to eccentricity e. $N_{Rd,fi}$ is the pure axial strength of column in fire as calculated previously.

3.7 SUMMARY

This chapter has provided an introduction of the composite column behaviour under both cold and fire conditions. This is then followed by detailed descriptions of appropriate design methods. Sources of design guidance are taken mainly from Eurocode 4, including Part 1.1 for cold design and Part 1.2 for fire safety design. These design methods have been derived from comprehensive experimental and theoretical studies. Due to complex behaviour of composite columns, design methods in these codes of practice are necessarily complicated. Hence to encourage more widespread use of composite columns, this chapter has also introduced some simplified design methods. For fire safety design, alternative simple design methods have been provided separately for protected and unprotected concrete filled columns.

Composite columns have been in existence for a considerable time and their behaviour appears to have been well researched and understood. However, there are still a number of areas where further research is desirable. In particular, there is a lack of information on the following aspects.

1. Experimental and theoretical studies on composite columns using high strength concrete, particularly composite columns under combined axial compression and bending.
2. Load transfer through connections to composite columns.
3. Simplified design methods for composite columns under realistic fire conditions.

3.8 ACKNOWLEDGEMENT

Dr. Wang would like to thank Dr. Allan Mann of Babtie, Allott and Lomax for thoroughly reading and commenting on this chapter.

3.9 REFERENCES

ACI (1995), *Building Code Requirements for Reinforced Concrete and Commentary, ACI 318-95*, American Concrete Institute, Detroit, Michigan

AISC (1996), *Load and Resistance Factor Design Specification for Structural Steel Buildings*, American Institute of Steel Construction, Chicago

ASTM (1990), *Standard Methods of Fire Endurance Tests of Building Construction and Materials, ASTM E119-88*, American Society for Testing and Materials, Philadelphia

Bergmann, R., Matsui, C., Meinsma, C. and Dutta, D. (1995), *CIDECT design Guide for Concrete Filled Hollow Section Columns Under Static and Seismic Loading*, Verlag TüV Rheinland

BSI (1979), *British Standard 5400: Steel, Concrete and Composite Bridges, Part 5: Code of Practice for Design of Composite Bridges*, British Standards Institution

BSI (1987), *British Standard 476, Fire Tests on Building Materials and Structures, Part 20: Method for Determination of the Fire Resistance of Elements of Construction (General Principles)*, British Standards Institution

BSI (1990), *British Standard BS 5950: Structural Use of Steelwork in Building, Part 8. Code of Practice for Fire Resistant Design*, British Standards Institution

BSI (2000), *British Standard BS 5950: Structural Use of Steelwork in Buildings, Part 1. Code of Practice for Design in Simple and Continuous Construction Hot Rolled Sections*, British Standards Institution

CEN (1992a), *DD ENV 1993-1-1, Eurocode 3: Design of Steel Structures, Part 1.1: General Rules and Rules for Buildings*, Commission of European Community

CEN (1992b), *Eurocode 4: Design of Composite Steel and Concrete Structures, Part 1.1, General Rules and Rules for Buildings*, Commission of European Community

CEN (1994), *Eurocode 4, Design of Composite Steel and Concrete Structures, Part 1.2: Structural Fire Design*, Commission of European Community

Kimura, M., Ohta, H., Kaneko, H., Kodaira, A. and Fujinaka, H. (1990), Fire resistance of concrete-filled square steel tubular columns subjected to combined loads, *Takenaka Technical Research Report*, **43**, 47–54

Kodur, V.K.R. (1999), Performance-based fire resistance design of concrete-filed steel columns, *Journal of Constructional Steel Research*, **51**, 21–36

Matsui, C., Mitani, I., Kawano, A. and Tsuda, K. (1997), AIJ design method for concrete filled steel tubular structures, *ASCCS Seminar*, Innsbruck

Oehlers, D.J. and Bradford, M.A. (1995), *Composite Steel and Concrete Structural Members, Fundamental Behaviour*, Pergamon

O'Shea, M.D. and Bridge, R.Q. (1999), Design of circular thin-walled concrete filled steel tubes, *ASCE Journal of Structural Engineering*, **126(11)**, 1295–1303

Roik, K. and Bergmann, R. (1990), Design method for composite columns with unsymmetrical cross-sections, *Journal of Constructional Steel Research* (Special issue on composite construction), 153–168

Roik, K. and Bergmann, R. (1992), Chapter 4.2: Composite columns *in Constructional Steel Design: An International Guide*, ed. Dowling, P.J., Harding, J.E. and Bjorhovde, R., Elsevier Applied Science

Structural Stability Research Council (SSRC 1998), *Guide to Stability Design Criteria for Metal Structures*, ed. Galambos, T.V., Fifth Edition, John Wiley & Sons

Uy, B. (1998), Concrete-filled fabricated steel box columns for multistorey buildings: behaviour and design, *Progress in Structural Engineering and Materials*, **1(2)**, 150–158

Uy, B. (1999), Strength of concrete filled steel box columns incorporating local buckling, *ASCE Journal of Structural Engineering*, **126(3)**, 341–352

Uy, B. and Das, S. (1999), Bracing of thin walled steel box columns during pumping of wet concrete in tall buildings, *Thin-Walled Structures*, **33**, 127–154

Uy, B. (2001), Strength of short concrete filled high strength steel box columns, *Journal of Constructional Steel Research*, **57**, 113–134

Wang, Y.C. (1999a), The effects of structural continuity on the fire resistance of concrete filled columns in non-sway frames, *Journal of Constructional Steel Research*, **50**, 177–197

Wang, Y.C. (1999b), Experimental study of high strength concrete filled steel columns, *Proceedings of the 2nd International Conference on Advances in Steel Structures*, Hong Kong

Wang, Y.C. (2000), A simple method for calculating the fire resistance of concrete-filled CHS columns, *Journal of Constructional Steel Research*, **54**, 365–386

3.10 NOTATIONS

A	area
d	depth of concrete in compression
D	diameter or dimension of steel tube
e	eccentricity
E, E_T	modulus of elasticity at 20 °C and elevated temperature
E_c	long term design modulus of elasticity of concrete
E_{cd}	effective design modulus of elasticity of concrete

Composite Columns

E_{cm}	secant modulus of elasticity of concrete
$(EI)_e$	effective rigidity of cross-section
$(EI)_{fi}$	effective rigidity of cross-section in fire
$f_{ck}, f_{ck,T}$	design characteristic strength of concrete at 20 °C and elevated temperature T
$f_{sk}, f_{sk,T}$	design strength of reinforcement at 20 °C and elevated temperature T
$f_y, f_{y,T}$	design strength of steel at 20 °C and elevated temperature T
f_1	multiplication factor
FR	fire rating
h_n	distance between plastic neutral axis under pure bending and centroid
I	second moment of area
k	magnification factor for column bending moment
L_e	effective length of column
$L_{e,fi}$	effective length of column in fire
L	column length
M	bending moment in column
$M_{max,Rd}$	maximum bending moment capacity of composite cross-section
$M_{pl,Rd}$	moment capacity of composite cross-section under pure bending
M_{Sd}	bending moment resistance of column
M_1, M_2	column end moments
N_c	$= \tfrac{1}{2} N_{c,Rd}$
$N_{c,Rd}$	total axial resistance of concrete in composite cross-section
N_{cr}	Euler buckling load of column
N_D	$= N_{c,Rd}$
$N_{G,Sd}$	permanent design load
$N_{pl,Rd}$	squash load of cross-section
N_{Rd}	design strength of column
$N_{Rd,e}$	design strength of column with eccentricity
$N_{Rd,fi}$	design strength of column in fire
$N_{Rd,e,fi}$	design strength of column with eccentricity in fire
$N_{Rd,pl}$	design strength of column with regard to plastic centroid
N_{Sd}	design axial load of column
P	column axial load
T	temperature
W_p	plastic modulus

Greek letters

χ	column strength reduction factor
χ_{el}	column strength reduction factor with regard to elastic centroid
χ_{pl}	column strength reduction factor with regard to plastic centroid
δ	column lateral deflection
δ_0	column initial deflection
ε	$= \sqrt{235/f_y}$
γ	ratio of column end bending moments
γ_m	partial safety factor for concrete modulus of elasticity ($=1.35$)
η_1, η_2	multiplication factors for concrete confinement
φ	modification factor for modulus of elasticity of material in fire
$\bar{\lambda}$	column slenderness ($= \sqrt{N_{Rd}/N_{cr}}$)
μ	non-dimensional bending moment resistance

Subscripts

a	steel
c	concrete
fi	fire
s	reinforcement
y,z	axis of bending

CHAPTER FOUR

Instability and Ductility

Alan R. Kemp

4.1 INTRODUCTION AND ELASTIC BUCKLING THEORY

Modes of elastic and inelastic buckling are considered in this chapter. These modes include local buckling of web and flange elements, lateral buckling of columns, and lateral-torsional and distortional buckling of beam cross-sections. The benefits of ductility are emphasized throughout this chapter. The influence of composite action on global frame behaviour is discussed separately in Chapter 7. As in steel design, the components of flexibility of the end connection are important to frame instability.

Section 4.1 describes the basic elastic theory applicable to local and overall buckling of plain steel and composite beams, and composite columns. The distinction between elastic and inelastic behaviour is clarified in the classification of ductility given in section 4.1.1. The subsequent sections 4.2 and 4.3 extend the elastic approach to identify the ultimate resistance of composite columns and continuous composite beams respectively. In the final section 4.4, attention is given to the ductility exhibited by continuous composite beams in regions of negative bending.

Instability of steel-concrete composite beams differs from the behaviour of plain steel beams in a number of important respects, including:

- In positive-moment regions, which are propped during construction, the concrete slab restrains the compression flange against local and lateral buckling, as discussed in sections 4.1.3 and 4.1.4.
- The steel section in composite beams should be inherently ductile to offset the strain-weakening behaviour of concrete in compression in positive-moment regions, and to allow for considerable redistribution of moment in continuous beams from internal supports to adjacent midspan regions, as described in section 4.3.1 (Figure 4.3).
- Inelastic local buckling in negative bending is modelled in section 4.3.2 by a strain limit at the centre of the compression flange at which a full wavelength buckle forms in the yielded region, allowing for moment gradiant (Figure 4.3) and a bi-linear moment-curvature relationship (Figure 4.4).
- The torsional restraint provided to the steel section by the transverse stiffness of the slab induces a distortional mode of buckling in the negative region of continuous beams, which is considered for slender beams in

section 4.3.3 (Figure 4.2) and for compact beams in sections 4.3.4 and 4.3.5. Typical distortions of the compression flange are shown in Figures 4.5 and 4.6.
- A limit-states criterion for ductility, described in section 4.4, is used to represent interactive, inelastic local and lateral buckling (Figure 4.6) for both rigid and semi-rigid end connections. An example is provided in section 4.4.5 (Figure 4.7). The level of the plastic neutral axis and thickness of the end plate (Figure 4.10) have an important influence on available ductility of composite beams and connections.

Composite columns are covered comprehensively in Chapter 3, where the instability aspects are part of the column design equations. Attention is given in section 4.1.6 to the definition of appropriate elastic properties for assessing the elastic critical buckling strength of composite columns. The second-order loss of resistance of beams and columns in Eurocodes 3 (1992) and 4 (1992) are described for completeness in section 4.1.7.

4.1.1 Classification of Beam and Column Elements

In modern design codes structural engineers are given the opportunity to analyse ultimate load effects in continuous beams either by plastic analysis or elastic analysis, with or without moment redistribution. They may determine the ultimate moment resistance using either rigid-plastic stress blocks or elastic stress limits. To clarify these options a classification has been introduced into European, North American and other codes as shown in Table 4.1.

It is apparent from Table 4.1 that the terminology used in different codes is somewhat confusing and the Eurocode definitions of Classes 1 to 4 will be used in this chapter. It is acknowledged that some inconsistencies may exist between the codes in terms of the definitions in Table 4.1.

Two generic groups of structural elements will also be used in this chapter to distinguish between broad subdivisions of ductility:

- "Slender" elements: Class 3 and 4 sections in which elastic analysis and elastic stress distributions less than or equal to yield are adopted in design.
- "Compact" elements: Class 1 and 2 sections in which elastic analysis (with considerable moment redistribution) or plastic analysis (for Class 1 sections), and plastic stress-blocks are adopted in design. For analysis based on uncracked concrete, Eurocode 4 (1992) allows 40% redistribution of moments for Class 1 sections and 30% for Class 2 sections.

Instability and Ductility

Table 4.1 Classification of flexural ductility of beam elements

Element Type	Method of analysing load effects	Method of calculating moment-resistance	AISC LRFD (1986)	Euro-code 3 (1992)	CSA 16.1 (1989)
Compact	plastic analysis or elastic analysis with large redistribution	yielded stress blocks	Compact	Class 1	Class 1 Plastic
	elastic analysis with limited redistribution	yielded stress blocks		Class 2	Class 2 compact
Slender	elastic analysis with limited redistribution	elastic stress limited to yield stress	non-compact	Class 3	Class 3 non-compact
	elastic analysis with no redistribution	elastic stress < yield stress	slender	Class 4	Class 4 slender

In building structures it is common to consider I-beams satisfying the requirements of Class 1 or 2, whereas in bridge structures non-uniform I-beams or girders satisfying Class 3 or 4 requirements, possibly with variable yield strength, are more common. In most design codes the classification of ductility in Table 4.1 is defined primarily on the basis of the slenderness for local buckling of the compression flange, b_f / t_f, (b_f and t_f are the width and thickness of the flange) and of the portion of web in compression, h_c / t_w, (h_c and t_w are the depth in compression and thickness of the web), adjusted for yield strength.

As described in section 4.4.3, test evidence indicates that for compact elements the most important parameter is the lateral slenderness ratio, kL / i_z, in negative bending, (kL is the effective length and i_z is the radius of gyration about the minor axis). Lateral buckling is dealt with separately in American and European design codes and not as part of the classification of ductility. This creates difficulties because lateral-torsional buckling is a property associated with a member or portion of a member, whereas local buckling is a property linked only to a cross-section. For this reason reference in this chapter will be made to compact "elements" (Class 1 or 2) as distinct from slender "elements" (Class 3 or 4). It is implicit in the classification of Table 4.1 that a Class 1 or 2 element should be able to achieve and maintain its fully-plastic moment resistance during the process of moment redistribution.

The same classification of ductility is used for both composite and plain steel elements (Table 4.1) and is based on local buckling properties, although tests indicate that lateral buckling is more important to ductility.

4.1.2 Elastic Local Buckling

When a plate element is subjected to direct compression, bending, shear or a combination of these, local buckling of individual elements of the cross-section may occur before the complete member becomes unstable or achieves its ultimate resistance. Attainment of the local buckling strength of such an element often requires distortion of the cross-section, redistribution of longitudinal stress across the width of the section and development of transverse membrane stresses. As a result local buckling resistance depends not only on plate slenderness, but also on yield strength, initial imperfections and residual stresses. This resistance may exceed or be less than the theoretical elastic buckling resistance of a perfect plate. The elastic local buckling strength of Class 3 or 4 plate elements will be considered first. Extension of this theory to inelastic local buckling will be developed in sections 4.2 and 4.3.

Bryan (1891) provided an analysis of the elastic local buckling stress σ_{cloc} of a long rectangular plate with simple supports on all four edges and subjected to a uniform longitudinal compressive stress. This expression has been generalised to cover varying boundary conditions, plate geometries and load distributions by the introduction of a buckling coefficient, k, as follows:

$$\sigma_{cloc} = k \frac{\pi^2 E}{12(1-\nu^2)(b/t)^2} \qquad (4.1)$$

in which (b / t) is the width-to-thickness ratio, and E and ν are the elastic modulus and Poisson's ratio of the plate. Minimum values of the buckling coefficient k are given in Table 4.2 for different boundary conditions and stress distributions. For a rolled-steel I-section in flexure, local buckling may develop in the outstand width of the flange, b = 0.5b_f (with k = 0.425), or within the depth of the web, d, (k = 4.0 under pure compression or k between 4 and 23.9 under combined compression and bending).

Equation 4.1 may be rearranged to identify the maximum value of plate slenderness (b / t)$_m$ which is predicted to achieve a buckling stress equal to the yield stress (σ_{cloc} = f_y), but not maintain this stress under additional strain:

$$(b/t)_m^2 = k \frac{\pi^2 E}{12(1-\nu^2)f_y} \qquad (4.2)$$

The width-to-thickness ratios obtained from this equation are given in Table 4.2. Comparative values are also provided of the local buckling limits for

Class 2 rolled I-sections in Eurocode 4 (1992). Equation 4.2 provides an unsafe estimate of Class 2 behaviour because it does not allow for local imperfections and residual stresses, or for maintaining the yield stress at levels of compressive strain required to develop the stress-block moment resistance.

Table 4.2 Local buckling limits at yield ($\varepsilon=\sqrt{275/f_y}$)

Type	Portion of cross-section	k	$(b/t)_m$ or $(d/t)_m$ Equation 4.2	Eurocode Class 2 limit
Rolled I-section beam	Flange outstand	0.425	17.1ε	10ε
	Web in uniform compression	4.0	52.5ε	35ε
	Web in flexure	23.9	128ε	77ε
Partially encased I-section*	Flange outstand			20ε
Concrete-filled rectangular tube	Tube in uniform compression	10.3 (Uy & Bradford)	84.3ε	48ε
Concrete-filled Circular tube	Tube in uniform compression			$77\varepsilon^2$

* Encased between flanges on both sides of web (Lindner and Budassis, 2000); no local buckling if fully encased.

For steel I-sections fully-encased by concrete, no verification of local buckling is required, but for partially-encased I-sections, in which the concrete is located between the flanges on both sides of the web, the increased flange outstand given in Table 4.2 is appropriate. For concrete-filled hollow sections, the concrete resists inward local buckling of the steel and achieves an enhanced strength due to the restraint provided by the hollow section. Eurocode 4 (1992) identifies that local buckling may be neglected if the limits for concrete-encased or concrete-filled steel sections in Table 4.2 are not exceeded. Unlike the United States code, Eurocode 4 does not provide guidance on the loss of column strength if these limits are exceeded.

Classic elastic buckling formulas such as Equations 4.1 and 4.2 and the ductility classification of Table 4.1 provide a basis for assessing complex local buckling behaviour of steel sections.

4.1.3   Design Considerations for Elastic Local Buckling

The "effective width" approach is a useful semi-empirical method of allowing for local buckling. In relatively slender plates under uniform edge compression in the longitudinal direction, the stress distribution is not uniform across the width of the plate with larger stresses associated with the stiffer portions of the plate adjacent to the longitudinal edges. The effective width represents the fraction of the width located adjacent to these boundaries which, if stressed at yield, will produce the same force as the actual stress distribution across the full plate width.

An example of the use of this effective width concept that has been widely adopted in design codes is the case of a Class 3 web and Class 1 or 2 flange of an I-section or plate girder in negative bending. Such a web may be replaced by two equal depths of web measured from the root of the compression flange and from the adjusted plastic neutral axis as shown in Figure 4.1, which is extracted from Eurocode 4 (1992). Each of these widths of web is approximately equal to half the maximum width of a Class 2 web in compression in Table 4.2, and is stressed at yield in calculating the ultimate resistance moment.

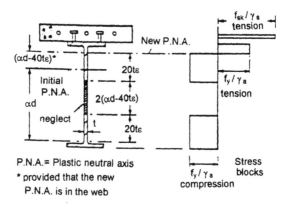

Figure 4.1 Effective width of web

The concept of effective width of slender plates is useful for designing composite beams with a Class 1 or 2 flange and a Class 3 web.

4.1.4 Elastic Lateral-Torsional Buckling of Beams

In positive-moment regions, composite beams which are propped during construction, are restrained by the concrete slab against lateral-torsional buckling. In beams which are unpropped during construction, the steel element is designed as a plain steel beam with appropriate lateral restraints under self-weight and construction loads.

In negative-moment regions of continuous composite beams, however, local and lateral instabilities of the steel web and bottom compression flange represent important potential modes of failure. Lateral buckling in negative bending involves distortional out-of-plane bending of the web, because the top flange is laterally and torsionally restrained by the concrete slab. It is reasonable to assume that the shear connection between the steel section and the concrete slab is sufficiently strong to ensure monolithic behaviour between the top flange and the slab. This assumption is justified in the absence of vertical web stiffeners, but in deep stiffened girder sections in bridge structures special consideration may be required. For relatively slender elements Johnson and Fan (1991) have indicated that not only the lateral, torsional and distortional stiffnesses of the steel section influence the resistance, but also the rotational stiffness of the transverse concrete slab.

The classic, elastic, lateral-torsional buckling resistance of an undistorted, plain steel beam, defined in the terms given in Eurocode 3 (1992), is as follows:

$$M_{cr} = \frac{k_c C_4}{L} \sqrt{EI_z[GI_t + (\frac{\pi}{k_w L})^2 EI_w]} \qquad (4.3)$$

In this equation the first term in the square brackets reflects the torsional-flexural stiffness of the section and the second term the warping stiffness; E and G are the elastic and shear moduli respectively; I_t, I_w and I_z are the St-Venant torsional inertia, the warping constant and the second moment of area about the minor axis respectively, and k_w is the effective length factor in warping (k_w is normally taken as 1.0). The factor k_c is a non-dimensional geometric parameter allowing for the positions of the centroid, centre of area and shear centre of the steel section. The factor C_4 allows for the loading arrangement, the ratio of end moments and the effective length factor k in lateral buckling. Eurocode 3 (1992) provides a development of this formula to allow for the level at which load is applied and different sizes of the compression and tension flanges.

In negative-moment regions of continuous beams the classic, elastic, lateral-torsional buckling equation provides a starting point for assessing flexural instability.

4.1.5 Elastic Lateral-Torsional Buckling of Partly Encased Steel Sections

Lindner and Budassis (2000) have investigated lateral-torsional buckling of steel I-sections in which the area between the flanges is encased in concrete on both sides of the web, but no concrete slab exists. Various bonding elements are considered such as open stirrups through holes in the web, headed studs, bars through holes in the web, and pure steel beams.

In this partly encased situation, warping restraint, as defined in the second term of Equation 4.3, is neglected and the buckling expression becomes:

$$M_{cr} = \frac{k_c C_4}{L} \sqrt{EI_z GI_t} \qquad (4.4)$$

Lindner has tested various assumptions, comparing the contribution of the partially cracked concrete to the torsional and lateral stiffness of the section, with the results of 22 experiments. Appropriate predictions for design use are found to coincide with the assumption that the concrete compressive stress block contributes to the moment resistance and to the torsional and lateral stiffness, but the concrete should be neglected in tension.

The critical buckling moment of partly-encased I-sections may be assessed by allowing for the concrete compressive stress block, but neglecting concrete in tension.

4.1.6 Elastic Buckling of Composite Columns

The elastic critical buckling load N_{cr} of a concrete-encased or concrete-filled composite column of effective length kL is given by the Euler buckling strength with an effective flexural rigidity $(EI)_e$. This buckling load is considered in section 3.2.2 of Chapter 3. It is determined as follows in Eurocode 4 (1992) and applies to cross-sections in which local buckling is avoided at yield by satisfying the limits for composite tubes given in the last column of Table 4.2:

$$N_{cr} = \frac{\pi^2 (EI)_e}{(kL)^2} = \frac{\pi^2 (E_a I_a + 0.8 E_{cd} I_c + E_s I_s)}{(kL)^2} \qquad (4.5)$$

in which I_a, I_c, and I_s are the second moments of area of the steel section, the uncracked concrete in compression and the reinforcement about the axis of buckling respectively, and E_a, E_{cd}, and E_s are the moduli of elasticity of the steel section, the concrete and the reinforcement. The effective contribution of the concrete to the flexural rigidity of the column, $0.8E_{cd}I_c$, contains the factor 0.8 obtained from test calibrations and includes a factored design value of the short-term secant elastic modulus, $E_{cd} = E_{cm}/1.35$. Limits are also given in this code for the lateral slenderness ratio, above which a defined provision should be made for creep and shrinkage of concrete.

The elastic critical buckling strength of composite columns is based on a combined flexural rigidity of the steel section, the concrete encasement or filling, and the reinforcement, with an adjustment factor for the concrete.

4.1.7 Design Buckling Resistance of Beams and Columns

Eurocodes 3 (1992) and 4 (1992) characterise the second-order loss of strength of beams and columns by a reduction factor χ. This factor is related to the non-dimensional slenderness ratio λ and the imperfection constant α by the following expressions:

$$\chi = [\phi + (\phi^2 - \lambda^2)^{0.5}]^{-1} \quad \text{but } \chi \leq 1 \tag{4.6}$$

$$\phi = 0.5 [1 + \alpha(\lambda - 0.2) + \lambda^2] \tag{4.7}$$

Four buckling Curves a, b, c and d with values of $\alpha = 0.21$, 0.34, 0.49 and 0.76 respectively, allow for lack of straightness, residual stresses and inelastic material behaviour.

For lateral-torsional buckling of beams, the slenderness ratio λ and design value of the moment resistance $M_{b.Rd}$ are given by Equations 4.6 to 4.9:

$$\lambda = \lambda_{LT} = (M_{pl} / M_{cr})^{0.5} \tag{4.8}$$

$$M_{b.Rd} = \chi M_{pl.Rd} \tag{4.9}$$

in which M_{cr} is the unfactored, elastic, lateral-torsional buckling moment given by Equation 4.3, M_{pl} is the unfactored, plastic moment resistance and $M_{pl.Rd}$ is the plastic moment resistance including partial material factors of $\gamma_a = 1.1$ for structural steel, $\gamma_c = 1.5$ for concrete and $\gamma_s = 1.15$ for reinforcing steel. The appropriate buckling curve is Curve a for rolled sections and Curve c for welded sections. Although $\chi \leq 1$ when $\lambda \geq 0.2$, Eurocode 4 (1992) allows $\chi = 1$ for $\lambda \leq 0.4$ to account for the benefit of the distortional nature of buckling in the negative region of continuous composite beams.

For lateral buckling of columns or struts, the slenderness ratio λ and design value of axial resistance $N_{b.Rd}$ are given by Equations 4.6, 4.7, 4.10 and 4.11:

$$\lambda = (N_{pl} / N_{cr})^{0.5} \qquad (4.10)$$

$$N_{b.Rd} = \chi N_{pl.Rd} \qquad (4.11)$$

in which N_{cr} is the unfactored axial force at which elastic buckling occurs as given by Equation 4.5, N_{pl} is the unfactored, plastic axial resistance and $N_{pl.Rd}$ is the plastic axial resistance including the same partial material factors as for beams. The appropriate buckling curve is Curve a for concrete-filled tubes, Curve b for concrete-encased steel sections buckling about the major-axis and Curve c for similar sections buckling about the minor-axis.

Further information on the design strength of composite columns is given in sections 3.2 and 3.3 of Chapter 3.

Eurocodes 3 (1992) and 4 (1992) provide a simple multiple-curve model for second-order loss of resistance of beams and columns.

4.2 ULTIMATE RESISTANCE OF COMPOSITE COLUMNS

4.2.1 Introduction

Composite columns have been used for over 50 years. Initially steel columns were encased in concrete to provide fire protection and the contribution of the concrete to the strength of the column was neglected. Early research indicated that, if adequate stirrups are provided, the concrete would contribute both to the strength and stiffness of the column. More recently the composite action of structural steel and reinforced concrete has been extended to include not only encased steel columns but also concrete-filled rectangular and circular tubes.

Composite columns represent an efficient form of construction because they take advantage of the properties of each component material. The axial and flexural strength and stiffness each include components contributed by the structural steel, the concrete and the longitudinal reinforcement. The detailed behaviour differs, however, between concrete-filled, hollow columns, where the concrete is constrained laterally by the steel section, and concrete-encased steel sections in which only the stirrups provide constraint, and spalling of the concrete may occur.

Leon and Aho (2000) have conducted an extensive study of the results of over 1100 tests on concrete-encased and concrete-filled composite columns. This study includes a comparison between the test measurements of ultimate strength, and the capacity predicted in European and United States design codes. They conclude that ultimate stress-block calculations, with an adjustment factor for confinement of the concrete of the type used in Eurocode 4 (1992), represent the best match to the test data. Interaction curves provide the most effective means of assessing the resistance of a column under combined axial compression and uniaxial bending. Roik and Bergmann (1992) propose the use of 4 or 5 significant points to approximate the full interaction curve.

4.2.2 Design Considerations for Composite Columns

The general form of the Eurocode provisions for lateral buckling of beams and columns is described in section 4.1.7. The plastic axial resistance of a short composite column ($N_{pl.Rd}$ in Equation 4.11), as defined in Eurocode 4 (1992), is given in section 3.2.1 of Chapter 3.

4.3 CONTINUOUS COMPOSITE BEAMS

4.3.1 Efficient Use of Composite Beams

Increasing attention is being given to the use of continuous composite beams in buildings for spans exceeding 9m or for heavily loaded members. A more general description of continuous composite beams is given in section 2.6 of Chapter 2. Relative to steel beams, the redistribution of moments and therefore the ductility required for an efficient design of composite beams is particularly demanding for the following reasons:

1. Elastic distributions of moment generally give higher values at internal supports than midspan regions, whereas resistances are greater in positive midspan regions.
2. The shape factor for composite beams is high (1.25 to 1.35) and this benefit is only utilised if stress-block analysis is permitted to determine the resistance.
3. To limit cracking of the concrete slab, longitudinal reinforcement is provided in the region adjacent to internal supports: this reinforcement together with relatively light end-plates may be utilised to develop negative resistance moments at these supports equal to about 40% to 80% of the moment resistance in positive bending at the centre of the span.

For building structures the inefficiency of a design based on elastic analysis makes it essential that composite beams are designed with resistances based on stress-block evaluations and with load effects assessed using plastic analysis or elastic analysis with considerable moment redistribution. For bridge girders this problem is overcome by providing heavier sections and larger resistance moments in negative bending adjacent to internal supports than in positive moment regions. This chapter will concentrate on potential inelastic instability of compact, continuous beams in building structures.

An important change of approach has taken place in recent years to improve the efficiency of composite beams. By adopting simple, semi-rigid end-connections and providing limited longitudinal reinforcement in the slab, simply-supported beams are replaced at little extra cost by continuous beams, with important stability and ductility implications. In such continuous beams, limit states of lateral buckling and local buckling of the lower flange and adjacent web may be important at internal supports. Inelastic moment redistribution creates a complex interaction in these zones of the beam.

Plastic analysis or redistribution of elastic moments in continuous beams and frames in buildings has always included the expectation that at each notional plastic hinge forming part of the collapse mechanism, the cross section will possess sufficient ductility to allow for the required plastic hinge rotation without loss of moment resistance. In the case of continuous composite beams with semi-rigid joints this expectation becomes a requirement because of the considerable redistribution of moments required. A more detailed description of frame behaviour and the influence of end connections is given in Chapter 6.

Extensive moment redistribution or plastic analysis is required in continuous composite beams due to the low ratio of negative to positive moment resistance and the high shape factor. This is particularly the case with semi-rigid end connections.

4.3.2   Inelastic Local Buckling of Flanges and Webs

Composite beams in negative bending differ from plain steel sections in that the location of the plastic neutral axis is above mid-depth of the steel section, to balance the tension force in the reinforcement. This has a detrimental effect on buckling of the web in compression, which is considered in the design codes by adjusting the value of k (see Table 4.2). It also reduces the inelastic curvature (strain gradient) for a particular critical strain in the compression flange. To account for the effect of the web on local buckling of the compression flange, Stowell (1950), Lay and Galambos (1967) and Southward (1969) adopted as their model a flange plate restrained against buckling by a central torsional spring, reflecting the contribution of the web. The buckling equation used by these authors may be expressed in the following form:

$$\varepsilon_{cloc} = \frac{12}{Eb_f^3 t_f} \left[\frac{Gb_f t_f^3}{3} + EI_w \left(\frac{2n\pi}{L_{cloc}} \right)^2 + K \left(\frac{L_{cloc}}{2n\pi} \right)^2 \right] \quad (4.12)$$

in which ε_{cloc} is the critical buckling strain at the center of the wave length, b_f and t_f are the overall width and thickness of the compression flange, L_{cloc} is the full wavelength of the local buckle, n is an integral number equal to unity for the first mode, I_w is the warping constant of the flange and K is the torsional stiffness of the web. The first term of this equation reflects the torsional stiffness of the compression flange, the second term represents the warping of the cross-section and the third term allows for the torsional (spring) stiffness of the web. The first term is identical to Equation 4.1 for k = 0.425, outstand width, b = 0.5b_f, G = E/2(1 + v) and v = 0.3, but expressed in terms of strain rather than stress.

If Equation 4.12 is differentiated with respect to L_{cloc} and equated to zero:

$$L_{cloc} = 2n\pi \left(\frac{EI_w}{K} \right)^{0.25} \quad (4.13)$$

Lay and Galambos (1967) substituted this result into Equation 4.12 and obtained an expression in terms of the flange and web properties. Southward (1969) pointed out that the resulting expression was dependent upon whether the web is restraining the flange elastically or plastically, or even being restrained itself by the flange (in which case it would provide negative stiffness). Essentially Equation 4.13 indicates that the minimum buckling stress occurs at a wavelength at which the second and third terms make an equal contribution. To eliminate the uncertainties of the web contribution the author equated the second and third terms to obtain the following expression, which includes the important influence of the yielded length, L_{cloc}, of the full-wave buckle:

$$\varepsilon_{cloc} = \frac{4G}{E} \left(\frac{t_f}{b_f} \right)^2 + \frac{96\pi^2 I_w}{b_f^3 t_f L_{cloc}^2} \quad (4.14)$$

For predominantly elastic local buckling of Class 3 or 4 elements, the second term in Equation 4.14 is negligibly small, associated with a large buckling wave-length, L_{cloc}, and the behaviour is dominated by the first torsional term. The precise nature of local buckling is complicated by the interaction of the web and flange slendernesses and by the moment gradient in the buckled region.

Inelastic behaviour has normally been considered by adopting a strain-hardening modulus, E_{sh}, and shear modulus, G_{sh}, based on plasticity laws. Southward (1969) evaluated the merits of incremental theories incorporating the tangent modulus, compared to deformation theories incorporating the secant modulus. He concluded that the results of inelastic plate buckling tests are more accurately reflected by deformation theories. He proposed an inelastic shear modulus, $G_{sh} = E_{sh}/3$, and warping constant, $I_w = b^3{}_f t^3{}_f / 144$, which differs from the incremental assessment of Lay and Galambos (1967). Using these values of E_{sh} and G_{sh} the author has identified the following expression for critical local buckling strain, ε_{cloc}, from Equation 4.14, which is defined in strain terms irrespective of the material properties (Kemp and Nethercot, 2001):

$$\varepsilon_{cloc} = 1.33(t_f/b_f)^2 + 6.6(t_f/L_{cloc})^2 \tag{4.15}$$

in which L_{cloc} is the full wavelength of the inelastic local buckle. In continuous composite beams inelastic local buckling is concentrated immediately adjacent to internal supports in regions of high moment gradient. At local buckling the yielded length, L_p, is equal to the full wavelength, L_{cloc}, and is bounded by the support on one side and by the transition point to elastic behaviour on the other. The critical strain, ε_{cloc}, is assumed to be the strain at the centre of this inelastic region.

An estimate of the critical average strain in the compression flange in inelastic local buckling is given by Equation 4.15.

4.3.3 Elastic Distortional Buckling

In slender sections, such as those used in continuous composite bridge girders, lateral buckling in negative bending is commonly restrained by torsional and / or lateral bracing to the bottom compression flange. Illustrations of the effectiveness of this bracing are given by Galambos (1998) and Yura (1993). In unpropped beams in which no precautions are taken during casting of the concrete to restrain the top compression flange in positive bending, lateral buckling occurs in the same way as a plain steel section.

In other cases of negative bending of continuous composite beams in building structures, the restraint provided by the slab to the top flange induces a distortional mode of buckling in composite beams, representing a rather fundamental difference from the lateral-torsional mode of plain steel sections, as illustrated in Figure 4.2. No closed-form solution is available for this distortional mode of buckling in negative bending. One of the first analytical studies of this behaviour appears to be that of Hamada and Longworth (1974) using finite-element analysis of local flange buckling. Bradford and Trahair (1981) subsequently developed a beam- or line-type element which provides improved computational efficiency. They concluded that distortional buckling modes are intermediate between conventional local and lateral-torsional buckling modes.

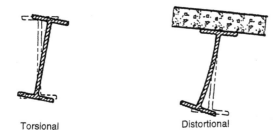

Figure 4.2 Lateral-torsional and distortional modes of buckling

Bradford and Johnson (1987) conducted finite-element studies on the buckling of unstiffened, fixed-ended, composite beams for an extensive range of relatively slender ratios of web depth-to-thickness, flange width-to-thickness and unbraced length-to-flange-width. They concluded that web depth-to-thickness ratio was the most important parameter influencing distortional lateral buckling. Weston *et al.* (1991) conducted a similar parametric study on sections with relatively slender webs and long spans, but included lateral slenderness ratio of the bottom flange as a variable. They observed three modes of failure: local web buckling, lateral buckling and a combination of the two modes. They concluded that both web slenderness, h / t_w, and lateral slenderness of the compression flange, kL / i_z, are important parameters.

Tests by Johnson and Fan (1991) on the negative-moment region of both T and inverted U arrangements of composite beams, falling on the boundary between Class 2 and 3 sections, illustrate the interaction between local and lateral buckling. It is apparent that inverted U-frames, comprising two steel beams as the legs of the U-frame and the transverse concrete slab as the cross connection, provide enhanced lateral buckling stiffness compared to the behaviour of a similar T-beam in isolation. They proposed an elastic U-frame model in which the rigid-web buckling formula of Equation 4.3 is adjusted to allow for distortion of the section and restraint to lateral buckling provided by the transverse concrete slab between beams. The observed distortional moment capacities are compared with the theoretical approaches of Bradford and Johnson (1987), Weston *et al.* (1991), the SCI (1988) and Eurocode 4 (1992). All four methods underestimate the average moment resistance in the tests by a factor of between 0.51 and 0.63. They observed a complex interaction between local and distortional buckling with the transition from symmetrical local buckling to S-shaped lateral buckling taking place at or near maximum load.

The modified expression for the critical buckling moment proposed by Johnson and Fan (1991) is the same as that subsequently adopted in Eurocode 4 (1992). The warping stiffness term, EI_w, in Equation 4.3 is replaced by $k_s(k_w L / \pi)^4$, in which k_s is the transverse flexural stiffness provided by the slab and web of the I-section to lateral distortion and is given in flexibility terms by:

$$(1/k_s) = [(a/X_1 EI_{sl}) + (4(1-v^2)h_s / Et_w^3)] \qquad (4.16)$$

in which EI_{sl} is the cracked transverse flexural rigidity of the concrete slab, a is the spacing of the composite beams, h_s is the depth between centres of the flanges, and $X_1 = 2$ for a simply-supported slab and $X_1 = 4$ for a slab continuous over a number of steel beams. The first term is the reciprocal of transverse stiffness provided by the slab and the second term is the reciprocal of lateral stiffness provided by the web. The resulting expression for the critical moment is then obtained from Equation 4.3 as:

$$M_{cr} = \frac{k_c C_4}{L} \sqrt{EI_z \left(GI_t + \left(\frac{k_w L}{\pi}\right)^2 k_s \right)} \qquad (4.17)$$

Dekker and Kemp (1998) have proposed an alternative spring model to calculate equivalent geometric properties in Equation 4.3 for $I^*_w = C_1 I_w$, $I^*_z = C_2 I_z$, and $I^*_t = C_3 I_t$. The equivalent properties I^*_w, I^*_z and I^*_t represent the change of stiffness due to distortional lateral buckling in steel and composite sections. Bradford and Trahair (1981) and Weston et al. (1991) have shown that distortional buckling is most severe in specimens with small span-to-depth ratios, L/h, large web slendernesses, h_s/t_w, and stocky flanges, (low b_f/t_f). In the extreme case, the effective section properties would tend towards those applying to the compression flange acting as an isolated strut.

The ratios of observed lateral buckling moment to plastic or yield moment in 19 analytical tests on relatively slender composite beams were initially evaluated by Weston et al. (1991) and subsequently compared by Dekker et al. (1995) to the theoretical models proposed by Weston et al. (1991), Bradford and Johnson (1987), Svensson (1985), unfactored Eurocode 4 (1992) and Dekker's spring model. The Weston et al. (1991) finite element results are considered to be the most accurate to use as a basis of comparison. It is apparent that Dekker's spring model compares favourably with these results and that other models are quite conservative.

A number of models have been proposed for assessing the predominantly elastic, distortional buckling moment of relatively slender composite beams in negative bending.

4.3.4 Effect of Yielding on Effective Length of Compact Beams

In developing lateral buckling theory for inelastic, compact composite and steel beams, emphasis is frequently placed on elastic stiffness properties which are adapted to determine the elastic critical buckling moment (see section 4.3.3). This elastic moment is then used to calculate the buckling strength of the beam on the basis of semi-empirical inelastic strength curves such as those defined in section 4.1.7. Considering the typical two-span subframe of Figure 4.3, this approach may appear to be justified in compact composite beams in negative bending, because lateral buckling is concentrated in the short length L_i where the bottom flange is in compression. Furthermore the yielded region is only a small proportion of this negative-moment length L_i.

However, this interpretation ignores the considerable loss of stiffness in the yielded region of the beam where the moments are largest. If the moment-curvature curve for composite or steel beams in negative bending is represented simply by the bi-linear relationship shown in Figure 4.4, it becomes apparent that the loss of stiffness in the yielded region is considerable. Intuitively (see Kemp and Dekker, 1991) an element of length, L, and uniform flexural rigidity, EI, subject to a linear, elastic moment gradient from M_{min} to M_m may be considered to be approximately equivalent to an element with an effective length of L_{eff} and uniform inelastic flexural rigidity, $E_{sh}I$, subjected to a uniform moment, M_p, where:

$$L_{eff} = L\sqrt{(E_{sh}/E)(0.6M_m + 0.4M_{min})/M_p} \qquad (4.18)$$

This formulation may be applied to the length L_i in negative bending in Figure 4.3, comprising firstly the elastic component of length L_e (between moments of zero and nM_p) and secondly the inelastic component of length, L_p, (between moments of nM_p and M_m), to give an approximate effective length, $L_{eff} = kL_i$, under uniform moment, M_p, and inelastic flexural rigidity, $E_{sh}I$:

$$kL_i = L_e\sqrt{0.6n(E_{sh}/E)} + L_p\sqrt{(0.6M_m + 0.4nM_p)/M_p} \qquad (4.19)$$

The idealized, bi-linear, moment-curvature relationship in Figure 4.4 represents a useful simplification for considering inelastic local and lateral buckling of compact sections. The author has shown (see Kemp *et al.*, 2001) that test measurements on steel I-beams are accurately represented in fibre-element models by such bi-linear relationships with n = 0.9. A similar model applies to composite beams in which the concrete is neglected in tension.

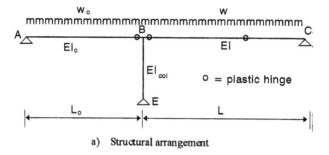
a) Structural arrangement

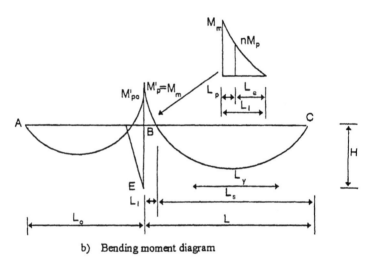
b) Bending moment diagram

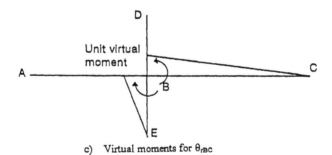
c) Virtual moments for θ_{rBC}

Figure 4.3 Typical two-span subframe

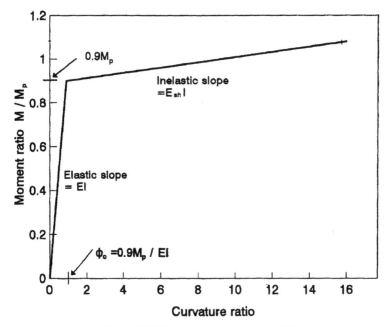

Figure 4.4 Bi-linear moment-curvature relationship

As an example, consider the case in which the pivot point in the bilinear moment-curvature relationship in Figure 4.4 occurs at n = 0.9, the maximum moment reaches $M_m=1.1M_p$, the minimum elastic moment is zero at the next lateral restraint and $E_{sh}/E = 0.01$. The length of the inelastic zone of negative bending, L_p, which represents less than 20% of the length L_i, would then contribute 75% of the total effective buckling length $kL_i = 0.24L_i$ of an equivalent element with uniform inelastic flexural rigidity of $E_{sh}I$ (or $kL_i = 2.4L_i$ of an element with uniform elastic flexural rigidity of EI). This explains the very close spacing of lateral restraints to the yielded portion of the compression flange required in design to ensure Class 1 plastic behaviour, and the potential inaccuracy of relying on elastic properties in stability assessments of compact beams under moment gradient.

Yielding of the bottom compression flange and adjacent web significantly changes the relative contribution of the different components of length and stiffness influencing inelastic lateral buckling. An effective length factor may be used to allow for the inelastic flexural rigidity and moment gradient in negative bending.

4.3.5 Inelastic Distortional Buckling of Continuous Compact Beams

At the low slenderness ratios at which inelastic lateral buckling occurs, Lay and Galambos (1965) recognized that the second term of Equation 4.3, reflecting warping stiffness of the flange, is much more important than the first term reflecting torsional stiffness, which may conservatively be discarded without significant loss of accuracy. If this essentially elastic equation is expressed not in terms of critical buckling moment, M_{cr}, but rather in terms of critical in-plane curvature represented by $\phi_{clat} = M_{cr} / EI_y$, the following expression is obtained:

$$\phi_{clat} = \frac{M_{cr}}{EI_y} = \left(\frac{\pi}{kL}\right)^2 \sqrt{\frac{C_1 I_w C_2 I_z}{I_y^2}} \qquad (4.20)$$

In this expression I_w is the warping constant of the plain steel section and I_y and I_z are the second moments of area about the major and minor axes of the steel section respectively. The effective length factor in lateral buckling, k, which is given by Equation 4.19 for negative bending adjacent to supports, replaces the effective length factor (π / C_4) in Eurocode 4. This effective length refers to a length under uniform moment of M_p (determined from plastic stress blocks) and a uniform strain-hardening modulus of E_{sh}. The corresponding critical moment is determined from Figure 4.4 for the assessed critical curvature ϕ_{clat}. For plain steel I-sections having relatively stocky flanges and thin webs, and no concrete slab, the values of C_1 and C_2 used for calculating the effective warping constant I^*_w and lateral stiffness I^*_z respectively, are close to unity (Dekker et al., 1995).

An important difference between the inelastic lateral buckling resistance of a plain steel beam and a composite beam in negative bending, is derived from the increase in warping stiffness caused by the shift in position of the shear centre from the centroid of a doubly-symmetrical plain steel I-section to the top flange of a composite beam which is restrained by the slab. This increases the warping stiffness from $I_w = I_z(h_f / 2)^2$ for a steel section to $I^*_w = I_{zf} h_f^2$ for a composite section (h_f is the depth between the centre of the flanges and I_{zf} is the second moment of area of the compression flange about the minor axis). Allowing for this shift in the shear center and the increase in lateral and torsional stiffness of the tension flange, the coefficient C_1 which is applied to the normal warping stiffness of a plain I-section, approaches a value of 4.0. This corresponds to an effective length factor due to distortional buckling, $k_d = [1/(C_1 C_2)]^{0.25}$ for a composite beam of approximately 0.7, which is considered further in section 4.4.3. The critical strain in the compression flange is obtained from Equation 4.20 as $\varepsilon_{clat} = \phi_{clat} h_c$ (where h_c is the depth of the centre of the compression flange below the plastic neutral axis).

Linked to studies of the Autostress method used in North America, White and Barth (1998) developed empirical equations for modelling moment resistance and inelastic rotation of sections with Class 1 or 2 flanges, Class 3 or

4 webs and closely spaced lateral restraints. Their results compare favourably with finite-element analyses and experimental results, and appear to be more accurate than the AISC LRFD code (1986). They confirm that the predicted design moment is achieved for the sections and note that failure is dominated by buckling of the compression flange.

Composite beams subject to inelastic negative bending exhibit distortional buckling in which the warping stiffness is the dominant component. The critical in-plane curvature for such a composite beam may be assessed using a spring model and adjusted effective length.

4.4 DUCTILITY CONSIDERATIONS FOR COMPACT BEAMS

4.4.1 Interaction of Inelastic Local and Lateral Buckling of Compact Elements

As indicated in section 4.3.4, a considerable loss of stiffness of the compression flange in lateral bending occurs due to yielding of the flange immediately adjacent to the internal support and local buckling within this yielded region. Consequently large lateral curvatures of this flange are concentrated in this locality as shown in Figure 4.5. Lay and Galambos (1967) originally recognised that local buckling of Class 1 or 2 beam elements is confined to a yielded and partially strain-hardened length L_{cloc} which is long enough to accommodate the full wavelength of the local buckle. On the other hand, lateral buckling occurs over an effective length, kL_i, in which L_i is the length over which the bottom flange is in compression between the section of maximum negative moment at the internal support and the adjacent point of inflection. An adjustment should be considered to allow for the incomplete lateral restraint of a compression flange at a point of inflection, (such as an increase of the elastic length of the negative moment region L_i by one beam depth), but this is less significant once yielding of a compact element occurs within the length L_i (see Equation 4.19).

In cases where inelastic lateral buckling is largely restrained, considerable ductility is evident before strain-weakening of the steel section occurs. This has been demonstrated in experiments conducted by Dekker (1989) on steel T-sections where both lateral buckling and web buckling could not develop, and in experiments on steel I-sections by Byfield and Nethercot (1998) in which closely spaced lateral restraints were provided in the yielded region to prevent lateral buckling.

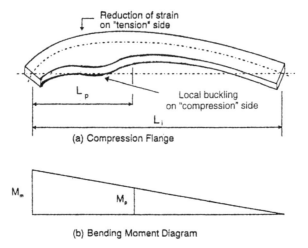

(a) Compression Flange

(b) Bending Moment Diagram

Figure 4.5 Interactive, inelastic, local and lateral buckling

Figure 4.6 Lateral and local buckling of compression flange (inverted beam in negative bending)

The transverse strain gradient across the compression flange caused by lateral displacements and the associated differential local buckling between the flange outstands provide a simple explanation of strain-weakening associated with lateral buckling. On the tension or outer fibre of the lateral buckle in Figure 4.5 the compressive strain is increasing due to in-plane bending but reducing due to lateral buckling. Once the net strain on this outer fibre is reducing elastically, the corresponding stress will fall rapidly below the yield stress and a loss of moment resistance will occur. Local buckling of the flange on the compression side of the lateral buckle will also contribute to strain weakening. Both of these effects are apparent in Figure 4.6 which shows the inelastic buckling under linear moment gradient of the compression flange of a steel I-section. This behaviour is apparent in numerous tests, not only from strain measurements, but also from the flattening of the local buckle on the tension side of the lateral buckle as shown in Figures 4.5 and 4.6.

Within this explanation of the cause of strain weakening there are, however, two modes of failure. The first mode occurs in Grade S275 steel with a lateral slenderness ratio L_i / i_z in the range of about 60 to 90, corresponding approximately to Class 2 behaviour. In this case, the section of maximum lateral deflection of the compression flange is located beyond the plastically yielded length L_p in Figure 4.5. The maximum in-plane moment M_m at which distortional buckling occurs is just above the plastic resistance moment M_p and reflects a yielded length L_p which is less than that required for local flange buckling. Consequently local flange buckling in the plastic region does not have a large influence on the maximum moment resistance achieved. Local web buckling may reduce the resistance to lateral buckling. Once strain-weakening commences, lateral deflections increase rapidly and the zone of yielding extends so that a local buckle will develop as a result of lateral buckling. The permanent deformation after unloading may overemphasize the influence of local flange buckling on failure, particularly because lateral deflections are largely elastic in this range of lateral slenderness ratio.

The second mode occurs with lateral slenderness ratios L_i / i_z less than about 60, corresponding approximately to Class 1 behaviour. In this case the section of maximum lateral deflection occurs within the yielded length L_p in Figure 4.5, which is long enough to accommodate the full wavelength for combined local flange and web buckling. As a result and as illustrated in Figure 4.6, the local buckle in the yielded flange, on the compression side of the lateral buckle, acts as a partial hinge in the vicinity of maximum lateral deflection and contributes directly to limiting the maximum moment resistance. The amplitude and wavelength of the local flange buckle is influenced by local web buckling, so that a fully interactive mode of local flange and web buckling and lateral-torsional buckling may occur.

4.4.2 Limit state of ductility

A broad spectrum of researchers have recognised that the most appropriate approach for classifying sections and elements of members according to Table 4.1 is to quantify the required and available ductility. This particularly applies to Class 1 or 2 sections where inelastic behaviour is inherent in the classification. Kemp and Dekker (1991) have proposed a simple limit states criterion for ductility which specifies that the available inelastic rotation at each notional plastic hinge, θ_a, (resistance) should exceed the required inelastic rotation, θ_r, (action effect) at the same notional hinge. A typical form of this criterion with partial factors included is as follows:

$$\theta_a / \gamma_a \geq \gamma_r \theta_r \qquad (4.21)$$

in which γ_a and γ_r are appropriate partial material and load factors for ductility, to allow for the mode of failure and uncertainties in the assessment of θ_a and θ_r. Due to the many conservative assumptions made in the assessment of the required rotation, the product $\gamma_a \gamma_r$ will probably lie in the range 1.5 for ductile modes to 3.0 for brittle modes of failure.

This formulation differs considerably from that adopted by Nethercot *et al.* (1995), Li *et al.* (1995) and others in the following respects:

1. The required rotation θ_r of the notional hinge in Equation 4.21 refers only to inelastic rotation implied by the process of moment redistribution or plastic analysis. Nethercot *et al.* (1995) and Li *et al.* (1995), on the other hand, consider the required rotation to be made up of an "elastic" component (associated with moment redistribution and concrete cracking) and a "plastic" component (reflecting the effect of inelastic behaviour in the positive-moment region). The advantage of the approach proposed in Equation 4.21 is that θ_r is a pure load effect and is independent of material properties such as steel yielding or concrete cracking. It is equal to the notional hinge rotation for plastic analysis or for moment redistribution relative to a conventional linear elastic analysis with rigid joints and uniform elastic properties, and has zero value if there is no moment redistribution.
2. The available rotation θ_a of the notional hinge on the other hand is the sum of the effects of inelastic behaviour of the component materials in positive and negative-moment regions of the beam (including concrete cracking and the total rotation within the end connection). The alternative approach only includes the rotation within the end connection as the available rotation, and all other materials-related effects such as steel yielding in positive bending and concrete cracking are included in the required rotation. This approach highlights the connection behaviour, but neglects, for example, yielding of the steel beam adjacent to the connection.

To some degree these differences are a matter of terminology. In the subsequent discussion, however, the required inelastic rotation at notional hinges is taken to result from a full or partial plastic redistribution of moments relative to the results of a conventional, linear, elastic analysis (neglecting rotation of the end connection and cracking of the concrete). The limit-states criterion in Equation 4.21 is satisfied by providing a beam with sufficient available inelastic rotation at each notional hinge. This is achieved by yielding of the steel section and reinforcement, crushing or cracking of the concrete, rotation of the end connection and possibly flexibility of the shear connectors. This available inelastic rotation θ_a therefore includes all the components of rotation at a notional hinge, which are not considered in a conventional elastic analysis. Yielding of the steel and crushing of the concrete in positive bending, produce a negative contribution to the available rotation of the support hinges.

Modelling approaches are described subsequently for quantifying the required and available inelastic rotations in this equation. By keeping the load-dependent items and material-dependent items on separate sides of the equation, a clearer picture of the design requirement is provided. Specific attention is given to the following important concepts:

1. Notional plastic hinges concentrated at sections of maximum moment are used to compare the available and required components of inelastic rotation on a consistent basis and these hinge rotations may be assessed simply using virtual work methods (see Kemp, 1999).
2. By considering the flexibility of these notional hinges, inelastic rotations due to yielding of the steel section, rotation within the end connection and cracking of the concrete may be combined in a logical way.

4.4.3 Available Inelastic Rotations for Compact Composite Beams

Introduction

The total available inelastic rotation θ_a of a notional plastic hinge at an internal support in a composite beam with semi-rigid end connections is made up of the sum of a number of components. These components are yielding of the steel and cracking of the concrete in the negative moment region, $\theta'_{ay} + \theta_{acr}$, total rotation of the end connection, θ_{acon}, and yielding of the steel and crushing of the concrete in the positive-moment region, θ_{ay}, as follows:

$$\theta_a = \theta'_{ay} + \theta_{acr} + \theta_{acon} - \theta_{ay} \tag{4.22}$$

These components are considered separately in the following sections and will be illustrated in the example structure of Nethercot *et al.* (1995) shown in Figure 4.7.

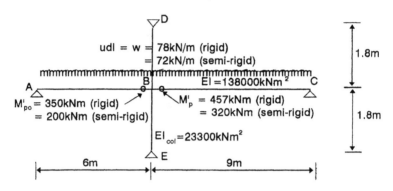

Figure 4.7 Example structure (Nethercot *et al.*, 1995)

Yielding of Steel Sections, θ'_{ay}, and Cracking of Concrete, θ_{acr}, in Negative Bending with Rigid End Connections

Thick extended end-plates may be designed at internal columns to support the full negative-moment resistance of the adjacent steel beam. Under these conditions the rotation of the end connection is normally small and is often ignored. At the end-plate the full steel section is effective in both tension and compression. The ultimate resistance moment and associated neutral axis depth are determined from conventional stress-block analysis for negative bending.

Inelastic flexural tests on compact steel or composite beams in negative bending are most commonly conducted on inverted double-cantilever specimens with an upward central load as shown in the inset of Figure 4.8. Lateral deflection is prevented at the central load and the two outside supports, but there is no restraint to rotation about a vertical axis at these points. Each half of the specimen represents the region of a continuous composite beam between an internal support and the adjacent point of inflection. The inelastic rotation at the notional plastic hinge at the internal support may be assessed approximately from the inelastic deflection at the end of each cantilever. Typical relationships between the central moment and the in-plane end rotation for Class 1 to 4 elements are shown in Figure 4.8.

Kemp (1996) has evaluated interactive lateral and local buckling in 58 double-cantilever tests on compact plain steel sections conducted by Adams *et al.* (1965), Lukey and Adams (1969), Kemp (1985,1986), Kuhlmann (1989) and Dekker *et al.* (1995). Some of these tests simulate the significant influence of coincident axial compression force on the steel section, reflecting the effect of tension reinforcement in a composite beam in negative bending.

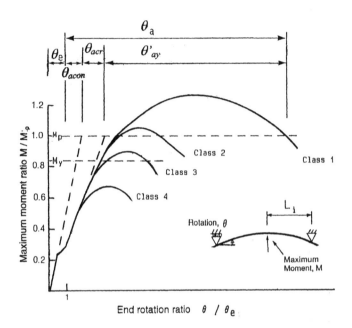

Figure 4.8 Moment-rotation curves

Observed inelastic rotation capacity, r_a, in these tests is expressed as the ratio of inelastic end rotation, θ'_{ay}, to the hypothetical, elastic end rotation at moment M'_{ps}, given by $\theta_{es}=0.5M'_{ps}L_i/EI_s$ as defined below. It is plotted in Figures 4.9a, 4.9b and 4.9c as a function of the following parameters:
(a) Flange slenderness ($b/t_f)\varepsilon_f$ in Figure 4.9a ($\varepsilon=\sqrt{250/f_y}$),
(b) Web slenderness ($h_c/t_w)\varepsilon_w$ in Figure 4.9b (h_c is web depth in compression),
(c) Effective lateral slenderness ratio λ_e (in the range $25 < \lambda_e < 140$) in Figure 4.9c and defined by:

$$\lambda_e = K_f K_w (L_i / i_{zc}) / \varepsilon_f \tag{4.23}$$

in which
$K_f = (b/t_f)/10\varepsilon_f$ is the flange factor for flange outstand b and thickness t_f in the range $0.7 < K_f < 1.5$,
$K_w = (\alpha d/t_w)/35\varepsilon_w$ is the web factor for web of depth d and thickness t_w in the range $0.7 < K_w < 1.5$,
α = ratio of web depth in compression to total web depth,
i_{zc} = radius of gyration of portion of the web and flange in compression,
L_i = length between zero and maximum negative moment in Figure 4.3,

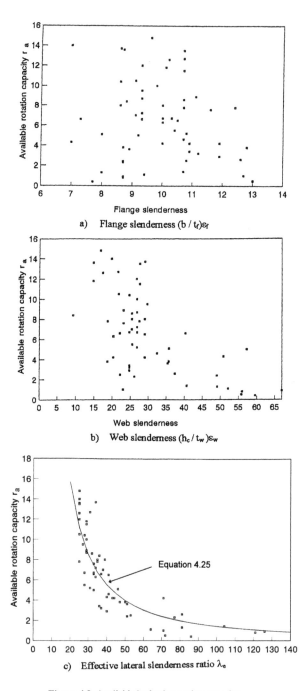

Figure 4.9 Available inelastic rotation capacity

M'_{ps} = plastic moment resistance of the steel section in negative bending,
EI_s = flexural rigidity of the steel section.

It is apparent from Figure 4.9 that the local buckling parameters, K_f and K_w, have a smaller and more random effect on the achievement of ductility than the lateral buckling parameter. The effective lateral slenderness ratio, λ_e, reflecting interactive lateral and local buckling (Equation 4.23 and Figure 4.9c), provides the best model of the behaviour. This figure also contains a proposed prediction for the available inelastic rotation, θ'_{ay}, of a plain steel section given by:

$$\theta'_{ay} = 1.5\theta_e(60/\lambda_e)^{1.5}/\alpha = 1.5\,(0.5M'_{ps}L_i/EI_s)\,(60/\lambda_e)^{1.5}/\alpha \qquad (4.24)$$

As this is an empirical model based on test results, it includes provision for moment gradient and localized yielding discussed in section 4.3.4.

This expression for available inelastic rotation of plain steel sections may be used in the negative moment region of continuous composite beams with rigid end connections if account is taken of the following differences in behaviour between steel and composite sections:

1. Unlike steel beams in pure flexure, the steel sections of composite beams in negative bending are subjected to a compressive axial force required to balance the tensile force in the reinforcement. This has the effect of increasing the depth of web in compression, which has a weakening influence on both local and lateral buckling. In addition, for a particular critical strain in the compression flange of the steel or composite beam, associated with local and lateral buckling, the strain gradient or curvature will reduce as the height of the plastic neutral axis increases, as shown in Figure 4.10. Consequently the inelastic rotation in a composite beam is reduced relative to the plain steel section as the neutral axis moves up the section. This is accounted for by the factor α in Equation 4.24 and the definition of K_w.
2. The concrete slab restrains the top flange torsionally and laterally producing distortional buckling as discussed in section 4.3.5. This is considered by introducing a distortional restraint factor k_d in Equation 4.23, which is equal to 1.0 for a steel section and 0.7 for a composite beam (see Dekker *et al.*, 1995, and section 4.3.5):

$$\lambda_e = K_f K_w k_d (L_i/i_{zc})/\varepsilon_f \qquad (4.25)$$

3. The in-plane elastic flexural rigidity of composite beams is not uniform over the length of the beam due to cracking of the concrete in negative moment regions at moment M'_{cr}. This aspect appears to provide less benefit than has been assumed in design codes and is allowed for separately by the

following approximate expression for available inelastic rotation due to cracking, θ_{acr}, which is a component part of Equation 4.22:

$$\theta_{acr} = 0.5L_i \ [(M'_p)^2 - (M'_{cr})^2][(1.0/EI_{cr}) - (1.0/EI)]/M'_p \qquad (4.26)$$

in which EI and EI_{cr} are the uncracked and cracked flexural rigidity of the composite beam in negative bending.

4. As buckling occurs in the steel section, the elastic and inelastic rotations of the steel section, θ_{es} and θ'_{ay}, in Equation 4.24 apply to both plain steel sections and composite beams. This produces a larger available rotation capacity, r_a, due to the reduced elastic rotation, θ_e, of composite beams.

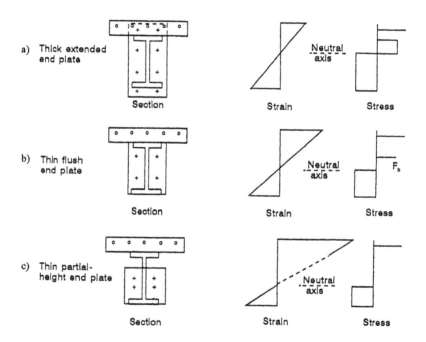

Figure 4.10 Effect of level of plastic neutral axis

In addition the ratio of ultimate moment resistance in negative and positive bending, M'_p/M_p, is smaller in composite beams than steel beams. The required redistribution of moment and therefore the required inelastic rotation at a notional support hinge in a continuous composite beam will be larger than for a pure steel beam, but the length of the region over which lateral buckling occurs in Figure 4.3 will be proportionately less. This is particularly the case with semi-rigid end connections discussed in the next section.

Couchman and Lebet (1996) have proposed a design method for compact Class 1 and 2 members in which they compare the available percentage moment redistribution (corresponding to the available rotation capacity, r_a) with the required percentage moment redistribution obtained from two generic diagrams reflecting a range of spans, loadings and moment ratios. They assessed the models proposed by Spangemacher and Sedlacek (1992), Johnson and Chen (1991) and Kemp and Dekker (1991) for determining available rotation capacity and concluded that the latter model, described in Equations 4.23 and 4.24, provides the best results in terms of their criteria of accuracy and conservatism.

In a parallel study Axhag (1995) derived empirical formulae for inelastic rotation and slope of the falling branch of the moment-rotation curve based on test measurements of specimens with Class 1 or 2 flanges and Class 3 or 4 webs. Closely spaced lateral restraints are again required to avoid premature failure.

Based on interactive lateral and local buckling observed in numerous tests on plain steel sections, a model has been proposed in Equations 4.23 to 4.26 for inelastic rotation of continuous steel and composite beams in the negative moment region with rigid end connections.

Yielding of Steel Sections, θ'_{ay}, and Cracking of Concrete, θ_{acr}, in Negative Bending with Semi-Rigid End Connections θ_{acon}

Semi-rigid end connections comprising either a flush end-plate or a partial-height end-plate over the lower portion of the beam are considered to be the most effective means of achieving ductile continuity in continuous beams, as demonstrated by Xiao et al. (1994), Ahmed and Nethercot (1998) and Leon (1998). Important components of available rotation at semi-rigid end connections, as defined in Equation 4.22 are due to the total rotation of the connection, θ_{acon}, as well as yielding of the steel beam and reinforcement, θ'_{ay}, and cracking of the concrete, θ_{acr}, in the adjacent portion of the beam.

In partial-height end-plate connections located over the lower portion of the steel section as shown in Figure 4.10c, the tension at ultimate load is transmitted almost entirely by the longitudinal reinforcement in the slab, due to the flexibility of the end plate. In flush end-plate connections shown in Figure 4.10b tension is also transmitted through the bolted end-plate. The tension force per bolt required to produce a yield-line mechanism in the thin end-plate, F_b, is assumed to act at the level of the top line of bolts as shown in Figure 4.10b and is conservatively estimated as:

$$F_b = 4.5 \, t_p^2 \, f_{yp} \qquad (4.27)$$

in which t_p and f_{yp} are the thickness and yield stress of the end plate. If the moments on either side of the internal column are unequal, the difference in moment is transferred to the column.

A design manual published by the Steel Construction Institute (SCI, 1988) describes seven main modes of failure at internal beam-column connections (reinforcement in tension, end plate in bending or bolts in tension, column web crushing, column web buckling, beam flange in compression, shear panel failure and column flange bending). Apart from reinforcement fracture, these modes reflect types of premature failure, which should be avoided by appropriate detailing of the composite connections. If this is done and reinforcement fracture is prevented, local buckling of the compression flange in negative bending will be the critical mode of relatively ductile failure. Lateral buckling is unlikely due to the limited length of yielding, the small depth of web in compression and the correspondingly lower strains in the compression flange (see Figure 4.10). This replaces the interactive mode of local and distortional buckling obtained with rigid end connections discussed in the previous section.

Based on an analysis of 22 tests on flush and partial-height end-plate connections conducted by Anderson and Najafi (1994), Aribert and Lachal (1992), Benussi et al. (1989), Brown and Anderson (2001), Kemp et al. (1995), Li et al. (1996) and Xiao et al. (1994), the following recommendations have been made by Kemp and Nethercot (2001) to reduce the possibility of reinforcement fracture in semi-rigid connections:

1. In the tests all but one of the failures due to reinforcement fracture involved the use of flush end-plates which were at least 15mm thick. The effect of a thick end-plate is to substantially increase the bolt force, F_b, in Equation 4.27 and Figure 4.10b and therefore to cause the steel section immediately adjacent to the end-plate to act in both tension and compression. This raises the plastic neutral axis, which limits the length over which yielding and strain hardening of the reinforcement occurs and thus reduces the length of the inelastic region adjacent to the connection. It is recommended that the thickness of the end-plate should not exceed the thickness of the compression flange of the steel section.
2. Unless strain-hardening of the reinforcement occurs, the moment resistance will not increase much above the value at first yield and the region of inelastic curvature will be small and concentrated. To ensure that the plastic neutral axis does not lie above the mid-depth of the steel section, it is recommended that for partial-height end-plates the area of longitudinal reinforcement should not exceed 40% of the area of the steel section for Grade S275 (or 50% for Grade S355). For flush end-plates these limits should be reduced to allow for a conservative estimate of the bolt force, F_b, from Equation 4.27. This is consistent with the design proposal given by the SCI (1988), although their maximum limit is based on column buckling.

Instability and Ductility 113

3. To achieve a moment of resistance sufficiently in excess of the moment at first yield, the area of reinforcement provided should be at least that required to yield the bottom flange and part of the adjacent web. To ensure that at least 20% of this web area yields in compression, it is recommended that a minimum area of high tensile longitudinal reinforcement be provided equal to the larger of 1% of the slab area or 25% of the area of the steel section for Grade S275 (or 35% for Grade S355). This is consistent with the proposal by Anderson *et al.* (1998) and the SCI (1988) to achieve a suitable distribution of cracking in the slab. In addition adequate ductility of the reinforcement should be specified, with smaller diameter bars at closer spacing being used where practicable.

If premature modes of failure including reinforcement fracture are avoided, Kemp and Nethercot (2001) have proposed a simple set of expressions for assessing the ultimate moment resistance and associated available inelastic rotation at internal supports in cases where the connections comprise flush or partial-height end-plates. These inelastic rotations include the effects of yielding of the adjacent steel section, cracking of the concrete and total rotation of the end connection. The analysis of these test results indicate that an available inelastic rotation, $\theta'_{ay} + \theta_{acr} + \theta_{acon}$, of at least 0.03 radians should be expected if the above precautions are followed with semi-rigid end connections.

The mode of failure at semi-rigid end connections differs from that at rigid end connections. If precautions are taken to avoid local failures within the end connection and fracture of the reinforcement, a ductile mode of local buckling failure is achieved with a minimum level of available inelastic rotation of approximately 0.03 radians.

Yielding of Steel and Crushing of Concrete in Positive Bending θ_{ay}

The last component of inelastic rotation in Equation 4.22, θ_{uy}, produces a negative contribution to the total available rotation at the support, because this inelastic behaviour in the midspan region increases the required rotations at the notional hinges in the negative-moment region. This applies whatever end connection is provided. Importantly, Li *et al.* (1995) have shown that this negative contribution to ductility is brought within manageable limits if the maximum positive moment used in the design is limited to 95% of the plastic moment resistance M_p. Assuming this limit on positive moment, θ_{uy} may be assessed by the following expression for a beam under uniformly distributed load:

$$\theta_{ay} = 0.175 L_y \left[\frac{0.0035}{h_n} - \frac{M_p}{EI} \right] \left[\frac{M_m - M_y}{M_p - M_y} \right]^3 \qquad (4.28)$$

in which the upper surface strain in the concrete is limited to 0.0035, h_n is the neutral axis depth from the upper surface of the concrete in positive bending, $(0.0035/h_n)$ is the curvature at the plastic moment M_p, (M_p/EI) is the elastic component of this curvature, M_m is the maximum design moment ($<0.95M_p$) and L_y is the yielded length in the sagging moment region.

A simple function of the inelastic positive moment (Equation 4.28) provides a basis for assessing the inelastic rotation in the positive-moment region during redistribution of moment from internal supports to adjacent sections of maximum positive moment.

4.4.4 Required Inelastic Rotation

The required ductility is assessed in terms of concentrated rotations at notional plastic hinges. This process commences in continuous beams as moment is redistributed from critical support sections to adjacent sections of high moment in midspan regions where the design resistance is not achieved initially. It may be limited by a defined proportion of moment redistribution, or embrace full plastic analysis.

A generic frame arrangement in Figure 4.3a is considered for the purpose of assessing the required inelastic rotation, θ_r, in Equation 4.21 at critical notional hinges at internal supports in continuous beams. The frames are assumed to be braced and subject to static loading. The span lengths are measured to the column face and the width of the column is not considered to have a significant effect on the required rotation of a notional hinge.

It is assumed that no moments are generated at external supports. Elastic-perfectly-plastic behaviour is assumed for redistributing moments or for plastic analysis. The assumed distribution of moment at ultimate load is shown in Figure 4.3b and the analysis is based on the formation of plastic hinges at the locations shown in Figure 4.3a with the sagging moment hinges being the last to form.

The required notional hinge rotation, θ_{rBC}, at B in the critical span BC is determined from a flexibility analysis or by integrating the product of the virtual moments in Figure 4.3c and the distribution of elastic curvature due to the applied load which is obtained from the moments in Figure 4.3b, as described by Kemp (1999). The virtual moments in Figure 4.3c correspond to the difference in end rotation at end B between the span BC and the column BE, reflecting the required hinge rotation, θ_{rBC}, in member BC. The resulting rotation compatibility equation at B is given by:

$$\theta_{rBC} = \theta'_{BC} - \theta_{BC} - \theta_{BE} \qquad (4.29)$$

Instability and Ductility 115

in which θ'_{BC} is the rotation at end B of member BC due to the applied load in this critical span if the ends of member of BC are simply supported, (for example for a uniformly distributed load w, $\theta'_{BC} = wL^3/24EI$ and for a central concentrated load W, $\theta'_{BC} = WL^2/16EI$); θ_{BC} is the rotation at end B due to the end moment M'_p at B in BC, given by $\theta_{BC} = M'_p L/3EI$, and θ_{BE} is the rotation of the column or columns due to the unbalanced moment, $(M'_p - M'_{po})$, at the end B of the two beams BC and BA, given by:

$$\theta_{BE} = \frac{(M'_p - M'_{po})}{\Sigma(3EI/H)_{col}} \tag{4.30}$$

Substituting all of these expressions into Equation 4.29 gives:

$$\theta_{rBC} = \theta'_{BC} - \theta_{BC} - \theta_{BE} = \theta'_{BC} - \left(\frac{L}{3EI}\right)M'_p - \frac{(M'_p - M'_{po})}{\Sigma(3EI/H)_{col}} \tag{4.31}$$

The other terms are defined in Figure 4.3 and the subscript "o" refers to the outside spans.

4.4.5 Example

The non-symmetrical frame of Figure 4.7 has been evaluated by Nethercot *et al.* (1995). It will be used as an example to illustrate the limit state of ductility as applied to a continuous frame in which the adjacent beams and the end-plate connections at B to beams BC and BA are assessed to be capable of resisting the following two cases of ultimate moments:

- Rigid, extended end-plate: 457 kNm at B in BC and 350 kNm at B in BA,
- Semi-rigid, flush end-plate: 320 kNm at B in BC and 200 kNm at B in BA.

The distributed load is adjusted from Nethercot's example to give a maximum positive moment, M_m, of less than 95% of M_p. The loads, dimensions, properties and moment resistances are given in Figure 4.7 and summarised in Table 4.3. The available and required inelastic rotations at the notional hinge at B in member BC are calculated in Table 4.3 from the preceding theory. The small secondary adjustments (Kemp, 1999) for equivalent notional hinges have been omitted for clarity. It is apparent that the limit states Equation 4.21 is satisfied for both the rigid and semi-rigid end connections with an adequate margin of safety.

Table 4.3 Example of limit states equation of ductility

		Equation	Rigid end connections at B	Semi-rigid end connections at B
Basic	M'_p, M'_{po} (kNm)		457, 350	320, 250
data	M_p, M_m (kNm)		614, 578	614, 577
	M_y, M_{ps} (kNm)		434, 330	434, 330
	EI, EI_{cr} (10^3kNm^2)		138, 65.6	138, 65.6
	EI_s, EI_{col} (10^3kNm^2)		44.1, 23.3	44.1, 23.3
Distributed load, w (kN/m)			78	72
Available inelastic rotation				
Effective	L_i, L_y (m)		1.30, 3.84	0.99, 3.98
slenderness	K_f		0.728	0.728
ratio λ_e	K_w		1.17	0.7
	λ_e	(4.23)	25	25
θ'_{ay} (rad.)		(4.24)	0.0322	Total
θ_{cr} (rad.)		(4.26)	0.0022	$\theta'_{ay} + \theta_{acr} + \theta_{acon}$
θ_{con} (rad.)			0.0	= 0.030
$-\theta_{ay}$ (rad.), h_n (m)		(4.28)	−0.0075, 0.133	−0.0076, 0.133
$\theta_{aBC} = \theta'_{ay} + \theta_{acr} + \theta_{acon} - \theta_{ay}$		(4.22)	0.0269	0.0224
Required inelastic rotation				
$\theta'_{BC} = wL^3 / 24EI$ (rad.)			0.0172	0.0158
θ_{BC} (rad.)			0.0099	0.0070
θ_{BE} (rad.)		(4.30)	0.0014	0.0015
$\theta_{rBC} = \theta'_{BC} - \theta_{BC} - \theta_{BE}$		(4.29)	0.0059	0.0073

4.5 REFERENCES

Adams, P.F., Lay, M.G. and Galambos, T.V., 1965, Experiments on high-strength steel beams. *Welding Research Council*, Bulletin No. 110, pp. 1–16.

Ahmed, B. and Nethercot, D.A., 1997, Prediction of initial stiffness and available rotation capacity of major-axis flush end-plate connections. *Journal of Constructional Steel Research*, **41** (1), pp. 31–60.

AISC LRFD, 1986, *Specification for structural steel buildings*. (Chicago: American Institute of Steel Construction).

Anderson, D.A. and Najafi, A.A., 1994, Performance of composite connections: major-axis end-plate joints. *Journal of Constructional Steel Research*, **31** (1), pp. 31–58.

Anderson, D.A., Aribert, J-M. and Kronenberger, H-J., 1998, Rotation capacity of composite joints. In *Proceedings COST-CI Conference*, (Liege: European Commission), pp. 177–186.

Aribert, J.M. and Lachal, A., 1992, Experimental investigation of composite connection and global interpretation. In *Proceedings COST-CI Conference on Semi-Rigid Joints*, (Strasbourg: European Commission), pp. 158–169.

Axhag, F., 1995, Plastic design of composite bridges allowing for local buckling. *Technical Report*, (Lulea: University of Technology).

Benussi, F., Puhali, R. and Zandonini, R., 1989, Semi-rigid joints in steel-concrete composite frames. *Construzioni Metallique*, **5**, pp. 1–28.

Bode, H. and Kronenberger, H-J., 1996, Behaviour of composite joints and their influence on semi-rigid composite beams. In *Proceedings of the Engineering Foundation Conference on Composite Construction III*, Irsee, Germany, (New York: Engineering Foundation), pp. 766–779.

Bradford, M.A. and Trahair, N.S., 1981, Distortional buckling of I-beams. *Journal of the Structural Division, ASCE*, **107** (ST2), pp. 355–370.

Bradford, M.A. and Johnson, R.P., 1987, Inelastic buckling of composite bridge girders near internal supports. *Proceedings of the Institution of Civil Engineers, Part 2*, **83**, pp. 143–159.

Brown, N.D. and Anderson, D.A., 2001, Structural properties of composite major-axis end-plate connections. *Journal of Constructional Steel Research*, **57** (3), pp. 327–349.

Bryan, G.H., (1891), On the stability of a plane plate under thrusts in its own plane, *London Mathematical Society*, **22**.

Byfield, M.P. and Nethercot, D.A., 1998, An analysis of the true bending strength of steel beams. *Proceedings of the Institution of Civil Engineers*, **128**, pp. 188–197.

CAN/CSA - S16.1-M89, 1989, *Limit states design of steel structures*. (Ontario: Canadian Standards Association).

Couchman, G. and Lebet, J-P., 1996, A new design method for continuous composite beams. *Structural Engineering International*, **6** (2), pp. 96–101.

Dekker, N.W., 1989, The effect of non-interactive local and torsional buckling on the ductility of flanged beams, *The Civil Engineer in South Africa*, **31** (4), pp. 121–124.

Dekker, N.W. and Kemp, A.R., 1998, A simplified distortional buckling model for doubly-symmetrical I-sections. *Canadian Journal of Civil Engineering*, **25**, pp. 1–10.

Dekker, N.W., Kemp, A.R. and Trinchero, P., 1995, Factors influencing the strength of continuous composite beams in negative bending. *Journal of Constructional Steel Research*, **34** (2–3), pp. 161–186.

Eurocode 3, 1992, *Design of steel structures, Part 1: General rules and rules for buildings*. (Brussels: European Committee for Standardisation).

Eurocode 4, 1992, *Design of composite steel and concrete structures, Part 1-1: General rules and rules for buildings*. (Brussels: European Committee for Standardisation).

Galambos, T.V., 1998, *Guide to stability design criteria for metal structures, Fifth Edition*. (New York: John Wiley).
Hamada, S. and Longworth, J., 1974, Buckling of composite beams in negative bending. *Journal of the Structural Division, ASCE*, **100** (ST11), pp. 2205–2219.
Johnson, R.P. and Chen, S., 1991, Local buckling and moment redistribution in Class 2 composite beams. *Structural Engineering International*, **4**, pp. 27–34.
Johnson, R.P. and Fan, C.K.R., 1991, Distortional lateral buckling of composite beams. *Proceedings of the Institution of Civil Engineers, Part 2*, **91** (3), pp. 131–161.
Kemp, A.R., 1985, Interaction of plastic local and lateral buckling. *Journal of Structural Engineering, ASCE*, **111** (ST10), pp. 2181–2196.
Kemp, A.R., 1986, Factors affecting the rotation capacity of plastically designed members. *The Structural Engineer*, **64B** (2), pp. 28–35.
Kemp, A.R., 1996, Inelastic local and lateral buckling in design codes. *Journal of Structural Engineering, ASCE*, **122** (4), pp. 374–382.
Kemp, A.R., 1999, A limit-states criterion for ductility of Class 1 and Class 2 composite and steel beams. In *Proceedings of the 6th International Colloquium on Stability and Ductility of Steel Structures*, Timisoaro, Romania, Elsevier, pp. 291–298.
Kemp, A.R. and Dekker, N.W., 1991, Available rotation capacity in steel and composite beams. *The Structural Engineer*, **69** (5), pp. 88–97.
Kemp, A.R. and Nethercot, D.A., 2001, Required and available rotations in continuous composite beams with semi-rigid connections. *Journal of Constructional Steel Research*, **57** (4), pp. 375–400.
Kemp, A.R., Dekker, N.W. and Trinchero, P., 1995, Differences in inelastic properties of steel and composite beams. *Journal of Constructional Steel Research*, **34** (2), pp. 187–206.
Kemp. A.R., Byfield, M.P. and Nethercot, D.A., 2001, Effect of strain hardening on flexural properties of steel beams. Submitted for publication to: *The Structural Engineer*.
Kuhlmann, U., 1989, Definition of flange slenderness limits on the basis of rotation capacity values. *Journal of Constructional Steel Research*, **14**, pp. 21–40.
Lay, M.G. and Galambos, T.V., 1965, Inelastic steel beams under uniform moment. *Journal of Structural Engineering, ASCE*, **91** (6), pp. 67–93.
Lay, M.G. and Galambos, T.V., 1967, Inelastic beams under moment gradient. *Journal of the Structural Division, ASCE*, **93** (ST1), pp. 381–399.
Leon, R.T., 1998, Composite connections. *Progress in Structural Engineering and Materials*, **1** (2), pp. 159–169.
Leon, R.T. and Aho, M., 2000, Towards new provisions for composite columns. In *Proceedings of the Engineering Foundation Conference on Composite Construction IV*, Banff, Canada, (New York: Engineering Foundation).
Li, T.Q., Choo, B.S. and Nethercot, D.A., 1995, Determination of rotational

capacity requirements of steel and composite beams. *Journal of Constructional Steel Research*, **32** (4), pp. 303-332.

Li, T.Q., Nethercot, D.A. and Choo, B.S., 1996, Behaviour of flush end-plate composite connections with unbalanced moment and variable shear/moment ratios. *Journal of Constructional Steel Research*, **38** (2), pp. 25-198.

Lindner, J. and Budassis, N., 2000, Lateral torsional buckling of partially encased composite beams without concrete slab. In *Proceedings of the Engineering Foundation Conference on Composite Construction IV*, Banff, Canada, (New York: Engineering Foundation).

Lukey, A.F. and Adams, P.R., 1969, Rotation capacity of wide flanged beams under moment gradient. *Journal of the Structural Division, ASCE*, **95**, pp. 1173-1188.

Nethercot, D.A., Li, T.Q. and Choo, B.S., 1995, Required rotations and moment redistribution for composite frames and continuous beams. *Journal of Constructional Steel Research*, **35** (2), pp. 121-164.

Roik, K. and Bergmann, R., 1992, Composite columns. In *Constructional steel design*, edited by Dowling, P., (New York: Elsevier Science Publishers), pp. 443-470.

SCI, 1988, Joints in steel construction: composite connections. *Publication No. 213*, (Ascot: Steel Construction Institute).

Southward, R.E., 1969, Local buckling in Universal sections. *Internal Report, University of Cambridge Engineering Department*.

Spangemacher, R. and Sedlacek, G., 1992, Zum nachweis ausreichender rationsfahigkeit von fliessgelinken bei der anwendung des fliessgelenkverfahrens. *Stahlbau*, **61**, pp. 329–339.

Stowell, E.Z., 1950, Compressive strength of flanges. *National Advisory Committee for Aeronautics*, Technical Note 2020, United States of America.

Svensson, S.E., 1985, Lateral buckling of beams analysed as elastically supported columns subject to a varying axial force. *Journal of Constructional Steel Research,* **5** (3), pp. 179-194.

Weston, G., Nethercot, D.A. and Crisfield, M.A., 1991, Lateral buckling in continuous composite bridge girders. *The Structural Engineer*, **69**, (5), pp. 79-87.

White, D.W. and Barth, K.E., 1998, Strength and ductility of compact-flange I-girders in negative bending. *Journal of Constructional Steel Research*, **43** (3), pp. 241-280.

Xiao, Y., Choo, B.S. and Nethercot, D.A., 1994, Composite connections in steel and concrete. I. Experimental behaviour of composite beam-column connections. *Journal of Constructional Steel Research*, **31** (1), pp. 3-30.

Yura. J.A., 1993, Fundamentals of beam bracing. In *Proceedings of Steel Structures Research Committee Conference on Is Your Structure Suitably Braced*, Milwaukee, p. 20.

CHAPTER FIVE

Composite Floors

J. Buick Davison

5.1 INTRODUCTION

Steel decking was first used to support a concrete floor in the 1920s. Loucks and Gillet described a steel-deck system in a patent filed in 1926. In this early development, the steel deck provided all the structural resistance, concrete was added to give a level surface and provide fire resistance. The use of a steel deck was attractive to contractors as it served as permanent formwork and a construction platform, and was thus an efficient alternative to reinforced concrete floors. Soon other advantages of steel decking floors became apparent including its comparatively light weight and the use of the troughs in the decking as ducts for wiring. By 1938 engineers were using a non-composite cellular floor system (known as the *keystone beam* because of the dovetail shape of the steel cross section) in two and three storey industrial buildings. The first composite slabs i.e. concrete reinforced by the steel deck, appeared in the 1950s. The first was a product known as *Cofar*, which was a trapezoidal deck section with cold drawn wires welded transversely across the deck troughs. Friberg (1954) analysed the system as a traditional reinforced concrete slab and found good correspondence between predicted strengths and experimental tests.

In 1961 the Inland-Ryerson Company produced a trapezoidal metal deck with indentations rolled into the profile to achieve horizontal shear transfer between the concrete and steel. The floor deck, known as *HiBond*, was the forerunner of modern composite decks which use embossments to effect an efficient concrete to metal interaction. By 1967 a number of manufacturers were producing composite metal decks which were examined by Bryl. In reviewing Bryl's work, Schuster (1976) notes that, in addition to providing a sound understanding of the behaviour of the various systems tested, the report also explained the many cost-saving advantages of composite slabs and predicted that they would be widely exploited in the future, opening up wider markets to steel frames. This prediction has proved correct. As each deck manufacturer was developing a product by extensive independent research, the need for one standard design specification became apparent. In 1967 the American Iron and Steel Institute initiated a research project at Iowa State University to develop a design approach for composite slabs. This research is briefly outlined by one of the co-investigators (Schuster, 1976). Design recommendations were developed (Porter and Eckberg, 1976) which formed the basis for the American Society of Civil Engineers standard on composite slabs (ASCE, 1985) and also form the basis of other national standards (British Standards Institution, 1994; CEN, 2001).

The increased popularity of steel framed construction over the last two decades is in part due to the advantages arising from the use of composite floors.

Although composite floors are most closely associated with multi-storey office buildings, they are also used in renovation projects (where the low self weight of the floor is advantageous), car parks, warehouse and storage buildings (heavy point loads and wheel loads from fork-lift trucks may require special attention), housing and community service buildings.

The design of composite floors requires particular consideration of the construction sequence. The metal decking must be sufficiently strong and stiff during construction as it is required to support the wet concrete. Once the concrete has hardened, the deck acts as all or part of the reinforcement to the slab and acts compositely with the concrete to support imposed live loading. After briefly considering the advantages of composite floors and current practice, these two conditions will be discussed in the sections "Behaviour as Formwork" and "Composite Behaviour".

5.2 CURRENT PRACTICE

5.2.1 Advantages of Composite Floors

Composite floors are quick to construct and provide a cost-effective alternative to conventional cast in-situ reinforced concrete slabs or precast units. The steel deck forms a quick to erect working platform. Because of its low weight and ease of handling, a team of four men can place up to 400 m^2 of decking in a day, weather permitting as windy conditions can make handling difficult and rain makes the decks slippery. Many of the profiles can be tightly packed on top of each other so large areas of flooring, up to 1500 m^2, can be transported on a single lorry. Once the decking is secured in place and edge trims fitted, the deck forms permanent formwork for the concrete slab. Depending on the span of the slab and depth of the metal deck profile, temporary props are often not required which considerably speeds up the construction process.

The steel deck acts as the tensile reinforcement for the composite slab. In many cases there is no need to fix bar reinforcement to resist positive (sagging) moments although mesh reinforcement is necessary to resist shrinkage or temperature movements and assist in improving fire resistance.

The shape of the deck reduces the required volume of concrete by as much as 0.04 m^3/m^2 of floor, equivalent to a reduction in self-weight of 1.0 kN/m^2. Composite slabs are usually more shallow than conventional reinforced concrete slabs which leads to a reduction in the overall construction depth.

In addition to these major advantages other attractions can be identified. Steel decks are manufactured under strict factory controlled conditions resulting in greater accuracy in construction. The deck is a good vapour barrier. The soffit remains clean during construction and may be left exposed. Colour coated sheets may be used but these can prevent the use of through-deck stud welding. Many decks incorporate dove-tail slots into which hangers, secured by special wedge shaped nuts, may be easily inserted. Holes in composite floors may be readily formed during concreting, by using a foam block to create the void, or by diamond cutting through the concrete.

5.2.2 Deck Types

Deck profiles are usually 45 to 80 mm high with troughs spaced at 150 to 300 mm. (Much deeper decks and more widely spaced troughs are used in shallow floor systems, these are discussed in section 5.9). Decks fall into one of two types – re-entrant (commonly known as dovetail) and trapezoidal. Figure 5.1 shows typical examples. Profiles are cold rolled from steel sheet of 0.9 to 1.5 mm thickness with yield strengths between 280 and 350 N/mm^2. For most applications, where the risk of corrosion is limited, galvanising to a thickness of 275 g/m^2 is usually specified. In more aggressive environments, precoating of the steel with polyester or polyvinylidene flouride paints may be necessary and in very corrosive atmospheres site applied protective treatments may be required.

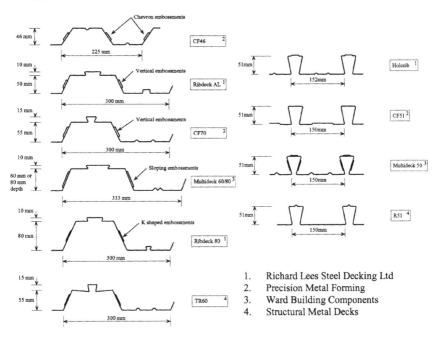

1. Richard Lees Steel Decking Ltd
2. Precision Metal Forming
3. Ward Building Components
4. Structural Metal Decks

Figure 5.1 Common deck profiles (after Couchman *et al.*, 2000)

In many cases the ability of the deck to support the loads arising during construction determines the maximum span. For this reason it is often advantageous to specify lightweight concrete (wet density 1850–1950 kg/m^3). Spans for conventional shallow decks are typically in the range of 3 to 4 m; deep decks can span more than 6 m.

5.2.3 New Types Under Development

Over recent years a demand has developed for decks which can span further. Deep decks of about 250 mm are now available and used in shallow floor systems. This will be addressed in section 5.9.

Harada (1998) reported a method to strengthen conventional decking at the construction stage by combining it with lightweight trusses fabricated from reinforcing bars and connected to the deck by a mechanical device. The combined arrangement increases the spanning capability of the deck under construction loads from 3.3 m to 6.0 m. After the concrete has hardened the truss acts as additional reinforcement for the slab.

Widjaja and Easterling (2000) reported a study of testing and analytical work on two deep profiles capable of spanning 20'(6.1 m). They noted that the volume of concrete used was similar to that used for a conventional slab and therefore the span was increased without a significant increase in weight. Deflections, during construction and in service, are likely to control the design and vibration characteristics were identified as requiring further investigation.

5.3 BEHAVIOUR AS FORMWORK

5.3.1 General

During construction the deck must support the weight of wet concrete and construction operatives. For unpropped construction this often constitutes the critical loading condition for the metal decking. The profiled steel sheeting is subjected to bending and shear and, due to the slender nature of the cross-section, is prone to local buckling. Rolled grooves and embossments tend to stiffen the flanges and webs of the cross-section but nonetheless buckling prior to yield is likely to occur and hence reduce the strength and stiffness of the deck. An effective width approach may be used but the design procedure involves iteration as the ineffective area of the compression flange increases under increasing bending moment, thus lowering the neutral axis of the profile and increasing the extreme fibre stresses. Because of the inherent difficulties and conservatism in analytical methods, many manufacturers have conducted tests to more accurately predict the performance of their decks.

5.3.2 Design Loading

At the ultimate limit state, loads arising from the weight of the wet concrete and steel deck, construction loads (operatives and equipment) and any 'ponding' effects (increased depth of concrete due to deflection of the sheeting) should be considered. Although construction loads account for the weight of the operatives and concreting plant, and also allow for any impact or vibration that may occur during construction, they are not necessarily sufficient for excessive impact (for example, by dropping concrete from a skip too high above the deck) or heaping concrete, or pipeline or pumping loads. British practice (BSI, 1994a) currently specifies the construction load as 1.5 kN/m² applied uniformly over the slab. Eurocode 4 (CEN, 2001) recommends a uniform characteristic loading of 0.75 kN/m² over the entire slab but locally increased to 1.5 kN/m² in any area of 3 m by 3 m (or span, if less). These loads should be placed to cause the maximum bending moment and/or shear as shown in Figure 5.2.

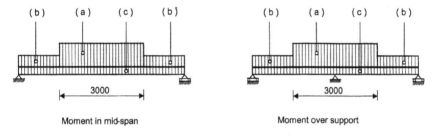

(a) Concentration of construction loads 1.5 kN/m²
(b) Distributed construction load 0.75 kN/m²
(c) Self weight

Figure 5.2 Critical load arrangements for sheeting acting as shuttering

In addition to the loads specified above, consideration should be given to the ability of the deck to resist concentrated point loads. The European Convention for Constructional Steelwork (ECCS, 1995a) recommends that without the concrete the decking should be shown by tests or calculation to be able to resist a characteristic load of 1 kN on a 300 mm square area.

If the central deflection of the sheeting under its own weight plus that of the wet concrete, calculated for serviceability, is less than one tenth of the slab depth the effect of ponding may be ignored in the design of the steel sheeting. Above this limit ponding should be allowed for. Eurocode 4 suggests assuming in design that the nominal thickness of the concrete is increased over the whole span by 70% of the central deflection.

5.3.3 Analysis

Continuous decking may be analysed using elastic analysis with the flexural stiffness determined without consideration of the variation of stiffness due to parts of the cross-section in compression not being fully effective. This approach is relatively conservative. Figure 5.3 shows a deck continuous over two spans. The maximum negative (hogging) moment ($0.125\ wL^2$) must be less than the resistance of the effective cross-section, $W_{eff}p_y$. As the effective plastic modulus for positive bending is greater than that for negative bending (as a result of the relatively narrow top flange), it is implied that the strength of continuous sheeting is lower than that for the simply supported case where the mid-span moment is $0.125\ wL^2$. Tests demonstrate that there is some redistribution of moment in the elastic range from the highly stressed support regions to the mid-span because of the variation in stiffness with moment. As shown in Figure 5.3, a steel deck has some limited post-buckling moment-rotation capacity and some residual negative moment resistance exists as the limiting mid-span moment is reached. Mullet (1998) comments that spans 10 to 15% greater than those predicted by elastic analysis are possible and for this reason many manufacturers prepare load-span tables on the basis of tests rather than analysis. To obviate the need for expensive testing of multi-span decks, Davies and Jiang (1997) have proposed a design procedure based on the formation of a pseudo-plastic collapse mechanism. The method requires the moment-rotation characteristic at the internal support (which may be determined by test or finite element computation) and the positive bending moment capacity from a simply supported test, or by calculation or finite element analysis.

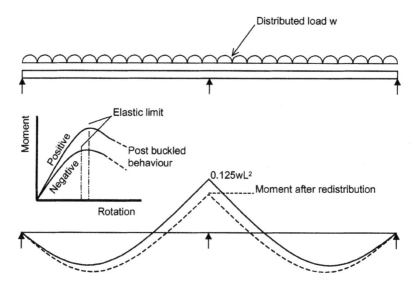

Figure 5.3 Redistribution of moment in continuous decking

5.3.4 Design Strength by Calculation

Strength calculations for metal decks are complicated by the necessity to allow for the effects of local buckling. Codes of practice for the design of profiled steel sheeting (BSI, 1995; BSI 1996) base the moment capacity on the attainment of a limiting compressive stress equal to the yield strength of the steel in the effective cross section. Figure 5.4 illustrates the effective cross sections of a variety of profiles. The presence of stiffeners in the flanges and webs of a cross section considerably enhances performance and guidance on how to account for their beneficial effect is available in codes of practise. Notwithstanding the availability of such calculation methods, many manufacturers prefer to test their products rather than rely on these approximate and necessarily conservative methods.

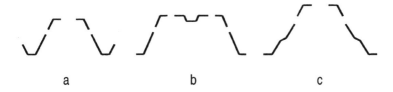

Figure 5.4 Effective cross section of a sheeting profile with (a) unstiffened trapezoidal profile (b) multiple-stiffened flange (c) stiffened web

Many profiles used in composite slabs have indentations rolled into the webs and top flange to improve the shear bond between the deck and in situ concrete. Davies and Jiang (1997) investigated the effect of these indentations on the bending strength of steel sheeting. They found that indentations in the webs reduced the bending strength by about 3% but those in the compression flange caused reductions of approximately 10%.

5.3.5 Design Strength by Testing

As analytical methods are conservative, design standards permit the use of testing for determining the performance of steel decking. To produce empirical design data a manufacturer would have to conduct a comprehensive test series. This would include tests to determine the moment capacity at midspan (where the shear force is negligible), the load capacity at an internal support (where the effects of moment and support reaction interact), shear at an end support and the resistance to local loads. At least four tests of each type would usually be required. Tests would normally be conducted on a complete as rolled width of decking. If this results in a free edge in compression, a part of a corrugation on one or both longitudinal edges should be removed. Uniformly distributed loads may be applied by a pressure bag, vacuum chamber, gravity loads or four equal line loads (applied to the troughs of the corrugations). Details of typical testing arrangements are contained in BS5950 Part 6 (BSI, 1995).

5.3.6 Serviceability Considerations

It is necessary to check that the residual deflection after the concreting operation has been completed is not excessive. Both EC4 and BS5950 recommend that the deflection of the sheeting under its own weight plus the weight of wet concrete, but excluding the construction load, should not be greater than L/180 (or 20 mm in the BS) where L is the effective span between supports, be they permanent or temporary. This limit may be increased if a greater deflection will not impair the performance of the floor and the additional weight of concrete due to ponding is taken into account. The British Standard recommends that where the deflection exceeds one tenth of the slab depth the additional weight of concrete due to the deflection of the sheeting should be taken into account in the self-weight of the slab. Where the soffit of the slab is to remain visible or where services are to be fixed to the underside of the slab it may be necessary to reduce the suggested limits.

Decking is usually installed over a minimum of two spans and continuity will reduce the mid-span deflection. ECCS (1995a) suggests that the central deflection of the side span of continuous decking be estimated from Equation (5.1):

$$\delta = \frac{(2.65 g_a + 3.4 g_c) L^4}{384 EI} \qquad (5.1)$$

where g_a and g_c refer to the self-weight of the decking and the weight of the wet concrete respectively.

For continuous decking it is also recommended to check (ECCS, 1995a) that under *characteristic* loading the combination of bending moment and support reaction causes no plastic deformation. The suggested limits are:

$$M_{S,ser} \leq 0.9 \, M_{Rd} \qquad (5.2)$$

$$R_{S,ser} \leq 0.9 \, R_{Rd} \qquad (5.3)$$

$$M_{S,ser} / 0.9 \, M_{Rd} \leq 1 \quad \text{if} \quad R_{support} / 0.9 \, R_{Rd} \leq 0.25 \qquad (5.4)$$

$$\left(\frac{M_{S,ser}}{0.9 M_{Rd}}\right)^2 + \left(\frac{R_{S,ser}}{0.9 R_{Rd}}\right)^2 \leq 1.25 \qquad (5.5)$$

where $M_{S,ser}$ and $R_{S,ser}$ are the bending moment and support reaction due to construction loads, without partial load factors (i.e. $\gamma_G = \gamma_Q = 1.0$) and M_{Rd} and R_{Rd} are the design moment and support reaction resistance for the serviceability limit state (i.e. $\gamma_M = 1.0$). Davies and Jiang (1997) used computational results to show that this interaction is quite conservative and that for low support reactions the failure moment can be increased above the ultimate resistance moment in pure bending. The suggested reason for this is that the complicated stress combination arising in the deck from the interaction of bending moment and reaction improves the stress distribution, with respect to longitudinal stresses, when compressive stress and shear stress are relatively small. If the shear stress is greater than half the yield stress the moment of resistance decreases rapidly due to either web crippling or shear buckling.

The design process for steel decking as formwork, including the important checks to be undertaken at the ultimate and serviceability limit states, is summarised in Figure 5.5.

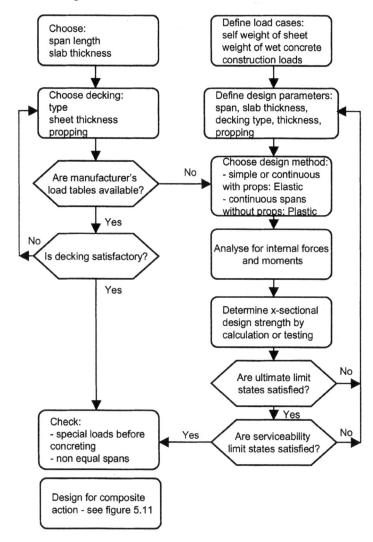

Figure 5.5 Design process for metal decking as formwork (based on ECCS, 1995a)

5.4 COMPOSITE BEHAVIOUR

5.4.1 Failure Modes

Once the concrete has hardened the steel deck and concrete combine to form a single structural unit, the composite slab. The response of a composite slab to load is analogous to that of a conventional reinforced concrete slab with an important proviso – the bond between the steel deck and concrete may not be fully effective and longitudinal slip may occur before the steel deck yields. As a result, two primary failure modes are possible; flexural failure and shear-bond failure.

It is instructive to consider the behaviour of a simply supported composite slab as shown in Figure 5.6. A composite slab bears on two external supports and is loaded symmetrically with two loads, P, applied at ¼ and ¾ of the span. A typical load-deflection curve, P-δ, is illustrated in the figure. The behaviour of the slab depends on the effectiveness of the steel to concrete connection, which is a function of the profile shape and embossment type and pattern.

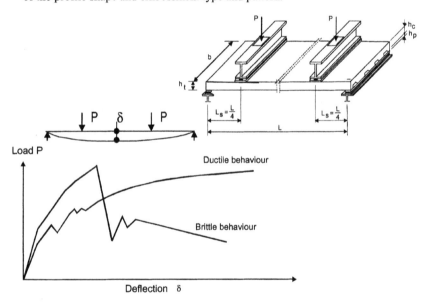

Figure 5.6 Composite slab test with typical load-deflection responses

As load is applied to the slab the initial behaviour is elastic as the concrete is uncracked and the composite action between the concrete and steel is complete. As further load is applied, the concrete below the neutral axis cracks, reducing the slab stiffness and increasing deflections. At this stage the adhesion between the sheet and the concrete is still capable of transferring the shear force between the cracks.

As further load is applied slip may occur between the concrete and steel deck as the shear-bond between them is exceeded. Two modes of behaviour may then follow:

1. Brittle behaviour in which slip causes a sudden decrease in load carrying capacity as the surface bond is broken. The extent to which the load reduces is dependent on the effectiveness of the mechanical embossments. This load reduction is not due to concrete cracking; it arises from relative slip between the concrete and steel. As the slab is deflected further the resistance to load is increased slightly but the mechanical means by which shear is transferred does not equal that arising from surface adhesion.
2. Ductile behaviour, in which case the mechanical shear connection is capable of transferring the shear force until failure occurs. This may be flexural or by longitudinal shear.

Eurocode 4 (CEN, 2001) defines the behaviour as ductile if the failure load exceeds the load causing first recorded end slip (greater than 0.5 mm) by more than 10%. If not, the behaviour is classed as brittle. Whether a brittle or ductile mode of failure occurs depends on the characteristics of the steel-concrete interface and has to be determined by tests. Slabs with open trough profiles tend to experience a more brittle behaviour, whereas slabs with re-entrant trough profiles exhibit more ductile behaviour. However, profiled decking producers ameliorate brittle behaviour with several mechanical means, such as embossments or indentations and the use of dovetail forms. It should be noted that the behaviour described above refers to simply supported slabs acting non-compositely with the supporting beams. Shear connectors between beam and slab also influence the failure mode. Where shear studs are provided to ensure composite action between the beam and slab the anchorage provided by the studs will enhance the longitudinal shear capacity and hence the load carrying capacity of the slab.

Failure of a simply supported slab may occur in one of three ways at locations along the beam as shown in Figure 5.7.

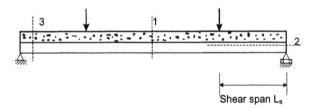

Figure 5.7 Location of critical sections in a simply supported composite slab

1. Failure type 1 when the bending resistance of the slab is exceeded at the point of maximum applied moment, section 1. This is generally the critical mode for moderate to high spans with a high degree of interaction between the steel and concrete.

2. Failure type 2 is due to exceeding the ultimate longitudinal shear load resistance at the steel concrete interface. This happens in section 2 along the shear span L_s.
3. Failure type 3 occurs when the applied vertical shear near the support (section 3) is greater than the shear resistance of the slab. This is only likely to be critical for deep slabs over short spans and subject to heavy loads.

5.4.2 Analysis for Internal Forces and Moments

Composite slabs may be analysed by elastic, rigid-plastic or elastic plastic methods. Elastic analysis may be used for both the serviceability and ultimate limit states and is the simplest method. The effects of longitudinal slip, buckling of the deck or yielding are ignored. Concrete cracking may also be ignored, in which case the support bending moments at the ultimate state may be redistributed by up to 30%. Cross-sectional properties can be considered constant and uniform in both positive and negative moment regions.

Plastic methods may be used only at the ultimate limit state. Although composite slabs are usually continuous over a number of spans, it is permissible to consider a slab as a series of simply supported spans with failure occurring either by slip between the decking and the concrete or by formation of a plastic hinge at the mid-span. Mesh reinforcement should be provided over intermediate supports to control cracking. If a collapse mechanism involving plastic moments at the supports is assumed, the cross-section must have sufficient rotation capacity. For spans less than 3 m, reinforced over the supports with reinforcing bars (not wire mesh), the rotation capacity may be assumed to be sufficient.

It should be noted at this point that many manufacturers of steel decking provide design charts, in the form of span-load tables, which means that analysis of the slab is often unnecessary. Detailed calculations are likely to be needed only when the design tables do not cover the required loading arrangements, span length or support conditions.

5.4.3 Bending Resistance

Either the steel sheeting yielding in tension or the concrete reaching its resistance in compression determines the sagging bending resistance. Where reinforcing bars have been placed in the troughs, usually to increase the fire resistance, these may be taken into account in calculating the composite slab resistance.

American, British and European codes (ASCE, 1985; BSI, 1994; CEN, 2001) all assume idealised rigid plastic stress-blocks as the basis for determining the resistance at the ultimate limit state. Material strengths are divided by a partial safety factor the value of which varies slightly between codes. Mesh reinforcement, or tension reinforcement for hogging bending, is usually in compression under sagging bending and is generally neglected when evaluating the sagging bending resistance.

The plastic neutral axis will most commonly lie above the steel decking as illustrated in Figure 5.8.

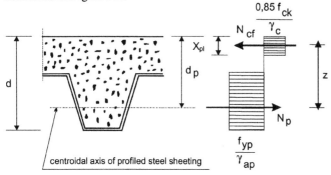

Figure 5.8 Stress distribution for sagging bending – plastic neutral axis above the steel deck

The resistance of concrete in tension is ignored. The tension force N_p in the steel sheeting is the product of the effective area A_{pe} (ignoring the galvanising thickness and the width of embossments and indentations) and the design strength of the steel sheet $f_{yp,d}$ ($=f_{yp}/\gamma_{ap}$). Equating this force with the compression force in the concrete over the width b of the cross-section and depth x_{pl} gives:

$$N_{cf} = bx_{pl}\frac{0,85\,f_{ck}}{\gamma_c} = A_{pe}f_{yp,d} \tag{5.6}$$

Equilibrium gives x_{pl} as:

$$x_{pl} = \frac{A_{pe}f_{yp,d}}{0,85\,b\,\dfrac{f_{ck}}{\gamma_c}} \tag{5.7}$$

and the design resistance moment is:

$$M_{ps.Rd} = A_{pe}\frac{f_{yp}}{\gamma_{ap}}(d_p - \frac{x_{pl}}{2}) \tag{5.8}$$

If the plastic neutral axis intercepts the steel sheeting (which is not a common situation), the calculation is slightly more complicated as part of the steel sheeting is in compression. For simplification, the concrete in the ribs as well as the concrete in tension are neglected. The stress diagram is shown in Figure 5.9 and may be considered in two parts. First, the equilibrium of the resistance of the concrete slab over the depth h_c (force N_{cf}) with part of the decking in tension (force N_p) in the steel sheeting. The lever arm z may be approximated as:

$$z = h_t - 0.5h_c - e_p + (e_p - e)\frac{N_{cf}}{A_{pe}f_{yp}/\gamma_{ap}} \qquad (5.9)$$

Second, the moment generated by the remaining part of the steel deck in tension and the upper section in compression, M_{pr}, called the *reduced plastic moment* is given by the expression:

$$M_{pr} = 1.25\, M_{pa}\left(1 - \frac{N_{cf}}{A_{pe}f_{yp}/\gamma_{ap}}\right) \le M_{pa} \qquad (5.10)$$

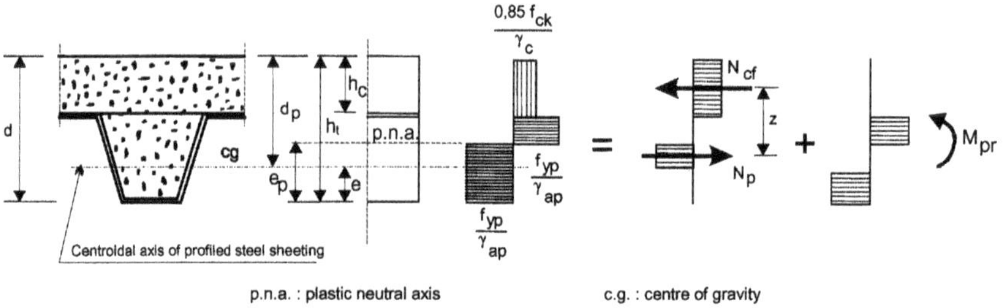

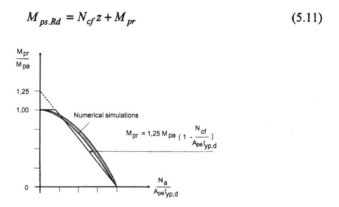

Figure 5.9 Stress distribution for sagging bending – plastic neutral axis in the steel deck

The bending resistance is then:

$$M_{ps.Rd} = N_{cf}z + M_{pr} \qquad (5.11)$$

Figure 5.10 Experimental relationship between M_{pa} and M_{pr} (after Stark and Brekelmans, 1996)

The approximate formula for M_{pr} (Equation 5.10), the reduced plastic moment of the steel sheeting, which is based on M_{pa}, design plastic resistant moment of the effective cross-section of the sheeting, is a result of numerical studies conducted on 8 steel profile types by Stark and Brekelmans (1996). Figure 5.10 shows the calibration of the expression with the analyses.

The hogging bending resistance is determined as an under-reinforced concrete slab neglecting the contribution of the steel sheeting and taking the width of concrete in compression as the average width of the concrete ribs over one metre.

Figure 5.11 summarises the steps required in the design of a composite slab.

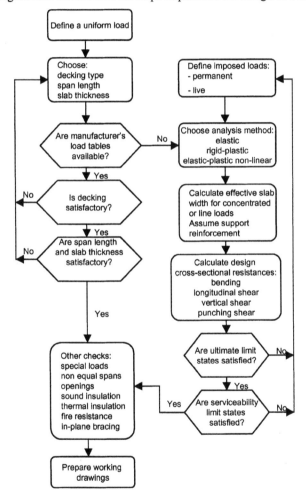

Figure 5.11 Flowchart of the design process for a composite slab (based on ECCS, 1995a)

5.4.4 Testing for Shear Interaction

Many codes use an empirical approach to assess the effectiveness of the shear interaction developed between by a particular steel deck profile. The basis of this approach is in work conducted by Porter and Ekberg (1976) at Iowa State University. Schuster (1976) reports that at the time in the United States and Canada each deck manufacturer was developing products by extensive independent research. Most of these tests were proprietary and therefore not reported or made generally available (Easterling and Young, 1992). In order to gain a fundamental understanding of the behaviour of composite slabs, and from this develop a standard design specification, the American Iron and Steel Institute commissioned an extensive research programme. Porter and Ekberg tested over 350 simply supported composite slabs (using two point loading as shown in Figure 5.12) and established that shear bond failure was the most likely mode of failure for most steel decks. During this work Schuster (1970) developed an expression for the longitudinal shear strength which was reported by Porter and Ekberg (1976) and became the basis of the American Society of Civil Engineers standard on composite construction (1984).

Figure 5.12, based on Porter and Ekberg, explains the approach. V_e is the experimental ultimate shear resulting from the application of an experimental failure load P_e.

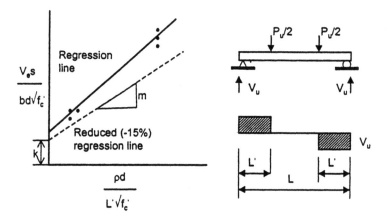

Figure 5.12 Shear-bond failure relationships (based on Porter and Ekberg, 1976)

The test slabs are cast on a single width of the deck and span between pinned and roller supports. This is an experimentally convenient set up but not representative of a composite slab in service, particularly one in which the slab acts compositely with the beams. Earlier a number of potential failure modes were described (see section 5.4.1) but for this test arrangement the most likely is shear bond failure in which the connection at the steel to concrete interface degrades over the shear span (the length of slab between the concentrated load and the support). As illustrated in Figure 5.13, for very long test specimens it is possible that flexural failure may be critical and for very short shear spans vertical shear may control.

Composite Floors

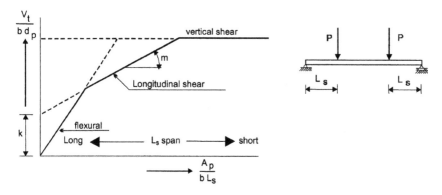

Figure 5.13 Relationship between failure mode and span

A minimum of three tests is conducted on each of two spans. The maximum vertical shear load is plotted against the shear span (the distance from the concentrated load to the support) and a regression analysis conducted to provide the best linear relationship as shown in Figure 5.12. The equation of the line is then given by:

$$\frac{V_u s}{bd\sqrt{f_c'}} = \frac{m\rho d}{L'\sqrt{f_c'}} + k \qquad (5.12)$$

where V_u is the calculated ultimate shear capacity, ρ is the ratio of the steel deck cross-sectional area to the area of concrete, A_s/bd (d is the effective depth of the slab measured from the top of the concrete to the centroid of the steel deck) and the other symbols are defined in Figure 5.12.

The reduced regression line shown in Figure 5.12 is a 15% reduction in the slope (m) and intercept (k) of the original regression analysis. This reduction is to ensure that lower-bound values are obtained. Once m and k have been determined these factors may be used in Equation 5.12 to predict the ultimate shear capacity of simply supported slabs of other configurations, span and slab thickness, cast on steel decks identical to those tested. Changes in the steel deck shape or embossment pattern would require further tests.

The s in Equation 5.12 was introduced by Porter and Ekberg to account for differences in the spacing of shear transferring devices, such as transverse wires, which may in rare cases vary even though the profile remains the same. Where the shear transfer is through embossments s may be taken as one. As this is the usual situation, the term s does not appear in codes of practice even though based on this work. The factors m and k reflect the contribution to shear transfer of mechanical interlock and friction respectively. The chemical bond between the deck and concrete is not reliable and should be broken before the test by cyclic loading of the slab.

According to Eurocode 4, the maximum design vertical shear $V_{t.Sd}$ for a width of slab b is limited due to the longitudinal shear resistance to $V_{L.Rd}$ given as :

$$V_{L.Rd} = b.d_p (m \frac{A_p}{bL_s} + k) \frac{1}{\gamma_{VS}} \tag{5.13}$$

where γ_{VS} is a partial safety factor equal to 1,25.

Figure 5.14 shows the results from tests conducted in accordance with Eurocode 4 recommendations. Notice that this code does not consider the influence of concrete strength (f_c' does not appear in the expressions used for the ordinate or abscissa) as the limited range of strengths used for buildings has only a minor effect. Also, EC4 defines the characteristic values as 10% lower than the minimum recorded for each group tested at the two shear spans and uses these values to define the straight relationship from which the slope and intercept may be determined.

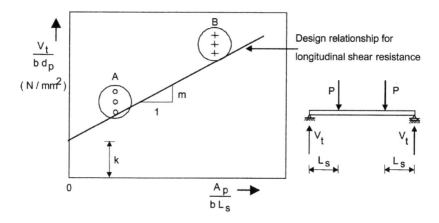

Figure 5.14 Derivation of m and k from test data in Eurocode 4

The shear span L_s depends on the type of loading and may be found by equating the area under the shear force diagram for a particular loading case with that due to a symmetrical two point load system applied at a distance L_s from the supports. For example, for a uniform load over a simply supported span, L_s equals L/4. Where the composite slab is designed as continuous, Eurocode 4 permits the use of an equivalent simple span taken as the distance between points of contraflexure but for end spans the full exterior span length should be used in design. If the longitudinal shear resistance of the slab is not sufficient, it can be increased by the use of some form of end anchorage, such as studs or local deformations of the sheeting.

5.4.5 Partial Shear Connection Method

Daniels and Easterling (1997) observed that the design capacity of many composite slabs is governed by longitudinal shear failure yet experiments on real floor slabs suggest that this type of failure will rarely be critical. Existing design methods are therefore quite conservative. This is because the experimental tests on which design methods are based are very much a worst case and do not reflect the common support arrangements encountered in practice. An alternative method, partial strength design, has been developed which has the advantage that additional parameters, such as end anchorage and reinforcement may be included. Although less empirical than the m-k method, it still requires data from full-scale tests and may only be applied to slabs exhibiting ductile behaviour.

Partial strength design of slabs follows the same principles as partial strength design of beams (see Chapter 2). To start, consider the two extreme conditions of zero and full interaction. If a simply supported slab, with no anchorage at the ends, were tested the minimum moment capacity (corresponding to the zero interaction case) would be the design resistance of the deck alone, M_{pa}. The maximum capacity would occur with full interaction and would be equal to the flexural resistance of the composite section, $M_{pl.Rd}$, achieved when either the concrete or the steel reaches its maximum capacity in compression or tension respectively. At intermediate levels of interaction, longitudinal shear resistance limits the force generated in the steel or concrete. The capacity of the partially connected slab may then be calculated as:

$$M_{Rd} = N_c z + M_{pr} \tag{5.14}$$

M_{pr} is the bending capacity of the deck in the presence of a tensile load and may be obtained from Figure 5.10. Note that in partially connected slabs the deck will not have yielded and therefore the reduced bending resistance of the steel deck contributes to the capacity of the composite slab. The significance of this term in the equation increases for a given deck depth as the overall composite slab thickness reduces. Because the degree of longitudinal shear interaction controls the magnitude of the force that can be mobilised in the concrete or steel, N_c, the shear interaction must be determined. This may be achieved by full scale testing of composite slabs in a very similar way to the tests used for the m-k method.

Figure 5.15 shows a plot of resistance moment against degree of interaction, expressed as the ratio of the axial force in the slab N_{cf} to the ultimate longitudinal shear force, $b\tau_{u.Rd}$ where b is the slab width and $\tau_{u.Rd}$ is the ultimate shear stress. If the shear stress is sufficiently large that the full axial strength of the concrete or steel may be mobilised the slab has full interaction. From tests the maximum bending moment, M_{test}, at the critical cross-section (under one of the two point loads) can be obtained. If this moment is plotted on Figure 5.15 the corresponding degree of interaction, η_{test}, can be deduced, from which the bond strength can be calculated as:

$$\tau_u = \frac{\eta_{test} N_{cf}}{b(L_s + L_o)} \qquad (5.15)$$

where L_s is the length of the shear span and L_o is the length of any overhang.

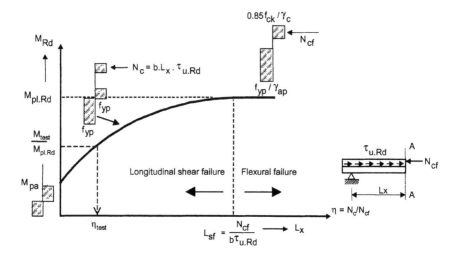

Figure 5.15 Determination of the degree of shear connection

Statistical analysis of the test results and application of a partial safety factor gives a design value of τ. (EC4 suggests using 90% of the minimum test value as the characteristic value and a partial safety coefficient of 1.25.) Once τ has been determined the value of N_c may be calculated as:

$$N_c = \tau_{u.Rd} bL_x + \mu R \leq N_{cf} \qquad (5.16)$$

the term μR is the frictional force generated at the support by the reaction R. The importance of including this friction term was explained by Bode and Minas (1997) and the consequent improvement in the prediction of the longitudinal shear capacity was demonstrated in a series of tests on simply supported composite slabs by Calixto *et al.* (1998).

Figure 5.16 shows the partial interaction curve with design values for materials and geometrical properties. This diagram may be used as a design tool with design bending moment diagrams superimposed on the resistant moment diagram – Figure 5.17. The abscissa is length, L_x. To plot the resistant diagram on the same plot the scale on the horizontal axis is given by the fictitious length L_{sf}, which would be required to give full interaction. This may be calculated as:

$$L_{sf} = \frac{N_{cf}}{b\tau_{u.Rd}} \qquad (5.17)$$

where N_{cf} is the minimum of either the design resistant force of the concrete slab of depth h_c or of the steel sheeting:

$$N_{cf} = \min\left(\frac{0{,}85 f_{ck} b h_c}{\gamma_c}; \frac{A_p f_{yp}}{\gamma_{ap}}\right) \qquad (5.18)$$

Where a simply supported slab has stud anchors attaching it to the supporting beams (as is often the case) the length L_{sf} required to develop full interaction is reduced as the anchors provide a very effective means of transferring shear between the concrete and deck (Jolly and Lawson, 1992). If the design strength of the end anchorage is V_{ld} the revised value of L_{sf} is given by:

$$L_{sf} = \frac{N_{cf} - V_{ld}}{b\tau_{u.Rd}} \qquad (5.19)$$

and the interaction diagram is moved in the L_x direction by $-V_{ld}/b\tau_{u.Rd}$ as shown in Figure 5.16.

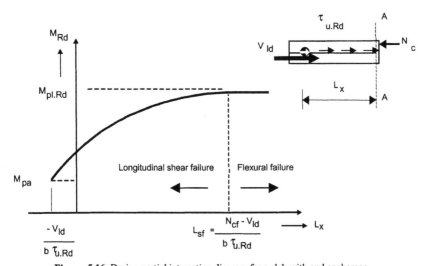

Figure 5.16 Design partial interaction diagram for a slab with end anchorage

For any cross-section of the span, the design bending moment M_{Sd} cannot be higher than the design resistance M_{Rd}. For slab spans where $L_x \geq L_{sf}$, the shear connection is full, so the bending resistance (flexural failure mode) is critical. For $L_x < L_{sf}$, the shear connection is partial, so the longitudinal shear resistance is critical.

Figure 5.17 illustrates the principle behind the partial connection method but it should be noted that the determination of the point of intersection of the two curves would require a computer programme because the partial interaction curve is not described by a mathematical function.

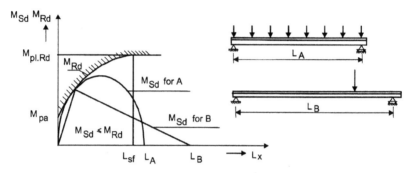

Figure 5.17 Slab design with a partial interaction diagram

Tenhovuori and Leskela (1998) considered the effect of slab depth and shear span on the value of bond strength, τ_u. Using FE analyses they observed that bond strength increases with slab depth and decreases as the shear span increases. Figure 5.18 shows this variation with a plot normalised to the τ_u value found for the minimum depth and shear span dimensions of a slab, namely 120 and 350 mm.

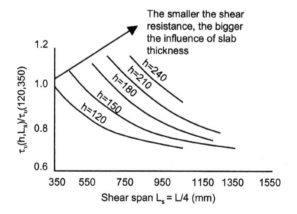

Figure 5.18 Increase in shear resistance with slab thickness (after Tenhovuori and Leskela (1998))

In tests for τ_u (and for m-k) shallow slabs yield the minimum values and therefore tests should be conducted on the minimum depth of slab to be used in practice. This contradicted the guidance given in clause 10.3.1.1(3) of EC4 (CEN, 1992) which stated that test results could be applied to slab thicknesses smaller than tested. This statement has subsequently been removed from EC4 (CEN, 2001). When conducting tests for τ_u Tenhovuori and Leskela recommend a minimum of three tests and that the largest possible shear span be used whilst still maintaining a longitudinal shear failure.

Many composite slabs have reinforcing bars placed in the troughs to improve fire resistance. This reinforcement may be accounted for in the partial connection method by adding its contribution to the moment resistance of the composite slab thus:

$$M_{rd} = N_c z_1 + M_{pr} + N_{as} z_2 \qquad (5.20)$$

where $N_{as} = A_s f_{sd}$ and z_2 is the depth from the centre of the concrete in compression to the level of the reinforcing bars.

Bode et al. (1996) briefly explain how the partial interaction method may also be used for the design of plastically analysed continuous slabs. The process is iterative and may be used with different profile geometry, degree of shear interaction and levels of reinforcement over the support.

5.4.6 Developments in Partial Connection Methods

A number of researchers have made progress towards developing design methods that are not empirically based and therefore do not require expensive full scale testing. Some of these analytical methods may also include for the effects of sagging reinforcement, which is often provided for fire resistance, end anchorage from shear studs required for composite action between the slab and beam, continuous spans and reinforcement in the hogging regions.

Easterling and Young (1992) compared the results of a number of experimental tests on 3-span continuous slabs with two analytical approaches to determine the capacity of the slab. The first calculated the strength as an under-reinforced concrete slab and the second based the moment capacity on elastic analysis of a cracked section limiting the deck to first yield at the extreme fibre. Both methods assume full composite action can be achieved. The tests were designed to consider the effects of end-span details and a number of edge details were investigated. None of the tests had additional reinforcement of any kind. The authors found that the strength based on elastic analysis of a cracked section with simply supported boundary conditions underestimated the strength in all cases. Strength calculations based on an under-reinforced concrete slab were greater than the test results in all cases except for the single test which used shear studs at the span ends. This test reached the predicted ultimate moment even though the strain gauge readings indicated that the entire cross-section had not yielded, as is assumed in the analysis. In order to achieve ductile behaviour and mobilise the full moment capacity of the cross-section, the authors observed that it is necessary to have properly secured end stops (to prevent the steel deck from slipping beneath the slab) and adequate attachment of the steel deck to the support beams by stud connectors. These details prevent longitudinal slip from being the limiting failure mode and therefore permit the strength of the slab to be calculated without recourse to test data. Use of an average of the cracked and uncracked second moments of area of an all steel transformed section gave slightly larger deflections than the test results at service load levels but was considered sufficiently accurate for design purposes.

Daniels and Crisinel (1993a) reported a new design procedure that uses slip characteristics derived from shear bond tests with a non-linear numerical model. The advantages of this approach are the reduced costs of testing for shear bond, compared with full-scale slab testing, and the versatility of numerical modelling which allows the effects of many more variables, for example end anchorage, sagging moment reinforcement and slab continuity, to be considered. The procedure also permits an estimate to be made of the minimum level of shear interaction necessary for satisfactory performance. If improved interaction is shown to reduce service load deflections or increase moment capacity, shear-bond testing of modified deck embossments can be economically undertaken without the necessity to conduct full-scale composite slab tests.

Figure 5.19a shows the details of a pull-out test arrangement devised to investigate the behaviour and strength of the shear connection at the deck to concrete interface. The test specimen must be long enough to contain a representative number of embossments but not so long that the force required to cause slip causes the deck to yield or that non-linear stress distribution is developed; a concrete block 300 mm long with 100 mm of deck projecting is recommended. Transverse loads are required to simulate the self-weight of the concrete and should remain constant throughout the test.

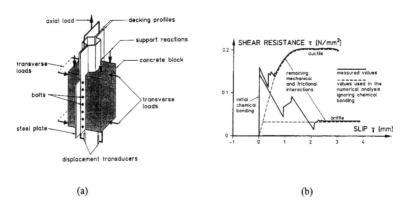

Figure 5.19 Shear-bond pull-out test specimen and typical results (after Daniels and Crisinel (1993a))

Figure 5.19b is a plot of typical measured values of shear resistance against slip. The plots clearly show the three shear transfer mechanisms. Initially the response is very stiff as the chemical bond between the deck and concrete has to be overcome. Next mechanical and frictional interactions transfer the shear. Mechanical interaction is developed by the embossed patterns or by re-entrant profiles. Frictional interaction is proportional to the applied normal force at the concrete-deck interface. Brittle behaviour occurs if the mechanical and frictional shear resistances are much lower than the initial chemical bonding. This may be the case in trapezoidal profiles with ineffective patterns of embossments. When testing full-scale slabs cycling loading prior to destructive testing breaks the chemical bond. In shear bond tests the chemical bond is measured but then ignored in analysis as shown by the dotted line representing the assumed behaviour for analysis.

Similar component tests are required to determine the strength of end anchorage details. Figure 5.20 illustrates the test arrangement. The deck is 450 mm long and concreted over a length of 300 mm. Shear studs are welded through the deck on to a short length of rolled section. The distance of the studs from the edge of the slab should be in accordance with minimum values used in practice. As load is applied through the lower rolled section it is important that this is adequately secured.

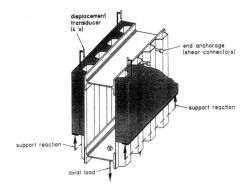

Figure 5.20 Pull-out test for end anchorage arrangements (after Daniels and Crisinel (1993a))

A non-linear finite element model uses the test load-slip behaviour to model the partial interaction resulting from embossments and end anchorage. In a companion paper Daniels and Crisinel (1993b) compared the analysis with test data. They found reasonable and conservative predictions of composite slab behaviour and maximum capacity in cases ranging from simple spans without end anchorage or additional reinforcement to those involving continuity, end anchorage and reinforcement both in the span and over the supports. Mid-span deflection rather than maximum load-carrying capacity was the limiting design criterion for many of the slabs investigated.

Patrick and Bridge (1994) proposed a similar combination of numerical analysis and component testing. Shear connection performance is measured using a slip block test (Patrick, 1990). The test is performed on a small block of composite slab (typically 300 mm long) subjected to both a vertical clamping force and horizontal force to shear the concrete from the metal deck. After breaking the adhesion bond, the clamping force is varied while continuously measuring the longitudinal slip. By plotting the relationship between the horizontal and vertically applied forces, the coefficient of friction and the contribution to slip resistance arising form mechanical interlock may be determined. These characteristic properties of the slab are then used in a partial strength design method to predict the moment resistance of the composite slab at critical sections along the span. A design manual explaining the method has been published (BHP, 2000) which gives clear guidance on the application of the method. The effects of end anchorage and conventional reinforcement in the slab can be included. This approach is being used as the basis for one of the strength design methods to be included in a new Australian Standard.

Veljkovic (1998) reported a parametric study of the behaviour of composite slabs under different load arrangements. It was shown that good correlation between full-scale tests and FE analysis could be obtained using longitudinal shear properties determined from slip block tests as proposed by Patrick and Bridge. For slabs that fail by longitudinal shear, load bearing resistance was seen to increase as the loading becomes more uniform. The partial shear connection design method of EC4 is only applicable for slabs which fail in a ductile manner. Veljkovic proposed modifications to the method which would remove this restriction.

5.5 DYNAMIC BEHAVIOUR

The relatively lightweight nature of composite floors and their essentially one way spanning behaviour makes them more vulnerable to vibration problems than heavier two-way spanning reinforced concrete slabs. Although vibration of floors can arise from external sources such as road and rail traffic the most usual source of dynamic excitation is pedestrian traffic. Chien and Richie (1984) explain the fundamental background to this problem and provide a brief commentary on Appendix G of the Canadian Standard (CSA, 1994) which is a guide for assessing floor vibrations. Floor vibrations can be categorised as continuous – caused by periodic forces of machinery, vehicles or human group activities like dancing, gymnastics or aerobics – or transient, arising from impulses caused by footsteps or other impacts. Vibrations caused by footstep impact are of particular importance in lightly damped floors such as open plan offices without partitions, shelves or file storage. Guidance on how to assess the likely vibrational behaviour of floors in steel-framed buildings is provided by Wyatt (1989). The approach is to divide the floor into bays (areas of floor substantially participating in the motion) and estimate the natural frequency of vibration based on a simple empirical formula. Depending on whether the natural frequency is above or below 7 Hz (defining floors of high or low natural frequency), two expressions are given for calculating a response factor which may then be compared with acceptable values according to the function of the floor. Where the response factor exceeds the acceptable value the floor mass may be increased to reduce the vibration. For metal deck composite slabs the relative contribution of the slab to the overall vibration of the floor system (slab and supporting beams) will usually be low. However, if the flexibility of the slab is larger than the flexibility of the floor beams (i.e. self-weight deflection of the slab is greater than that of the beam) and the natural frequency of the floor is found to be less than 7 Hz, Wyatt recommends that a detailed analytical method be employed to take proper account of the variations of deflection of the slab from point to point in both directions.

The behaviour of composite slabs subject to continuous vibrations, from machinery for example, is greatly affected by the mechanical bond arising from embossment patterns in the deck. Chien and Richie (1993) cite a number of Canadian research studies on this topic one of which (Temple and Abdel-Sayed, 1979) reported the results of a series of 19 tests comparing the behaviour of composite slabs subject to static and dynamic loading. Eight of the twelve composite slabs subjected to dynamic loading withstood 2,000,000 cycles of loading. Of the remaining four, three failed by fatigue of the metal deck. In these cases the shear span was larger than in other tests resulting in greater tensile stresses in the deck. The researchers concluded that special attention should be paid to cases in which the tensile stresses in the deck are relatively high. The sharp corners of the embossments were considered a contributory factor to the fatigue failure. As the behaviour of a particular deck under repeated loading is likely to be affected by the type and pattern of embossments deck manufacturers should be asked for information on the dynamic performance of their products in such situations.

Later research by Krige and Mahachi (1995) studied the effect of load amplitude on the fatigue strength of composite slabs with re-entrant profile decks. They found that with a central point load applied to the slab the maximum load that can be applied, irrespective of the minimum load, in order to sustain repeated loading is 50% of the static maximum. For two point loading the endurance to repeated loading is slightly improved and may be found from a design chart in the form of a modified Goodman diagram.

5.6 CONCENTRATED LOADS AND SLAB OPENINGS

5.6.1 Concentrated Loads

Composite flooring in multi-storey warehouse buildings in which forklift trucks are used should be checked for the effects of heavy moveable point loads. After selecting a floor for the appropriate uniformly distributed design loading, the resistance of the slab to punching shear and local shear transfer of point loads should be assessed. Eurocode 4 defines the critical perimeter as shown in Figure 5.21 and refers the designer to Eurocode 2 for the calculation of the punching shear resistance.

The width of slab, b_{em}, considered to be effective for checking bending and longitudinal shear in simply supported slabs is defined in EC4 as:

$$b_{em} = b_m + 2L_p[1-(L_p/L)] \leq \text{slab width} \qquad (5.21)$$

where L_p is the distance from the centre of the support of the load to the nearest support and L is the span length.

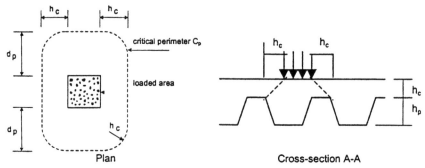

Figure 5.21 Critical perimeter for punching shear

For vertical shear the effective width, b_{ev}, is given as:

$$b_{ev} = b_m + L_p[1-(L_p/L)] \leq \text{slab width} \qquad (5.22)$$

Although a composite slab may have sufficient bending resistance to sustain the moments arising from a uniformly distributed load, the high longitudinal shear forces that occur when a point load is placed close to a support may cause shear bond failure between the concrete and metal deck. It is therefore recommended (AD150, 1993) that where the moment capacity of the slab is inadequate to resist the bending moments arising from local point loads, additional reinforcement be placed over the supports. The reinforcement, a minimum of 0.5%, should be placed in the top of the slab (to control cracking) and extend into adjacent spans for 25% of the span. Where the loads are exceptionally high it may be necessary to design the slab as a continuous reinforced concrete slab with conventional bar reinforcement throughout and ignore the contribution of the steel deck.

5.6.2 Slab Openings

Holes for the passage of services through composite slabs may be formed most easily by boxing out the required shape, using pressed metal, timber or polystyrene

Composite Floors

formwork, prior to concreting and trimming the metal deck by burning or cutting after the concrete has gained 75% of its strength. Cutting holes through hardened concrete is not recommended as this may damage the bond between the deck and concrete, particularly if percussive methods are used. Small holes may be formed without the need for extra reinforcement in the slab. Larger holes will require the slab to be strengthened, either by the provision of extra reinforcement or trimming steelwork. Deck manufacturers are usually able to provide detailed guidance on the size of holes that can be accommodated in their products. For shallow decking Couchman *et al.* (2000) suggest that typically holes up to 300 mm square will not require additional reinforcement; between 300 and 700 mm square the hole will need to be reinforced by additional bar reinforcement in the slab and above this size permanent trimming steelwork will be necessary. They also suggest that where specific guidance on the provision of extra reinforcement is not available from a manufacturer that a load path as shown in Figure 5.22 be assumed.

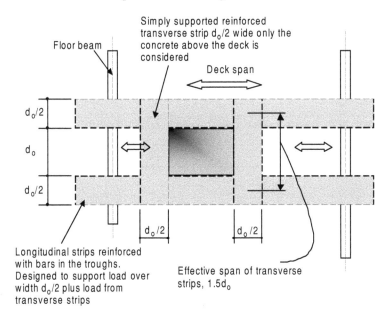

Figure 5.22 Assumed load path around large holes

5.7 FIRE RESISTANCE

When a composite slab is subjected to fire the metal deck, which is usually unprotected, heats up rapidly and loses strength and stiffness. Tests reported by Cooke *et al.* (1988) showed that simply supported slabs have inherent fire resistance. Composite deck slabs were shown to be adequately strong in fire if reinforced with mesh. The performance of slabs in fire may be determined by test or more usually by calculation methods which conservatively ignore the residual strength of the deck and rely only on the contributions of the concrete slab and reinforcement provided for this purpose. The section may be analysed as a reinforced concrete slab with the concrete and reinforcement strength reduced in line with increasing temperature. In addition to satisfying strength requirements, it is essential that slabs also provide adequate insulation and maintain the integrity of fire compartments. Providing a minimum depth of slab satisfies insulation requirements. The UK code (BSI, 1994a) specifies the depth of normal weight and lightweight concrete required above trapezoidal profiles and the overall depth required for slabs cast on re-entrant, dovetail, decks (provided the width of the dovetail is not more than 10% of the spacing). Non-structural screeds may be included in the depth. Eurocode (CEN, 1994) expresses the minimum in terms of an average depth given by two empirical formulae defined by the profile geometry and concrete cover. For lightweight concrete the minimum depths may be reduced by 10%. Only 20 mm of screed may be included in the minimum depth.

Newman (1991) has prepared a simplified design method based on test results. For slabs which are continuous over at least one support and reinforced with a single layer of mesh, tables of the type shown in Table 5.1 are presented for trapezoidal and dovetail decks. Many manufacturers also provide data sheets of this type.

Table 5.1 Simplified design for trapezoidal decks

Maximum span of continuous slab (m)	Minimum dimensions to obtain 60 minutes fire resistance (mm)			Mesh Size mm^2/m
	Deck thickness t (mm)	Overall slab depth D_s		
		NWC	LWC	
2.7	0.8	130	120	142
3.0	0.9	130	120	142
3.6	1.0	130	120	193

Alternatively, the hogging and sagging moment of resistances of a composite slab in fire may be calculated using empirically based design methods which account for the degradation of concrete and steel strengths with temperature (Bailey et al., 1999). These methods, which neglect the contribution of the deck and require additional reinforcing bars, are applicable to single and continuous spans. Purkiss (1996) provides a useful comparison of the UK and Eurocode methods and Lawson and Blaffart (1996) give a concise summary of the Eurocode approach. By assuming a simple plastic collapse mechanism the load to cause failure may be obtained from:

$$M_{sag} + 0.5\left[1 - \frac{M_{hog}}{WL}\right] \geq \frac{WL}{8} \quad \text{for end spans} \quad (5.23)$$

$$\text{and} \quad M_{sag} + M_{hog} \geq \frac{WL}{8} \quad \text{for internal spans} \quad (5.24)$$

To determine the sagging strength, EC4 uses the concrete strength at ambient temperature, as it is on the cooler top side of the slab, but reduces the strength of the reinforcement depending on the temperature. The temperature may be estimated using a parameter z which is in turn a function of the position of the bars within the troughs, as shown in Figure 5.23.

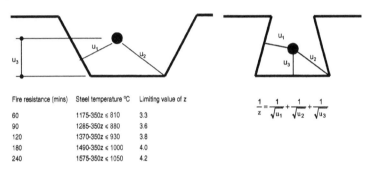

Figure 5.23 Calculation of reinforcement temperatures in trapezoidal and dovetail profiles with normal weight concrete using the parameter z

The reduced strength of the reinforcement may then be found from a strength reduction-temperature relationship as shown in Figure 5.24. The moment resistance is then calculated in the usual way.

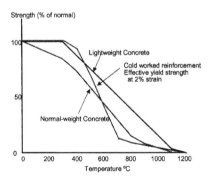

Figure 5.24 EC4 strength reduction for normal weight siliceous concrete, lightweight concrete and steel reinforcement at elevated temperatures

In order to calculate the resistance in hogging it is necessary to define the temperature profile through the slab. Temperature profiles for normal weight concrete flat slabs are given in EC4 Part 1-2 (CEN, 1994) and illustrated in Figure 5.25. These may be related to the composite slab by considering it as a flat slab with an effective thickness (calculated in the same way as for checking the minimum thickness required for insulation). From the temperature, the strength reduction factors for concrete and the embedded steel may be found, see Figure 5.24. The reduced compressive strength of thin (say 10 mm) layers of the concrete may be determined and the neutral axis depth found by equating the tensile force in the reinforcing bars with the compression in the concrete. Taking moments about the neutral axis position gives the hogging moment resistance.

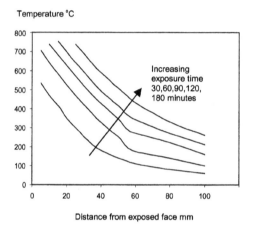

Figure 5.25 Temperature distribution through a concrete slab

Although a considerable number of fire tests have been conducted on composite slabs, Lawson and Blaffart (1996) note that to be realistic such tests should be performed on continuous slabs as the continuity effect provided by reinforcement greatly increases the fire resistance of slabs. Results from full-scale fire tests on an eight storey steel framed building have shown that composite steel deck flooring systems play a very significant role in the performance of steel buildings in fire (Bailey and Moore, 2000a). The development of in-plane forces as a result of large displacements enhances the load carrying capacity of the slab. Bailey and Moore (2000b) propose that this inherent extra capacity be utilised in a new design approach to fire design of composite frames.

5.8 DIAPHRAGM ACTION

5.8.1 Temporary stage

During construction the steel decking is often assumed to provide adequate lateral bracing to resist in-plane forces arising from wind loading. The ability of the decking to function as a stressed-skin diaphragm is dependent on the fixing details in place at the time of the applied loading. Initially the deck will only be secured to the beams with shot fired pins. Through deck welding of shear connectors is likely to occur soon after but there could be a delay of a few days. This is when the building is at its most vulnerable. Some guidance on an approximate method of assessing the diaphragm action of steel decking during construction is available (AD175, 1995). Where 4 mm shot fired pins are used at 300 mm spacing, the recommended horizontal distance between points of effective vertical bracing during construction are between 2 and 4 times the floor depth on plan. The value depends on the height of the building, whether it is clad during construction and the presence or not of fasteners along the seams of the sheets. Where a project requires a more detailed analysis a stressed-skin design method, such as BS5950 part 9, may be employed (BSI, 1994b).

5.8.2 Permanent stage

In many multi-storey buildings the floor is required to act as a stiff plate and transfer horizontal loads to the parts of the frame designed to provide lateral resistance. Easterling and Porter (1994) conducted the most comprehensive study of the ability of composite slabs to function as diaphragms. Prior to this limited tested data was available, the most useful being four full-scale tests conducted by Davies and Fisher (1979). Easterling and Porter tested to destruction 32 full-size composite diaphragms to provide the background for the development of design provisions. The tests were designed to consider the important parameters, steel deck shape and thickness, type and number of connectors (spot welds or stud shear connectors), slab aspect ratio, edge member size and concrete thickness. In all the tests only the deck reinforced the slabs; no mesh or reinforcing bars were provided. In-plane loads are introduced from the steel frame to the composite slab through narrow regions near the edges of the diaphragm (known as edge zones). If spot welds are used to connect the deck to the steel frame, the load path is through the welds into the deck and into the concrete through the deck-concrete interface. Where headed shear studs are used, the load has a more direct path through from the steel beams into the concrete.

The concrete part of a composite slab is considerably stiffer than the metal deck and it is therefore the concrete that primarily supports the load (providing all parts of the load transfer path are adequate to introduce the forces into the concrete). The shear resistance of the concrete slab alone may be assessed from concrete design codes and compared with the induced shear force. In most practical cases the concrete in the composite slab will have adequate shear resistance to resist the forces induced by wind loading. If this is not the case, the concrete thickness may be increased and additional reinforcement added to resist the horizontal diaphragm stresses.

5.9 SLIM FLOOR DECKING

5.9.1 Developments in Slim Floor Systems

The 1980s saw dramatic increases in the market share for steel framed multi-storey office buildings in Finland and Sweden (Latter, 1994). This was mainly due to the development of floor systems which contained the structural floor beams within the depth of the concrete slab. These floor systems offer a number of advantages over conventional slab-over-beam composite construction (Lu and Makelainen,1996). A reduction in overall structural depth reduces the total height of the building, perhaps permitting the construction of an extra floor in a given height, and reduces the cost of cladding the building. Slim floor systems have a flat soffit which makes installation and later replacement of services much easier. Inherent fire resistance of slim floor systems is significantly improved as only the bottom flange of the steel section is exposed and, depending on the required fire resistance period, fire protection may not be necessary.

Initially, fabricated beams in a "top hat" cross-section were used (ECCS, 1995b). Later, built up asymmetric beams were produced from the upper half of an I beam with a welded bottom plate (the ARBED integrated floor beam) or an H section welded to a bottom plate (the British Steel Slimflor beam (Mullet, 1992)). In each case the floor slab was usually a hollow-core slab or in-situ concrete cast on concrete planks. These are heavy and difficult to manoeuvre and an alternative deep metal deck capable of spanning 6 m unpropped, was developed by Precision Metal Forming Ltd (Mullet and Lawson, 1993). In 1995 British steel launched a rolled asymmetric beam designed to be used with a revised and slightly deeper decking in a combined slim floor system called *Slimdek* (Lawson *et al.*,1997) as illustrated in Figure 5.26.

Figure 5.26 Installation of deep decking (notice the closure pieces fixed to the rolled steel beams)
Photograph courtesy of Corus plc

Interest in slim floor systems is spreading world-wide. Qeiroz *et al.* (1998) reported results of a full-scale test conducted in Brazil on a slim steel-concrete composite floor using conventional 75 mm deep steel decking supported on the bottom flange of steel beams 200 to 250 mm deep.

5.9.2 Steel Decking in Slim Floor Systems

Steel decking for slim floor construction is typically 200–225 mm deep and 1.0–1.25 mm thick. The decks are cold-rolled from galvanised strip into units 600 mm wide. Each unit forms a single trough and clip together to form a completed rib (see Figure 5.27). End diaphragms fixed to the lower flange of the beam help to stabilise the deck and prevent concrete flowing out under the deck during construction. If lightweight concrete is used the slab weight can be as little as half that of an in-situ flat slab of the same depth (Lawson and Leskela, 1996).

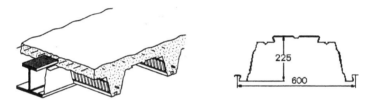

Figure 5.27 Cross section through SD225 deep decking

Research to support the development of the deep decking used in the Slimdek system was conducted at a number of centres throughout Europe. Lawson *et al.* (1999) report that research into the behaviour of deep decks under construction loading was conducted at British Steel's Welsh Technology centre. Tests were conducted using vacuum pressure and also by loading directly with jacks. Spans of up to 6.5 m are possible with lightweight concrete, reducing to 6 m when normal weight concrete is used. If propped construction is used spans of up to 9 m are possible (Lawson *et al.*, 1997).

5.9.3 Composite Action in Deep Decks

Composite deep deck slabs differ from conventional shallow composite floors in two fundamental ways. In conventional slabs, bar reinforcement in the deck troughs is not usually necessary except where fire considerations, heavy loading or large openings control the design. In deep deck composite slabs bars are placed in the troughs to enhance the moment resistance in both normal and fire conditions. Secondly, the overall depth of a deep deck composite slab may in some cases be dictated not by structural and fire resistance requirements but the necessity to provide sufficient depth of concrete over the top flange of the integral floor beam.

Tests conducted at TNO in the Netherlands showed that the bending resistance of a composite slab is significantly enhanced by the addition of bar reinforcement in the troughs. The resulting moment resistance may be determined by adding the capacity calculated for an equivalent reinforced concrete slab to the moment resistance of the composite slab limited by shear-bond resistance. This approach was used to verify the capacity of slabs cast on a 225 mm deep deck and tested at the Universities of Kaiserslautern (in Germany) and Salford (in the UK) and was found to give good results (Lawson et al., 1999). The re-entrant shapes embodied in the profile were found to produce excellent interlock between the profile and steel deck. Longitudinal shear bond strengths ($\tau_{u \cdot Rd}$, as described in section 5.4.5) of 0.23 to 0.37 N/mm^2 were achieved in the tests from which suggested design values of 0.25 N/mm^2 and 0.2 N/mm^2 were proposed for normal and lightweight concrete respectively. These tests were followed by a full-scale test on a double bay floor assembled with 280 mm deep rolled asymmetric beams (ASBs) and a 300 mm thick composite slab cast on a 225 mm deep deck. The test was devised to investigate the structural performance of a complete *Slimdek* system, particularly the combined action of the beam and slab and the deflection and vibration performance. Tests on airflow and heat transfer were also conducted. The failure load (at a total deflection of span/50) was some 50% greater than the required design value. The natural frequency of the bare floor system was measured as 6.4 Hz.

As the performance of composite slabs is based on test data, for practical design of deep deck composite slabs manufacturers provide load-span tables and software. Composite action is usually more than adequate for normal imposed loads and therefore the construction stage condition, where the metal deck acts alone, is generally critical. Detailed practical advise on the use of deep deck composite floors, including guidance on the provision of holes to accommodate services, is contained in Couchman et al. (2000).

5.9.4 Fire Resistance of Slim Floor Decking

A principal advantage of slim floor systems is the inherent fire resistance which is a result of the encasement of the steel beams. A number of experimental test programmes (Lawson et al., 1999) and numerical studies (Ma and Makelainen, 2000) have investigated the behaviour of the complete slim floor arrangements and confirmed their excellent performance in fire. The metal deep deck behaves in a similar way to that in a shallow composite slab with the steel being largely ineffective at high temperatures and its contribution to flexural strength is ignored in fire. The reinforcing bars already present in the troughs of deep composite decks to enhance the moment resistance in service conditions may be used to provide bending strength under fire conditions. A minimum thickness of slab is required to meet insulation requirements, as is the case for shallow composite slabs. Guidance on the performance of a deck in fire is most likely to be derived from tests and be made available by the manufacturer. The performance under fire conditions may also be assessed by calculation using reduced material strengths at elevated temperatures. Bailey *et al.* (1999) note that tests have shown the method of predicting the temperature of reinforcing bars according to EC4 leads to unconservative (cooler) estimates when applied to deep decks. They propose that the design temperature for the reinforcement in SD225 steel decks be taken as 398 °C, 598 °C and 586 °C for fire resistance periods of 60, 90 and 120 minutes respectively, with the centre of the bars at 70 mm from the bottom of the profile for 60 and 90 minutes but 100 mm for 120 minutes.

5.10 REFERENCES

AD150, 1993, Composite floors – wheel loads from forklift trucks. *Advisory Desk, New Steel Construction,* **1(7)**, December, p. 37.

AD175, 1995, Diaphragm action of steel decking during construction, *Advisory Desk, New Steel Construction,* **3(4)**, August, pp. 44–45.

American Society of Civil Engineers, 1985, Standard for the Structural Design of Composite Slabs, ANSI/ASCE 3-91.

Bailey, C.G., Newman, G.M. and Simms, W.I., 1999, *Design of steel framed buildings without applied fire protection,* The Steel Construction Institute, Publication No. 186.

Bailey, C.G. and Moore, D.B., 2000, The structural behaviour of steel frames with composite floorslabs subject to fire: Part 1: Theory. *The Structural Engineer, Journal of the Institution of Structural Engineers,* **78(11)**, pp. 19–27.

Bailey, C.G. and Moore, D.B., 2000, The structural behaviour of steel frames with composite floorslabs subject to fire: Part 2: Design. *The Structural Engineer, Journal of the Institution of Structural Engineers,* **78(11)**, pp. 28–33.

BHP Structural Steel, 2000, Design of composite slabs for strength, *Composite Structures Design Manual DB3.1,* BHP Co. Pty Ltd, Australia.

Bode, H., Minas, F. and Sauerborn, I., 1996, Partial connection design of composite slabs. *Structural Engineering International, Journal of the International Association for Bridge and Structural Engineering,* 1, pp. 53–56.

Bode, H. and Minas, F., 1997, Composite slabs with and without end anchorage under static and dynamic loading. *Composite Construction – Conventional and Innovative, IABSE Conference Report,* pp. 265–270.

British Standards Institution, 1994a, BS5950: Structural use of Steelwork in Building Part 4: Code of practice for design of composite slabs with profiled steel sheeting, London.

British Standards Institution, 1994b, BS5950: Structural use of Steelwork in Building Part 9: Code of practice for stressed skin design, London.

British Standards Institution, 1995, BS5950: Structural use of Steelwork in Building Part 6: Code of practice for design of light gauge profiled steel sheeting, London.

British Standards Institution, 1996, ENV 1993-1-3: Eurocode 3: Design of steel structures Part 1.3, General rules – supplementary rules for cold formed thin gauge members and sheeting, London.

Canadian Standards Association, 1994, Limit States Design of Steel Structures, CAN/CSA-S16.1-94, Rexdale, Ontario, 1994.

Calixto, J.M., Lavall, A.C., Melo, C.B., Pimenta, R.J. and Monteiro, R.C., 1998, Behaviour and strength of composite slabs with ribbed decking. *Journal of Constructional Steel Research,* **46(1-3)**, Paper No. 110.

CEN (European Committee for Standardisation), 1992, ENV 1994-1-1: Eurocode 4: Design of composite steel and concrete structures Part 1.1, General rules and rules for buildings, Brussels.

CEN (European Committee for Standardisation), 1994, ENV 1994-1-2: Eurocode 4: Design of composite steel and concrete structures Part 1.2, Supplementary rules for structural fire design, Brussels.

CEN (European Committee for Standardisation), 2001, ENV 1994-1-1 Draft No. 2: Eurocode 4: Design of composite steel and concrete structures Part 1.1, General rules and rules for buildings, Brussels.

Chien, E.Y.L. and Richie, J.K., 1984, *Design and Construction of Composite Floor Systems,* (Toronto: Canadian Institute of Steel Construction).

Chien, E.Y.L. and Richie, J.K., 1993, Composite floor systems – a mature design option. *Journal of Constructional Steel Research,* **25**, pp. 107–139.

Cooke, G.M.E., Lawson, R.M. and Newman, G.M., 1988, Fire resistance of composite deck slabs. *The Structural Engineer, Journal of the Institution of Structural Engineers,* **66(16)**, pp. 253–261,267.

Couchman, G.H., Mullett, D.L. and Rackham, J.W., 2000, *Composite slabs and beams using steel decking: best practice for design and construction,* (Ascot: The Metal Cladding and Roofing Manufacturers Association Technical Paper No. 13 and The Steel Construction Institute Publication No. P300).

Daniels, B.J. and Easterling, W.S., 1997, Shortcomings in composite slab design codes. In *Proceedings of an Engineering Foundation Conference.* Germany, edited by Buckner, C.D and Shahrooz, B.M., (American Society of Civil Engineers), pp. 370–379.

Daniels, B.J. and Crisinel, M., 1993a, Composite slab behaviour and strength analysis. Part I: Calculation procedure. *ASCE Journal of Structural Engineering,* **119(1)**, pp. 16–35.

Daniels, B.J. and Crisinel, M., 1993b, Composite slab behaviour and strength analysis. Part II: Comparison with test results and parametric analysis. *ASCE Journal of Structural Engineering,* **119(1)**, pp. 36–49.

Davies, J.M. and Jiang, C., 1997, Design procedures for profiled metal sheeting and decking. *Thin Walled Structures,* **27(1)**, pp. 45–53.

Easterling, W.S. and Young, C.S., 1992, Strength of composite slabs. *ASCE Journal of Structural Engineering,* **118(9)**, pp. 2370–2389.

ECCS (European Convention for Constructional Steelwork), 1995a, Design guide for composite slabs, ECCS publication No. 87.

ECCS 1995b, Multi-storey buildings in steel – Design guide for slim floors with built-in beams, ECCS publication No. 83.

Friberg, B.F., 1954, Combined form and reinforcement fo concrete slabs. *Journal of the American Concrete Institute,* **50**, May, pp. 697–716.

Harada, M., 1998, Small art extends the range of the steel deck market. *Journal of Constructional Steel Research,* **46(1-3)**, Paper No. 118.

Jolly, C.K. and Lawson, R.M., 1992, End anchorage in composite slabs: an increased loadcarrying capacity. *The Structural Engineer, Journal of the Institution of Structural Engineers,* **70(11)**, pp. 202–205.

Krige, G.J. and Mahachi, J., 1995, Dynamic behaviour of composite floors, *Journal of Constructional Steel Research,* **34**, pp. 249–269.

Latter, R., 1994, The European market for constructional steelwork. *New Steel Construction,* **2(5)**, pp. 17–19

Lawson, R.M. and Blaffart, H., 1996, Fire safe design of composite decks. *Structural Engineering International, Journal of the International Association for Bridge and Structural Engineering,* **2**, pp. 119–122.

Lawson, R.M. and Leskela, M., 1996, Slim floor construction. *Structural Engineering International, Journal of the International Association for Bridge and Structural Engineering,* **2**, pp. 122–126.

Lawson, R.M., Mullet, D.L. and Rackham, J.W., 1997, *Design of asymmetric Slimflor beams using deep composite decking,* SCI publication 175, (Ascot: Steel Construction Institute).

Lu, X. and Makelainen, P., 1996, Slim floor developments in Sweden and Finalnd. *Structural Engineering International, Journal of the International Association for Bridge and Structural Engineering,* **2**, pp.127–129.

Mullet, D.L., 1992, *Slim floor design and construction,* SCI publication 110 (Ascot: Steel Construction Institute).

Mullet, D.L. and Lawson, R.M., 1993, *Slim floor construction using deep decking,* SCI publication 127, (Ascot: Steel Construction Institute).

Mullet, D.L., 1998, *Composite floor systems,* (London: Blackwell Science Ltd).

Newman, G.M., 1991, *The fire resistance of composite floors with steel decking,* SCI publication 056, (Ascot: Steel Construction Institute).

Patrick, M., 1990, A new partial shear connection strength model for composite slabs. *Steel Construction Journal, Australian Institute of Steel Construction.* **24(3)**, pp. 2–17.

Patrick, M. and Bridge, R.Q., 1994, Partial shear connection design of composite slabs, *Engineering Structures,* **16(5)**, pp. 348–362.

Porter, M.L. and Ekberg, C.E., 1976, Design recommendations for steel deck floor slabs. *ASCE Journal of the Structural Division,* **102(11)**, pp. 2121–2136.

Purkiss, J.A., 1996, *Fire safety engineering design of structures,* (Oxford: Butterworth-Heinemann).

Queiroz, G., Pimenta, R.J., Calixto, J.M. and da Mata, L.A.C., 1998, A new type of slim floor. *Journal of Constructional Steel Research,* **46(1–3)**, Paper No. 111.

Schuster, R.M., 1970, *Strength and behaviour of cold-rolled steel-deck reinforced concrete floor slabs,* PhD Thesis, Iowa State University.

Schuster, R.M., 1976, Composite steel-deck concrete floor systems. *ASCE, Journal of the Structural Division,* **102(5)**, pp. 899–917.

Stark, J.W.B. and Brekelmans, J.W.P.M., 1996, Plastic design of continuous composite slab. *Structural Engineering International, Journal of the International Association for Bridge and Structural Engineering,* **1(96)**, pp. 47–53.

Temple, M.C. and Abdel-Sayed, G., 1979, Fatigue experiments on composite slab floors. *ASCE Journal of Structural Division,* **ST7**(July), pp. 1435–1444.

Tenhovuori, A.I. and Leskela, M.V., 1998, Longitudinal shear resistance of composite slabs. *Journal of Constructional Steel Research,* **46(1-3)**, Paper No. 319.

Veljkovic, M., 1998, Influence of load arrangement on composite slab behaviour and recommendations for design. *Journal of Constructional Steel Research,* **45(2)**, pp. 149–178.

Widjaja, B.R. and Easterling, W.S., 2000, Developments in long span composite slabs, *American Institute of Steel Construction, Engineering Journal,* **37(2)**, pp. 73–81.

Wyatt, T.A., 1989, *Design guide on the vibration of floors,* (Ascot: Steel Construction Institute).

CHAPTER SIX

Composite Connections

David B. Moore

6.1 INTRODUCTION

A composite connection may be defined as a joint between members where one or more of the members is a composite section (usually the beam) and the reinforcement connecting the members is intended to contribute to the resistance, stiffness and ductility of the joint.

Using this definition the connections in a variety of different forms of construction can be considered and designed as composite connections. These include the beam-to-column connections in a steel frame where the columns are non-composite (bare steel, 'I', 'H' or box section) and the beams are designed to act compositely with the floor slab. In this type of structure the floor slab may consist of any of the following forms of floor construction:

- concrete acting compositely with a profiled metal decking
- an insitu flat slab
- a flat slab constructed from pre-cast concrete units.

Other forms of composite connection exist and include slimfloor connections, connections between composite beams and composite columns and between composite beams. They also include the more exotic connections that are sometimes used in earthquake regions of the world.

Different types of bare steel joints can further increase the range of possible composite connections. For example, the following different types of bare steel connection can be used:

- end-plate connection (Figures 6.1b and 6.1d)
- fin plate connection (Figure 6.1c)
- fully-welded connection
- flange and web cleat connection (Figure 6.1a)
- contact plate.

Figure 6.1 shows some of these details used as part of a composite connection. With such a large variety of composite connections it is not surprising that well-established design rules only exist for a limited range of connection types.

Over the past twenty years the behaviour of composite connections has been extensively examined but the majority of this work has concentrated on composite connections between steel columns and composite beams with simple end-plate steel connections. Other forms of composite connection have been studied but the design methods are judged to be not sufficiently well established and therefore have not been included in this publication. Furthermore, no attempt is made in this publication to provide a complete set of design rules for all forms of composite connection.

Figure 6.1 (a) Cleated connections; (b) full-depth end-plate; (c) fin plate connections; (d) partial depth end-plate. Reproduced with kind permission of Building Research Establishment Ltd.

Consequently, the methods presented in this chapter place the emphasis on composite connection with traditional forms of steel joints based on the methods and procedures developed for end-plate connections in Eurocodes 3 [1] and 4 [2]. Where, alternative approaches are available, such as those described in the BCSA/SCI publication 'Joints in Steel Construction—Composite Connections' [3] these are included.

Most design procedures for composite connections assume that the forces are carried by a combination of the bare steel connection and the reinforcement. Such an approach assumes that the reinforcement is properly designed and detailed to transfer the forces from one member to another. Where the reinforcement is not continuous or where only a brittle welded mesh is provided the capacity should be based only on the behaviour of the bare steel connection. Later in this chapter the effect that continuous, embedded reinforcement has on the stiffness, strength and ductility of composite connections is discussed in detail.

The design procedure presented in this chapter is very general and can be used to design composite connections with other types of bare steel joint (fully welded, flange cleat, contact plate, etc.) provided that the rules are used with care and any additional modes of failure are taken into account.

The method given in Eurocodes 3 and 4 also includes a system for classifying connections based on their moment resistance and rotational stiffness. This in turn is related to a classification system for framed structures. Such a procedure allows the design of the connection to be consistent with the engineer's assumptions regarding the structural behaviour of the frame. This approach is described and the basis for the stiffness and strength limits between pinned, semi-rigid and rigid connections is presented.

This chapter starts by describing the structural model used to determine the behaviour of composite connections and presents the advantages and disadvantages of using different types of bare steel connections.

6.2 TYPES OF COMPOSITE CONNECTIONS

Figure 6.2 shows a typical composite connection, which consists of a steel joint and a concrete slab with continuous tension reinforcement across the joint. In this joint the reinforcing bars and the upper part of the steel connection provide the tensile resistance and, provided there is no axial force in the beam, these forces are balanced by compression forces between the lower part of the beam's steel section and the column. In composite connections the vertical shear resistance of the slab is small and difficult to calculate. It is therefore assumed that the vertical shear resistance of the connection is provided by the steel connection alone.

The type of steel connection used can have a significant influence on both the buildability of the structure and the performance of the composite connection. While the choice of steel connection is usually based on simplicity, duplication and ease of fabrication, its influence on the behaviour of the composite connection must also be considered. A range of typical beam-to-column composite connections with different steel joints is shown in Figure 6.3. Similar configurations can be used for beam-to-beam joints. The influence that each of these has on the behaviour of the complete composite connection is briefly described below.

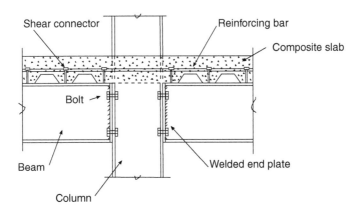

Figure 6.2 Typical composite connection

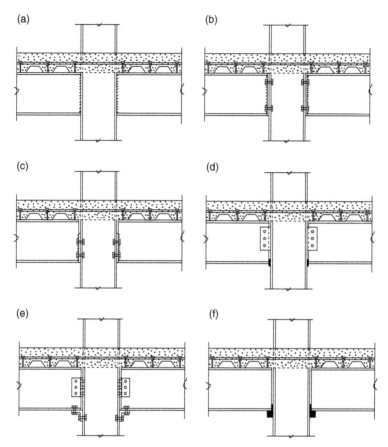

Figure 6.3 (a) Welded connection; (b) full depth end-plate connection; (c) partial depth end-plate connection; (d) fin-plate connection with connect plate; (e) cleated connection; (f) contact plate

Site welded connections, as shown in Figure 6.3a, can be used as part of a composite connection. They provide a high degree of strength and stiffness. However they are expensive to fabricate and their resistance may be limited by local instability of the compression flange or by excessive deformation caused by column flange bending if horizontal stiffeners are not provided. Furthermore, the welds must be designed using a high partial safety factor to prevent brittle failure. This type of connection is rarely used but with careful planning there is no reason why they should not be used.

End-plate connections like those shown in Figures 6.1b, 6.1d, 6.3b and 6.3c are commonly used as part of a composite joint. These can be flush, extended or partial depth and consist of a single plate fillet welded to the end of the steel beam and site bolted to either a supporting beam or column. This type of steel connection is relatively inexpensive but has the disadvantage that there is no room for site adjustment. Consequently, overall beam lengths need to be fabricated within tight limits, although 'finger shim' packs can be used to compensate for fabrication and erection tolerances. The contribution of the end plate to the overall behaviour of the composite joint will be influenced by the yielding of the column flange and/or end-plate and by local instability of the column web.

Another advantage of using an end-plate is that the overall strength, stiffness and ductility of the composite connection can be adjusted by varying the arrangement and properties of the end-plate and its components.

Fin plate connections like the one shown in Figure 6.1c are often used in practice because they are cheap to fabricate and simple to erect. However, fin plates do not provide the same degree of continuity as end-plate or welded connections. Furthermore, the fin plate must carry the compressive forces alone and this can result in lateral-torsional buckling and a reduction in both moment capacity and ductility. As an alternative a contact plate (like the one shown in Figure 6.3f) can be placed between the column flange and beam compression flange to enhance the compression capacity of the connection.

Cleated connections like those shown in Figures 6.1a and 6.3e can be used. In this connection the bottom flange cleat carries the compression. However, its effectiveness can be reduced by bolt slip caused by clearance holes. Tightening the bolts to increase the friction between the steel cleat and bottom flange of the beam can prevent this.

Contact plate connections or boltless connections like those shown in Figure 6.3f are sometimes used in continental Europe. By welding a plate to the column they ensure a direct transfer of the compression forces without slip.

6.3 DESIGN PRINCIPLES

Eurocodes 3 and 4 and the UK's national steel code, BS5950 [4], give three approaches for the design of a structure in which the behaviour of the connection is fundamental. In the Eurocodes these methods are defined as *Simple design*, *continuous design* and *semi-continuous design*. Elastic, plastic and elastic–plastic methods of global analysis can be used with any of these three approaches and Table 6.1 shows how the joint classification, the type of framing and the method of global analysis are related.

Simple design is based on the assumption that the beams are simply supported and implies that the connections must be sufficiently flexible to restrict the development of end fixity. When using this approach the connections are classified as nominally pinned no matter what method of global analysis is used. Early research [5] on pinned connections and their effects on structures

Table 6.1 Methods of frame design

Method of global analysis	Classification of joint		
Elastic	Nominally-pinned	Rigid	Semi-rigid
Plastic	Nominally-pinned	Full-strength	Partial-strength and ductile
Elastic–Plastic	Nominally-pinned	Rigid and full-strength	Semi-rigid and/or partial strength
Type of framing	Simple	Continuous	Semi-continuous

showed that most so-called pinned connections are capable of developing some moment capacity. More recently, tests on composite connections [6] have shown that when even a small amount of reinforcement is placed in the concrete the stiffness of the connection can be significant and in some cases approach that of a so-called rigid connection.

If the continuous approach is adopted the type of connection used will depend on the method of global analysis. When elastic analysis is used the joints are classified according to their stiffness and rigid connections must be used. When plastic analysis is used the connections are classified according to their strength (moment capacity) and full-strength connections must be used. The term full-strength relates the strength of the connection to that of the connected beam. If the moment capacity of the connection is higher than that of the connected beam then the connection is termed full-strength. The purpose of this comparison is to determine whether the joint or the connected member will limit the resistance of the structure. If the elastic–plastic method of global analysis is used then the connections are classified according to both their stiffness and strength and rigid, full-strength connections must be used. These connections must be capable of carrying the design bending moment, shear force and axial load while maintaining the original angle between the connected members. While continuous design can produce economies in beam size with respect to simple design most of these savings are offset by the need to supply joints with adequate rigidity.

While the simple method ignores stiffness and the continuous method only allows full-strength connections, the semi-continuous method accepts the fact that most practical connection are capable of providing some degree of stiffness and their moment capacity may be limited.

Once again the type of connection used will depend on the method of global analysis. When elastic analysis is used the connections are classified according to their stiffness and semi-rigid connections can be used. It should be noted that the term semi-rigid is a general classification and can be used to encompass all connections including nominally pinned and rigid connections. If plastic global analysis is used the connections are classified according to their strength. Connections that have a lower moment capacity than the connected member are termed partial-strength. In this case the connection will fail before the connected member and must therefore possess sufficient ductility to allow plastic hinges to form in other parts of the structure. Where the elastic–plastic method of global analysis is used the connections are classified according to both their stiffness and strength and semi-rigid, partial strength connections are used.

Figures 6.4 and 6.5 show the moment-rotation curves for a number of different types of composite connection. From these figures it is clear that connections 1, 2 and 4 can be classified as full-strength because their moment capacity exceeds that of the connecting beam. The moment capacity of connections 3 and 5 is lower than the connecting beam and these can be classified as partial strength connections. Similarly, connections 1, 2, 3 and 4 can be classified as rigid while connection 5 is semi-rigid. Finally, connection 6 is a nominally pinned connection.

The traditional system of classifying a connection only allows two types of connections (pinned and rigid) and it is relatively straightforward to use engineering judgement to classify a connection as either pinned or rigid. For an extended system such as the one described above that includes 'semi-continuous' connections it is preferable to use the structural properties of the connection as

Composite Connections 167

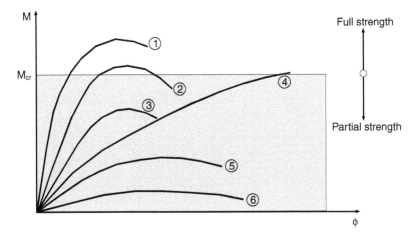

Figure 6.4 Classification by strength

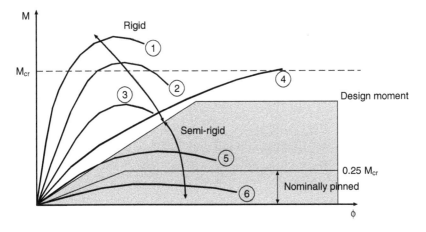

Figure 6.5 Classification by rigidity

a way of classifying the joint. To do this a set of limits must be developed to define the boundaries between the three types of connection in the stiffness and strength domains. Such a classification system has been developed in Eurocode 3 [1] and Eurocode 4 [2], and is described in the next section.

6.4 CLASSIFICATION OF COMPOSITE CONNECTIONS

In Eurocode 3 [1] a connection can be classified by its stiffness, by its strength and by a combination of both its stiffness and strength. Figure 6.5 illustrates the three ways in which a connection can be classified by its stiffness and Figure 6.4 illustrates the classification by strength. These figures also show the limits given in Eurocode 3 for a braced frame. Using its stiffness together with these limits a connection can be classified as pinned, rigid or semi-rigid. Similarly a connection can be classified using its strength as pinned, full-strength or partial strength. A fuller description of a connection's behaviour can also be obtained by classifying by both stiffness and strength. Such a classification leads to connections, which are pinned, rigid/full-strength, rigid/partial strength and semi-rigid/partial strength.

To classify a connection by its stiffness Eurocode 3 compares the initial stiffness of the joint, $S_{j,ini}$, against the following limits:

Rigid if $S_{j,ini} > 8EI_b/L_b$
Semi rigid if $0.5EI_b/L_b < S_{j,ini} < 8EI_b/L_b$
Pinned if $S_{j,ini} < 0.5EI_b/L_b$

In these inequalities EI_b and L_b are the flexural rigidity and length of the connected beam respectively.

Bijlaard et al. [7] established the stiffness boundary between rigid and semi-rigid connections by investigating the relationship between connection stiffness and the Euler buckling load for a single-bay, single-storey frame. In arriving at the boundary between these two connection types Bijlaard et al. assumed that a semi-rigid connection could be considered as rigid provided the difference between the Euler buckling load of a single-bay, single-storey frame with semi-rigid connections ($F_{cr,sr}$) and the Euler buckling load of a similar frame with true rigid connections ($F_{cr,r}$) was less than 5%. This criteria is expressed mathematically in the following expression:

$$\frac{F_{cr,sr}}{F_{cr,r}} \geq 0.95$$

By adopting this approach Bijlaard et al. were able to devise a classification system that is based on EI_b/L_b only. While such a system is easy to use, during its introduction into Eurocode 3 it attracted the following criticism:

- When compared to the stiffness limits given in some European national standards (most of which are based on custom and practice) the above limits appear to be conservative.
- The classification system given in Eurocode 3 can be applied to any steel/concrete composite framed structure but as the limits have been determined on the basis of a single-bay, single-storey frame the accuracy of their application to multi-bay, multi-storey frames is questionable.
- The stiffness boundaries between connection types have been determined on the basis of the ultimate limit state and on the assumption that a difference of 5% between the performance of a frame with rigid and semi-rigid connections is small and can be neglected. However, this does not necessarily mean that the differences at the serviceability limit states, where displacements of the structure are more important, are equally small and can be neglected. Clearly, when deriving classification criteria both serviceability and ultimate limit states should be considered.

At the time of writing some of these issues are being investigated by a number of European groups. The results from these studies may be used to develop a more generally acceptable classification system for future versions of Eurocodes 3 and 4.

Although the method in Eurocode 3 was developed for use with bare steel connections, with a few modifications it can be applied to composite connections. For example when classifying the stiffness of a composite connection it is necessary to decide whether to use the cracked or the uncracked properties of the connected composite beam. To identify which properties to consider it is necessary to examine the basis on which the above limits were determined. The boundary for a rigid connection was determined by examining the failure of a column in a simple frame. On the point of failure the connection will provide restraint to the column. Therefore the moments at the connection will reverse and the whole of the connected beam will be in sagging bending. In this situation the whole of the concrete will be in compression and therefore the uncracked properties of the composite beam should be taken.

For classification by strength the resistance of the connection is compared with the capacity of the connected beam. The main aim of this parameter is to identify which of the two members will fail first (the connection or the beam). Therefore the capacity of the composite beam should be taken as its capacity in hogging bending.

Just because a connection has adequate stiffness and strength does not mean that it is satisfactory for all situations. Where partial-strength joints are used in combination with plastic analysis there is a requirement for rotational capacity. This is needed to permit the redistribution of moments. The limits to redistribution are dependent on the local-buckling classification (class 1 plastic, class 2 compact, etc.) of the beam section in hogging bending. The rotational capacity required to permit redistribution of moments and the rotation capacity available from composite connections are discussed later in this chapter.

From the above discussion it is clear that three parameters are required to classify a connection; its stiffness, moment capacity and rotational capacity. The following sections describe in detail the methods than can be used to calculate each of these three characteristics for a composite connection.

6.5 CAPACITY OF COMPOSITE CONNECTIONS

The structural action of a composite joint with a steel end-plate connection is shown in Figure 6.6. The tensile resistance is provided by the reinforcement in the concrete slab and the upper part of the steel connection. Provided there is no axial force in the beam, these tensile forces are balanced by compression forces between the lower part of the beam's steel section and the column. By splitting the connection into its tension, compression and shear zones, the resistance of the complete connection can be determined by assembling the resistances of the individual components in such a way that the following three basic requirements are satisfied:

- The internal forces are in equilibrium with each other and with the external forces applied to the connection.
- The internal forces never exceed the resistance of the components.
- The maximum deformation capacity of the components is never exceeded.

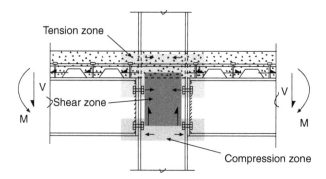

Figure 6.6 Typical composite connection showing force transfer by the various components

The following sections describe the background and methods used to calculate the resistance of a composite connection in the tension, compression and shear zones.

6.5.1 Tension Zone

Tests on composite connections have shown that any of the following four components may control the tension capacity of the connection:
- The tension capacity of the reinforcement.
- The shear connection between steel beam and concrete.
- The tension capacity of the steel components.
- The longitudinal shear force between the slab and the steel beam.

6.5.1.1 Tension Capacity of the Reinforcement

Yielding of the Reinforcement

The reinforcement in the concrete slab has a significant effect on the stiffness, strength and ductility of a composite connection. To see how the reinforcement affects each of these parameters Figure 6.7 compares the stress–strain behaviour of a reinforcing bar with that of an embedded reinforcing bar. It can be clearly seen that the un-embedded bar exhibits bi-linear behaviour while the curve for the embedded reinforcement shows a quite different stress–strain history. Initially, the curve is steeper because at this stage the concrete is uncracked and both the reinforcement and the concrete are acting together to support the applied tension. At a strain of $\Delta\varepsilon_{sr}$ the concrete begins to crack and at the cracks the load is carried by the reinforcement. As the load is increased this leads to further cracking of the concrete and a reduction in the stiffness of the member.

It is interesting to examine the distribution of forces between the concrete and the reinforcement either side of a crack. Figure 6.8 shows the distribution of strain in the concrete and reinforcement either side of a single crack. At the crack the force is carried entirely by the reinforcement but further away from the crack a bond is developed between the concrete and the reinforcement until

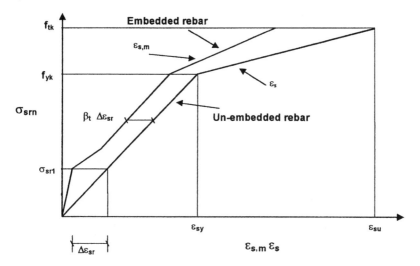

Figure 6.7 Comparison between the stress strain curves for an un-embedded reinforcing bar and an embedded bar

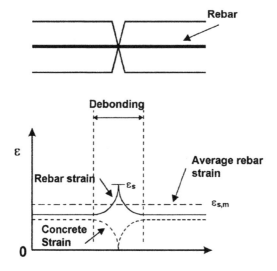

Figure 6.8 Distribution of strain in the concrete and reinforcement either side of a crack

eventually both concrete and steel act together. This effect is known as 'tension stiffening' and has the effect of reducing the deformation capacity of the member. A comparison between the embedded and un-embedded reinforcement at fracture shows the influence that tension stiffening can have on deformation capacity. Its effect is to cause rupture of the reinforcement at low levels of ductility. This is because the ultimate strain in the reinforcement can only be achieved at the cracks.

From the above discussion it is apparent that the tensile capacity of embedded reinforcement is limited by the ultimate capacity of the reinforcing bars. Therefore the potential resistance of the reinforcement in tension is given by the following expression:

$$P_{reinf} = f_y A_{reinf} / \gamma_m$$

where
f_y is the design yield strength of reinforcement
A_{reinf} is the area of reinforcement within the effective width of the slab
γ_m is the partial safety factor for reinforcement (usually taken as 1.05)

Local Crushing of the Concrete

Under unbalanced loading the slab on the less heavily loaded side can bare against the column and failure can occur by local crushing of the concrete. This may limit the capacity of the longitudinal reinforcement and result in premature failure. To investigate concrete crushing a truss model was developed to simulate the behaviour of a double-sided connection subject to unbalanced loads. This model is shown in Figure 6.9.

This model shows that the force in the longitudinal reinforcement is limited by the strength of that area of concrete, which bares against the column. By using this model it can be shown that the area of reinforcement must not exceed the following expression:

$$A_{reinf} \leq \frac{0.6 b_e d_s}{\mu} \frac{f_{cu}}{f_y}$$

where
b_c is the width of the column
d_s is the depth of the slab
f_{cu} is the cube strength of the concrete
μ is a function of the difference in moments applied either side of the connection and is given by the following expression:

$$\mu = 1 - \frac{M_{low} h_{r1}}{M_{high} h_{r2}}$$

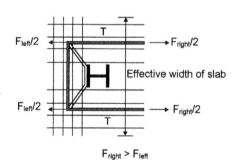

Figure 6.9 Truss model for the behaviour of a composite connection under unbalanced moment

In this expression the parameter are defined as follows:

M_{low} is the smaller applied moment
M_{high} is the larger applied moment
h_{r1} is the lever arm of the reinforcement on the high moment side
h_{r2} is the lever arm to the reinforcement on the low moment side

The bearing of the concrete against the column creates transverse tension in the concrete slab, which has to be resisted by transverse reinforcement. This additional reinforcement supplements any transverse reinforcement present and prevents longitudinal shear failure of the composite beam. Once again the truss model can be used to determine the following expression for the required area of transverse reinforcement:

$$A_T \geq \frac{0.35 \mu A_{reinf}}{\{e_T/e_L - 0.3\}}$$

where
$e_T = 0.2 b_c$
$e_L = 3.0 b_c$

6.5.1.2 The Shear Connection Between the Steel Beam and Concrete Slab

Composite action between the steel beam and the concrete slab is achieved by attaching shear connectors to the top flange of the steel beam. These connectors transfer the shear forces between the steel beam and the concrete slab. However, the degree of shear connection can have a significant influence on the stiffness, strength and ductility of a composite connection.

Full shear connection means that the shear connectors distributed along the beam are sufficient to develop the maximum bending capacity of the composite section. Partial shear connection occurs when there are insufficient shear connectors to carry the total forces between the steel beam and concrete slab. This results in a reduction in the moment capacity of the composite section. With regard to the connection, a reduction in the number of shear connectors reduces the contribution that the reinforcement can make to the capacity of the connection. This changes the distribution of forces through the connection and can also change the mode of failure.

To illustrate this Figure 6.10 shows the moment-rotation characteristics for four similar connections with different degrees of shear connection [8]. In each test the steel connection consisted of a double web cleat and a bottom flange cleat. Full-shear connection was used in test D2 while in tests D3 and D4 the shear connection was reduced to 75% and 50% respectively. D1 is a test on the bare steel connection and is included to illustrate the advantages of using composite connections.

In the early stages the relationship is dominated by the slip between the cleats and the beam web. After this there is a marked increase in the stiffness of the connection. By comparing tests D2, D3 and D4, it is clear that the increase in stiffness after slip is directly related to the degree of shear connection. It is also clear that the moment capacity of the connection increases with increase in shear connection. The changes in shear connection also have an influence on the mode of failure. In test D2 failure occurred when the reinforcement fractured while in tests D3 and D4 failure was by fracture of the shear connectors. This latter mode of failure also limited the rotation capacity of the connection.

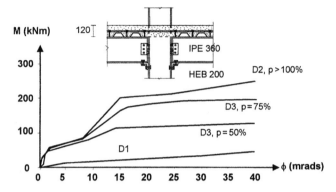

Figure 6.10 Moment rotation characteristics of four connections with different degrees of shear connection

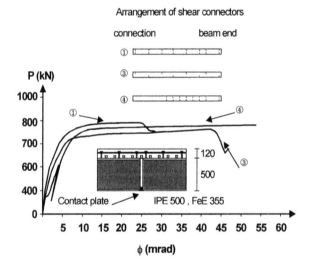

Figure 6.11 Load rotation characteristics for three connections with different arrangements of shear connectors

The behaviour of the connection can also be influenced by the arrangement of shear connectors close to the joint [9]. This is best illustrated by considering tests on a series of connections in which the arrangement was changed. Figure 6.11 shows load rotation characteristics of two interconnected floor beams where the main supporting beam is underneath. In all these tests the beam size and the amount and distribution of reinforcement remained the same. Test 3 had 70% partial shear interaction and the shear connectors are distributed uniformly along the beam. Test 4 had the same level of partial shear interaction but the shear connectors are located towards the end of the beam furthest from the connection. Both tests failed by fracture of the reinforcement.

Comparison between tests 3 and 4 show that test 4 failed at a higher rotation. This is because the greater distance between the joint and the first shear connector leads to a more uniformly stressed tensile band in the reinforced concrete slab. Therefore higher ductility is obtained from the concrete slab which leads to an increase in the rotation capacity of the composite connection.

In design guides the two effects discussed above are usually taken into account by imposing the following restrictions:

1. There are sufficient shear connectors to develop the full tensile strength of the reinforcement. This eliminates failure of the shear connectors as the governing mode and the associated reduction in rotation capacity.
2. To improve the rotational capacity of the connection the first shear connector is usually positioned at least 100 mm from the face of the column. This ensures that the reinforcing bars are strained over a substantial length.

6.5.1.3 The Tension Capacity of the Steel Components

Figure 6.3 shows a number of composite connections with a range of different steel components. Of the following different types of steel joint shown only three, the welded connection, the joint with flush or extended end-plate and the joint with partial depth end-plate have steel components that may contribute to the tension capacity of the composite connection:

- joint with welded connections
- joint with end-plate connections (flush or extended)
- joint with partial depth end-plate connections
- joint with fin-plate
- joint with cleated connections
- joint with contact plate.

For the remainder the tension capacity is based on the tension capacity of the reinforcement alone. However consideration most also be given to the resistance of these connections in the shear and compression zones.

The method used to determine the resistance of a bare steel end-plate connection is based on the design method given in Eurocode 3: Part 1.8 Joints (EC3: Part 1.8) [10]. This method assumes that the connection transmits moment by coupling tension in the reinforcement and the bolts with compression in the bottom flange. The previous section described the method for calculating the potential resistance of the reinforcement and described how partial shear connections and the distribution of the shear connectors might influence this. This section will concentrate on the methods and techniques used to calculate the potential resistance of the bolts and the end-plate/column flange in the tension region.

In the UK it has been traditional practice to assume a triangular (or linear) distribution of bolt forces in a connection. The method described in EC3: Part 1.8 uses a plastic distribution of bolt forces. These two approaches are shown in Figure 6.12. In the traditional approach the center of compression is assumed to be in line with the compression flange of the beam, with the bolt row furthest from the center of compression attracting the highest tension. One disadvantage of this method is that to achieve the assumed bolt force distribution the designer often has to stiffen the column flange.

In the Eurocode no assumption is made about the distribution of the bolt forces. Instead each bolt row is allowed to attain its full potential resistance. This model relies on adequate ductility of the connecting part in the uppermost bolt row to develop the design strength of the lower bolt rows. To ensure

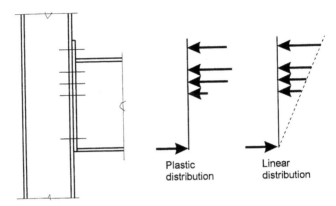

Figure 6.12 Comparison between linear and plastic bolt force distributions

adequate ductility of the bolt rows in the steel part of a composite connection the BSCA/SCI publications 'Joints in Steel Construction—Composite Connections' [3] places a limitation on the distance a bolt row is above the plastic neutral axis. The limit is 200 mm and has been derived from test results. For bolt rows that are less than 200 mm above the neutral axis a triangular limit is imposed to establish their resistance. For example the resistance of a row 150 mm from the neutral axis would only provide three quarters of its full potential resistance.

The resistance of each bolt row in the tension zone is determined from a consideration of the following modes of failure:

- bolt tension
- end-plate bending
- column flange bending
- beam web in tension
- column web in tension.

In addition to the above the adequacy of the flange to end plate welds and the web to end plate welds in tension should also be considered. Figure 6.13 illustrates a typical weld failure in an end-plate connection.

The approach described in EC3: Part 1.8 is used to determine the capacity of the bolts, the end-plate in bending and the column flange in bending. This approach is based on many years of experimental and theoretical research on end-plate connections. The results of which suggest that an equivalent tee-stub with an effective length of L_{eff} can be used to model the tension region of either the column flange or end-plate. Figure 6.14 shows how an equivalent tee stub can be used to model the failure of an end-plate. A full description of the effective lengths used for different bolt configurations is beyond the scope of this book and the reader is referred to either EC3: Part 1.8 [10] or the BCSA/SCI design guide on Joints in Steel Construction—Moment Connections [11].

In a tee stub, failure can occur by one of three mechanisms depending on the relative stiffness of the flange and the bolts. These mechanisms are usually referred to as complete yielding of the flange (Mode 1), simultaneous bolt fracture and yielding of the flange (Mode 2) and bolt fracture (Mode 3). These modes are shown diagrammatically in Figure 6.15 while Figures 6.16 and 6.17 are pictures of the first two modes of failure. Figure 6.15 also shows the prying forces (Q) that

Figure 6.13 Typical weld failure in a composite end-plate connection. Reproduced with kind permission of Building Research Establishment Ltd.

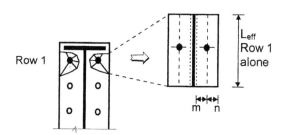

Figure 6.14 Equivalent tee-stub

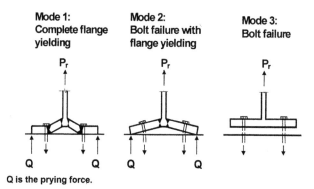

Q is the prying force.

Figure 6.15 Failure mechanisms of a Tee stub

Figure 6.16 Complete flange yielding. Reproduced with kind permission of Building Research Establishment Ltd.

develop at the edges of the T-stub and which increase the forces in the connecting bolts. The equations for calculating each of these modes of failure are given below:

MODE 1 COMPLETE FLANGE YIELDING

The potential resistance of either the column flange or end-plate, P_r, can be determined from the following expression:

$$P_r = \frac{4M_p}{m}$$

where
M_p is the plastic capacity of the equivalent tee-stub representing the column flange or end-plate
m is the distance from the bolt centre to a point 20% into either the column root or end-plate (Figure 6.14).

MODE 2 BOLT FAILURE AND YIELDING OF THE FLANGE

The potential resistance of the column flange or end-plate in tension is given by the following expression:

$$P_r = \frac{2M_p + n(\Sigma P_t')}{m + n}$$

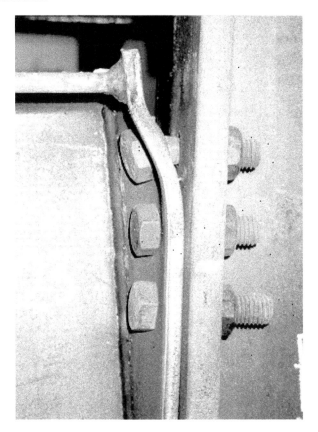

Figure 6.17 Combined bolt fracture and yielding of the end-plate. Reproduced with kind permission of Building Research Establishment Ltd.

where
$\Sigma P_t'$ is the total tension capacity of all the bolts in the group
n is the effective edge distance (see Figure 6.14).

MODE 3 BOLT FRACTURE

The potential resistance of the bolts in the tension zone is given by the following expression:

$$P_r = \Sigma P_t'$$

The equations for modes 1 and 2 do not explicitly include prying action (Q—see Figure 6.15) nor are any equations given to calculate its value. This is because prying action is implicit in the expressions for the calculation of the effective length L_{eff}. The principal author of this method, Zoetemijer [11], addresses the problem of prying action and developed three expressions for the equivalent length of an unstiffened column flange. Each of these expressions takes into account a different level of prying action.

For prying force = 0.0 $L_{eff} = (P + 5.5m + 4n)$
For maximum prying action $L_{eff} = (P + 4m)$
For an intermediate value $L_{eff} = (P + 4m + 1.25n)$

where
P is the bolt pitch.

Zoetemijer [12] explains that the first expression has an inadequate margin of safety against bolt fracture while the margin of safety in the second is too high. He therefore suggests using the third expression, which allows for approximately 33% prying action. This approach simplifies the calculations by omitting complicated expressions for determining prying action.

6.5.2 Compression Zone

The compression resistance of most of the steel details shown in Figure 6.3 is based on a consideration of the following modes of failure:

- the compressive resistance of the beam's lower flange and web
- the compressive resistance of the column's web.

In addition to these modes of failure consideration must be given to bolt slip in the flange cleat connection and lateral–torsional buckling of the fin plate.

A number of different types of column can be used in composite connections. These can be bare steel columns, concrete filled steel columns or cased columns. When bare steel columns are used the behaviour of a composite connection in the compression zone is similar to that of a bare steel connection and consideration should be given to the above modes of failure.

The effect of the concrete in concrete filled and cased columns is to provide additional strength to the column, which may be limited, by the compressive strength of the concrete. Therefore when using this type of column the compressive resistance of the steel column need not be considered.

6.5.2.1 Compression Resistance of the Beam's Lower Flange and Web

Compression failure of the beam's lower flange and web is shown in Figure 6.18. The method given in Eurocodes 3 and 4 for calculating this mode of failure assumes that the compression resistance of the beam is given by the following expression:

$$F_{c,fb,Rd} = M_{c,Rd} / (h - t_{fb})$$

where
h is the depth of the connected beam's steel section
$M_{c,Rd}$ is the moment resistance of the steel section, reduced to allow for shear
t_{fb} is the thickness of the connected beam's steel flange.

If the bottom flange alone is assumed to resist the compressive forces then the above method implies that stresses greater than yield are possible. However, discussions between writer and the authors of this method revealed that the stresses in the beam are assumed to be limited to the yield strength of the steel.

The SCI's publication on Composite Connections [3] presents two alternative methods to the approach given in Eurocode 3. These methods take advantage of

Figure 6.18 Buckling of the beam's compression flange and web. Reproduced with kind permission of Building Research Establishment Ltd.

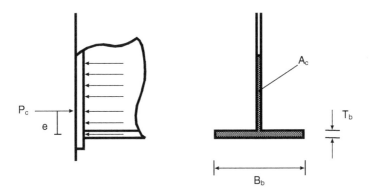

Figure 6.19 Beam flange in compression

strain hardening and allow the stresses in the beam to exceed yield. These methods are illustrated in Figure 6.19. In the first method it is assumed that only the bottom flange carries the compression in the beam while in the second method the compression zone is allowed to spread up the beam and into the web. In the second approach the centre of compression will move from the centre of the flange into the web. This approach is usually used in cases with either high moments or high moments combined with axial load.

Beam Flange in Compression—Method 1

The resistance of the beam flange is given by the following expression:

$$P_c = 1.4 p_{yb} T_b B_b$$

where
p_{yb} is the design strength of the beam
T_b is the beam flange thickness
B_b is the breadth of the beam's flange.

The 1.4 factor in front of this expression accounts for two effects. Firstly, it allows for the spread of compression into the beam's web and secondly it accounts for possible strain-hardening of the steel in the compression flange.

Beam Flange in Compression—Method 2

In this approach the resistance of the beam's flange is given by the following expression:

$$P_c = 1.2 p_{yb} A_c$$

where
A_c is the area in compression shown in Figure 6.19.

It should be noted that in this method the 1.4 factor is reduced to 1.2 since the contribution of the web is now taken into account directly. It should also be noted that the centre of compression is now at the centroid of the area A_c and the lever-arm of the bolts and the longitudinal reinforcement is reduced accordingly. Changing the position of the centre of compression will also affect the moment capacity of the connection and an iterative calculation becomes necessary.

6.5.2.2 Compression Resistance of Column Web

In many composite designs it is common for the column web to be loaded to such an extent that it governs the design of the connection. However, this can be avoided by either choosing a heavier column or by strengthening the web with one of the compression stiffeners shown in Figure 6.20.

Eurocode 3 gives the following expressions for calculating the resistance of an unstiffened steel column web subject to compression:

$$F_{c,wc,Rd} = \omega b_{eff,c,wc} t_{wc} f_{y,wc} / \gamma_{M0}$$

but

$$F_{c,wc,Rd} \leq \omega \rho b_{eff,c,wc} t_{wc} f_{y,wc} / \gamma_{M1}$$

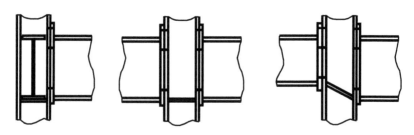

Figure 6.20 Typical compression stiffeners

where
ρ is a reduction factor for plate buckling
ω is a reduction factor to allow for the possible effects of shear in the column web
$b_{eff,c,wc}$ is the effective width of the column web in compression
t_{wc} is the thickness of the column web
$f_{y,wc}$ is the design strength of the column web
γ_{M0} and γ_{M1} are material partial safety factors.

This method allows for the effects of shear in the column web and uses the factor ω to reduce its compressive resistance. The resistance of the column web is further reduced to allow for the effects of the longitudinal compressive stress in the column. This is achieved by multiplying the values of $F_{c,wc,Rd}$ by the reduction factor k_{wc}. This factor takes the following values:

- When the longitudinal stress in the column, $\sigma_{com,Ed}$ is less than half the yield stress of the column web.

$$k_{wc} = 1.0$$

- When the longitudinal stress in the column, $\sigma_{com,Ed}$ is greater than half the yield stress of the column web.

$$k_{wc} = 1.25 - 0.5\sigma_{com,Ed}/f_{y,wc}$$

The BCSA/SCI publication 'Joints in Steel Construction—Moment Connections' [11] presents an alternative approach for calculating the compressive resistance of an unstiffened column web in compression. This method is simpler to apply than the method given in Eurocode 3 as it does not include the reduction factors for shear and longitudinal compressive stresses. In this approach the resistance of an unstiffened column web subject to compressive forces P_c is given by the smaller of the expressions for either column web crushing or column web buckling.

Column Web Crushing

The resistance of the column web to crushing is based on an area of web calculated by assuming the compression force from the beam's flange is dispersed over the length shown in Figure 6.21.

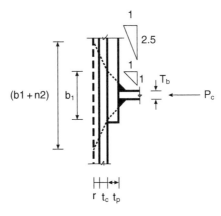

Figure 6.21 Force dispersion for web crushing

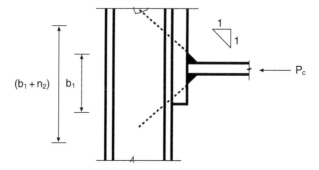

Figure 6.22 Length for web buckling

From this the resistance is given by the following expression:

$$P_c = (b_1 + n_2) t_c p_y$$

where
b_1 is the stiff bearing length based on a 45° dispersion through the end-plate from the edge of the welds
n_2 is the length obtained by a 1:2.5 dispersion through the column flange and the root radius
t_c is the thickness of the column web
p_y is the design strength of the column.

Column Web Buckling

The resistance of the column web to buckling is based on an area of web calculated assuming the compression force from the beam's flange is dispersed over a length shown in Figure 6.22.
From this the resistance of the column web to buckling is given by the following expression:

$$P_c = (b_1 + n_1) t_c p_y$$

where
b_1 is the stiff bearing length shown in Figure 6.22
n_1 is the length obtained by a 45° dispersion through half the depth of the column
p_y is the compressive strength of the column web taken from BS5950: Part 1: 1990, Table 24 with $\lambda = 2.5 d / t_c$
d is the depth of web between fillets.

6.5.3 Shear Zone

The column web panel must be designed to resist the resulting horizontal shear forces. To calculate these forces the designer must take account of any connection to the opposite column flange. In a single sided connection with no axial force

Composite Connections

the resultant shear force will be equal to the compressive force in the beam. For a symmetrical two-sided column connection with balanced moments the resulting shear force will be zero. However, in the case of a two sided connection subject to moments acting in the same sense the resultant shears will be additive. For any connection the resulting shear force can be calculated from the following expression:

$$F_v = M_{b1}/Z_1 - M_{b2}/Z_2$$

where
M_{b1} and M_{b2} are the moments in connections 1 and 2 (hogging positive)
Z_1 and Z_2 are the lever-arms for connections 1 and 2.

Eurocode 3 gives the following expression for calculating the shear resistance of an unstiffened column web: ,

$$V_{wp,Rd} = \frac{0.9 f_{y,wc} A_{vc}}{\sqrt{3}\gamma_{M0}}$$

where
A_{vc} is the shear area of the column.

The 0.9 factor is a general coefficient to allow of the interaction between axial compression in the column and shear.

As an alternative to the above method the BCSA/SCI publication 'Joints in Steel Construction—Moment Connections' [11] provides a more familiar expression that does not allow for the interaction between shear and axial compression. In this method the resistance of an unstiffened column web panel in shear is given by the following expression:

$$P_v = 0.6 p_{yc} t_c D_c$$

where
p_{yc} is the design strength of the column
t_c is the thickness of the column web
D_c is the depth of the column.

The webs of most UC sections will fail in panel shear before they fail in either bearing or buckling and therefore most single sided connections are likely to fail in shear. The strength of a column web can be increased by either choosing a heavier column section or by using one of the shear stiffeners in Figure 6.23.

6.6 DUCTILITY OF COMPOSITE CONNECTIONS

The discussion in Section 6.5.1.1 indicated that the ductility of a composite connection is influenced by the deformations that can be achieved in the reinforcement over the crack lengths in the concrete.

To illustrate this behaviour consider the three stress strain curves shown in Figure 6.24. These curves are for three similar reinforced concrete members

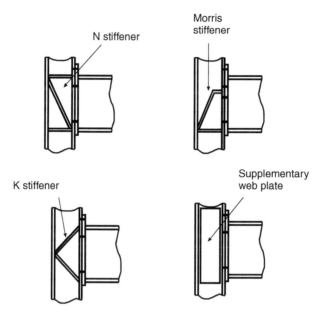

Figure 6.23 Typical shear stiffeners

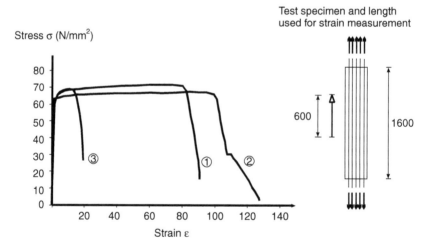

Figure 6.24 Influence of percentage reinforcement on ductility

subject to applied tension [13]. In test 1, 0.6% reinforcement is used and this is achieved by using 10 mm diameter bars with 12.0% strain at fracture. In test 2 a similar level of reinforcement is used but in this case 8 mm diameter bars are used with 15.2% strain at fracture. The final test incorporates 0.2% reinforcement by using 8 mm diameter bars with 15.2% strain at fracture. Comparing tests 1 and 2 it can be seen that their ductility at fracture is quite different. Test 1 fails at a strain of approximately 80 while test 2 fails at a strain of about 100. The reduced ductility of test 1 is due to the difference in the elongation at fracture of 10 mm and 8 mm diameter bars used in tests 1 and 2 respectively. This shows that by using bars with higher levels of elongation at fracture higher levels of ductility can be achieved. Comparing tests 2 and 3 shows that increasing the percentage of reinforcement also increases the ductility of the member. In general the rotation capacity of a composite connection increases as the area and ductility of the reinforcement increases. This is because the level of strain, at which the reinforcement fails, allowing for tension stiffening, also increases.

When semi-continuous framing is used with plastic global analysis the connections must be partial strength with adequate rotational capacity. To ensure that the connection has the required rotational capacity it is necessary to specify a minimum area of reinforcement and a minimum percentage elongation of the reinforcement at fracture.

The BCSA/SCI's design guide on Composite Connections [3] lists minimum areas of reinforcement required for compact (Class 2) and plastic (Class 1) composite connections. These limits are based on beam size, the grade of steel, the properties of the reinforcement and the type of connection (compact or plastic) and are presented in Table 6.2.

Minimum areas of reinforcement are given in Table 6.2 for reinforcement that is capable of achieving 5% and 10% elongation at maximum force. The value of 5% elongation at maximum force is in line with BS 449 [14] and BS EN 10080 [15]. Minimum areas of reinforcement are also given for 'high ductility' bars, which are capable of achieving 10% elongation at maximum force. These values are given because manufactures are able to produce reinforcement with higher levels of ductility than those quoted in BS 449 and BS EN 10080. The higher levels of elongation are also given because Eurocode 4 includes the requirement that for plastic global analysis only reinforcement of 'high ductility' as defined in Eurocode 2 should be used. This increase in ductility has considerable advantages as it permits the designer to use less reinforcement.

The above prescriptive approach is normally used to ensure that composite connections have adequate ductility, however, other performance-based approaches

Table 6.2 Minimum areas of reinforcement (mm^2) for Compact (Class 2) and Plastic (Class 1) connections

Type of joint	Type of steel	Elongation of rebar (%)	Beam depth (mm)							
			203	254	305	356	406	457	553	610
Class 2	S275	5	500	500	500	500	500	600	750	1150
		10	500	500	500	500	500	550	650	800
	S355	5	500	500	500	500	500	600	750	1150
		10	500	500	500	500	500	550	650	800
Class 1	S275	5	500	500	500	650	1100	1450	1800	3000
		10	500	500	500	500	500	600	750	1150
	S355	5	500	500	600	1400	2100	3100	—	—
		10	500	500	500	500	650	900	2000	2850

are available. These approaches require the designer to calculate both the required rotational capacity (i.e. the rotational capacity required by the connection to allow plastic hinges to form in other parts of the structure) and the connection's available rotation capacity. By comparing these two values it is possible to demonstrate whether or not the chosen composite connection has adequate ductility.

The next two sections describe the methods than can be used to calculate both the required and available rotation capacities for composite connections.

6.6.1 Required Rotation Capacity

The required rotation capacity for composite construction has been investigated by a number of authors [16,17]. The method used in the SCI publication—'Joints in Steel Construction—Composite Connections' [3] was developed by Li and is described below.

In the method proposed by Li, the rotations required at the end of the beams are divided into elastic and plastic components. These rotations are expressed as integrals of the elastic and plastic curvatures that occur in the span of the supported composite beam. The elastic and plastic components of curvatures in the beam are shown schematically in Figure 6.25. This figure clearly demonstrates the importance of plastic curvatures in the beam on the rotation capacity required from the composite connection. Indeed, Li et al. [18] showed that the required connection rotations depend on the following six parameters:

1. The beams' grade of steel. The higher the grade of steel the greater the strain and therefore curvature needed to attain yield.
2. The depth of the beam. The curvature needed to attain a given yield strain increases as the depth of the beam increases.
3. The ratio of connection and span moments. As this ratio decreases (i.e. the connections becomes a simple support) the end rotation increases.
4. The design mid-span moment. Less strain and therefore less curvature is required in the span to achieve $0.9M_p$ than to achieve M_p.

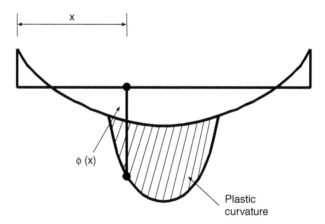

Figure 6.25 Elastic and plastic components of curvature in a beam

5. The type of loading. For a single point load at mid-span, plasticity is confined to a short length of beam either side of the point of maximum applied moment. For two point loads at third span positions, plasticity occurs over the central third of the span. Integrating plastic curvatures over this length means that both the plastic rotation requirements and the required rotational capacity of the connections are particularly high for beams subject to two point loads.
6. The span of the beam.

To calculate the required rotation capacity for all types of composite structure is clearly a very daunting task. However, based on the above findings practical limits can be imposed on the structure and its loading that will enable designers to determine the range of required rotation capacities. For example the SCI design guide for composite connections imposes limits on the span to depth ratio, the minimum allowable support to span moment ratio and on the design moment of the beam.

For a beam subject to a uniformly distributed load, a span to depth ratio of 25 is imposed. Whereas for a beam subject to two point loads at third span positions a more onerous limit of 20 is imposed. In the case of the ratio of support to span moment a minimum limit of 0.3 is imposed. This means that the composite connection must have a moment capacity of at least 30% of the sagging moment capacity of the composite beam. The final limit on beam design moment requires more consideration. Li *et al.* [18] calculated that if the design moment is reduced from $0.95M_p$ to $0.9M_p$ then the required rotation capacity of the composite connections is reduced by 30%. Couchman's [18,20] work on required rotation supports Li's findings and demonstrated that a reduction in required rotation can be obtained if the design moment of the beam is further reduced to $0.85M_p$. Using the charts developed by Li *et al.* [18] Couchman was able to derive Table 6.3 below which gives the required connection rotations to achieved span moments of $0.95M_p$, $0.9M_p$ and $0.85M_p$ for beams subject to either a UDL or two point loads.

The above discussion is based on the assumption that the composite beam is propped during construction. As the majority of composite construction is unpropped it is necessary to consider the effect that propping has on the required rotation capacity of the connections. In unpropped construction the dead load is supported by the bare steel and as a consequence greater strains and therefore curvatures are developing in the beam. These curvatures in turn lead to an increased demand in the required rotation of the connections because as explained above the end rotations are integrals of the curvature over the span of the beam. The effect of propping on the required rotation was investigated by Anderson and Najafi [21]. They concluded that the required rotations for an unpropped beam designed for $0.85M_p$ are similar to those for a propped beam

Table 6.3 Required connection rotation (μRads)—propped construction

Type of load	Span/depth	Mid-span moment		
		$0.95M_p$	$0.9M_p$	$0.85M_p$
UDL	25	40	28	20
Third points	20	46	32	23

Table 6.4 Required connection rotation—unpropped construction

Type of load	Span/depth	Mid-span moment		
		$0.95Mp$	$0.9Mp$	$0.85Mp$
UDL	25	56	39	28
Third points	20	64	45	32

designed for $0.9Mp$. This is similar to saying that the required rotations for unpropped construction are 40% larger than those required for propped construction. Based on this simple result Couchman was able to develop the required rotations for unpropped construction. These are given in Table 6.4.

6.6.2 Available Rotation Capacity

Numerous researchers have investigated the rotation capacity of composite connections and they have identified many sources of ductility some of which are included the following list:

- elongation of the reinforcement,
- deformation of the shear connectors,
- localized plastic deformation of the lower part of the beam's steel section,
- slip between steel components,
- buckling,
- elongation of bolt holes (this depends on the type of steel connection).

For a composite connection the design moment of resistance is calculated assuming that the reinforcement in the slab yields. A minimum elongation of the reinforcement can therefore be calculated from the ductility properties of the embedded reinforcement. The deformation of the shear connectors has been widely studied [22,23,24,25,26] and has been shown to enhance the rotation capacity of a composite connection. Yielding of the lower part of the beam's steel section provides a further source of deformation within the connection but this is generally small and is not included here.

The remaining sources of deformation are either unreliable or difficult to calculate with any certainty. For example, the slip between steel components is erratic [27] and therefore unreliable, the extent of buckling deformations depends on imperfections whose magnitudes are unknown to the designer and the deformations due to elongation of the bolt holes are dependent on whether or not the bolts slip into bearing. For these reasons only the first three sources of deformation provide reliable and predictable deformation capacity within the connection.

Based on the above arguments an empirical model for predicting the available rotation capacity of a composite connection was developed by Anderson, Aribert and Kronenberger [28]. In this model the available rotation capacity depends on the following sources of deformation:

- Elongation of the reinforcement.
- Slip at the interface of the steel and concrete as a result of the deformation of the shear connectors.

Composite Connections

Assuming that rotation takes place about the underside of the steel section, the available rotation capacity, θ, can be calculated from the following expression:

$$\theta = (\Delta_{us})/(D+D_r) + (s^{(B)})/D$$

where
Δ_{us} is the allowable elongation of the embedded reinforcement
$s^{(B)}$ is the interface slip
D is the depth of the steel beam
D_r is the distance from the top of the steel beam to the centreline of the reinforcement.

The above expression is consistent with the observation made by Najafi [29] that the rotation capacity of a connection increases as the area of the reinforcement increases. This is because as the area of reinforcement is increased the longitudinal force in the reinforcement also increases leading to an increase in the number of shear studs and a longer length of strained reinforcement.

For practical connection details the elongation of the embedded reinforcement makes the most important contribution to the connection's available rotation capacity and the next section describes a method for calculating the elongation of embedded reinforcement.

6.6.2.1 Elongation of the Embedded Reinforcement

Figure 6.7 compares the stress–strain behaviour of embedded reinforcement with the corresponding curve for bare reinforcement. As explained in Section 6.5.1.1 due to the effect of tension stiffening the embedded reinforcement has a lower overall ductility than bare reinforcement. Consequently the ultimate average strain is calculated from the following expression:

$$\varepsilon_{sm} = \varepsilon_{sy} - \beta_t \Delta_{sr} + \delta(1 - \sigma_{sr1}/f_{ys})(\varepsilon_{su} - \varepsilon_{sy})$$

where
Δ_{sr} is the increase of strain in the reinforcement at the crack when cracking first occurs
σ_{sr1} is the stress in the reinforcement in the crack, when the first crack has formed
β is taken as 0.4 for short term loading
δ is taken as 0.8 for high ductility bars.

Equations to calculate Δ_{sr} and σ_{sr1} are given in the CEB-FIP Model code [30]. However, these equations are only valid for concrete members in tension. The concrete slab of a composite connection is subject to combined tension and bending and therefore these two parameters must be calculated taking account of this combined action. Δ_{sr} and σ_{sr1} may therefore be determined from the following expressions:

$$\sigma_{sr1} = f_{ctm} k_c (1 + \rho E_s/E_c)$$

$$\Delta_{sr} = \sigma_{sr1}/E_s - f_{ctm} k_c/E_c = f_{ctm} k_c/(E_s \rho)$$

where
$\rho = A_s/A_c$ and $k_c = 1/(1+d/(2z_0))$
d is the thickness of the concrete flange excluding any ribs
z_0 is the vertical distance between the centroids of the uncracked, unreinforced concrete flange and the uncracked unreinforced composite section calculated using the modular ratio for short-term effects E_a/E_{cm}.

Before the extension of the reinforcement can be calculated the length of reinforcement over which the strain, ε_{sm}, acts must be determined. Expressions for calculating this length are given below.

If $\rho < 0.8\%$ then $\quad \Delta_{us} = 2L_t \varepsilon_{sm}$
If $\rho \geq 0.8\%$ and $a < L_t$ then $\quad \Delta_{us} = (h_c/2 + L_t)\varepsilon_{sm}$
If $\rho \geq 0.8\%$ and $a > L_t$ then $\quad \Delta_{us} = (h_c/2 + L_t)\varepsilon_{sm} + (a - L_t)\varepsilon_{sm}$

In these expressions a is the distance from the face of the column to the first shear connector along the beam. h_c is the depth of the column section in the direction parallel to the longitudinal reinforcement.

L_t is the introduction length shown in Figure 6.9 and Hanswille [31] developed the following expression for calculating the introduction length:

$$L_t = (k_c f_{ctm} \phi)/(4\tau_{sm}\rho)$$

where
ϕ is the diameter of the reinforcement
τ_{sm} is the average bond stress along the introduction length.

In situations where low amounts of reinforcement are used ($\rho < 0.8\%$) a single crack generally forms in the concrete at the connection. The reinforcement will yield at this location and further deformation of the beam will lead to fracture of the reinforcement at this location. In this situation $L_t < h_c/2$.

For higher levels of reinforcement ($\rho > 0.8\%$) the elongation length of the reinforcement increases but is limited by the location of the main crack and the position of the first shear connector. The second expression for Δ_{us} represents the situation where the first shear connector is located within the introduction length while the third expression for Δ_{us} is used when the distance between the first shear connector and the face of the column is larger that the introduction length.

6.6.2.2 Deformation Capacity Due to Slip at the Steel/Concrete Interface

The deformation capacity due to slip at the steel/concrete interface has been investigated by Aribert [32]. As part of this work Aribert develop a simple expression for calculating the stiffness of the shear connection between the steel section and the reinforced concrete slab. This expression is given below:

$$K_{sc} = Nk_{sc}/((\beta - ((\beta-1)/(1+\alpha))(h_s/d_s))$$

where
N is the number of shear connectors in the hogging region
k_{sc} is the stiffness of a single shear connector. This is determined by test but for a 19 mm diameter headed stud it is usually taken as 100 kN/mm²
d_s is the distance between the line of action of the tension resistance of the slab to the centroid of the beam's steel section
h_s is the distance from the reinforcement to the centre of the steel's bottom flange.

$$\alpha = E_a I_a / (d_s^2 E_s A_s)$$
$$\beta = [(1+\alpha) N k_{sc} l d_s^2 / (E_a I_a)]^{0.5}$$

where
l is the length of the beam in hogging bending. This may be assumed to be equal to $0.15 L_b$.

Aribert also recognized that a typical moment-rotation curve for a composite connection could be represented by the tri-linear curve shown in Figure 6.26 and that point A on this curve corresponded to the yielding of the shear connection between the steel section and concrete slab. By assuming that the initial linear-elastic portion of the curve was valid upto a maximum load of $0.7P_r$ in the heaviest loaded shear stud Artibert developed the following equations for calculating the end slip at point A, $S^{(A)}$, and the force in the reinforcement at point A $F_s^{(A)}$:

$$S^{(A)} = 0.7 P_r / k_{sc}$$
$$F_r^{(A)} = S^{(A)} K_{sc}$$

It is unfortunate that Aribert chose to use similar symbols for the stiffness of a single shear connector and the stiffness of the shear connection as this can cause confusion. For clarification the first expression uses the stiffness of a single shear connector while the second expression uses the stiffness of the shear connection.

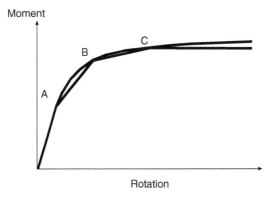

Figure 6.26 Tri-linear representation of the moment-rotation curve for a composite connection

Table 6.5 Comparison between test data and predictions for available rotation capacity

Source	Test details		Mode of failure	Rotation capacity		$\dfrac{\text{Test}}{\text{Calc.}}$
	Beam	Rebar		Test	Calc.	
Brown [34], 4	UB457	4T16	Rebar	35	33	1.06
Najafi [35], S8F	UB305	8T12	Rebar	35	33	1.09
Najafi [35], S4F	UB305	4T12	Rebar	27	22	1.23
Kronenberger [36], S2-1	IPE500	2.42%	Rebar	23	23	1.00
Kronenberger [36], S2-2	IPE500	1.45%	Rebar	22	23	0.96
Kronenberger [36], S2-3	IPE500	2.42%	Shear	41	39	1.05
Kronenberger [36], S2-4	IPE500	2.42%	Rebar	52	53	0.98
Kronenberger [36], S5-2	IPE300	1.54%	Rebar	44	44	1.00

Point B on the curve corresponds to the attainment of the maximum force in the reinforcement. For a connection with full shear interaction this can be calculated from the following expression:

$$F_s^{(B)} = A_s f_y$$

where
A_s is the area of reinforcement
f_y is the strength of the reinforcing steel.

As a result of the elastic–plastic behaviour of the shear studs the slip at point B is greater than the slip at point A and Aribert established the relationship between the slip at these two points as follows:

$$S^{(B)} = 2 S^{(A)} (F_s^{(B)}/F_s^{(A)})$$

6.6.2.3 Validation

The method for calculating the available rotation capacity of a composite connection has been validated by comparison against available test data on beam-to-beam and beam-to-column connections from a number of test laboratories. Couchman prepared Table 6.5 as part of the validation for the SCI's design guide on composite connections.

The comparisons made in Table 6.5 show good agreement with the test results. The ratio of the test value to calculated value of rotation capacity varies between 0.98 to 1.23.

6.7 STIFFNESS OF COMPOSITE CONNECTIONS

Calculating the stiffness of any connection can be a difficult process. For this reason the BSCA/SCI publication 'Joints in Steel Construction—Composite Connections' takes a more pragmatic approach and lists a number of simple rule-of-thumb guidelines which, if followed, will in most circumstances ensure that frame design assumptions are not violated. These simple rules result in a composite connection than can be considered to be rigid. Research on composite connections has shown that even when a small amount of reinforcement is used the connection can be considered to be rigid at serviceability limit-state.

Eurocode 3: Part 1.8 [10] incorporates an alternative method for calculating the stiffness of bare steel connections based on work carried out initially by Zoetemeijer [12] and more recently by Jaspart [33]. This method uses the 'component approach' in which the rotational response of the joint is determined from the mechanical properties of the different components (end-plate, cleat, column flange, bolts, etc.) within the connection. The advantage of this approach is that the behaviour of any joint can be calculated by decomposing the connection into its components (end-plate, cleat, column flange, bolts etc).

6.7.1 Basic Model

The procedure is to derive the rotational stiffness of a connection from the elastic stiffnesses of its component parts. The elastic stiffness of each component is represented by a spring with a force-deformation relationship of the form given below:

$$F_i = E \, k_i \, w_i$$

where
F_i is the force in component i
E is the elastic modulus of steel
k_i is the stiffness of component i
w_i is the spring deformation of component i.

These components are then added together using a simple spring model. The method is best illustrated by a simple example. The spring model shown in Figure 6.27 is for an end-plate connection. In this example the stiffness of the connection can be determined from the stiffness of the following components:

Unstiffened column web panel in shear (w_1)
Unstiffened column web in compression (w_2)
Column web in tension (w_3)

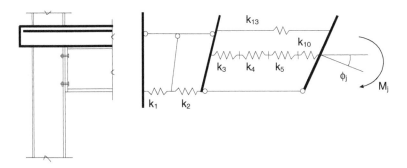

Figure 6.27 Stiffness model for a composite connection with a steel end-plate

Column flange in bending (w_4)
End plate in bending (w_5)
Bolts in tension (w_{10})

The initial stiffness of the connection is given by the following expression:

$$S_{j,ini} = \frac{M_j}{\phi_j}$$

where
M_j is the moment acting on the connection
ϕ_j is the rotation of the connection.

In this example the moment, M_j, acting on the connection is assumed to be equal to the product of the tension and compression force and the distance, z, between the centre of tension and the centre of compression. The rotation of the connection is simply calculated by adding together the deformation of each component taking into account the direction of deformation. For this example the rotation is given by:

$$\phi_j = (w_1 + w_2 + w_3 + w_4 + w_5 + w_{10} + w_{13})/z$$

Substituting the expressions of the moment and rotation into the equation for the initial stiffness gives the following expression:

$$S_{j,ini} = \frac{M_j}{\phi_j} = F_z/[(w_1 + w_2 + w_3 + w_4 + w_5 + w_{10} + w_{13})/z]$$

By replacing the deformations, w, with the force-displacement expression the following general formula can be derived for the initial stiffness of any connection:

$$S_{j,ini} = \frac{Fz^2}{\frac{F}{E}\left(\frac{1}{k_1} + \frac{1}{k_2} + \frac{1}{k_3} + \frac{1}{k_4} + \frac{1}{k_5} + \frac{1}{k_{10}} + \frac{1}{k_{13}}\right)}$$

This can be further reduced to the following familiar expression given in EC3: Part 1.8 [10]:

$$S_{j,ini} = \frac{Fz^2}{\Sigma(1/k_l)}$$

This expression can be used to calculate the initial stiffness of any connection provided that the stiffnesses of its component parts are known. Jaspart [33] and Yee et al. [34] have derived stiffness expression for the components of end-plate and cleated connections. A list of these components together with their stiffnesses is given in Table 6.6. For a detailed derivation of the stiffness equations the reader is referred to references [33] and [34] or to the paper by Weynand et al. [35].

Table 6.6 Stiffness coefficients for the basic components of a steel connection

Component	Parameter	Stiffness coefficient*
Unstiffened column web panel in shear	k_1	$\dfrac{0.38 A_{vc}}{\beta z}$
Unstiffened column web in compression	k_2	$\dfrac{0.7 b_{eff,c,wc} t_{wc}}{d_c}$
Column web in tension	k_3	$\dfrac{0.7 b_{eff,c,wc} t_{wc}}{d_c}$
Column flange in bending	k_4	$\dfrac{0.85 l_{eff} t_{fc}^3}{m^3}$
End plate in bending	k_5	$\dfrac{0.85 l_{eff} t_p^3}{m^3}$
Flange cleat in bending	k_6	$\dfrac{0.85 l_{eff} t_a^3}{m^3}$
Bolts in tension	k_{10}	$\dfrac{1.6 A_s}{L_b}$

Note
*Definitions are given in EC3: Part 1.8 [13].

6.7.2 Stiffness of Composite Components

For the approach described above to be applied to composite connections it is necessary to identify the additional components that contribute to its stiffness and to derive simple expressions for calculating these stiffnesses. Research has shown that the components that have a major influence on the stiffness of a composite connection include:

- The longitudinal reinforcement.
- The deformation of the shear connection at the interface between the concrete slab and the steel beam.

Expressions for each of these components are developed in the following two sections.

6.7.2.1 Longitudinal Reinforcement

When a moment is applied to a composite connection a gap develops between the face of the column and the edge of the concrete slab. This gap is a result of the elongation of the longitudinal reinforcement. For a double sided connection under balanced moments the elongation of the reinforcement and therefore the stiffness of each connection is based on a length equal to half the depth of the column, $h_c/2$ and is given by the following expression:

$$k_{13} = \frac{2 A_l}{h_c}$$

where
A_l is the area of the longitudinal reinforcement
h_c is half the depth of the column.

This coefficient represents the component stiffness at the face of the connection. Most simple analytical models assume that the joint occurs at the intersection of the two connected members. When this type of nodal model is used the above stiffness must be modified by a transformation factor which compensates for the additional member flexibility introduced by the model. Extensive research has been carried out by Tschemmernegg *et al.* [36,37] and the following modified form of the above equation has been developed.

$$k_{13} = \frac{2A_l}{h_c(1 - k_{trans})}$$

where
k_{trans} is the transformation factor and is given by:

$$k_{trans} = \frac{A_l}{A_{l,beam} + 0.64\, t_{fb} b_b}$$

t_{fb} is thickness of the beam flange
b_b is the breadth of the beam flange.

In the case of unbalanced moments or a single sided connection the flexibility of the joint results in a gap developing between the face of the column and the concrete on the more heavily loaded side. To determine the initial stiffness for this type of connection Tschemmernegg *et al.* [37] developed a very sophisticated spring model. This model is shown in Figure 6.28. This model was used to develop a redirection factor, K_μ that modifies the stiffness component for the balanced case. For unbalanced moments the stiffness component is given by the following expression:

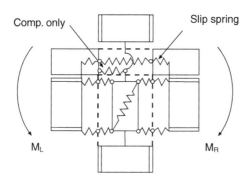

Figure 6.28 Tschemmernegg's spring model for a composite connection

For $M_{right} > M_{left}$

$$k_{11,left} = \frac{2A_l}{h_c(-0.5k_{trans})}$$

$$k_{11,right} = \frac{2A_l}{h_c(1 + K_\mu - 0.5k_{trans})}$$

where
K_μ is the redirection factor and is given by the following expression

$$K_\mu = \nu\mu \, (77\mu^2 - 13.8\mu + 8.9)$$

$\nu = 1.0$ for a composite slab
$\nu = 1.4$ for a solid slab.

6.7.2.2 Shear Connection

Aribert [32] recognized that slip between the interface of the concrete slab and the steel beam can have a significant influence on the stiffness of a composite connection. Furthermore, this effect can be accounted for either by partial-interaction analysis of the composite beam or by making an allowance in the stiffness model for the connection. As designers are unlikely to choose the partial-interaction analysis approach a simple method was developed by Aribert [32] which modifies the stiffness of the connection. In this method the stiffness of the shear connection, K_{sc}, is evaluated in the hogging moment region using the approach outlined in Section 6.6.2.2. This factor is then used to reduce the effective stiffness coefficient k_{13} for the reinforcement in tension.

6.8 SUMMARY

It has long been recognized that the strength and stiffness of traditional forms of composite construction can be enhanced if the connections at the ends of the beams are also designed to take into account the interaction between the concrete slab and the components of the steel joint. If this approach is adopted some of the benefits that can be realized include:

- reduced beam depths, which can facilitate service integration and
- reduced deflections at serviceability.

The behaviour of composite connections has been extensively examined over the past twenty years but the majority of this work has concentrated on composite connections between steel columns and composite beams with simple end-plate steel connections. Other forms of composite connection have been studied but the design methods are judged to be not sufficiently well established. The scope of this chapter is therefore limited to composite connections where the connection between the main steel members is an end-plate. However, many of the checks described are general and can be applied to other connection types.

In this chapter the behaviour of a composite connection is split into the following three main properties:

- strength
- ductility
- stiffness

and the influence that the type of connection, its components and the embedded reinforcement has on each of these properties is presented. Most of the design methods described are based on the recommendations given in Eurocodes 3 and 4. However, where alternative methods are available these have been included. From the discussion it is clear that the type, size and location of the additional reinforcement embedded in the concrete slab immediately above the steel joint has a significant influence on the connection's strength, stiffness and ductility.

For each of the above properties the main modes of failure and design checks are described in detail and photographs showing some of the modes of failure have been included to aid understanding. These checks are based on the latest research and practical aspects associated with current fabrication and erection techniques. Where appropriate the design methods are compared with available experimental data to demonstrate the validity and accuracy of the approach.

The successful performance of all composite structures depends as much on the connections as it does on the size and shape of the members. Therefore as designers move towards ever more economic structures they must be aware of the consequences of design decisions and the effect that connections details have on the force distribution and the overall behaviour of the structure. The details in this chapter are provided to help the designer understand the consequences of his/her decisions and to enable them to design safe and economic structures.

6.9 REFERENCES

1. Eurocode 3: Part 1.1: Design of Steel Structures: General Rules and Rules for buildings. ENV 1993-1-1, CEN, Brussels, 1992.
2. Eurocode 4: Part 1.1: Design of Composite Structures: General Rules and Rules for Buildings, ENV 1994-1-1, CEN, Brussels, 1994.
3. BCSA/SCI, Joints in Steel Construction—Composite Connections, The Steel Construction Institute, Publication Number 213, 1998.
4. BS 5950 : 1, Structural Use of Steelwork in Building—Part 1: Code of Practice for Design—Rolled and Welded Section, BS5950-1, BSI, London, 2001.
5. Wilson, W.M. and Moore H.F., 'Tests to determine the rigidity of rivited joints of steel structures', University of Illinois, Engineering Experimental Station, Bulletin 104, Urbana, USA, 1917.
6. Xiao, Y., Choo, B.S. and Nethercot, D.A., 'Composite connections in steel and concrete I. Experimental behaviour of composite beam-column connections', J. Constructional Steel Research, vol. 31, 1994, pp. 3–30.
7. Bijlaard, F.S.K. and Steenhuis, M., 'Prediction of the influence of connection behaviour on the strength, deformations and stability of frames by classification of connections'. Proceedings of the Second AISC and ECCS Workshop on Connections, Pittsburgh, USA, 1991, pp. 307–318.

8. Aribert, J.M., Lachal, A., Muzeau, J. P. and Racher, P. 'Recent tests on steel and composite connections in France', COST C1, Proceedings of the Second State of the Art Workshop, Prag, 1994.
9. Bode, H. and Kronenberger, H.J., 'Behaviour of composite joints and their influence on semi-continuous beams', Proceedings of the Engineering Foundation Conference Composite Connection III, Irsee, 1966.
10. Eurocode 3: Design of Steel Structures Part 1.8: Design of Joints, prEN 1993-1-8: 20xx, Preliminary draft, European Committee for Standardization, September 2000.
11. BCSA/SCI, 'Joints in Steel Construction—Moment Connections', The Steel Construction Institute, Publication Number 207/95, 1995.
12. Zoetemeijer, P., 'A design method for the tension side of statically loaded bolted beam-to-column connection'. Heron 20, No.1, Delft University, Delft, The Netherlands, 1974.
13. Bode, H., Ramm, W., Elz, S. and Kronenberger, H-J., 'Composite Connections—Experimental Results', Semi-rigid structural connections, IABSE Colloquium Istanbul, 1996.
14. British Standards Institution, BS 4449: 1988: 'Specification for carbon steel bars for the reinforcement of concrete', BSI, 1988.
15. British Standards Institution, BS 10080: 1996: 'Steel for the reinforcement of concrete. Weldable ribbed reinforcing steel B500. Technical delivery conditions for bars, coils and welded fabric', BSI, 1996.
16. Li, T.Q. 'The analysis and ductility requirements of semi-rigid and composite frames', University of Nottingham, Ph.D. Thesis, 1992.
17. Nethercot, D.A., Li, T.Q. and Choo, B.S., 'Required rotations and moment redistribution for composite frames and continuous beams', Journal of Constructional Steel Research, Vol. 35, No. 2, 1995, pp. 121–163.
18. Li, T.Q., Nethercot, D.A. and Lawson, R.M., 'Required rotation capacity of composite connections', Journal of Constructional Steel Research, Vol. 56, No. 3, Nov. 2000, pp. 151–174.
19. Couchman, G., 'Design of continuous composite beams allowing for rotation capacity', Thesis No. 1308 (1994), Lausanne, EPFL 1995.
20. Couchman, G. and Lebet, A., 'A new design method for continuous composite beams', Structural Engineering International, May 1996.
21. Anderson, D. and Najafi, A.A., 'Ductile steel-concrete composite joints', Composite Construction—Conventional and Innovative, Innsbruck 1997, Conference report.
22. Johnson, R.P. and Anderson, D., 'Designers' Handbook to Eurocode 4', Thomas Telford, London, 1993.
23. Xiao, Y., Choo, B.S. and Netercot, D.A., 'Composite connections in steel and concrete I: Experimental behaviour of composite beam-column connections', Journal of Constructional Steel Research, 31, 1994, pp. 3–30.
24. Aribert, J-M, Lachal, A., Muzeau, J-P. and Racher, P., 'Recent tests on steel and composite connections in France', COST C1 Semi-rigid behaviour of civil engineering structural connections. Proceedings of the Second State of Art Workshop, Wald, F. (ed.), Prague, 1994, pp. 61–74.
25. Ren, P. and Crisinel, M., 'Effect of reinforced concrete slab on the moment-rotation behaviour of standard steel beam-to-column joints: experimental study and numerical analysis', COST C1 Semi-rigid behaviour of civil engineering structural connections. Proceedings of the Second State of Art Workshop, Wald, F. (ed.), Prague, 1994, pp. 175–194.

26. Tschemmernegg, F., 'Deformation of semi-rigid composite joints', COST C1 Semi-rigid behaviour of civil engineering structural connections. Proceedings of the Second State of Art Workshop', Wald, F. (ed.), Prague, 1994, pp. 195–207.
27. Davison, J.B., Lam, D. and Nethercot, D.A., 'Semi-rigid action of composite joints', The Structural Engineer, 68(24), 1990, pp. 489–499.
28. COST C1, 'Composite steel-concrete joints in braced frames for buildings', Anderson, D. (ed.), published by the European Commission, Brussels, 1996.
29. Najafi, A.A., 'End plate connections and their influence on steel and composite structures', Ph.D. Thesis, University of Warwick, 1992.
30. CEB FIP Model Code, Design Guide 1990, Thomas Telford.
31. Hanswille, G., 'Cracking of concrete: mechanical models of the design rules in Eurocode 4', Composite Construction in Steel and Concrete, ASCE, New York, 1997, pp. 421–433.
32. Aribert, J.M., 'Influence of slip of the shear connection on composite joint behaviour', Third International Workshop on Connections in Steel Structures, University of Trento, May 1995.
33. Jaspart, J.P., 'Etude de la semi-rigidite des noeuds poutre-colonne et son unfluence sur la resistance et la stabilite des ossatures en acier', Ph.D. Thesis, University of Liege, Belgium, 1991.
34. Yee, Y.L. and Melchers, R.E., 'Moment rotation curves for bolted connections', Journal of the Structural Division, ASCE, Vol. 112, ST3, March 1986, pp. 615–635.
35. Weynand, K., Jaspart, J.P. and Steenhuis, M., 'The stiffness model of revised annex J of Eurocode 3', Connections in Steel Structures III, Behaviour, strength and design, Proceedings of the Third International Workshop held in Trento, May 1995, pp. 29–31.
36. Tschemmernegg, F. and Huber, G., 'Joint transformation and the influence for global analysis', Paper T6, COST/ECCS TC11 Drafting Group for Composite Connections, University of Innsbruck, 1995.
37. Tschemmernegg, F., Huber, G. and Pavlov, G., 'Tension region in the panel zone of a composite joint'. Paper T4 COST/ECCS TC11 Drafting Group for Composite Connections, University of Innsbruck, 1995.

CHAPTER SEVEN

Composite Frames

Graham H. Couchman

7.1 INTRODUCTION

7.1.1 What is a composite frame

Composite beams are widely used. Their benefits are recognised, design principles well understood, and authoritative design and construction best practice guidance is available. Composite connections are currently much less common, although we can expect to see this situation change with time now that authoritative guidance is available (SCI and BCSA 1998).

The combination of composite beams and connections with steel or (rarely) composite columns is what we shall define as a composite frame. The connections may provide full continuity between the beams and columns, in what is traditionally known as either continuous or rigid construction. Alternatively, they may only provide semi-continuity, which may result in a more economical framing solution.

Composite frames may be either braced or unbraced. In a braced frame the continuity between the beams and columns allows the sagging moments in the beams to be reduced (Figure 7.1). In an unbraced frame the primary role of the continuity between the members is to provide a means of resisting horizontal loading (Figure 7.2).

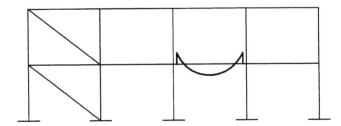

Figure 7.1: Role of continuity in a braced frame

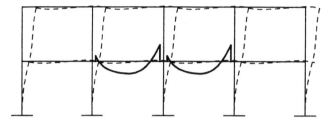

Figure 7.2: Role of continuity in an unbraced frame

7.1.2 Components and their role

Beams and connections

The nature of the beams in a composite frame is dictated by the use of composite connections to the columns. So that connections with reinforcement remote from the beam bottom flange (to provide adequate stiffness and strength) can be adopted, the beams must be of a 'traditional' downstand nature. Standard details are only currently available (SCI and BCSA 1998) for connections to in-situ concrete with steel deck composite slabs, but provided adequate reinforcement location and anchorage can be achieved there is no reason why precast planks could not also be considered. It will be appreciated later that appropriate connection performance is most easily ensured by imposing quite stringent detailing requirements, and the best way of satisfying these requirements is to adopt standard connections.

Although there are numerous composite beam options available for use in simple construction, connections have not yet been developed that would allow many of these to be incorporated in a continuous (or semi-continuous) composite frame. The importance of the detailing on the connection performance can be appreciated by referring to Figure 7.3. This figure shows typical (simplified bi-linear) moment-rotation responses for steel and composite connections. In addition to a response for a pinned connection, one curve indicates the performance that could be expected from a standard flush end plate bare steel connection, and a second curve indicates the performance that could be expected from a composite connection achieved by simply adding some slab reinforcement in the connection zone (whilst maintaining the same steelwork detailing). A line of constant moment on the figure indicates the moment resistance of the beam itself. The following information can be identified from the figure:

- It can readily be seen that the composite connection is both stiffer and stronger than the bare steel connection. This can only be achieved when sufficient reinforcement can be located and anchored in the slab. The use

of a steel deck based slab with in-situ concrete facilitates the incorporation of this reinforcement.
- Both bare steel and composite connections may be what is known as partial strength. This means that they have a lower moment resistance than that of the adjacent beam (which in the case of composite construction is the beam in hogging).

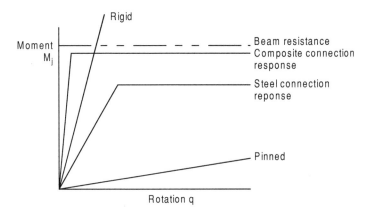

Figure 7.3: Simplified bi-linear moment-rotation curves for steel and composite connections

Comparisons of moment-rotation curves for connections to a number of different beams would also show that both the stiffness and strength of a connection increase with beam depth. This is true for bare steel and composite connections. Standard connections have not yet been developed that provide the levels of stiffness and strength that are required in an unbraced frame with relatively shallow steel beams, as might be used in slim floor construction for example.

It is also worth noting that the need to use downstand beams in a composite frame effectively precludes the use of deep decking. This may impose certain restrictions on the frame spacing that may be considered, which will be limited by the spanning capabilities of shallow decking.

In an unbraced frame the composite connections provide continuity, or 'frame action', between the beams and columns. In order to limit the horizontal sway of the frame under lateral loading the connections must possess a degree of stiffness. The importance of the connection stiffness may be appreciated by considering the fact that in the 'wind moment method', of which more later, the sway of a semi-continuous frame is assumed to be 50% greater than that of an equivalent continuous frame. Composite connections are generally 'rigid', and therefore relatively effective in limiting frame sway.

The (generally) partial strength nature of composite connections allows sufficient moment transfer to deal with the internal moments that arise due to the lateral loading on an unbraced frame. The partial strength of a connection is defined as its moment resistance divided by that of the adjacent beam. Composite connections typically possess high partial strength in hogging (80% or more of the beam strength). Care must be taken to avoid confusion over the definition of partial strength due to the fact that composite beams possess different resistances in hogging and sagging; the connections do not possess 80% of the considerably higher sagging resistance of the beams. It should also be noted that a composite connection in sagging, to which it may be subjected in an unbraced frame, is considerably less stiff and strong. The fact that the connections are generally partial, rather than full, strength has implications in terms of the moment redistribution that is needed to achieve an efficient composite frame design, as we shall see later.

In a braced frame moment resisting connections between the beams and columns attract hogging moment to the beam ends, thereby reducing the sagging moments that the beams must support. The use of connections that provide any reasonable degree of continuity, expressed in terms of the partial strength of the connection, can be beneficial. The stiffness of the connections is also beneficial in reducing beam deflections considerably. Clearly, however, the use of composite connections will allow more substantial reductions in sagging moments than when weaker bare steel connections are used. One of the great benefits of semi-continuous construction is that the beams and connections can be 'balanced'; savings in beam depth or weight can be weighed-up against connection costs to achieve an optimum solution for a given example. This is explained in more detail in Section 7.2.1.

Figure 7.4 shows a detail of a composite connection between a beam and column. It will be appreciated from this figure that the necessary attributes of stiffness and strength can be obtained without incurring high fabrication costs. The generation of tension forces in the reinforcing bars, which are remote from the connection's point of rotation, means that large moments can be generated without needing, for example, a costly haunch.

Columns

Composite columns are not common in the UK, although they have been used in situations where their particular attributes of high strength and relatively small size have proven beneficial. A particular issue with composite columns is the detailing of the connections to beams; standard composite connection details have not yet been developed for connections to filled hollow sections, and connections to the flanges of encased, or partially encased, open section columns require attention to the sequence of construction. Almost all composite frames in the UK adopt bare steel open sections, such as UCs, for the columns.

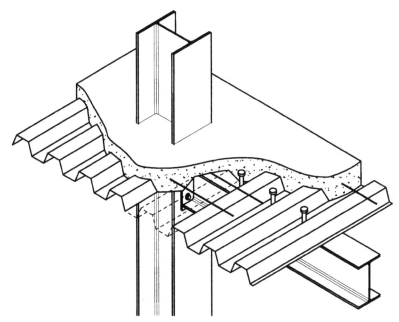

Figure 7.4: A composite beam-column connection

7.2 PRINCIPLES OF FRAME BEHAVIOUR

7.2.1 Braced frames

It should be noted that being braced does not necessarily make a frame 'non-sway'. Similarly, an unbraced frame is not necessarily 'sway'. A 'non-sway' frame is one for which the sway deflections are sufficiently small to make second order forces and moments negligible. Some configurations of bracing allow significant sway, so that the frame analysis must then consider second order effects; either explicitly or by using a simplification such as the amplified sway method (BS5950-1 1990). Alternatively, the design of the columns may allow for the second order effects by using effective lengths in excess of the system lengths. Guidance on how to determine whether a frame is 'sway' (which actually means sway sensitive) or 'non-sway' (not sway sensitive) is given in BS5950-1 (1990).

As discussed earlier, in a braced frame the use of partial strength connections is an economic way of reducing the sagging moments that the beams must resist. The economy results from the fact that the connections need only resist limited moments, and therefore costly fabrication can be avoided. The amount of moment to be transferred through the connections can be chosen to produce an optimum design. For example, a designer might wish to keep beam depth to a minimum in a building that is to be heavily serviced; minimising the structural floor depth maximises the depth available to accommodate the services. However, in order to minimise the beam depth it would be necessary to adopt composite connections with substantial reinforcement, and this would lead to increased connection costs. If minimising the floor depth was not particularly important, the use of weaker, less expensive connections would be more appropriate. One of the great benefits of semi-continuous construction is that the designer is not 'stuck' with one of the two extremes of simple or continuous construction, but is free to choose the degree of continuity that is most appropriate for a given frame.

Connections

A consequence of the connections possessing partial strength, rather than full strength, is that as load is applied to the frame the applied moments at the beam ends soon equal the moment resistance of the connections. The connections then plastify (assuming for simplicity that they possess an idealised elastic-plastic moment rotation response, as shown in Figure 7.3). As further load is applied plastic hinges may be assumed to form in the connections; the connections rotate, maintaining a constant level of moment, and redistribution takes place to increase the sagging moments in the beams. Hinges form in the connections rather than the adjacent beams because the connections are the 'weak link'. Moments in the beams in a braced frame are illustrated schematically in Figure 7.5. It will be appreciated from this figure that the amount of redistribution needed, particularly when the connections are relatively weak, is substantial (remembering that for a built-in beam the end moments are twice that at mid-span).

Connections that are assumed in the frame analysis and design to behave as plastic hinges must be physically capable of behaving in an appropriate manner; they must be able to rotate without loss of moment resistance. How much the connections in a frame need to be able to rotate is difficult to quantify. The amount of rotation required depends on variables such as the frame geometry, the nature of applied load (distributed or concentrated), and whether the beams are propped during construction. Quantifying required rotations has been the subject of a number of doctoral theses (Nethercot *et al.* 1995 and Couchman and Lebet 1996), but as yet no practical rules for everyday design have been developed. As an upper bound to the requirement, a connection capable of achieving 30 mrad of rotation may be deemed suitable to behave as a plastic hinge in a braced frame where the beams are designed assuming they achieve their plastic resistance moment $M_{pl.Rd}$. If the frame design is based on the attainment of less than the full plastic moment in the beams in sagging, then connection rotation demands are significantly reduced; the standard composite connections developed by The Steel Construction Institute and British Constructional Steelwork Association (SCI and BCSA 1998) were developed to provide enough rotation to allow sagging beam moments up to

$0.85M_{pl.Rd}$. The 15% reduction in beam moments allowed the connection rotation requirements to be reduced by around 50%, thereby permitting less restrictive detailing rules. When unpropped construction is adopted it may be necessary to limit the maximum stresses in the steel beams under construction loading to guarantee that the standard details are sufficiently ductile.

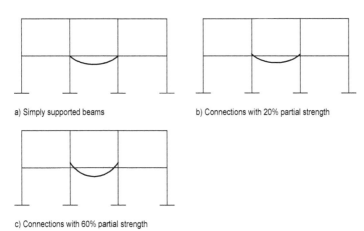

Figure 7.5: Beam moments in a braced frame

Various models have been proposed for determining how much connections are able to rotate, quantified in terms of available rotation capacity. The semi-empirical model developed by the COST initiative (European Commission 1996) may be considered to represent the current state-of-the-art and is based on a substantial number of tests. The available rotation capacity of a connection greatly depends on its detailing. The inclusion of a large amount of localised reinforcement is one of the most effective ways of achieving good rotation capacity. The reinforcement distributes cracking in the concrete, thereby reducing localised strains (that result in reinforcement failure) and allowing high 'average' strains to develop in the reinforcement. Appropriate connection detailing can be used to ensure that the average strain occurs over a substantial length, for example by placing the 'first' shear stud on the beam some distance (say 100 mm) from the column face. High strains over a substantial length result in large deformations of the reinforcement.

The reinforcement also helps to develop large compressive forces in the lower parts of the steel section, causing plastification which leads to large strains in the steel. Plastified regions of the steel section may also be assumed to extend over a significant length. Large tensile deformations in the reinforcement, and compressive deformations in the steel beam, equate to significant connection rotation (Figure 7.6). Figure 7.6 also shows that the slip which inevitably takes place at the steel-concrete interface helps in the attainment of high rotation capacity by reducing the strains that the reinforcement is subject to. In practice, by using standard connection details such as those developed by SCI and BCSA (1998), which are known to possess adequate rotation capacity, a designer can avoid the need to explicitly quantify the available rotation capacity of a given connection detail. It is worth noting that most tests measure rotation on the beam side of the connection. Any additional source of rotation on the column side, for example due to panel shear, may be beneficial in some applications.

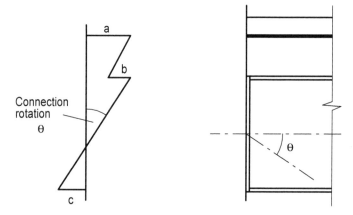

Deformations
a. Elongation of reinforcememt
b. Slip between steel beam and slab
c. Plastic compression of beam

Figure 7.6: Deformation and rotation in a composite connection

In order to permit the use of smaller beams than would be needed in simple construction, it is not sufficient for the connections to reduce the sagging moments applied to the beams. This would not address the issue of the greater flexibility that the 'smaller' beams would then possess. It is important that the connection strength is complemented by some stiffness; either 'semi-rigid' or 'rigid' details are needed. This stiffness provides rotational restraint at the beam ends and thereby reduces the beam deflections. Not a great deal of connection stiffness is needed to substantially reduce the beam deflections. Because a fully built-in beam deflects only one fifth as much as a simply supported beam, it is not necessary to move substantially away from a simple condition to gain significant benefits. Further information on the relationship between support restraint stiffness and beam deflections has been published by The Steel Construction Institute (Couchman 1997).

Composite Frames 211

There are, however, one or two complexities to be understood when considering the benefits offered by stiff connections. The first of these is a consideration of loading history. Figure 7.7 shows a beam with semi-rigid partial strength connections at both ends (these are represented in the figure by rotational springs). It is assumed that the beam is propped during construction and that, as will normally be the case, plastification of the connection occurs as gravity loading is applied. The figure also shows a plot of the moment-rotation characteristic for the left hand side connection. The following points plotted on this characteristic show the state of the connection:

1. after the application of dead load
2. after the first application of full imposed load
3. after the removal of this imposed load
4. after the re-application of some imposed load.

It will be appreciated from this simple, though representative, example that the connections only possess stiffness during the application of the dead load, and during general applications of imposed load. They do not possess stiffness during the first application of imposed load, at which stage the connection has attained its full moment resistance and is 'unwinding' as a plastic hinge with zero stiffness. Connection stiffness can therefore be relied upon to reduce beam deflections, and therefore occupant discomfort, during general applications of imposed load, but the designer should be aware that some initial rotations and deflections will take place.

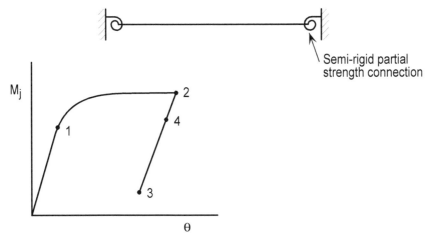

Figure 7.7: Effective connection stiffness at various stages of load application
(see text for key to points 1 to 4)

A further complexity when considering composite connections is that, if unpropped construction is adopted, the connections will be considerably less stiff during their initial (bare steel) state during construction. This will result in greater initial rotations and deflections than would occur with propped construction. For such cases it will be necessary to undertake a two-stage frame analysis, beginning with a model of the bare steel frame under construction loading, and using the rotations and deflections of this frame as a starting point for the analysis of the composite frame (with increased member and connection stiffnesses) under service loading.

Beams

The stiffness of a composite beam varies considerably according to whether the section is subject to hogging or sagging moments. In sagging the concrete slab is subject to compression (Figure 7.8a); design codes such as BS5950-3.1 (1990) give equations for calculating the stiffness of this 'uncracked' section, using the transformed section approach. For typical section sizes used in building applications the uncracked stiffness is up to 3.0 times the stiffness of the bare steel beam. In hogging the slab is subject to tension, and may be assumed to comprise the reinforcement alone (Figure 7.8b). Any stiffness the cracked concrete possesses is ignored for design purposes. The stiffness of the composite beam in hogging may be little more than that of the steel beam alone.

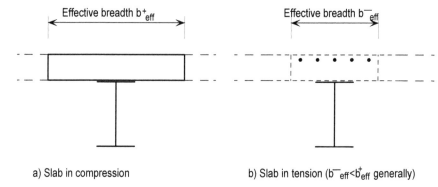

a) Slab in compression b) Slab in tension ($b^-_{eff} < b^+_{eff}$ generally)

Figure 7.8: Composite beam cross-sections a) sagging b) hogging

The substantial difference between the stiffness of a composite beam in hogging compared with its stiffness in sagging means that the relative lengths in hogging and sagging have a significant influence on the effective stiffness of a beam between its end points.

In a braced frame design the effective beam stiffnesses must be quantified in order to calculate beam deflections. Fortunately, this is not a particularly onerous task. The hogging moments at the beam ends may be taken as the connection strengths, and assumed not to vary. The sagging moments can be determined by considering the differences between the free bending moments and the hogging moments (Figure 7.9). Consideration of the geometry of the bending moment diagrams, which are parabolic for uniformly distributed loading (UDL), then allows the points of zero moment to be identified. Because the moment resistance of composite connections tends to be relatively small compared with that of the beams in sagging (less than 50%), the lengths of beam in hogging tend to be relatively small (typically less than 10% of the span at each end of the beam under UDL). For beams subject to concentrated loads the lengths in hogging may be greater, but can be easily determined from the straight line moment diagram. Knowing the lengths of beam in hogging and sagging it is a relatively easy task, for instance using a simple software model with elements of differing stiffness to represent the three regions (hog-sag-hog), to determine an effective stiffness for the beam.

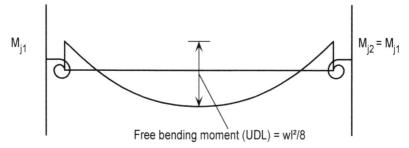

Figure 7.9: Distribution of hogging and sagging moments in a beam subject to UDL

Columns

In a continuous frame is it accepted practice to consider 'pattern loading' in order to identify the most onerous design condition. It is assumed that by removing imposed load from some of the beams the resulting unbalanced nodal moments may combine with the, admittedly reduced, axial forces to produce a critical combined load on the columns. How reasonable this assumption is will be discussed shortly, but its consequence is to increase the labour and complexity of continuous frame analysis considerably.

In a semi-continuous frame, here thinking about one with partial strength composite connections, the connections act as 'fuses' which can only transfer a limited amount of moment into the columns. The fact that the connections reach their moment capacity quite quickly as load is applied, and thereafter act as plastic hinges, has already been discussed.

For propped construction, in which the connections behave compositely under both dead and imposed loading, it may reasonably (and if anything conservatively) be assumed that the connections will reach 75% of their moment capacity under dead load alone. This assumption means that it is not necessary to analyse the frame explicitly under pattern loading in order to establish values of axial load and moment to be combined in a column check. The moment can be determined by considering the difference between the connection strengths either side of a column node, with one of these strengths simply reduced by 25%. This principle is illustrated in Figure 7.10, which shows the moment applied to a column for propped construction.

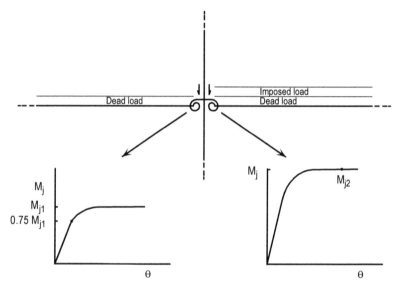

Figure 7.10: Column moments – propped construction

For unpropped construction it may be assumed that the bare steel connections reach their full moment capacity under dead load alone. Unbalanced moments due to pattern loading are then given by the difference between the bare steel connection capacity on one side of a node (the side which is assumed to be under construction) and the full composite connection capacity on the other side. This principle is illustrated in Figure 7.11.

The supposed need to consider pattern loading in order to identify the critical combination of axial load and moment on a column arises because an elastic analysis can show cases where removal of imposed load from some beams is detrimental in terms of the supporting column. According to BS5950-1 (1990) the unbalanced moment that arises at a node should be distributed between the members (normally two beams and an upper and lower length of column) meeting at that node. The amount of moment distributed to each member depends on its relative stiffness (EI/L). However, a real, as opposed to ideal elastic, column looses stiffness as it begins to fail in overall buckling. This loss of stiffness means that the column does not attract the moment predicted by an elastic analysis, which is rather shared between the other members meeting at the node. It may therefore be that the supposed danger of pattern loading is merely the result of over-simplistic assumptions. Considerable analyses and testing (Gibbons et al. 1993) have shown that the provision of beam connections that are capable of transferring moments may in fact benefit a column by providing a means of restraint once it starts to buckle, rather than hinder it by applying moment.

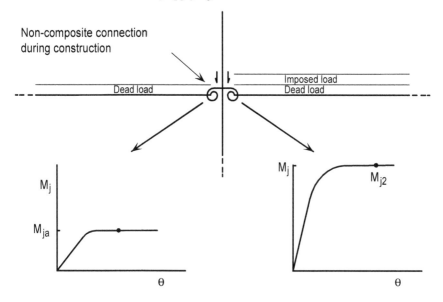

Unbalanced moment applied to column = $M_{j2} - M_{ja}$

Figure 7.11: Column moments – unpropped construction

A final point concerned with braced frames is the question of effective lengths that should be used when designing the columns. Composite connections, being rigid, may allow columns to be designed assuming an effective length factor that is less than 1.0. However, it should be remembered that the connections in their bare steel state, during construction, will only provide semi-rigid restraint and therefore a longer effective length may need to be considered.

7.2.2 Unbraced frames

In an unbraced frame the use of semi-rigid, partial strength connections to create semi-continuity is an economic way of resisting lateral loading by 'frame action'. The beams and their connections must be sufficiently strong and stiff to restrain the columns and prevent excessive frame sway. It will be appreciated from Figure 7.12, which shows schematically how the beams and columns deform and connections rotate as a frame sways, that the stiffness of all three of these frame components is important; the beams and columns must resist double curvature bending and the connections must resist rotation. It can also be appreciated from this figure that some base fixity is important, particularly given that first storey sway is often critical.

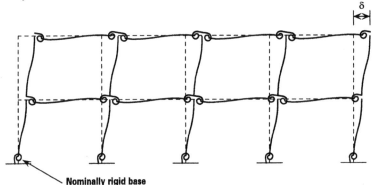

Figure 7.12: Deformations that take place as an unbraced frame sways

Economy can be achieved because the levels of strength and stiffness that are needed from the beam-to-column connections do not necessarily demand full continuity. Relatively simple composite connection details with low levels of fabrication can provide sufficient semi-continuity.

Connections

The behaviour of the connections is clearly very important to the behaviour of the frame. This relationship can best be understood by initially considering a simple model for the connection behaviour, and then having established a certain level of understanding addressing the implications of the simplifying assumptions. Figure 7.13 shows part of an unbraced frame, and includes a schematic moment-rotation response for two of the connections in that frame. Assuming that the beams are first subject to full dead and imposed loading, the partial strength connections may be assumed to have reached their moment capacity before the wind blows. They will be on the plateau of their moment-rotation response, as shown by the points A on the responses in the figure. If it is assumed that the wind subsequently blows

Composite Frames 217

from the left, the leeward connection (1) will unload, and will do so along the line A_1-B_1. The connection will therefore possess some stiffness (equal to its initial stiffness), and play a role in preventing the frame falling over due to the wind loading. The windward connection (2) will, however, continue to 'load', doing so along line A_2-B_2. This connection will not therefore possess any stiffness, and not contribute to the frame stability. When the wind subsequently blows from the right this connection situation will be 'reversed'; the windward connection will unload from point B2, the leeward connection will load along the line B1–A1.

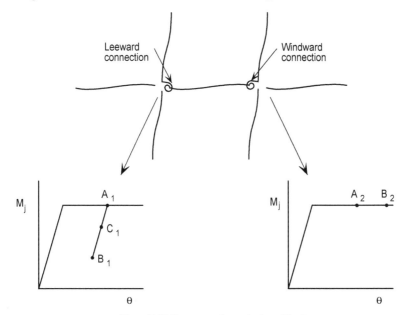

Figure 7.13: Frame swaying under lateral loads

Real frame behaviour is considerably more complex than suggested by the simple model considered in the previous paragraph. Levels of imposed load vary as specific loading is applied and removed, causing the connections to unload and reload. Wind loads vary in both magnitude and direction and this causes continual loading and unloading of the connections. The result of all this loading and unloading is that the frame will be subject to a phenomenon known as 'shakedown' An excellent, detailed description of the 'shakedown' of an unbraced frame has been published by Hughes (1999). The most important thing for a designer to appreciate about 'shakedown' is simply that the exact state of rotation (and associated moment) of the connections cannot be accurately predicted. All that can be said for certain is that the level of moment in the connections will lie somewhere between the limits of hogging and sagging capacity, and that there will be some residual rotation. Point C_1 on the moment-rotation curve for connection 1 might represent such a shakendown condition. Extensive studies (Hensman and Nethercot 2001, Brown *et al.* 1999) indicate that it is not necessary to take 'shakedown' into account in order to obtain a reasonable prediction of frame behaviour; a simple model can provide reasonable results. Designers should not therefore be overly worried by this complicated and apparently unpredictable phenomenon.

The connection behaviour itself is also considerably more complex than the simple model considered above. The use of substantial amounts of reinforcement to achieve adequate ductility means that composite connections possess reasonable levels of both strength and stiffness when subject to hogging moments (i.e. when the slab is in tension). However, when the slab is in compression composite connections may have strengths and stiffnesses which are only slightly in excess of those of the bare steel connection on which they are based. This 'unsymmetric' behaviour (Figure 7.14) has implications on the use of composite connections in unbraced frames; horizontal (wind) loads must be small relative to the vertical loads to ensure that variations in the applied connection moments (depending on the direction of the wind) are not excessive. The lack of 'resistance symmetry' possessed by the connections is not a problem provided that the range of applied moments does not exceed the resistance range (the absolute sum of the hogging and sagging resistances); because of 'shakedown' the effective origin of the moment-rotation response cannot be accurately predicted anyway. This limitation means that composite connections should not be used in high rise unbraced frames.

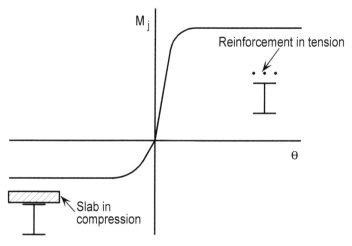

Figure 7.14: Moment resistance of a composite connection in hogging and sagging

Composite Frames

The 'frame-action' that an unbraced frame relies upon to resist horizontal loading is achieved by transferring moments between the beams and columns (Figure 7.15). In order to achieve this moment transfer it is necessary for the slabs to bear against the 'back' faces of the columns, and the lower beam flanges to bear against the 'front' faces. Limitations on a slab's ability to bear against the columns may govern the amount of moment that can be transferred. Although the concrete is confined and can therefore achieve quite high levels of stress, it only bears against a relatively small area of column flange. Appropriate design and detailing guidance has been produced by SCI and BCSA (1998).

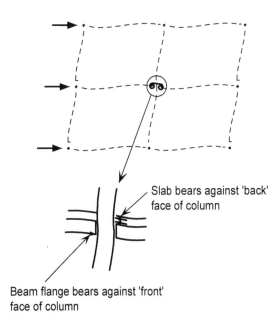

Figure 7.15: Mechanism for transfer of moments from beams to columns in an unbraced frame

Beams

For the design of an unbraced frame the stiffness of the beams must be quantified for several reasons:

- In order to determine the vertical beam deflections
- In order to calculate the horizontal sway of the frame – the stiffer the beams the more restraint is provided to the columns, which therefore sway less

- In order to calculate the distribution of moments around the frame – unless a simplified approach (such as the wind moment method, which is described later) can be adopted, the distribution of moments must be determined as a function of the relative stiffnesses of the frame members.

A problem with unbraced frames lies in establishing the lengths of beam in hogging and sagging. Because the beams and columns in an unbraced frame must resist horizontal loads as well as vertical loads, the connections load and unload as wind loads vary in magnitude and direction. This means that the hogging moments at the beam ends vary, and the lengths of beam in hogging also vary (Figure 7.16). As a result of these variations in moment distribution the effective stiffnesses of the beams also vary.

Bending moment diagram for a beam subject to UDL plus wind load from the left

Bending moment diagram for a beam subject to UDL plus wind load from the right

Figure 7.16: Variations in bending moments as a function of wind load direction

Despite this apparent complexity, simplified models to predict an acceptable constant effective beam stiffness that may be used in analyses have been developed:

- According to Leon *et al.* (1996), the effective second moment of area may be derived from those of the uncracked (I_1) and cracked (I_2) sections

  $I_{eff} = 0.4I_2 + 0.6I_1$ (when there is semi-continuity at both ends)
  or
  $I_{eff} = 0.25I_2 + 0.75I_1$ (when there is semi-continuity at one end)

- According to Hensman and Way (1999) a slightly more complicated formulation gives greater accuracy:

  $I_{eff} = (7.5I_1I_2^2)/(9I_2^2 + 2I_1I_2)$

Columns

The design of the columns in an unbraced frame must take into account so-called second order effects due to geometric non-linearity. These are also known as P-delta effects; a column node which sways by an amount delta will lead to a secondary moment in the column equal to the axial force P times the lever arm delta. Such second order effects may be allowed for in one of three ways:

- explicitly in a second order analysis (some software offers this capability, or an iterative procedure could be implemented)
- using a simple method to increase the moments and forces due to the first order loading
- using a simple method to decrease the resistance of the columns

The Amplified Sway method described in BS5950-1 (1990) provides an easy way of increasing the applied moments and forces to represent second order effects. The increase is simply a function of the frame's sway stiffness, which is quantified in terms of its sway. This method may be applied to 'orthodox' frames that possess a reasonable degree of sway stiffness.

A simple method of artificially decreasing the resistance of the columns, so that this resistance can be compared with the 'known to be too small' first order applied moments and forces, is to consider an effective length in excess of the system length. Procedures for determining appropriate values of effective length are fully described in BS5950-1 (1990). The length is a function of the restraint provided to the column by adjoining members. The so-called Wind Moment Method, which is described is Section 7.3.4, uses this approach for dealing with second order effects, and as a simplification adopts an effective length 1.5 times the system length, regardless of the frame details.

7.3 FRAME ANALYSIS AND DESIGN

In this section the analysis and design of braced and unbraced composite frames will be discussed. The frame at both the construction (steel) and final (composite) stages will be considered. Key points are discussed, including some in-depth guidance, but in the interests of concision reference is made to existing guidance where appropriate.

7.3.1 Braced frames

Construction condition

The guidance given below relates to frames in which the beams are unpropped during construction, so that the construction loads are applied to bare steel beams and connections. This is the most common practice for composite frame construction due to the speed and therefore economic benefits of the site process.

Composite connections are generally based on relatively thin flush end plate steelwork details. Such details are appropriate because:

- The lower part of the end plate provides a direct load path for the often substantial compression loads to transfer from the lower beam flange into the column
- The thin end plate behaves in a ductile manner, so that the connection can achieve high rotations without failure of any components
- The availability of capacity tables for standard details greatly simplifies the design process.

In their bare steel state such flush end plate connections are 'semi-rigid' and 'partial strength'. Typically their moment resistance is 40 to 50% of that of the adjacent bare steel beam. The frame must therefore be analysed under construction loading recognising these attributes of semi-rigidity and partial strength. They lead to what is known as a semi-continuous frame; the connections are neither stiff enough nor strong enough to provide full continuity between the beams and columns. The frame may be considered to be statically determinate if plastic hinges are assumed to form in the connections, so the beams and columns can be considered as separate members.

Hogging moments in the connections, equal to their strength, should be allowed for when determining the sagging moments in the beams. Depending on the construction details it may be necessary to design the beams considering them to be laterally unrestrained. Columns should be designed for unbalanced moments when the opposing connections at a node are of unequal strength. It has been demonstrated by both testing and analysis that unbalanced moments due to 'pattern loading' do not occur in a semi-continuous frame, as discussed earlier (Couchman 1997). The stiffness of the semi-rigid connections may be taken into account when calculating the beam deflections. For typical frames the deflections will be at least 30% less than those of simply supported beams (Couchman 1997). It is important to limit the deflections so that 'ponding' of the concrete is not excessive. 'Ponding' results in additional concrete being placed as the beams deflect. Design standards such as BS5950-3.1 (1990) recommend different deflection limits depending on whether or not 'ponding' is explicitly allowed for.

Full analysis and design procedures for a semi-continuous braced steel frame have been published by The Steel Construction Institute (Couchman 1997).

Final (composite) condition

The frame in its final condition remains semi-continuous. Although composite connections are generally 'rigid', full continuity cannot be achieved because they remain 'partial strength'. The composite frame may therefore also be analysed considering the beams and columns separately, with plastic hinges at the connections (Figure 7.17). Hogging moments equal to the connection strengths should be allowed for when determining the sagging moments applied to the beams.

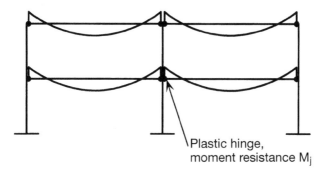

Figure 7.17: Moments in a braced composite frame

Column design is slightly more complicated than at the construction stage. The reasons why pattern loading may be ignored remain at least partly valid for the composite frame, if less justified by extensive testing and finite element modelling. The higher moments that can be transferred by the composite connections into the columns do however make it worthwhile considering other models. The connections may be pictured as 'fuses', which cannot transfer a level of moment greater than their capacity. Some simple rules can then be derived for determining the level of unbalanced moment that may be applied to an internal column, as discussed in Section 7.2.1.

The 'rigid' nature of the composite connections allows substantial reductions in the beam deflections compared with a simple frame; only bending of the supporting columns and adjacent beams will lead to deflections greater that those for a fully built-in condition. The contributions to deflection due to these members bending should be determined using standard procedures (Couchman 1997).

7.3.2 Procedures – braced frames

In the section which follows the key stages in the design procedures for a braced frame are given. The order in which the design checks should be considered in order to minimise wasted effort may vary according to spans, relative loading at different stages etc. It is anticipated that the practitioner will become familiar with the most effective order for a particular frame from experience. It is also worth emphasising that the most cost effective frame design is generally not the one that uses the least weight of steel; for example when column sections are being considered it may be less expensive to use a heavier section in order to avoid stiffening (CIMSteel 1995).

Construction condition

1. Size beams and connections for strength

    - Calculate the free bending moment (FBM) applied to a beam
    - Assuming that the connection will have a partial strength equal to 40% of the beam moment resistance, choose a beam with a moment resistance M_s equal to FBM/1.4 for an internal span with equal strength connections at either end. For an end span with a weaker connection at the perimeter column end the value 1.4 would reduce (Figure 7.18)

    A reduced beam resistance should be considered if the top flange is not laterally restrained to prevent lateral torsional buckling, although restraint is normally achieved by the provision of adequate deck fixings (Couchman *et al.* 2000).

    - Confirm details of a standard connection (SCI and BCSA 1998) that will provide at least 40% of the moment resistance of the chosen beam. If a stronger connection is chosen it might be possible to reduce the beam size in an iterative process.

a) Internal span   b) External span

Required beam moment resistance is M_a (construction stage)

Figure 7.18: Hogging, sagging and free bending moments in a beam (a) internal span (b) external span

2. Size columns for strength

    Note: This check will only be required in exceptional circumstances because the columns will normally remain non-composite at the final stage, at which time the applied loads will be considerably greater. Any column that is acceptable in service should therefore be adequate during construction.

    - Estimate a column size and check for overall buckling under combined axial load and unbalanced moment. Unbalanced moment will only occur if there is a one-sided connection, or opposing connections are of different strengths.
    - Confirm that the local capacity of the column is sufficient for any combined loading. For this check it is necessary to consider unbalanced moment due to pattern loading, although because the check is unlikely to be critical simplifying assumptions may be made (SCI and BCSA 1998).

3. Check beam deflections

    - Calculate beam deflections recognising the (initial) semi-rigid nature of the connections (Couchman 1997) and any permanent rotation that takes place as the connections reach their plastic limit.
    - Check beam deflections against agreed limits. Suggested values are given in numerous design codes and publications, for example Couchman *et al.* (2000), but the exact values should be chosen to suit the details of an individual structure.

Final condition

4. Size composite beams and connections for strength

    - Calculate the FBM applied to a beam.
    - Assuming that the composite connection strength in hogging will be α times the moment resistance of the steel beam (with $\alpha = 1.1$ for S275 beams and 0.9 for S355) from the FBM determine the sagging moment that the composite beam must resist (Figure 7.19).

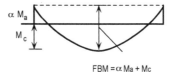

FBM = α Ma + Mc

α is a composite connection factor
   (1.1 for S275 beams or 0.9 for S355 beams)
Ma is the moment resistance of the bare steel beam
Mc is the required moment resistance of the
   composite beam in sagging

Figure 7.19: Hogging, sagging and free bending moments in a composite beam, internal span

    - Confirm that the steel beam size dictated by the construction condition is reasonable. This can be done using the Table 7.1, which relates the moment resistance of a typical composite beam to that of the steel section on which it is based.

Table 7.1: Ratios of moment resistances

Nominal depth of steel section	Slab depth	Ratio of composite: steel moment resistances (sagging)
300	80	1.7
	100	1.9
	120	2.1
400	80	1.5
	100	1.7
	120	1.9
500	80	1.4
	100	1.6
	120	1.7
600	80	1.4
	100	1.5
	120	1.6

- Carry out detailed design of the composite beam in sagging, choosing the decking, shear connector layout, transverse reinforcement etc. Remember that sufficient studs must be placed between points of zero and maximum sagging moment.
- Confirm that a standard composite connection detail is available that will provide the necessary strength. Iteration between beam (thinking about sagging resistance) and connection choices may be possible in order to give the most effective 'combined' resistance.
- Specify longitudinal reinforcement for the beam in hogging that will ensure the resistance is not less than say 1.25 times the connection moment resistance.

5. Size columns for strength

- Check assumed column sizes for overall buckling under combined axial load and unbalanced moment. Unbalanced moment may occur if there is a one-sided connection, opposing connections are of different strengths, or because pattern loading is present. For propped construction it may be assumed that under pattern loading the connection to the beam that is only subject to dead load will have attained 80% of its moment resistance. This should be balanced against an assumed 100% on the other side. If unpropped construction is adopted then under pattern loading one of the

connection strengths should be assumed equal to that of the steel connection alone, the other equalling its full composite resistance.
- Confirm that the local capacity of the column is sufficient. This is unlikely to be critical as in general the overall buckling resistance will be less than the local capacity.

6. Check beam deflections

- Under repeated applications of imposed load the connections will load and unload with a stiffness that may be taken as equal to that of their initial response. This stiffness will reduce the beam deflections considerably compared with a simple condition.
- Permanent rotations of the connections will take place on the first application of dead and imposed load as they first behave inelastically. This permanent rotation may need to be allowed for, depending on what is being checked (for example if deflections need checking to ensure that brittle components such as glazing are not damaged on the first application of imposed load).

7.3.3 Unbraced frames

Construction (steel) condition

The guidance given below relates to frames in which the beams are unpropped during construction and temporary bracing is not used to ensure frame stability. In its construction state a composite frame is semi-continuous due to both the 'semi-rigid' and 'partial strength' nature of the steel connections. Semi-continuity between the beams and columns makes an unbraced frame very difficult to analyse.

Unlike a braced frame it is not possible to assume at the ultimate limit state that the connections are carrying moments equal to their resistances, because as wind loads vary the connections will load and unload (see Section 7.2.2). The stiffnesses of the connections also vary with the loading. The frame analysis must therefore incorporate:

- connection stiffnesses that differ from those of the adjacent beams
- connection stiffnesses that vary according to the loading
- checks that elastic limits are not exceeded, which indeed may happen in the partial strength connections at relatively low levels of load.

Non-linear software may be used to include the connection moment-rotation characteristics explicitly. Software with this potential is, however, not currently available to most practitioners. Alternatively it may be possible to improvise using linear elastic software, albeit at the expense of considerable complexity (see Section 7.4.2). A third possibility for analysing and designing an unbraced steel frame is the so-called wind-moment method. However, the use of the method has only been justified for frames which satisfy certain conditions relating primarily to loads and geometry (Salter *et al.* 1999). Construction loading is outside the recognised scope of the method.

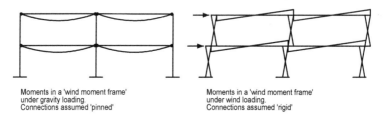

Figure 7.20: Assumptions of the wind-moment method

Final (composite) condition

Once the concrete reaches sufficient strength, and the frame may be considered to be composite, the rigidity of most composite connections means that at low levels of loading it can be analysed as a continuous structure. Clearly however there remains a need to check that none of the partial strength composite connections has reached its capacity, in which case an elastic analysis becomes invalid. Once some of the connections have started to behave in an inelastic way the analysis model must reflect these localised areas of substantially reduced stiffness.

Analysis options using software are as for the bare steel frame during construction; non-linear software that can model the connections explicitly, laborious and complex improvisation with linear software. To these options should now be added the wind-moment method. This method allows for the connection behaviour by assuming that they behave as nominal pins under gravity loading and as rigid connections under horizontal loading (Figure 7.20). Clearly the connections are not in reality 'schizophrenic', and able to modify their behaviour depending on which way the load is acting, but extensive finite element modelling has proven the validity of these apparently rash assumptions for quite a broad range of orthodox composite frames (Hensman and Way 1999). The method allows for second order effects by reducing the column resistances (using an effective length factor of 1.5). It allows for the flexibility of the composite connections that have gone beyond elastic behaviour by using a multiplier to increase sway deflections calculated for a rigid frame. Extensive details of the use and range of application of the wind-moment method for composite frames have been published by The Steel Construction Institute (Hensman and Way 1999), which includes the scope of frame types for which the method has been shown to be acceptable.

7.3.4 Procedures – unbraced frames

As described earlier the only practical way to analyse and design an unbraced semi-continuous frame under construction loading may be by using software. At the final, composite stage, it is possible to use either software or, for frames which are relatively orthodox (and therefore satisfy some scope limitations), the wind-moment method may be used. The wind-moment method is the only practical option for analysing and designing unbraced semi-continuous frames by hand. Extensive guidance on the use of the wind-moment method for composite frames is given by Hensman and Way (1999). The procedure is as follows:

1. Check that frame geometry and loading complies with the scope for which the wind moment method has been validated (Hensman and Way 1999).
2. Determine shear forces, bending moments and axial forces due to wind loads and notional horizontal loads.
3. Design composite beams for the ULS
   - Checking both hogging and sagging moment regions
   - Under gravity loading assume a nominal 10% end fixity due to the connections.
4. Check the composite beams are satisfactory for the SLS
   - Allow for increased deflections due to partial shear connection if necessary
   - Ignore the benefits of end restraint from the connections.
5. Size columns for the ULS
   - Use effective lengths of 1.5L for major axis bending (in the plane of the wind-moment frame) and 1.0L for minor axis bending (a limitation of the method is that nodal points must be held in position laterally by bracing).
6. Size connections for the ULS
   - Use capacity tables for standard connections (SCI and BCSA 1998) to compare with the applied moments for the various load combinations
   - Ensure that the areas of reinforcement required are within acceptable limits (SCI and BCSA 1998).
7. Check the horizontal sway of the frame
   - Calculate sway assuming rigid connections (i.e. a fully continuous frame)
   - Apply a sway amplification factor to allow for the connection flexibility.

7.4 DESIGN USING SOFTWARE

7.4.1 Braced frames

Software that could 'automatically' design a braced composite frame, which by definition would normally be a frame with semi-continuity between the beams and columns, would need the ability to model the partial strength attributes of the connections. This might be achieved by inputting non-linear moment-rotation curves for the connections, based on test results, or allowing for a simplified bi-linear connection response.

Software capable of modelling the frame during construction would need a similar capability, the only difference being a need to recognise the semi-rigid (rather than nominally rigid) initial part of the connection response.

Although software which is capable of considering non-linear connections currently exists, it is not generally available to practitioners. Some inaccuracies would also be expected if the software was not able to accurately model the load-unload behaviour of the connections (see Section 7.2.1). The author is not aware of any current software that offers this capability. Failure to model unloading accurately, whilst certainly not catastrophic, would lead to some overestimation of beam deflections on the 'nth' application of imposed load (which may be of interest in terms of occupant comfort).

The analysis model would also need to recognise the difference between the composite beam stiffnesses in hogging and sagging. This could be done using simplified effective stiffnesses for the complete spans (see Section 7.2.1).

For a designer 'limited' to using linear elastic software the semi-continuity of the frame could be allowed for in the following way:

- For checking the frame at the construction stage, model the semi-rigid nature of the connections (provided their stiffness is known) using dummy beam elements. Each dummy beam would be chosen so that the initial stiffness of the connection was equal to $EI/2L$ with E taken as the default steel value for simplicity, and L chosen as a nominal value such as 100 mm. An appropriate value of I can be readily determined (Figure 7.21).

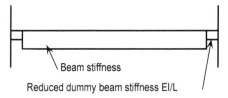

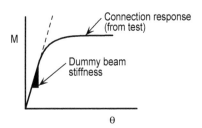

Figure 7.21: Modelling connection semi-rigidity using dummy beam elements

- For both the construction and final stage check it will also be necessary to monitor the connection moments as load increases. When the connection moment resistance is reached at any point the structural system should be changed by introducing a pin, with equal and opposite moments (equal to the connection strength) applied to the members either side of the pin. These measures model the loss of stiffness as the connection plastifies, but recognise its retained strength.
- For checking beam deflections on the 'nth' application of imposed load the dummy beams should be considered in order to give the right end restraint stiffness.

7.4.2 Unbraced frames

Software that could 'automatically' design an unbraced composite frame would need the ability to:

- Model the partial strength attribute of the connections (as for a braced frame).
- Allow for second order effects (geometric non-linearity as the frame sways).

Software capable of modelling the frame during construction would need a similar capability, the only difference being a need to recognise the semi-rigid (rather than nominally rigid) initial part of the connection response.

Software with these capabilities is not generally available to practitioners. Also, as stated above, software is not available which models the load-unload response of the connections accurately. An inability to do this, and model the associated 'shakedown', will affect frame sway predictions, although studies suggest the analysis may not be overly sensitive to this simplification (Hensman and Nethercot 2001).

As for braced frames the analysis model would also need to recognise the difference between the composite beam stiffnesses in hogging and sagging. For an unbraced frame the relative lengths in hogging and sagging vary with the applied load, further complicating predictions of beam stiffness. However, studies suggest that a simplified effective stiffness for the complete span may be used (see Section 7.2.2).

Including a realistic model for the base fixity is of importance for an unbraced frame because first storey sway is often critical. Guidance has been published by The Steel Construction Institute (CIMSteel 1995).

For a designer 'limited' to linear elastic software the following procedure is suggested to allow for the semi-continuity of the frame:

- Assume that the vertical loads are applied to the frame before the horizontal loads.

- Assume that the windward connections continue to load as the wind blows (with increased rotation but no change in moment), and that the leeward connections unload along a straight line parallel with the initial part of the moment-rotation curve.
- Model the windward connections as pins to represent zero stiffness, applying equal and opposite moments (equal to the connection strength) to the adjacent members to represent the connection resistance.
- Check that the range of applied moments in each connection due to wind loading does not exceed the sum of the hogging and sagging resistances of the connection. If this condition is not satisfied then under some load combinations both leeward and windward connections might be plastic, with negligible stiffness, so the frame would fall over. Ranges and sums should be compared, rather than looking at hogging and sagging in isolation, because 'shakedown' means that the 'starting point' is unknown (although it may be assumed that the frame will reach a suitable point).
- Allow for geometric non-linearity. This can be done using one of the accepted simplified methods, for example the amplified sway method (BS5950-1 1990) to increase forces and moments, or considering effective lengths greater than the system lengths to reduce column resistances.

For analysing the frame at the construction stage it will also be necessary to model the (semi-rigid) leeward connection stiffness using dummy beam elements (may only be necessary for the construction check), unless temporary bracing is used to stabilise the frame.

Clearly the analysis model would also need to allow for varying beam stiffnesses and the restraint offered by base fixity.

7.5 CONCLUSIONS

Making steel beams act compositely with the concrete slab they support is well recognised in terms of the stiffness and strength improvements that can be achieved compared with a non-composite solution using the same basic components. These benefits can be increased when the beams are used in conjunction with composite connections to form a composite frame. The columns in such a frame are generally bare steel, indeed the use of composite columns would complicate the connection detailing. The connections are used either to reduce the sagging moments in the beams, when the frame is braced, or to achieve frame action in order to resist horizontal loading on an unbraced frame.

In this chapter the principles of composite frame behaviour have been identified. Consideration has been given to the need for the frame to resist construction loading when in its bare steel state.

Step-by-step outline procedures have been given for the design of both braced and unbraced frames. These procedures would allow a designer to combine existing, generally codified, guidance to design a braced frame. Unbraced frame design is considerably more complex. The Wind-moment Method may be used as a means of effectively avoiding this complexity by applying simple assumptions and rules that have been shown to be valid for frames satisfying certain conditions. The design of unbraced composite frames that are outside the scope of the Wind-moment Method practically relies on the use of software. The attributes that appropriate software would need to possess have been identified in this chapter. Unfortunately, commonly available (linear elastic) software is not sufficient for the 'automatic' design of an unbraced composite frame. However, guidance has been included on how this type of software could be used in an interactive way so that the true stiffness and partial strength of the connections can be allowed for in the frame design.

7.6 REFERENCES

BS5950-1, 1990, *BS5950: Structural use of steelwork in buildings Part 1: 1990 Code of practice for design in simple and continuous construction: hot rolled sections*, BSI

Nethercot D A, Li T Q and Choo B S, 1995, *Required rotations and moment redistribution for composite frames and continuous beams*, Journal of Constructional Steel Research, Vol 35, No 2

Couchman G and Lebet J-P, 1996, *A new design method for continuous composite beams*, Structural Engineering International, May 1996

SCI and BCSA, 1998, *Joints in steel construction: Composite connections (P213)* The Steel Construction Institute and The British Constructional Steelwork Association

European Commission, 1996, COST C1 *Semi-rigid behaviour of civil engineering structural connections, Composite steel-concrete joints in braced frames for buildings*, EC

Couchman G H, 1997, *Design of semi-continuous braced frames (P183)*, SCI

BS5950-3-1, 1990, *BS5950: Structural use of steelwork in buildings Part 3: Design in composite construction, Section 3.1: 1990: Code of practice for the design of simple and continuous beams*, BSI

Gibbons C, Nethercot D A, Kirby P A and Wang Y C, 1993, *An appraisal of partially restrained column behaviour in non-sway steel frames*, Proceedings of the Institution of Civil Engineers, Structures and Buildings, February 1993

Hughes A F, 1999, *A fresh look at the wind moment method*, The Structural Engineer, Vol 77, No 16, 17 August 1999

Hensman J S and Nethercot D A, 2001, *Design of unbraced composite frames using the wind-moment method*, The Structural Engineer, Vol 79, No 11 5 June 2001

Brown N D, Anderson D and Hughes A F, *Wind-moment steel frames with standard ductile connections*, Journal of Constructional Steel Research, Vol 52

Leon R T, Hoffmann J J and Staeger T, 1996, *Partially restrained composite connections: A design guide*, AISC

Hensman J and Way A, 1999, *Wind-moment design of unbraced composite frames (P264)*, SCI

CIMSteel, 1995, *Design for manufacture guidelines*, SCI

Couchman G H, Mullett D L and Rackham J W, 2000, *Composite slabs and beams using steel decking: Best practice for design and construction (P300)*, SCI and MCRMA

Salter P R, Couchman G H and Anderson D, 1999, *Wind-moment design of low rise frames (P263)*, SCI

CIMSteel, 1995, *Modelling of steel structures for computer analysis*, SCI

Index

asymmetric section 45
axial force 169
axial resistance 54

biaxial bending 64
bisteel 27
bolts 111, 175
braced 203, 207, 221
brittle behaviour 131
buckling 19, 42, 51, 52, 81, 183

castellated beam 45
characteristic value 6
circular beam 45
classification 82, 167
cleats 161, 165
column 49, 90, 213
column buckling curve 56, 90
combined loading 58, 65, 75, 128
compact element 82
composite beam 3, 23, 24, 28, 39, 45, 91, 204, 212
composite columns 19, 49, 51, 82, 88, 90, 206
composite connections 20, 102, 161, 163, 204, 216
composite construction 1
composite floors 19, 121
composite slab 12, 27, 43
compression zone 180
concrete 5, 14, 28, 50, 69
concrete filled tube 49, 73, 85, 90
confinement 53
connections 20, 41, 161
construction loads 124
contact plate 161, 165
continuous beams 40, 81, 91, 114

cracking 17, 41, 91, 132, 170
creep 6, 55

deck 12, 24, 123, 153, 161, 205
deflection 17, 34, 161, 205
diaphragm action 153
distortional buckling 42, 94, 100
ductility 24, 38, 81, 83, 101, 104, 144, 166, 175, 185, 190, 209

eccentricity 55, 75
effective breadth 30, 42
effective length 71, 97, 176, 215
effective width 86
elastic neutral axis 28
endplate 161, 164
Euler load 54, 88

failure modes 130, 176
fin plate 161, 164, 165
fire resistance 45, 49, 68, 73, 132, 149, 156
flange 92, 103, 176, 180
floors 19
flush endplate 111, 222
formwork 124
frames 20, 208, 221
full interaction 24

high strength concrete 51, 59, 77

imperfections 65
inelastic buckling 81
interaction diagram 59, 63, 67, 139

lateral torsional buckling 42, 83, 87
lightweight concrete 123

load introduction 68
local buckling 53, 84, 86, 92, 103, 127, 169
longitudinal splotting 39

material properties 5
moment gradient 93
moment of resistance 32, 61, 82, 113, 139, 163
moment rotation curve 166, 205, 229

natural frequency 44, 146

partial connection 23, 33, 35, 139, 143
partial section analysis 31
partial strength 205, 208
pattern loading 214
Perfobond 10
permanent formwork 121
plastic centroid 58
plastic hinge 37, 92, 104
Plastic neutral axis 32, 60, 86, 135
plate element 84
Poisson's ratio 53
precast slabs 14, 122
primary bending 66
propped construction 213
push-off test 6, 14, 35

redistribution 81, 91, 126, 169
reinforcement 5, 41, 112, 132, 162, 187
rotation capacity 41, 104, 105, 113, 132, 169, 188, 208
rotational stiffness 163, 194

secant stiffness 40
second order bending 55, 65, 207
semi-continuous 166
semi-rigid connections 41, 111, 163, 168, 211
serviceability limit state 17, 23, 128, 219
shakedown 217
shear 39, 68, 170
shear bond 131

shear connection 3, 9, 35, 144, 173, 192, 199
shear connectors 6, 32, 35
shear deformation 30
shear interaction 136
shear lag 28
shear span 138
shear studs 6, 10, 24, 30
shear zone 184
shrinkage 6, 55
simple construction 70
slab openings 148
slender element 82
slenderness 56, 66, 83
slimflor 26, 45
slip 4, 192
slip block test 146
squash load 54, 59, 70
steel 5, 28, 32, 49, 59, 69, 109
stem-girder 45
stiffness 194, 210
stirrups 90
strain-weakening 81
stud anchors 141
sway 205

tee stub 176
tension zone 170, 175
testing (as basis for design) 127, 144
thermal expansion 72, 136
transformed section 34
transverse shear 24, 39

U-frame 95
ultimate strength 16, 31, 124
unbraced 203, 205, 216, 227
unpropped construction 17, 26, 124, 188, 209
unprotected columns 73
unsymmetrical section 50, 56, 67

vibration 17

warping constant 87, 93, 100
web 92, 103, 182
welded connection 164

For Product Safety Concerns and Information please contact our EU representative GPSR@taylorandfrancis.com
Taylor & Francis Verlag GmbH, Kaufingerstraße 24, 80331 München, Germany

www.ingramcontent.com/pod-product-compliance
Ingram Content Group UK Ltd.
Pitfield, Milton Keynes, MK11 3LW, UK
UKHW021441080625
459435UK00011B/331

Introduction

There has been a lot of confusion about Brigid, who is unique in still being venerated today in Ireland, Europe and the USA as a goddess, as a saint and a mixture of the two. As a goddess she is often confused with a British deity called Brigantia, but the two are only distantly related, and Brighid is a purely Irish goddess. This work includes original research which, for the first time and, probably uniquely, dates the 'conception' of a deity, explains why she was 'created' and by whom. It explains how this goddess 'became' a saint; and how events in first-century Britain were to lead to Ireland becoming unified into five provinces under a High King.

Some historians have dismissed the saint of the same name, the second most important Irish saint after St Patrick, as a fictitious character on the basis that she was merely a pagan goddess who was transmuted into a Christian saint. It was a Druidess who provided the link between the Goddess Brighid and St Brigid of Kildare, a real person, whose fame during her life was widespread, but whose reputation after her death grew over the next 1,400 years until her name was known throughout much of the western world and attracted many adherents to her cult.

Many of the characteristics of the goddess were later attributed to St Brigid, who is strongly connected with the fertility of crops, animals and humans, on the face of it an unlikely association for a virgin saint. Her special day, 1 February, the same as her goddess namesake, marks the beginning of spring in the Celtic calendar.

St Brigid probably has more traditional customs associated with her than any other saint, some of which date back to pre-Christian times. These were once practised in much of the western European Celtic world, and some, such as making Brigid's crosses are still carried out today. There were strong links between Somerset and Ireland from the ninth century and there is an ancient tradition that St Brigid visited the county in 488 AD and left behind a number of relics, a story first recorded in 1120, and the origin of this belief is explained for the first time.

St Brigid had a strong connection with a large part of the natural world, ranging from cattle to carrots, which is covered in this work, as is the story of her relics, which were once widespread. Some still survive today, one being carried in procession at the Irish National Pilgrimage held each July to honour her.

Today Brigid is still venerated and celebrated by people throughout the world and continues to be a bridge between the pagan and Christian worlds as she was during her lifetime. This, one of the most comprehensive works on Brigid, combines early Celtic history, archaeology, tradition and folklore to give a truly fascinating insight into all aspects of this intriguing goddess-saint.

One

Who was the Goddess Brigantia?

The People of Brigantia

The Goddess Brighid is mistakenly regarded by many people as being exactly the same as another goddess, usually referred to – at least in Britain – by the name Brigantia, the Latin form of the British name Briganti. *Brig* means 'High One', and she seems to have been worshipped in other parts of the Celtic world, sometimes under different names such as Brigindo and Brigandu. Brigantia was so important to one British tribe that they named themselves after her, being known collectively as the Brigantes, a name derived from the singular form *Brigans*, which comes from the Celtic root word *Brig*, in this case meaning 'The High Ones'. It was the political and military situation of this British kingdom in the first century AD that was to lead to the 'conception' of the Goddess Brighid.

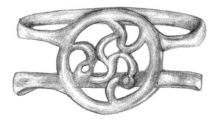

1 Bronze sword scabbard mounts found on Lambay Island, off the east coast of Ireland. The open-work mounts were first cast and then polished. Their style indicates they date to the early second half of the first century AD and were made in the British Brigantian kingdom.

Another Celtic tribe, the Brigantii, who lived near Bregenz in Austria, also took their name from Brigantia and, like their British counterparts, equated her during the Roman period, with the goddess Minerva. Celts quite often traced their tribal descent from a divine ancestor, and so it seems the British Brigantes were conforming to an accepted tradition, and it may be that their overall leader, whether a king or a queen, would also have claimed descent from this goddess.

The Brigantes were numerically the largest tribe in Britain, occupying what is now the modern six counties of northern England, stretching from coast to coast with their southern boundary probably running from the River Mersey to the River Humber, then curving southwards to include the Derbyshire Peak District, with their northern boundary probably extending beyond the line of Hadrian's Wall. Covering a huge area, this large tribe was divided into a number of sub-groups or minor kingdoms, which archaeology suggests probably numbered fifteen, occupying favourable valleys and areas of lighter soils but separated by areas of hill, mountain, moorland and bog.

The names of six of these sub-kingdoms are known from the Romano-British period: the Gabrantovices in North Yorkshire, the Setantii in Lancashire, the Textoverdi in the upper valley of the Tyne, the Lopocares around Corbridge, the Carvetii in the upper Eden Valley and the Latenses. While each sub-kingdom revered their own regional deities they were unified by the worship of one High Goddess, Brigantia, who took precedence over the local male gods, a system that was reflected in their social structure, having a High King or Queen over all their sub-kings. There is evidence that the concept of a Great Goddess or a Mother Goddess that was above all the lesser or regional deities was known in a large part of the Celtic world.

It is obvious from early literature that most of the Celts had this belief in a Mother Goddess who presided not only over mortals but over the gods themselves, and it was this High Goddess who nurtured the gods as well as the land, its crops and its animals, so affecting the fate of mankind. While the Mother Goddess was a more or less universal concept and a very ancient tradition, she was later credited with a variety of other roles which varied from region to region and tribe to tribe, giving her a very wide appeal as well as many names.

So what at first sight appears to be a number of individual goddesses are, in fact, variations or interpretations of one basic Mother Goddess. The importance of a High or Mother Goddess in Celtic society influenced its make up, giving women a role that was, in many cases, equal to or even superior to that of men. This concept was a novelty to the Romans, and the writer Tacitus (56–c.120 AD) mentions that 'they make no distinction of sex in their appointment of commander'. For example, females could be members of any of the three categories of Druids and they were often employed as ambassadors, women would fight as warriors, and they could be monarchs in their own right or, like Queen Boudicca of the Icini, could succeed to the throne on their husbands death and, in her case, take command of a force of warriors estimated to have eventually numbered 230,000 to resist the Romans.

Celtic women worried the Romans! Dio Cassius described Queen Boudicca as being 'huge of frame, terrifying of aspect and with a harsh voice; a great mass of bright red hair fell to her knees'. While Ammianus Marcellinus made the comment that

> a whole host of foreigners would not be able to withstand a single Gaul if he called to his aid his wife, who is usually very strong and has blue eyes; especially when swelling her neck, gnashing her teeth and brandishing her sallow arms of enormous size, she begins to strike blows mingled with kicks as if they were so many missiles from the string of catapults.

Following typical Celtic practices and the war-like nature of their society, tribes were not adverse to raiding each other, to 'acquire' cattle and horses as part of the heroic ideals that meant so much to the Celts. The Brigantes were mostly engaged in pastoral pursuits raising cattle, sheep and horses rather than the agricultural activities of tribes further to the south in Britain.

Julius Caeser said that the entire Celtic people were 'exceedingly given to religious superstition' but this simply indicates that they belonged to a very religious society. They saw many stones, streams, lakes, springs, groves and other features as being inhabited by a local deity or spirit, and were very aware of the existence of the Otherworld. They regarded some days as being 'lucky' and others 'unlucky', being guided in this matter by their priests, the Druids. Evidence from ancient Roman and Greek writers indicates that the Celts were very vain and restless, easily given to quarrelling, fond of boasting, and with a love of ornaments, especially the wearing of torcs, a distinctive type of neck ring, along with bracelets on their arms and wrists, all usually made of gold and of the finest workmanship. In 21 AD Strabo said that 'the whole race was war mad, both high spirited and quick for battle, although otherwise simple and not ill-mannered.'

The Brigantes were united under the protection of their powerful goddess, Brigantia, and even her name 'High One' suggests the important position she held in tribal tradition and beliefs. It is probable that in the early period, i.e in the centuries before the Roman invasion of Britain in 43 AD, their goddess may not have existed in representational form. At most there may have been crude wooden carvings that represented her 'spirit' rather than being a detailed carving of a human-like figure. Only a few such representations of pre-Roman Celtic deities carved in wood are known, and are rare survivors because of the perishable nature of the material. Detailed iconographic carving of deities was a Greek and Roman tradition that was alien to the Celts until the Roman conquest.

The Brigantes probably 'saw' their High Goddess and the evidence of her power in the landscape of their tribal territory, in the form of the great rivers, springs and the impressive, distinctive hills of the area. This does not mean that the rivers and hills themselves were worshipped, but that they represented the divinity of the goddess, especially in her most ancient role of a Mother or Earth Goddess. Such a view is not derived from modern fancy, but can be seen in the writings of ancient Irish mythology where rivers and hills are named after sacred beings whose legends they commemorate, and which preserve a very ancient tradition found all over the Celtic world. Two rivers, the Braint on the Island of Anglesey and the River Brent in Middlesex are said to derive their name from the Goddess Brigantia, although a similar word as a river name in British Celtic means to boil or froth, and is related to the Welsh *brydio*, which means the same. She was closely associated with water, and almost certainly such an important deity would have had other rivers named after her that have now been forgotten.

The Celts saw many deities as represented by woods, hills, streams and other natural features, before they were influenced by the Roman and Greek practice of making carved representations of their gods and goddesses. This explains the rarity of early representations of Celtic deities, and the reaction of the Gaulish King Brennus in the early fourth century BC, as described by Diodorus Siculus, when the king entered the temple at Delphi:

> Brennus, the king of the Gauls, on entering a temple found no dedications of gold and silver, and when he came only upon images of stone and wood, he laughed at them [the Greeks], to think that men, believing that gods have human form, should set up their images in wood and stone.

The religious observances of the Brigantes would have been conducted by Druids, and usually carried out in sacred groves of trees, particularly of oak, which they regarded as particularly sacred, but there were probably more conventional shrines, particularly near the sources of rivers as was the case in Gaul (France). There were three categories within the Order of Druids. The first were the Druids (*Druaid*) themselves who undertook training for twenty years and acted as priests and intercessors with the gods, as well as magicians, healers, herbalists, teachers, judges, lawyers and political advisors to the Celtic aristocracy.

The second category were the Bards (*Bhaírds*), who spent seven years practising composition before becoming a full Bard, and were poets and story tellers, keeping alive legends and folk stories, and formally praising their leaders and satirising their enemies. The third category were the *Vates* (or *Velitas*) who, after twelve years, were qualified as interpreters of sacrifices and foretellers of the future, and regarded as persons of general great learning. All members of the Order could marry and have children, and women could belong to the Order and be members of any of the three categories and, like the men, were taught the secret Druid language, known in Irish Gaelic as *béala na bhfile* (language of the poets).

The Isle of Anglesey, off the north Wales coast, was the centre of the Druid Order in Britain and a focus of resistance to the Roman invasion, harbouring many political refugees as well as being the 'grain store' of the Ordivici, a northern Welsh tribe that mounted very strong opposition to the Roman conquest of Wales. However, in 60 AD, Anglesey was attacked by the Romans under the command of Caius Suetonius Paulinus and the Druid headquarters was destroyed. This was graphically described by the Roman historian Tacitus in his *Annals XIV*.

Paulinus prepared a fleet of flat-bottomed boats to cross the Menai Straights, but as they approached the shore of Anglesey the Romans were confronted by an intimidating multitude on the shore. Black-clad Druidesses, long haired and brandishing torches, mingled with the warriors and the male Druids who, according to Tacitus, by 'lifting up their hands to heaven and pouring forth dreadful imprecations, so scared our soldiers by the unfamiliar sight, so that, as if their limbs were paralysed, they stood motionless and exposed to wounds'.

Then, 'urged by their generals and mutual encouragement's not to quail before a troop of frenzied women', they moved forward, cutting down all resistance. Tacitus then describes how 'a force was next set over the conquered and their groves, devoted to inhuman superstitions, were destroyed. They deemed it, indeed, a duty to cover their altars with the blood of captives and to consult their deities through human entrails.' The groves and shrines were destroyed and a fort for a Roman garrison was constructed.

While the Romans accepted many local religions all over their extensive empire they objected to Druidism not so much on religious grounds, as they were quite capable of equating their own gods with those of the Britons, and were to do so for more than 300 years, but for two main reasons. The first was because the Druids practised human sacrifice which the Romans abhorred, somewhat ironically given their acceptance of putting people in the amphitheatres to fight each other or wild animals and the practice of crucifying criminals. However, the second and more important reason why the Romans wanted to dismantle the Druid Order was because of their role as political advisors to the Celtic leaders, and its ability to unite the disparate Celtic tribes in opposition to the expansion of the Roman Empire. In an appeal to the gods for help before the attack on Anglesey the Druids consigned a great many valuable items to a lake on the island, today called Llyn Cerrig Bach (the Lake of Small Stones). Among the huge number of objects, such as swords, spears, daggers, shield and chariot fittings, horse harness, bronze cauldrons, a trumpet, iron currency bars and slave chains, were distinctively Brigantian items, suggesting that Druids from the Brigantian kingdom were present when the island was attacked.

Gaulish Druids were suppressed by the Emperor Tiberius in 21 AD and the whole Order was abolished in 54 AD, followed by the attack on their Anglesey headquarters in 60 AD.

Druidry did survive, however, but was stripped of its political powers. Individual Druids and Druidesses continued to act as intercessors with the gods, now given a Roman overlay, at many temple sites throughout Britain, also continuing their role of foretelling the future and practising healing using herbs. Druidesses do not seem to have attracted the same hostility from the Romans as the male members of their Order. In Ireland female Druids played an active role in prophesying the outcomes of battles and other events and influencing the aristocracy, while also practising their traditional role as healers.

The arts of healing and prophesy continued to be practised by a small number of individuals, mostly female, for centuries after the end of Roman Britain. These 'descendants' of the Druids were eventually to evolve into the 'cunning folk' who performed a vital role in many villages, using herbs, often coupled with 'magical formulae', for curing ills and other problems. They often acted as midwives and fortune tellers, right down to the twentieth century in some places.

It is probable that Cartimandua, whose name means 'Sleek pony', was already High Queen of the Brigantes in 43 AD when the Roman invasion started, and was one of the eleven British rulers who submitted personally to the Roman Emperor Claudius, on his sixteen-day visit to Britain in August of that year. She may have negotiated a treaty to make the Brigantes allies of the Romans at that time, with herself retaining her throne as a Client Queen. However, this treaty was certainly in place in 47 AD and meant that the Romans would not attack or occupy her kingdom and, indeed, would offer her support as an ally of Rome. In exchange she would not hinder their conquest of Britain outside the Brigantian kingdom.

So at this date the northern extent of Roman Britain lay on the southern boundary of the Brigantian kingdom. While their territory lay beyond the Roman Empire it had the advantage of being able to acquire Roman pottery and other desirable goods and send cattle and other items to the markets over the border. In 48 AD the Romans began the difficult task of subduing the Welsh tribes who were led by a very able war leader called Caractacus who had fled to Wales from south eastern Britain. He was the son of a powerful king and had been fighting the Romans since the start of the invasion. He was a very charismatic person who roused the Welsh tribes to actively resist the Roman advance, and Tacitus says that 'his many undefeated battles, even victories, had made him pre-eminent among British chieftains.'

The Roman campaign against the Welsh was in full swing when some form of unrest broke out among the Brigantes, possibly fermented by Brigantian Druids who would have been in close contact with the Druid headquarters on Anglesey. These Druids would have been greatly concerned at the Roman advance towards North Wales. The Roman Governor of Britain, Aulus Didius Gallus, broke off the campaign and, as the support of Queen Cartimandua was so important to protect his northern flank, sent military support to her aid. The unrest was dealt with, after which the campaign against the Welsh tribes recommenced.

By 51 AD Caractacus had moved his base of resistance to the territory of the Welsh Ordivices, occupying a fortress which, archaeological evidence suggests, was on the limestone spur of Llanymynech overlooking the western edge of the north Shropshire plain. At the western foot of the massif is a Roman campaign base not far from a Roman fort, built in the early first century AD, at Llansantffraid (whose church is dedicated to St Ffraid – the Welsh name for Brigid). However, after a difficult battle the Romans took this fort and Caractacus' wife, children and brothers were captured. Caractacus himself escaped and fled to the Brigantian kingdom where he planned to rally anti-Roman elements to himself and launch an attack on the Romans.

This situation presented a great danger for the pro-Roman Cartimandua because, if Caractacus succeeded in his plan, the Romans would attack Brigantian territory and she would lose her kingdom and power. Despite the military strength of the Brigantes, she seems to have been aware from the first year of the invasion that they would not be able to resist the might of the Romans. So to maintain her alliance she had Caractacus arrested and handed over to the Roman authorities, who sent him to Rome.

Because of the scattered nature of the Brigantes it was difficult for Queen Cartimandua to keep close control over her people, especially the sub-kingdoms in the west which lay closest to the Welsh tribes that were resisting the Roman advance. In addition the Welsh tribes were harbouring warriors who had fled from areas of the south of England that had been conquered by the Romans. Objections to Cartimandua's alliance with the Romans may also have been fanned by the Druids on Anglesey who must have been aware of the danger of being isolated from their sources of support in both the Brigantian kingdom and elsewhere, as well as seeing the danger to the Brigantian Druids.

Her action in handing over Caractacus also led to a split between the pro-Roman Queen and her anti-Roman consort, Venutius, who now began a resistance campaign against the Romans. Cartimandua then 'astutely trapped' all Venutius's relatives and held them hostage. Roman auxiliary forces were sent to her aid and to oppose the resistance of Venutius. For a while this restored the situation, although some time later, around 54 AD, the IXth legion, under the Command of Caesius Nasica, had to intervene again in the continuing struggle between Cartimandua and her consort as, according to Tacitus, Venutius organised outside support to his cause from enemies of Cartimandua who 'infuriated and goaded by fears of feminine rule, invaded her kingdom with a force of picked warriors.'

Venutius bided his time, but was not in a position to make further serious trouble even in 60 AD when the Icini, Trinovantes and other tribes, led by Queen Boudicca, began a rebellion in the south-east of England, presumably because Cartimandua was still strong enough to resist him and hold her kingdom together. However, in an effort to undermine him and win over his followers she divorced him and took a new consort, Vellocatus, a member of the aristocracy and formerly armour-bearer to Venutius. This tactic failed as Venutius gathered enough supporters to wage war on his wife who, no doubt once again called on the Goddess Brigantia to help her in her time of need. Her prayers were unanswered as Venutius captured her, but she managed to get a message to her Roman allies who once again had to come to her aid.

2 A beaded bronze collar found on Lambay Island. This is a distinctively Brigantian design of the second half of the first century AD and was brought to this Irish island by British Brigantians fleeing the Roman invasion of their kingdom.

A force of auxiliary infantry and cavalry was dispatched into the Brigantian kingdom by order of the Governor, M. Vettius Bolanus and, after some indecisive battles, managed to rescue Cartimandua, but were unable to defeat Venutius who, for the time being, was left in power. In effect the Romans had lost the Brigantes as an ally and client kingdom and were now left with an actively hostile tribe on their northern boundary. This stalemate ended in 71 AD when Q. Petillius Cerialis arrived in Britain as Governor with a new legion, the II Adiutrix, and ordered the XXth legion, with Agricola as its Legate, to concentrate on the subjugation of the Brigantes and, after a number of battles over a period of three years, their kingdom was conquered. The people of the Goddess Brigantia were now no longer members of an independent kingdom but had been absorbed into the Roman Empire by 74 AD.

Over a period of time, as the tribe became Romanised and no longer engaged in conflict, the perception of their goddess Brigantia began to change as her aspects as a warrior goddess became less relevant. She was still revered as a protective deity however, and retained her association with water and healing. The absorption of the Brigantes had far reaching and costly effects for Rome, since if the alliance could have been maintained the Romans would not have had to keep a strong military presence in the region of the Scottish lowlands and the strength of the Roman garrison in Britain could have been reduced once the conquest of Wales was completed.

It appears that a large number of the Brigantians left their kingdom around the time of its conquest by the Romans in the early 70s AD and settled in Ireland. The geographer Claudius Ptolemaeus (Ptolemy) compiled an eight-volume *Geography* around the middle of the second century AD. No original maps still exist, the information surviving as a series of latitudes, measurements and tables, and his Irish 'map' shows some fifty-five tribal and place names, with the south-east portion of Ireland being occupied by the Brigantes, who are also shown on the *Verona List*, c.312 AD.

On the island of Lambay, off the coast of Co Dublin, sometime before the 1860s, an iron sword and decorated gold band were discovered. The whereabouts of the sword is now unknown, but the gold band is in the National Museum of Ireland in Dublin. This measures 1.9cm in width and its present length is 21cm, but was originally longer. It bears two patterns, both incorporating wheeled crosses, and the few similar finds of such objects from Britain indicate a date in the second half of the first century AD and stylistically would suggest a Brigantian origin. Suggestions for its original use range from decoration for a wooden casket or a chariot, but most likely it was some form of headband or decoration on a head dress, perhaps worn by a Druid.

In 1927 a first-century BC/AD cemetery was uncovered on Lambay Island while work was being undertaken to construct a harbour. It is unfortunate that this was destroyed by workmen before a full archaeological examination could be carried out, but the skeletons were on their sides with their knees drawn up and seemed to have been laid in shallow pits cut into the clay and covered by clean silver sand. Crouched burials are rare in the Early Iron Age but a few are known from England, notably the warrior burials from Grimthorpe in Yorkshire – Brigantian territory.

One of these burials was obviously of an important man, as it was accompanied by a rich range of grave goods, including a sword, shield and other ornaments. The sword is not like the short Irish weapon used at that period, but had a heavy, long parallel-sided blade and a leather or wood scabbard with three bronze mounts, two of which had open-work ornamentation of a Brigantian style datable to the second half of the first century AD. The shield, probably of wood or wood and leather, had only its bronze central boss surviving.

Other finds from the cemetery included a jet and bronze bracelet, five Roman brooches, a beaded bronze collar, an iron mirror of northern British origin and various bits of decorated bronze sheet. These finds are unique in Ireland and all the items are 'foreign' in character. The beaded collar is notable as it is of a distinctive pattern confined to the British Brigantian kingdom, and again a date in the second half of the first century is generally assigned to these. It consists of two main parts, a plain strip a little larger than a semicircle, the circle being completed by a smaller length strung with eight large bronze beads alternating with seven bronze discs.

Dating evidence puts the burials at some time in the second half of the first century, suggesting these people were Brigantians fleeing the Roman attack on their kingdom. Since the Brigantian kingdom was independent until 74 AD, it is almost certain that among the people who left it for Ireland around the time it came under attack by the Romans were Brigantian Druids. They would have taken their regional religious practices and deities with them, along with their experience of confronting the Romans, and their idea that the way to deal with the Roman threat was by uniting the Celtic people, a process that British Druids had been engaged in at the time of the initial invasion of Britain.

The Goddess Brigantia under the Romans

The Romans generally allowed, and indeed encouraged, the continued worship of native gods and the use of ancient shrines. They brought these local deities into the realm of the official Roman pantheon of gods by equating them with whichever of the Roman deities the native ones attributes and qualities most resembled, or even sometimes equated them with 'local' deities from other parts of the empire.

Following the conquest of the Brigantian kingdom during a series of campaigns between 71 and 74 AD, the people still continued to venerate their High Goddess, Brigantia, though human nature would suggest that some may have regarded her as failing them in their hour of need as they tried, unsuccessfully, to resist the Romans. However, she had many attributes besides being a protective warrior deity that made her popular and was to ensure her survival during the Roman occupation.

During the Romano-British period when Brigantia is mentioned on dedications or shown as a carving, she was often associated with Victoria, 'Victory' or, more commonly, with the Roman goddess Minerva, a protective goddess among several other attributes she was credited with. The seven inscriptions found so far that mention Brigantia all cluster on the northern or southern limits of the Brigantian kingdom. The four dedications from the north of their territory, in the Roman frontier zone (Birrens, Brampton, Corbridge and South Shields) were all given by members of the army or military personnel, while the three from the south (Adel, Castleford and Greetland) are dedications by civilians. Also from Yorkshire is a dedication to *Deus Bregans* (the god Bregans) who, it has been suggested, may have been the consort of Brigantia.

There are clear indications that Brigantia also had other roles such as healing and a connection with water, which was an element associated with healing in the Celtic world. For example, from Yorkshire come two dedications to *Dea Nympha Brigantia* (Brigantia, Goddess of the water), showing that she was, in these cases, regarded as a water goddess. Water in the form of rivers, springs, lakes and wells was regarded by the Celts as being sacred and a channel via which to communicate with the gods and the Otherworld.

Such water sources were also regarded as having strong links with fertility as water is a vital requirement of life and essential for growing good crops, so providing an obvious connection to the Mother Goddess. Huge numbers of offerings have been found deposited in rivers, lakes, wells and bogs. It seems that wells and rivers often formed part of tribal or district boundaries, so it is not unreasonable to link a territorial goddess such as Brigantia with water. This belief of the sacredness of water persisted into Christian times and the remnants of such beliefs can still be seen today in the rite of baptism and the existence of many holy wells all over Britain and Ireland.

From Greta Bridge in Yorkshire an inscription reads *Deae numeriae Brig et Jan* (To the deities Brigantia and Janus), while another dedication bears the inscription '*Brigantiae s. [acrum] Armandus Architectus ex imperio, imp. I.* (Armandus Architect, by Imperial command, [made] this sacred offering to Brigantia in the first year of the Emperor)'. She is also referred to as *Victoria Brigantia* (Victorious Brigantia) on an altar that also mentions Titus Aurelius Aurelianus, a *Magister Sacrorum*, the only Romano-British mention of a 'Master of Sacred Ceremonies'. A centurion of the Second, Augusta Legion dedicated an altar to *Jupiter Dolichenus, Caelestis Brigantia* and *Salus*.

As the Romans stationed in Britain came from all over the Empire, so many overseas local gods were 'imported' and coupled with those of the official Roman pantheon, but could also be equated with a local god from another country, as is Brigantia in this case. Jupiter was the supreme ruler of heaven and earth and on the altar the inscription is coupled with Dolichenus, a Syrian god who was Lord of the Firmament; Caelestis was the daughter of Astarte, the mythical founder of Carthage, so may represent a protective deity which is why she is equated with Brigantia, but can also be equated with Juno, the consort of Jupiter and Queen of the gods; while Salus was a goddess of healing and so also linked to Brigantia.

What is interesting is the apparent equating of Brigantia with Juno in the form of Caelestis, which suggests the recognition that Brigantia, like Juno, was a goddess who presided over the other gods, a recognition that she was regarded as the 'High One', the meaning of her name. However, Brigantia is not put first on this inscription as in the official pantheon of Roman gods Jupiter was seen as the pre-eminent deity, with Juno second to him in importance.

There is a superb carving of Brigantia from Birrens (*Blatobulgium*), Dumfriesshire, the site of a Roman fort, which shows some of her complex attributes. In this she is depicted as the Roman goddess Minerva, holding a spear with a shield to her left and with the *gorgoneion*, the head of the Medusa (Latin) or Gorgon (Greek) worn round her neck. She is wearing a conical helmet which has a battlemented wall around its base indicating that she is a protective city goddess, presumably of the Brigantian tribal administrative capital of the Roman period, *Isurium Brigantum* (Aldborough, Yorkshire). She holds a globe, symbolic of victory, and behind her are the wings of Victory.

Below this figure is the inscription '*Brigantia Samaadvs*' whose exact meaning is unclear. Could it be a Celtic word, *Samhach* (peaceful) 'twisted' into a Latin form? This is now in the National Museum of Antiquities, Edinburgh. The Birrens representation shows Brigantia in the form of Minerva with whom she was often equated in the Roman period. However, without the inscription on the carving it would have been impossible to identify her as the Celtic goddess Brigantia. Minerva in the classical world does not have a water or healing connection, and only in her Celtic interpretation has she these associations.

Since Brigantia was equated in the British Celtic mind with Minerva, it is not unreasonable to suppose that many other Romano-British images of this Roman goddess actually represent Brigantia or whatever regional name this High Goddess was known by. In Celtic tradition, where deities are depicted in iconographic form, they are not usually named on an associated inscription. Inscriptions from East Gaul (France) indicate that Brigindo, the Gaulish name for Brigantia, was also a deity of healing, crafts and fertility. Since we know that Brigantia was closely associated with water and healing, it would seem that the equating of Brigantia with Minerva can be seen in a number of examples from Roman Britain. A silver gilt *patera* (a broad shallow dish used for pouring libations), from Capheaton, Northumberland, shows Minerva (Brigantia) presiding over springs and a temple.

Another example of linking a native goddess with a modified Minerva can be seen at Bath in Somerset, known in Roman times as *Aquae Sulis*, where there is an elaborate series of Roman baths and temple complexes. The natural hot spring there was a focus for the worship of the native goddess Sul or Sulis long before the Roman invasion, her name apparently meaning 'opening' or 'orifice', but she is also connected with the sun. From inscriptions it appears that Sulis was concerned with blessing, prophesying, healing and possibly childbirth.

The appeals to Sulis found on the so-called 'curse tablets' from Bath, lead sheets on which are inscribed appeals to the goddess for help, give a further insight into how her devotees viewed her. They show that she was also seen as a deity that could help in the recovery of lost property, particularly money. For example, one reads 'I have given the Goddess Sulis six silver pieces which I have lost. It is for the Goddess to extract it from the debtors Senicianus, Saturninus and Anniola…' while another tablet reads 'Whether pagan or Christian, whosoever man or woman, boy or girl, slave or free has stolen from me, Annianus, six *argenti* [silver coins] from my purse. You, lady Goddess, are to exact them from him…' On the other side of the tablet are eighteen names, presumably a list of the suspects which Annianus drew up to assist the goddess!

The Roman baths are dedicated to Sulis Minerva in her aspect of a deity responsible for healing and therapeutic treatments, and it is possible that Sulis is a south-west England version of Brigantia. The baths and surroundings became a centre for the veneration of a mixture of native, imported native and Roman gods and goddesses, but note that Sulis is given precedence in the coupling with Minerva. Sulis is also known from archaeological finds from Cirencester and Colchester, as well as from Rome, Hungary, France and Germany. While organised Druidism in Britain was destroyed in the first century AD, worship of the gods was still conducted by British priests similar to Druids but without the political influence they formerly had, and an inscription from Bath shows that 'priests of Sulis' were present at the temple.

Another example from Bath that links underlying Celtic beliefs with a Romanised expression of religion, as well as the importance of Minerva, can be seen in one of the most famous carvings from the Roman baths, the Gorgon head that once adorned the pediment of the temple of Sulis Minerva. The inspiration for this carving comes from an ancient Greek story that was also popular with the Romans and concerned the Gorgon's, three monstrous sisters with huge teeth like those of swine, claws for hands, and hair of writhing serpents. Their mere glance would turn people to stone.

One day King Polydectes sent Perseus, the son of Zeus (Jupiter), to cut off the head of one of the Gorgon sisters named Medusa who had lain waste the king's country. Perseus was a favourite of Athene (Minerva) who lent him her shield in his quest, while Mercury (Hermes) lent him his winged shoes giving Perseus the power of flight. Perseus approached Medusa as she slept and, taking care not to look directly at her, but guided by her image reflected in Athene's bright shield, cut off her head. He later gave it to Athene who fixed it in the middle of her shield.

Based on this story the carving at Bath should depict a female head; however, it clearly shows a male with flowing beard, hair and moustache. There is also a pair of wings on each side of the head, perhaps symbolic of victory, while the carved figures that support the head

Who was the Goddess Brigantia?

appear to be Victories depicted in the classical style. Other male 'Medusa' heads are known from Roman Britain, many of which are associated with healing springs.

There was a very strong cult of the head in the Celtic world, both in the pre-Roman period and throughout it. Classical writers described how Celts took the heads of enemies. Livy (59 BC–17 AD) wrote of Celtic warriors 'with heads hanging at their horses breasts, or fixed on their lances, and singing their customary songs of triumph'. While Strabo (c.60 BC–20 AD) says the heads were preserved and treated with great respect, and that sometimes the skulls were decorated with gold, and that the 'heads of enemies of high repute were then preserved in cedar oil and exhibited to strangers'. Heads were displayed in temples and houses.

There are many references to this custom of 'taking heads', which is backed up by archaeological evidence from Britain and Europe. A great many carved stone heads have been found, ranging from pebble size to boulders, along with a few wooden heads that by luck have survived. Heads are also very common as decoration on weapons, buckets, jewellery and other items during the Iron Age. The 'taking of heads' not only indicated military prowess in Celtic society, but was a way of venerating the head which was seen as a symbol of divinity and Otherworld powers, as well as being the part of the body in which they believed the 'spirit' of the person dwelt.

The veneration of the head was not unique to the Celts but was widespread among many different peoples and something that archaeological evidence shows goes far back into prehistory. The practice of 'taking heads' was generally viewed by the Romans with disgust and once Britain was settled after the conquest it was a practice that died out. However, the carving of male 'Medusa' heads in Roman Britain was a way of retaining the symbolism of the traditional divine head, although now thinly disguised as classical art. This devotion to the cult of the head may well be another reason why Minerva particularly appealed to the Romano-British who still venerated Brigantia in one of her forms.

3 A carving of the Goddess Brigantia from Birrens, Dumfriesshire. She is shown as the Roman goddess Minerva, but below is the inscription '*Brigantia Samaadvs*'.

Minerva was a protective warrior deity, often depicted with spear and shield, so in the Celtic mind her acceptance and display of a severed head from the brave warrior Perseus would have further endeared her to them as a worthy equivalent of the Goddess Brigantia. The classic story of Persius and Minerva says that the goddess fixed the head of Medusa to her shield, but carvings often show it either on her breastplate or, as appears to be in the case of the Birrens carving, worn round her neck.

As Minerva was equated with the native goddess, it suggests that most of Brigantia's attributes would be very similar, which gives some idea of the full range of attributes that she might have been credited with. Besides being a protective warrior deity, Minerva was regarded as a goddess of wisdom, who presided over both the ornamental arts and household crafts such as spinning and weaving, but was also concerned with agricultural matters. Brigantia was also a protective warrior deity and her distant origin as a Mother Goddess would link her to fertility, while her water association linked her to the healing arts. She was also probably connected to agriculture, particularly that involving cattle, derived from her veneration by a pastoral people, and which also links her to prosperity as cattle were an indicator of wealth in the Celtic world, hence the connection with money and its recovery at Bath where Brigantia/Minerva seems to be equated with Sulis.

As Britain became settled in the first forty years following the Roman invasion of 43 AD there was less need for a warrior deity, unless you were a member of the army, so Brigantia's other attributes, protection of a personal nature and particularly healing, became more emphasised among the civilian population. This appears to be the case through the whole of the Roman period in Britain until the official withdrawal of the Roman's from Britain in 410 AD. However, a new religion was beginning to spread and become more popular from the sixth century onwards. Christianity was actively being promoted by missionaries, many from Ireland and Wales, although a few Christians had certainly been present in Britain from the third century and possibly earlier.

Over the next three centuries there were major changes in Britain as various areas became settled by Saxons, Angles and Jutes. While they brought their own deities with them, the native population would have continued the veneration of their own gods and goddesses, although it may have been 'diplomatic' to equate them now to the deities of the new settlers! Christians would have continued the worship of their one God, and rapidly managed to convert many of the pagan settlers to their own religion. The Christians regarded the old gods and goddesses as 'demons', and discouraged their worship, although many pagan practices and customs, notably the time of year that ceremonial gatherings occurred and even the form they took, were adopted by the Christians, so making it easier for pagans to convert to the new religion.

So Brigantia, in her persona of the British Minerva, along with Sulis, disappeared, along with all the other traditional deities and many other aspects of Roman life by the end of the fifth century in Britain. She did not even survive in some altered and debased form such as a fairy, as was the case with a few of the other Celtic deities, for example Gwynn ap Nudd who, in medieval stories, is described as 'king of the fairies', but was originally a pre-Christian Celtic god of the Otherworld. So it is to Ireland that we have to look for the Goddess Brighid.

Two

The Conception of the Goddess Brighid

In Britain the Order of Druids was destroyed after the attack on their headquarters on Anglesey (*Mon Insulae* in Latin) in 60 AD. While Druids and Druidesses still survived as individuals, acting as intercessors with the gods, using herbs for healing, practising magic, foretelling the future and keeping alive stories, poems and songs, they no longer had the organisation or wielded the political influence they once had. However, Ireland was never part of the Roman Empire and the Order of Druids remained strong there until the coming of Christianity in the fifth century and they continued to play an important role in Irish society considerably later. There is also evidence that quite large numbers of Celts from the kingdom of the Brigantes, including Brigantian Druids, migrated to Ireland in the second half of the first century.

Early Irish writings often mention Druids, and many of the Irish legal tracts, known as the Brehon Laws, were based on those devised by the Druids. A number of folktales have come down to us that pre-date Christianity, although in some cases these have been given a Christian veneer. It is therefore fortunate that there exists in Ireland a considerable body of knowledge about the Celtic gods, goddesses and heroes which gives some idea of the beliefs of the British Celts up to the time of the Roman invasion.

In pagan Celtic Ireland there was a goddess called Brighid, daughter of the god Daghdha. However, she is often described as three goddesses or sisters all called Brighid. These three Brighids each had a specific sphere of influence, one for learning, poetry and protection, one for healing and one for metal working. As was common in Celtic beliefs, one goddess could be visualised as three, so the three sister goddesses called Brighid would also represent one goddess with three main aspects or spheres of influence. There is no direct mention of any warrior traits in Brighid or the Brighids, and at first sight this is surprising considering her parentage, since it is clear that her main attributes are derived from both her parents, but Brighid seems to have been conceived to represent a particular set of values that did not emphasise those of a war-like nature, neither aggressive or protective war, and her mother, Mór-Ríoghain, already existed as the war goddess of the Irish Celts.

Brighid was a goddess whose characteristics clearly represent those held in high regard by the Druids, so she can be regarded as the patron goddess of this priestly caste. Many of Brighid's attributes are also those that would appeal to women, and in Ireland Druidesses played an active role in influencing the Celtic nobility and prophesying the outcome of battles, and she is also connected with agricultural activities. The spur for the creation of the Goddess Brighid in Ireland was the arrival of Brigantian Druids between 71–74 AD, following the attacks on the Brigantian kingdom in Britain by Roman forces.

Brigantian Druids would have brought the veneration of Brigantia with them and, as they merged with Irish Druidry, may have also wanted to merge their High Goddess with the Irish equivalent, probably as part of a policy to unite the disparate Irish Druidry. Introducing Brigantia in the form that she was venerated in Britain would have been difficult as she was an 'alien' deity as far as the Irish were concerned, and she did not possess the range of attributes that was required to bring about the unification of all the Irish Druids and, with their encouragement and influence, the Irish tribes.

The influence of these Brigantians, and their numbers, must have been substantial since the geographer Ptolemy, in his *Geographia Claudii Ptolemai*, compiled in the second quarter of the second century AD, shows a number of Irish tribes and that an area of the south east of Ireland was occupied by Brigantes. However, the Mother Goddess of the Irish, Anu (sometimes called Danu), was primarily a fertility goddess. Anu is commemorated in the geographical feature known as the Paps (breasts) of Anu (*Dá chích Annan*), two distinctive mountains at Killarney in Co Kerry, that were regarded as the breasts of this Earth Mother, confirmation of the fertility with which Anu is connected. She is probably very ancient in origin going far back into prehistory when the main deity of the Celts and their ancestors was a female.

So, rather than try to modify this ancient Irish Mother deity who was already widely venerated, it seems that a new goddess was 'conceived' by some of the Irish Druids, under the influence of the Brigantian Druids. They were able to do this as the Celts did not regard

4 Map by Robert Morden, 1690, showing the main tribes of Ireland. This is based on information in the second-century AD *Geography of Ptolemy,* and shows the area of south east Ireland settled by Brigantians from Britain.

their gods as their creators but their ancestors, so it would not have been too difficult to 'find' a lost deity, and Pomponius Mela stated, *c.* 43 AD, that the Druids 'profess to know the will of the gods', so it would have been relatively easy for them to inform the other Druids and the people of this newly discovered or conceived goddess's desire to unite the tribes of the country.

This new deity also had a strong appeal to the Druidesses (*ban Druaid*) who, in Irish Celtic society were very influential, particularly in the art of foretelling the future, and many leaders would not do anything without consulting their seer. The Druids had immense power and influence with the tribal leaders, as is clearly shown by Dio Chrysostom (*c.*40–111 AD) who wrote: 'the kings were not permitted to adopt or plan any course, so that in fact it was these [Druids] who ruled and the kings became their subordinates and instruments of their judgement.'

Brighid (the High One) was the name of this new goddess, but see pages 44 and 66 for more information on her 'real' name. While she had a few characteristics in common with Brigantia, she had many more taken from two powerful and important existing Irish deities, the Daghdha and Mór-Ríoghain who, between them, combined the attributes most valued by the Druids. Brighid was also credited with fertility aspects, particularly those concerning livestock, something that would also have given this 'newly revealed' goddess a wide appeal to the ordinary people, most of whose lives were closely bound up with agriculture and its natural concern about the fertility of animals and crops. One of the few surviving stories about Brighid as goddess mentions three places, all of which are in the area of Ireland occupied by the Brigantes (see page 232).

While Brighid can be considered to be distantly related to Brigantia, she is primarily a new goddess conceived between 71–74 AD by the Druids, incorporating the attributes of two existing Irish deities who, the Druids felt, would be suitable 'parents' for Brighid, so making her acceptable to the Irish people. Further evidence for Brighid's relatively late 'conception' and that she was an imposed deity is suggested by the fact that there was already an older Irish deity, Goibhniú, whose name means 'smith', and was a smith god, while another, Dian Cécht, whose name translates as 'he who travels swiftly', was a god of medicine, both characteristics attributed to Brighid. See page 232–33 for an ancient story to account for her taking on the role of the smith god.

Lugh was a sun god but remained widely revered during Brighid's time as a goddess, unlike Goibhniú and Dian Cécht who rapidly faded away, suggesting that she was not primarily a sun deity as some writers have maintained, but probably had some responsibility for crop fertility, in which the sun plays a part. Further evidence for her late appearance is suggested by the fact that her main symbol, the St Brigid's cross (*Crois Bhríde*), see page 96, is not found on any Bronze Age pottery or stone monuments, or any of the ancient megalithic monuments associated with other Irish goddesses such as Bóann or Macha. Being able to date Brighid's 'conception' accurately makes her unique among ancient deities in being able to say when she came into existence.

Until the Order was destroyed in Britain by the Romans, the Druids were the most important of the three categories within the Order, although the Bards (*Bhaírds*, the poets and singers), and *Vates* (foretellers of the future), also played a major part in their religion and culture. Until the Roman invasion the Britons were divided into twenty-one large kingdoms, but in Ireland at that time the Celts consisted of many smaller independent tribes or petty kingdoms, one estimate putting these at as many as 150. Their tribal Chieftains would each have had Druids, Bards and *Vates* in their 'courts', but generally only serving the needs of that tribe (*tuath*) and venerating their particular chosen deities. In the fourth century the name for an Irish diviner, *Vates*, changed to *Filidh* and, as Christianity began to take hold in Ireland, the *Filidh* began to take on many of the roles of both the Druids and the Bards.

The influx of Brigantian Druids, with their policy of uniting both the Irish Druids and the Irish Celts, seems to have started the process of unifying these many different Irish tribes into larger groupings, and so a goddess to initially unite the Irish Druids would have made a good start. This policy would have been seen as essential by these Druids in the light of their experience of Roman expansion and invasion when they were in Britain. What the British Celts who had fled to Ireland could not have been sure of, is where the Roman expansion would stop, as the legions had already conquered all of France and Spain, part of Germany and most of Britain, and there must have been a real fear that Rome would turn its attention to Ireland, so a united people would be much more capable of repelling a Roman invasion.

Archaeology suggests that the intention in the 70s AD was indeed for the Romans to invade Ireland in a campaign to be launched from the city of Chester (Roman *Deva*). Troops would sail up the River Dee and across the Irish Sea for a landing on the east coast of Ireland. Chester would have become the capital of a Roman province consisting of northern Britain and Ireland, which explains the grand Roman buildings found at Chester that date to this early period. The Romans traded with Ireland and were in contact with Irish leaders, so undoubtedly rumours about an intended invasion would have been circulating in Ireland at that time.

What the Irish could not have known was that in 82 AD the Roman military commander, Gnaeus Julius Agricola, during his campaigning in Scotland, took a decision while at the Mull of Kintyre on the west coast of Scotland, that he would not cross to Ireland, just 17km (11 miles) away at that point. Despite the fact that, as the Roman historian Tacitus tells us, Agricola 'saw that Ireland…conveniently situated for the ports of Gaul might be a valuable acquisition', and tells how an Irish minor king, expelled by a dynastic struggle, had been received by Agricola in the hope of being able to use him in the future. Agricola estimated that Ireland could have been invaded and overcome by a single legion (10,000 men) and a moderate number of auxiliary troops. This clearly indicates what the Romans thought of the weakness of the disunited Irish tribes, as it had taken four legions (40,000 men) plus a large number of auxiliaries (cavalry, archers, engineers and other specialist units), to invade Britain.

There was an expansion of some Irish tribes and absorption of others and it was no doubt an unsettled time for many people. By the end of the fifth century however, this unification movement initiated by the Brigantian Druids 400 years earlier resulted in the creation of five kingdoms with a ruling monarch (*airdri cuicid*) for each, with individual tribes and tribal leaders under them, and a High King (*airdri ri Eireinn*) senior to the other four rulers. This system was prevailing in the British Brigantian kingdom before it was taken over by the Romans (see chapter one).

This new unifying goddess had to be given a suitable genealogy and, as a new deity, Brighid had to have a creditable Irish parentage using existing Irish deities. While Brigantian Druids were probably the prime movers in 'conceiving' this new goddess, Brighid was a purely Irish deity designed to appeal to both the Irish Druids and the people. Irish tradition says that Brighid's father was the Daghdha, the Good God, in the sense of technical skills rather than his morals, but he is also referred to as *Eochaid Ollathair* (the Great Father), *Daghdha Mór* (the Great Good God), *Ruadh Ró-Fheassa* (the Noble of Great Knowledge), and *Aodh Ruadh Ró-Fheassa* (Red Fire of All Knowledge). He was regarded as the greatest of the Irish gods. Ancient Irish stories also mention a deity called Daire, a name which means 'oak' as well as 'fruitful one', which seems to have been an alternative title for the Daghdha.

The Daghdha was huge, one account saying his face was 'broader than half a plain', while others say he was able to clear twelve plains in one night, cut the path of twelve rivers on

Brighid i.e. a poetess, daughter of the Daghdha. This is Brighid the female sage, or a woman of wisdom, i.e. Brighid the goddess whom poets adored, because very great and very famous was her protecting care. It is therefore they call her goddess of poets by this name. Whose sisters were Brighid the female physician [*ban líaigh* – woman of healing] Brighid the female smith [*ban goibnechtae* – woman of smith work]; from whose names with all Irishmen a goddess was called Brighid.

In this statement can be seen the triple aspect of the goddess so common in Celtic paganism and Roman Britain, along with many of the characteristics of Brighid. She had the attributes of wisdom and learning, was seen as a protective deity, as a goddess of healing, a goddess of smiths, suggesting a connection with the production of weapons as well as items such as tools, cauldrons, jewellery and other useful objects. Her fire connection, to judge by later traditions, also extended to the hearth and domestic fire. She was a goddess specially revered by the *filidh* who, by Christian times in Ireland, had taken on many of the roles of the Bards. They were concerned not only with poetry but story telling, tribal genealogy and wisdom in general.

By looking at the attributes of Brighid's parents, excluding their war-like ones, it can be seen why a fusion of the two into a single goddess would have appealed to the Druids, reflecting both their beliefs and their role within Celtic society. It can also be seen that many of these traits were later attributed to St Brigid because of her connection to the goddess.

Attributes of Brighid's parents that she 'Inherited'

Daghdha: the Good God	Mór-Ríoghain: the Goddess of War
Magic	Fertility
Healing	Foretelling the future
Knowledge	Animal husbandry
Producing an abundance of food	Association with fire
Control of the weather and environment, particularly in relation to crops	Protection of 'her' people
Fire	

The Goddess Brighid

Stories concerning the Irish gods and goddesses were perpetuated in an oral tradition for many centuries, and some of those about the Goddess Brighid persisted into Christian times long after she had been 'converted' to a saint. Some of the miracle stories told about St Brigid, which were recorded by monks in her *Lives* from the seventh to the tenth centuries, give a clue to some of the tales that must have originally been told about the goddess, even though these have not been directly preserved. So a few of the supposed miracles in the early accounts of St Brigid give an idea of what else the Goddess Brighid was almost certainly associated with, for example cattle and fire.

Brighid possessed two Royal oxen (the word can also indicate uncastrated bulls) (see page 231) which suggests a connection with cattle, while her mother, Mór-Ríoghain, had a herd of Otherworld cows. St Brigid was traditionally so closely involved with cows that it would strongly suggest that the goddess too was associated with this animal. Cattle not only

Brigid: Goddess, Druidess, Saint

5 Brú na Bóinne (Newgrange), Co Meath. A prehistoric tomb which, according to tradition, was built by the Goddess Brighid's father, the Daghdha, as both his palace and an entrance to the Otherworld. The structure, 85.3m (280ft) in diameter and 15.2m (50ft) high, is faced with white quartz stone.

indicated wealth in the Celtic world, but were an animal which played a prominent part in Druid sacrifices on their most important festivals.

While there is no direct mention of any fertility attributes in the case of the goddess, the fact that she is celebrated at *Imbolc*, a festival connected with the end of winter and the birth of new life, suggests she did have a fertility role, and her possession of two oxen may be significant. The fertility aspects of many recent customs to celebrate the saint's festival are quite clear and are almost certainly 'handed down' from the time of the goddess. She may also have had some form of sun association, suggested by the St Brigid's crosses (see page 96) which look like a sun symbol, if their origin goes back to the time of the goddess, and she may have inherited this aspect from her father who had responsibility for the weather in relation to crops. It is clear that Brighid's main attributes were very much those held dear by the Druid Order:

Learning

One of the prime roles of the Druids in Celtic society was the preservation of a wide range of knowledge of matters both spiritual and physical. They were responsible for the guarding and passing on of traditional knowledge along with the stories of their deities, and preserving tribal and family genealogies. The understanding of the stars and world around them as they interpreted it were also their provenance. Tradition credits one of the Brighids with the invention of the ancient Ogham alphabet, although other stories say it was invented by the god Ogma mac Elathan. It seems to have come into existence in the south of Ireland by the fourth century AD, and some scholars regard it as a sign language used by the Druids, and it is found on stones set up above burials.

Poetry

Poetry was very important in the Celtic world, and remained so until quite recent times. With its associated traditions of songs, storytelling and the keeping of genealogies and histories, poetry was one of the major roles of the Druid Order, and was used as a way of recording important events such as successful battles, and praising their chieftains and brave warriors. It was also used as a means of ridiculing and denigrating their enemies.

Healing

This was traditionally the province of women in many societies, including that of the Celts, being related to the use of herbs and plants in cooking as well as for healing and various 'magical' purposes. As one of the roles of Druids was that of physicians and healers, its not unreasonable to suppose that it was primarily the responsibility of the Druidesses within the Order, although many Druids are also recorded as well known healers.

Metal Working

The metal working attributed to Brighid by the ninth century Bishop Cormac mac Cuilleanáin is interesting, since the smith was regarded almost as a magician with the power to transform ore into weapons and other useful items. In many societies the smith was believed to be an intercessor with the gods and, until quite recently, the blacksmith was regarded as someone special in the village. In Ireland, at least in more rural areas, it was traditional to give the blacksmith a gift of the head of each cow or other animal killed for food, as well as gifts of oats or straw. The smith was believed to be able to cure diseases in both animals and humans, and often the water in which he quenched the hot metal played a part in the healing.

Related to Brighid's association with smiths is the connection with fire, not only in an industrial sense but in the domestic form of the hearth with all its connotations of the home and, by extension, various other domestic activities. As a saint, many of the customs associated with her, particularly in the Scottish islands, are closely connected with the hearth (see page 79). It should also be remembered that not only did fire play a major role in some of the most important festivals of the Druids, but it also represented both the sun and purity, aspects connected with both the Goddess Brighid and St Brigid.

Fertility

While only slight indirect evidence can be seen for this role as far as the goddess is concerned, many of the beliefs and the customs carried out in honour of Brigid as a saint are concerned with the fertility of crops and animals (see page 85). This was almost certainly one of the main roles of the goddess, particularly as her celebratory day marked the beginning of spring. In addition, the Druids undertook various ceremonies to encourage the fertility of the land, crops and animals, so this would have been an important attribute for a goddess who was especially revered by these Celtic priests.

Magic

The Druids used magic for a variety of purposes, and while no stories concerning this have survived that relate directly to Brighid, some of the early miracle stories concerning St Brigid are probably based on folk tales about the goddess which have been given a Christian veneer, and are indictors of some of the 'magical' acts she was credited with.

Three

The Goddess Brighid becomes a Saint

By the end of the fourth century there were Christian communities in Ireland, partly due to contact with Roman Britain in the form of trade for at least 200 years but also, since at least the middle of the fourth century, Irish raiding bands were bringing back slaves from Britain. Since Christianity was widespread there it was inevitable that some of these British slaves were Christians who would have had some influence on the Irish families to which they belonged.

There are a number of references in early fifth-century writings suggesting that Christianity in Ireland had recently increased. St Jerome says that 'even Britain, the province fertile in the breeding of tyrants, and the Irish peoples, and all the barbarian nations round to the very ocean, have come to know Moses and the prophets.' St Augustine said 'So it is that God's word has been preached not only on the continent, but even in the islands which are set in the middle of the sea; even they are full of Christians, full of the servants of God', while in another letter, referring to the 'islands of the ocean', he says there are no islands 'where the church is not.'

The official introduction of Christianity to Ireland started in 431 AD with the arrival of Palladius, who was ordained as first Bishop of Ireland by Pope Celestine, but, according to later accounts, his mission was not too successful and he later left to preach in Scotland. Another early Christian missionary was Sedulius, who may have been an Irishman but who had studied in Gaul before returning to his own country, and the names of other bishops active in Ireland appear in early fifth-century accounts, among whom were Secundinus, Auxilius and Iserninus.

However, it seems Christianity did not really began to make serious inroads among the Irish until St Patrick (whose real name was Succat; the name Patricius/Patrick means Nobleman) returned to Ireland. Which was, according to the latest authorities, in the mid-450s AD after having spent six years there some time before as a slave. At some point after this many of the attributes of the Goddess Brighid seem to have been transferred to a woman generally called today St Brigid of Kildare.

The Goddess Brighid was described by Bishop Cormac Mac Cuilleanáin, in his ninth-century *Sanas Cormac* (*Cormac's Glossary*) (see page 29). Brighid seems to have had a wide appeal in Ireland by the fifth century AD and was revered by many of the Irish Celts and

The Goddess Brighid becomes a Saint

6 Bronze figure of a Druid from Nevvy-en-Sullias, Loiret, France. A rare contemporary representation of a member of the Celtic priesthood.

Irish Druids, and it seems that many early Christians did not completely abandon belief in the old gods, goddesses and local spirits, which is not too surprising as these beliefs went back many generations. Secret veneration of the 'old ones' may also have been seen as a form of insurance!

Rather than trying to suppress pagan beliefs, it was common for the Christian church to adopt and adapt many pagan festivals as part of a deliberate policy of Christianising the pagans. They even re-dedicated pagan temples to the Christian God, so making it easier for pagans to convert. This policy was embodied in the instructions St Augustine received from Pope Gregory four years after he had begun his mission to England in 597 AD, but in practice it had already been happening for some time:

> Because they [the Anglo-Saxons] are accustomed to slay many oxen in sacrifices to demons, some solemnity should be put in the place of this, so that on the day of the dedication of the churches, or the nativity's of the holy martyrs whose relics are placed there, they may make for themselves bowers of branches of trees around those churches which have been changed from heathen temples, and may celebrate the solemnity with religious feasting.
>
> Nor let them now sacrifice animals to the Devil, but to the praise of God kill animals for their own eating, and render thanks to the giver of all things for their abundance, so that while some outward joys are retained for them; so they may the more easily respond to inward joys.

A further letter from Pope Gregory, sent to Abbot Mellitus in 601 AD just before his journey to England, shows the church's attitude to pagan temples and their reuse:

> When by God's help you come to our most reverend brother Bishop Augustine, I want you to tell him how earnestly I have been pondering over the affairs of the English – I have come to

the conclusion that the temples of the idols in England should not on any account be destroyed. Augustine must smash the idols, but the temples themselves should be sprinkled with holy water and the altars set up in them in which relics are to be enclosed. For we ought to take advantage of well built temples by purifying them from devil-worship and dedicating them to the service of the true God. In this way, I hope the people, seeing their temples are not destroyed, will leave their idolatry and yet continue to frequent the places as formerly, so coming to know and revere the true God.

So how did the change from goddess to saint occur? The early Irish Church had the same policy of adapting firmly held pagan beliefs and sites if at all possible, and the people kept alive stories about their traditional deities. In the early Celtic church saints were either real people who had founded churches or monasteries or, so tradition maintained, had performed miracles during or after their life. In a few cases they were symbolic replacements for their ancient deities and heroes as both were able to work magic/miracles. In Brigid's case a number of these strands came together to effect the change from goddess to saint. There are clues to this adoption not only in St Brigid's attributes, but particularly in one of the practices concerning the saint that may explain not only how the change occurred but even when the Goddess Brighid became 'Christianised'.

Gerald of Wales (c.1145–1223), real name Gerald de Barri, described in his work *Expurgatio Hibernica* written in 1185 how a fire had been kept continually burning at the church of St Brigid at Kill-dara (*Cill Dara*). This name means 'church of the oak tree', a tree sacred to the Druids, and a great oak by her church seems to have survived until the tenth century. This would seem to be a continuation of the way the Goddess Brighid was venerated at her cult-centre long before Christianity took hold in Ireland.

Originally a perpetual fire would have been maintained in honour of this goddess, which may suggest she was particularly associated with fire, symbolic of purity among other things. It was tended by Druidesses (*ban Druaid*) and, if the later practices were based on what was happening in pre-Christian times, then males would have been excluded from the focal centre of the temple, the fire shrine. It was sometimes the case in pagan religions that the chief priest or priestess of a temple or shrine would identify themselves with the deity they venerated, sometimes taking the deities name as part of their title. So, bearing in mind that *Brig* means 'High' or 'Exalted', the Chief Druidess could well have coupled her name with the title of this goddess.

Members of the Druid Order were knowledgeable in many fields, both religious and secular, were in contact with fellow scholars, spoke and wrote Greek and Latin, at least in the case of British and French Druids, and had the chance of travelling widely. With the spread of Christianity in Ireland the Druids started to lose their authority that for centuries had been unchallenged in their tribal societies. They were now confronted with a hard choice as to what to do next. While there was a degree of conflict between Christianity and Druidic beliefs and magic, at the same time there was not a huge difference between many of the beliefs of early Celtic Christianity and Druidism. While it was relatively easy for them to convert to the new religion, as undoubtedly many did.

It is obvious that Greek and Roman writers admired many of the traits of the Druids, particularly the sincerity of their beliefs and teachings. This was summed up by Diodorus Siculus (c.60–c.21 BC): 'They are of much sincerity and integrity, far from the craft and trickery of men among us, contented with homely fare, strangers to excess and luxury', and this is confirmed by other writers. Their philosophy was basically a moral system that stressed right (*fás*) from wrong (*neamhfás*), and distinguished those things that were lawful (*dlea-thach*) from those that were unlawful (*neamhdhlea-thach*), being enforced by a series of clearly stated taboos (*geasa*).

Christianity would have a natural appeal to many of them, with its message of the Son of God who came to earth to save mankind, who grew up to be a great teacher and preacher. He was a healer and prophet and finally ended his time on earth as a sacrificial victim, but who did not really 'die', and promised a life beyond death, beliefs close to those of the Druid's themselves. In addition the idea that deities dwelt in springs, wells, rivers, trees, mountains, stones, birds, animals and so on was easily altered to a belief that the one Christian God could be seen in all these aspects of the world around them.

Looking at the early *Lives* of the saints it can be seen how similar the Druids and these early Christians were in so many respects, including their ceremonies. Both baptised babies at which time their name was given; both had fires that were symbolically extinguished and re-lit at certain seasons; both effected cures using magic/miracles; both were able to transfer illness from humans to other objects; both were able to predict forthcoming events; both were teachers of the young; both led religious observances, both were advisors and counsellors to kings, and both were intercessors between mankind and the gods/God.

Many of the Celtic deities were worshipped in a triple form besides Brighid and her mother Mór-Ríoghain, so the Christian God, who took the form of the Father, Son and Holy Spirit was not a difficult concept for Celts to grasp. So the shift from worshipping many deities to venerating one God would not have been a difficult one to cope with, and some Druids, as natural leaders of religious observances, entered the church as priests. Others became scholars and worked as monks in the *scriptoria* of the monasteries working on the manuscripts with their beautiful illuminated pictures and decoration, and recording

7 Statue of St Brigid in the Market Square, Kildare, Ireland. She wears a simple hooded tunic of the type that Brigid probably wore, not the medieval-style nun's habit in which she is often depicted.

not only the Christian stories but those of an earlier age that had survived in an oral tradition. However, these new Christian priests still respected their former Druid 'brothers' who had not converted as is clear from early literature.

In the earliest surviving biography of a British Celtic saint, dating to the end of the sixth century AD, *A Life of Sampson*, his teacher is stated to have been Illtyd (*c*. 425–505 AD) who was 'by descent a most wise Druid'. While in the *Life* of the British Celtic saint Beuno it says that the last words that he uttered on his deathbed were that he saw 'the Holy Trinity, the saints and the Druids.' One of the most famous 'converts' was Crimthann (*c*.521–97) who, when he began his Christian ministry, took the name Columcille (Columba in English), and was later canonised as a saint, becoming one of the three most famous saints of Ireland and also much revered in Scotland. He was a member of an aristocratic family, but also a *filidh* [Druid bard], and had undergone training in this before becoming a Christian.

According to the *Life of Saint Columcille* when the saint landed on the Scottish island of Iona to establish a monastery, he found two Druids who claimed to be bishops and told Columba that they had already planted the Christian faith there. However, he believed that they had not been properly ordained and so ordered them to leave the island, which they did. This shows they had converted to Christianity even if Columba did not think their office was officially sanctioned. In some of the very early Celtic literature the word Druid is used when referring to a saint.

Druids were present at battles as advisors and to give an advantage to their own side using their magic. Priests and bishops, carrying relics of the saints, were also present at battles in Christian Ireland even as late as the fifteenth century to give an advantage to their own side by prayers and the use of relics to call on the power of the saints. The Druids used wands or staffs to direct their power and perform magic, while Christian saints in Ireland used their staff (*bachall*) as their main vehicle of power to convey the holy energy by which they brought about miracles. Several staffs survive encased in croziers as relics. The poets too continued to apply their skills, as poetry and storytelling remained popular in Celtic societies almost to the present day, keeping alive the tales from the 'old days' but Christianised and diluted for their new Christian audiences.

This conversion from one religion to another seems to have applied to the Druidesses of the temple of the Goddess Brighid. When they converted to Christianity it would have been quite natural for them to have wanted to continue their method of venerating their deity. However, they now maintained the perpetual flame in honour of the Christian God rather than the goddess, particularly as fire symbolism is very common in the Bible, and God's messenger, and even God himself, is visualised as a flame in Exodus 3.1–4:

> And the angel of the Lord appeared unto him in a flame of fire out of the midst of a bush: and he looked, and, behold, the bush burned with fire, and the bush was not consumed. And Moses said, I will now turn aside, and see this great sight, why the bush is not burnt. And when the lord saw that he turned aside to see, God called unto him out of the midst of the bush, and said, Moses, Moses. And he said, Here am I.

The power of healing, the attributes of wisdom and knowledge, the connection with fertility, agriculture and the keeping of animals, all aspects connected with the goddess, are also found in many parts of both the Old and New Testaments of the Bible. The church therefore actively encouraged the people to now associate them with the Christian God, and there has been an extraordinary symbiosis in Ireland between the ancient religion and Christianity.

The Goddess Brighid becomes a Saint

This can be clearly seen in the many religious customs that were once carried out and which, to some extent, still are.

However, folk tales connecting the Goddess Brighid with such aspects had been prevalent for generations and were widespread, so continued to survive but with a Christian veneer. The former Chief Druidess and now first Abbess of Kildare continued to wield considerable influence and power, and almost certainly was still carrying out customs and services that were concerned with healing, fertility and blessings as she did before her conversion. The Christianised tales about the Goddess Brighid became associated with this woman, particularly following her death.

The *Táin Bó Cuailgne* (*The Cattle Raid of Cooley*) written down in the seventh century, describes the clothing worn by a Druidess called Fidelma which may allow us to visualise what the Chief Druidess of Brighid, and the person on whom the early *Lives* of St Brigid was based may have looked like. Many illustrations of St Brigid show her wearing the clothing of a nun or abbess as worn from the medieval period, but the 'real' Abbess Brigid would have worn the type of clothes she had worn as a Druidess of Brighid, of a style and quality suitable for a woman of her status.

Most people visualise Druids as always wearing white robes, as they certainly did in Gaul (France) for one ceremony that involved the gathering of mistletoe. However, Druids also wore other ceremonial robes such as the 'bull's hide and a white speckled head dress with fluttering wings' worn by one Irish Druid who, on another occasion, was described as wearing a 'coloured cloak with gold ear rings'. So it is possible that Brigid, as Chief Druidess, might have worn clothes resembling those that Fidelma wore as described in the *Táin Bó Cuailnge*:

> She wore a speckled cloak fastened around her with a gold pin, a red embroidered hooded tunic and sandals with gold clasps. She had hair in three tresses; two wound upwards on her head and the third hanging down her back, brushing her calves. She had a light gold weaving rod in her hand, with gold inlay.

It is interesting to note that one of the relics of the saint kept at St Brigid's chapel at Beckery near Glastonbury, Somerset, and said to have been left by the saint herself, was a weaving tool. Was this a symbolic item as the description of the Druidess Fidelma specifically mentions this? Traditional stories and beliefs about St Brigid say she wore white, and one early *Life* says Bishop Mac Caille clothed her with a white cloak and placed a white veil over her head, and while this may simply have been symbolic of her purity and virginity, it may also be a memory of the actual colour of the clothing worn by the first nuns of her order. Perhaps, on becoming Christians, the Druidesses changed the colour of the clothing they had worn as servants of the goddess, possibly adopting a white hooded tunic (*cochal brat*) with a white or coloured cloak. One of the relics of St Brigid is a piece of her cloak kept in the Cathedral of St Sauveur, Bruges, in Belgium (see page 226) and her cloak is mentioned in some of the stories detailing the miracles attributed to her. In one case she used it to acquire land for her monastery (page 58) and in the other she hung it on a sunbeam by mistake! (page 82).

While it is difficult to know for certain what a fifth/sixth- century monk who was formerly a Druid would have worn, it was probably similar to images of priests and monks found on such items as reliquaries of the eighth and ninth century. These show them wearing an ankle length, belted tunic with an ankle length cloak, however, we do know what their hair style would have been. There are many examples from Irish sources which show that the male Druid's had a tonsure, something that is common in priests of many world religions. For example, Lucat Mael and Caplait, two Druids, were described as having their hair cut in the style known as *airbacc giunnae*, which means 'fence cut of the hair', that is it was cut from ear to ear leaving the front shaved, with a sort of 'fence' of hair between the shaved and unshaved area, and left long at the back.

As Christianity began to take hold in Ireland this style was retained as the tonsure of the Celtic Christian monk, despite opposition to it by St Patrick and was to become one of the causes of argument between the Roman and Celtic churches at the Assembly of Whitby in 664 AD. However, the Celtic tonsure continued to be worn by some Celtic Orders as late as the eighth century, and was apparently worn by one order of wandering monks, the Culdes (*Cele De* – Servant of God), in Scotland as late as the fourteenth century. An early nickname for an Irish Celtic cleric was *táilchenn*, 'adze – head', a reference to this Druidic-style of tonsure.

The monks who wrote the *Lives* in the early centuries of Christianity were themselves drawn from the local population; and while 'folk memory' can sometimes record an event with great accuracy from generation to generation, in other cases it can distort an event within a few years to something whose origin is almost unrecognisable. The ideas of the Irish people during the early Christian era were shaped by generations of pagan legends about their 'old' gods, many of which contained a large element of magic and tales of supernatural events, and these ideas were transferred to the stories about some of these early saints, including Brigid. While some of the stories are obviously based on incidents in the Bible, others, particularly those recorded before the tenth century, incorporate elements which are based on the life of the last Chief Druidess of Brighid and incorporate attributes of the goddess, being Christianised versions of stories about the Goddess Brighid that go back to her 'conception' in the first century AD.

Early Christian literature describes St Brigid of Kildare as a real person, an abbess, who founded a double monastery at Kildare about 470 AD and died about 525 AD. Although there are six early versions of the life of the saint surviving earlier than the tenth century, three only give a very short account of her early life, but three of the *Lives* written down in the eighth and ninth century, but almost certainly based on earlier manuscripts or oral stories, give considerably more detail of the saint's early life, and one in particular contains a lot of biographical information. Reading this it sounds like it is describing the events in the life of a real person in the fifth century before Christianity had got much of a hold in Ireland.

This account of Brigid's early life describes the close ties she initially had with Druidism. This version would be less acceptable for inclusion in many later *Lives* as Druids were often described as 'wizards' in a negative and derogatory sense after the tenth century. Later many people would have felt it would not have been suitable to have such a respected and widely venerated saint so closely associated with Druids, and it is these early *Lives* that are likely to be the most accurate. It is unlikely that their compilers would have made a special effort to show the saint was base born, and were reflecting what was generally believed about her at the time, when the wisdom of the Druids was still well known and respected. Even the *Rennes Dinnsenchus* from the twelfth-century *Book of Leinster*, which stylistically is eleventh-century but records an older, oral tradition, says Brigid was a *ban druaid* (Druidess) before she converted to Christianity.

So it seems that the Brigid described in some of these first versions of the *Life of Saint Brigid* was based on memories of the Chief Druidess of the Goddess Brighid's temple at Kildare, who either took the name Brighid or used it as part of her title, and later became a Christian and the first Abbess of what was to became a double monastery. This woman did

The Goddess Brighid becomes a Saint

not 'found' the monastery at Kildare but was responsible for 'Christianising' the pagan veneration at a cult-centre, and converting it to a monastery. However, it would have been after Brigid's death, when she became a 'larger than life figure', that folk stories and characteristics of the goddess became attributed to her, particularly as she had the same name or title as the goddess. The 'magic' in the tales now being 'miracles', the Christian term for similar events, and these intermingled with real events from her life.

In three of the early *Lives* hints about the career of this woman are given, from the signs that indicated she was to be a special person within the Order of Druids, which was foretold before her birth, to her conversion to Christianity. How she continued to live with her former Druidesses and Druids at the cult centre venerating the Christian God using practices much the same as they had with the Goddess Brighid, and how she went on to convert her former fellow Druids and the ordinary people. All this can be discerned when one of these accounts is examined in detail:

> One day her father, Dubhthach, Chieftain of Leinster, was driving in his chariot with his bondmaid, Broicsech, who was to be the mother of Brigid. They passed the house of a certain Druid, called Mathghean who prophesied that the bondmaid would give birth to a child of great renown. Because of the jealousy of his wife, Dubhthach sold the bondmaid to a *Bhaírd* [poet], who in turn then sold her to a Druid of Tirconnell. This Druid gave a great feast to which he invited the King of Conaill whose wife was also expecting a child. The Druid foretold that 'the child which should be born at sunrise tomorrow, and neither within the house nor without', would be greater than any other child in Ireland.
>
> That night the Queen gave birth, but her child was dead. At sunrise the bondmaid, Broicsech, bore a child with one foot inside the house and the other outside it. Broicsech's child was then brought into the presence of the Queen's child who, miraculously, was restored to life. This Druid travelled with Broicsech and her baby into Connacht when, in a dream, he saw 'three clerics in shining garments', who poured oil on the girl's head and thus completed the order of baptism in the usual manner, calling the child Brigid.
>
> When she was a grown girl the Druid and his wife went one day to the dairy where their daughter was helping her mother and demanded to have a great hamper, eighteen hands high to be filled with butter. The girl had only the making of one churning and a half but, at her prayer, the butter increased so that if the hampers of the men of Munster had been given to her she would have filled them all.
>
> Whereupon the Druid freed Brigid's mother, presented to Brigid the butter and the cows, which she gave to the poor and needy, received baptism and remained until his death in Brigid's company.

Examining the story with the possibility that this concerns the last Chief Druidess of the Goddess Brighid it can be seen that it tells of a Druid called Mathghean, who prophesied that the bondmaid Broicsech was going to give birth to a 'child of great renown', a special child, that is one chosen by the gods. Prophesying was one of the skills of the Order of Druids, usually carried out by the *Vates* and Mathghean seems to have been foretelling the coming of a great person as far as the Order was concerned. As the child is described as of 'great renown' not as a great warrior and, because of its mother's status, it could not be a renowned member of the aristocracy or a monarch.

Broicsech was then 'sold' to a poet, that is a Bard of the Druid Order, because of the jealousy of her master's wife, so continuing the Druid link. This Bard then 'sold her' to a Druid of Tirconnell. The Bard obviously did not 'keep' her very long, so is this symbolic of the importance of this child being recognised by all three categories of the order – *Vates*, Bards and Druids even before she was born? The Druid of Tirconnell also made a prophesy showing the great part this child would play in the Druid Order when he said the child would be born at sunrise, 'neither within the house nor without.'

The meaning of this is that the child would be in a position of liminality (Latin: *limen* – a threshold), which in Celtic tradition indicates that it would occupy a position between two worlds, usually referring to the world of mankind and the Otherworld. In this case it was to be between the worlds of paganism and Christianity, and this is reinforced by the time of birth at 'sunrise' so spanning the time between dark and light, suggesting the bringing of great enlightenment to a place or people who were in darkness – the knowledge of Christianity to the pagans. He also confirms that it will be 'greater than any other child in Ireland.'

Broicsech is now associated with a Druid who is obviously very important as he is in a position to throw a feast for a king. There are two pregnant women in the story, one a lowly bondmaid, the other a queen. While on the face of it this 'child of great renown' should be the issue of the king and queen, it seems that the Druid was actually foretelling the coming of one of the greatest leaders of the Druid Order, the Chief Druidess of the temple of their patron goddess, Brighid, who was later to convert much of the Druid Order and the Irish people to the new religion, Christianity.

Then comes the confirmation that this child was the special one whose coming was foretold by the Druids, when she showed the power of healing by restoring life to the stillborn child of the Queen just by being brought into the presence of the dead baby. Why was Broicsech's child 'brought into the presence of the queen's child'? Was this simply an early indicator of the power of the child, or could this indicate something deeper? The original plan may have been for Broicsech's child to be adopted by the queen to replace her dead baby, but that the restoring of life to the queen's child meant that Broicsech's baby, who already had her life mapped out by the Goddess Brighid, could continue along its pre-ordained path.

This early manifestation of the power of healing was something the child shared with the Goddess Brighid and confirmed her connection with the deity. It suggested this baby was later destined to actively serve the goddess in the most important position, as Chief Druidess of her cult-centre and, as Brighid was a patron goddess of the Druid Order, this would have been at the most important temple in Ireland. Broicsech and her baby then travelled with the Druid to Connacht where he had a dream that formed another prophesy. In this Broicsech's child had oil poured on her head by 'three clerics in shining garments', the number three being of great significance to the Celts. Was this foretelling the anointing of a great Druid leader, in much the same way that British monarchs are still traditionally anointed with oil at their coronations? The child was then either given the 'name' Brighid (High One) after the goddess, or perhaps it was being foretold that she would later take the name as part of a title indicating her future role as the most notable of the Chief Druidesses of Brighid's cult centre.

Druids did carry out baptisms, called in Irish *baisteadh gintlí* (baptism of a pagan), according to a number of Irish sources. Some people have argued that this ceremony was simply a parody of a Christian concept that the scribes attributed to the Druids, but the evidence for actual Druidic baptism is found in several sources. When the hero Conall Cernach was born, 'Druids came to baptise the child' and sang a ritual song over him. Ailill Ollamh of Munster was 'baptised in Druidic streams', and the practice was not confined to Ireland as the Welsh hero Gwri of the Golden Hair was baptised by Druids. While it seems that water was usually

The Goddess Brighid becomes a Saint

used in these rituals, the use of oils in the case of Brigid probably indicated her importance and points to her future as a great Druid leader.

Later in the account it says that 'the Druid and his wife went to the dairy where their daughter was helping her mother…' This shows the Druid adopted Brigid, a not uncommon practice among the early Celts who had a system of fosterage (*altram*), where children were placed with foster parents even if their blood relatives were still alive. There were a number of reasons for doing this, such as gaining advancement for a youngster, but most often it was done as a form of apprenticeship. Druids were very popular as foster parents, with the type of training and conditions being laid down in legal tracts known as the Brehon Laws. This clearly indicates that Brigid was undergoing training as a member of the Druid Order.

Then occurs a miracle where Brigid uses her powers (or those of the goddess?), to greatly increase the yield of the cattle, so important to Celtic wealth, perhaps suggesting fertility aspects of the goddess coming into the story as well as a link with the goddess's mother

8 A statue of St Brigid outside the church of St Bridget and St Oliver, Inishmore, Aran Islands, off the west coast of Ireland. The inscription reads: 'May the cloak of Saint Brigid be on [protect] the young people of Aran. St Brigid's Day 1995.'

Mór-Ríoghain who had a herd of Otherworld cattle. This follows a Celtic tradition of notable beings increasing the yield of food. Brigid is then presented with the butter and cows which she gives to the poor and needy, so we may now have indicators of the changeover point from paganism to Christianity. The Goddess Brighid has fertility aspects and probable involvement with cows that she 'inherited' from her mother, and an overall protective role, now in its Christianised form, symbolised as assisting the poor and needy. As a saint she was often invoked for her ability to protect the family, herds, crops and the house, particularly in Ireland and Scotland.

So it seems the Chief Druidess of the goddess's temple, and presumably her fellow Druidesses, become Christians but, despite this, continued their veneration and practices in a very similar way to before. However, they now did it in honour of the Christian God who is regarded as one or three (Father, Son and Holy Spirit) instead of the Goddess Brighid, a deity who was also visualised as one or three. If the generally accepted date for the 'founding' of the double monastery at Kildare is correct, it indicates that the conversion of the Druidesses took place in 470 AD, which would also be the date when the Goddess Brighid's temple became a Christian church.

As the Druid in the story does not receive baptism till after Brigid was obviously Christianised, could this indicate that she was not able to persuade the male Druid's to follow her example till some time later when they saw that things had not changed too much? Although it might simply indicate that other Irish Druid's became Christians some time after those at Kildare were converted. Christianity and Druidism existed side by side for a long time in Ireland. For example, Diarmaid mac Cearrbheoil, High King of Ireland 544–565 AD, had both Druids and Christian priests at his court as advisors, and took counsel from both equally.

The newly baptised Druid in the story 'remained until his death in Brigid's company', suggesting that the former Druidesses and Druids continued to occupy the cult-centre of the goddess. This became a double monastery (*Monasteria duplicia*) for males and females, now called monks and nuns but which took the form of linked but separate communities, following Christian practice. The former Chief Druidess of Brighid presided over both communities, as is stated in all accounts of the life of St Brigid, continuing to reflect her importance in this religious establishment. This followed the Celtic practice of giving equal status to women, and sometimes even a superior role in some religious contexts, particularly in the case of abbesses. When the first *Lives* of St Brigid were written it would be natural to base it on the life of the last Chief Druidess of the goddess's temple and the first Abbess of the church, who was so intimately connected both with the goddess but played such an important part in early Irish Christianity. So it is her details that have come down to us in this version as the 'real' St Brigid and provides the link between a pagan deity and Christian saint.

This pre-Christian attitude to female equality carried forward into the early Celtic church where initially some women even celebrated Mass as well as priests. However, this practice was abhorrent to the Roman church which engaged in a long struggle for supremacy over the Celtic church. The fact that women were celebrating the Mass came to the attention of Rome around 520 AD when three Roman bishops wrote a letter to Lovocat and Catihern, two Breton priests: 'you celebrate the divine sacrifice of the Mass with the assistance of women to whom you gave the name of *conhospitae*… We are deeply grieved to see an abominable heresy… Renounce these abuses…'.

Further evidence that women played an equal part in Irish Celtic society into early Christian times, is indicated by the large numbers of female Celtic saints in comparison to those in other European countries. This was to gradually change to a patriarchal society, which was to be reinforced and firmly established when Roman Christianity replaced that of the Celtic church from the seventh century. In the popular mind of the early Irish Christians the essential characteristics of their saints was not so much moral goodness but the mysterious power to work miracles, and although many of the stories concerning Brigid do mention her goodness and virginity, they are outweighed by stories connected with wisdom, cattle, prosperity, protection and healing, the attributes of Brighid as a goddess.

One ancient story about St Brigid, also noted in *The Book of Lismore*, concerns the supply of a love potion, not really a Christian activity, but one that was carried out by Druids:

> One day a man came to Brigid to tell her of his distress about the state of his marriage, as his wife would have nothing to do with him and was threatening to leave him. He therefore asked the Abbess for a spell to make his wife love him again. Brigid blessed some water and gave it to him, instructing him to sprinkle over their house, over their food and drink, and over their bed. After he returned home he carried out Brigid's instructions and his wife's attitude completely changed and she became exceedingly loving and wanted to be with him all the time.

As this last Chief Druidess of the Goddess Brighid's temple was instrumental in bringing Christianity to this area, building on the work of conversion started by St Patrick about twenty years earlier, and further spreading the new religion, her role would have been both influential and a very important one. Her 'story' would be remembered, but with elaboration, changes and additions as time passed and her fame spread after her death. It would have been expected that the Chief Druidess would have changed her name upon conversion to Christianity, and may well have done so, but her former name or title was so ingrained in local folk memory that it was still used for her, and passed into the histories of the saint, the earliest of which was written well over 100 years after her death.

Many new stories about St Brigid's supposed miracles were added to her repertoire over the next few centuries as the appeal and veneration of the saint spread and became entangled with the real life of Brigid, the first Abbess and last Chief Druidess of the Kildare site. We are fortunate that Irish monks recorded the '*Lives*' of many of the early saints, but they were not too interested in accuracy since they wanted to promote the saint who was often the founder of their own monastery, so increasing its status. They also had a tendency to attribute miracles found in the Bible to their 'own' saint, and often included folk tales about them if it added weight to their work. During the medieval period St Brigid become very popular, and veneration of her became widespread in Ireland, Britain and Europe which led to the appearance of many more stories about her life inspired by both the Celtic medieval imagination and the romantic view so popular at that period, especially regarding the miracles she was said to have performed. So, in a sense, the Goddess Brighid, some of the folk stories about her, many of her characteristics and some of the customs associated with her, survived her Christian 'conversion' and lived on, initially in her last Chief Druidess/first Abbess of Kildare, and later in the beliefs of the many people who have venerated to the present day, either as a saint, a goddess or a blend of both.

The second Abbess after the death of Brigid was Darlughdacha, whose name means 'Daughter' (*Dar*) of Lugh plus a personal name, *dacha* or *dorcha* meaning dark one, probably a reference to her hair colour. Lugh was a Celtic god who was depicted as a young and handsome warrior, a sun god who was credited with a wide range of attributes. This suggests that the second Abbess came from the Luigne tribe who took their name from the god Lugh, and may perhaps, have been a former Druidess of the temple of the Goddess Brighid. Cill Dara was situated in the kingdom of the Luigne so Brigid's family was also a member of this tribe, and the Chief Druidess's 'real' name would probably have been Darlugh [personal name] brighid. Darlughdacha only held the position of Abbess for a year before her death.

In the earliest *Life of Saint Brigid*, written by Cogitosus c.650 AD, the original of which is housed in the Dominican Friary at Eichstatt in Bavaria, he only briefly mentions her parentage, saying that St Brigid was born in Leinster, of parents belonging to the 'tribe of Eochaid'. 'Her father was Dubhthach, her mother Broicsech.' The Eochaid that he mentions were the Eochaid Fúath nAirt, the ancestors of the Fothairt, a group of noble families with interests in the Kildare area and other parts of Leinster. The earliest surviving non-religious reference to St Brigid dates to the eighth century and is found in a document that explains the origin of the Fothairt sept (called a clan in Scotland i.e. extended family group), in which she is stated to have been a member of this sept, and is described as 'truly pious Brig-eoit' and another 'Mary'.

The rapid and early development of her cult was closely connected with the rise to power of a new sept, the Uí Dhúnlainge, in Leinster, whose leader, Faolán mac Colmáin seized the kingship of the province about 633 AD. He was married to a woman of the Fothairt and his brother, Áed Dub mac Colmáin, was Bishop of Kildare. So it would have been advantageous to stress the importance of St Brigid and the fact that he was related to her, and many of the early abbesses of Kildare were members of the Fothairt. Indeed family control of the senior executive office in the church, abbots and abbesses, was enshrined in Irish law and was justified if the office holder was *fine érlama*, that is related to the founder of the church or saint, or was related to the landowner who endowed the church or monastery. So this may give a further clue to the true genealogy of the last Chief Druidess of the temple of Brighid at Kildare.

It was not unusual for membership of the Druid Order to be hereditary, and this tradition continued after the conversion to Christianity of the Chief Druidess of Brighid who became the first abbess of Kildare. While Brigid may not have been a hereditary Druid, having achieved her high status within the Druid Order by study and her role being 'foretold', her Christian successors to the post of abbess at Kildare were all related to her for a long time. This was a powerful position as the abbess was not only in effective control of the monastery and nunnery at Kildare, but other Christian communities as well. She was also effectively Governor of the major secular city that seems to have developed on the outer monastic precinct by the seventh century.

Reliable records of St Brigid's abbess successors begin in the early seventh century with the death of Gnáthnat in 690 AD, and while her family name is not given, her successor, Sébdann filia Chuirc, was a member of the Ui Chuirc, a sub-group of the Fothairt family to which St Brigid belonged. This claim is reinforced as almost all the Abbesses of Kildare seem to have been members of one or other branches of the Fothairt family, with only one break between 690 AD and 1016 when the last Fothairt abbess died. All used the title of *Abatissa* (Abbess) or *Dominatrix* (Governor), until the appointment of Cobflaith ingen Duib-duin (died 916 AD) who took the title of *Comarba Brigte* (successor of Brigid). The next two Abbesses reverted to the simple title of *Abatissa* but *Comarba Brigte* was used again by Muirenn ingen Chongalaig (died 979 AD) and by all abbesses who came after her, although her immediate successor, Eithne ingen Uí Suairt was the last member of the Fothairt family to hold the post. Although her successors continued to use the title *Comarba Brigte* it was in a spiritual sense as they were no longer actually related to St Brigid. Even Cormac Ua Cathasaig (died 1146), Archbishop of Leinster, used that title for a while, the abbess of the time using the title *Banchomarba Brigte*, 'female successor of Brigid'.

In many respects pagan elements at what was formerly the cult-centre of the goddess, and now the double monastery of Kildare, seemed to have continued for at least another 750 years. The most notable aspect was the tending of a perpetual fire as described by Gerald of Wales (see page 78). One place where the veneration of the goddess rather than the saint seems to have continued until the nineteenth century was in the parish of Knockbride in Co Cavan, Ireland. There are a number of Brighid/Brigid connections in the district, and even the parish is named after a 'sacred' hill called Knockbride, which means Hill of Brigid. On top of this hill is a well, an ancient fort and two lakes called Upper Loughbride (Upper Brigid's Lake) and Lower Loughbride (Lower Brigid's Lake). Many archaeological finds have been made in the area, including a carved Celtic stone head which was believed locally to be that of St Brigid, but whose exact whereabouts are now unknown. This stone head was passed down in a local family from one generation to the other, and was revered as something very special connected with St Brigid. However, it is uncertain whether the locals regarded this head as representing the Goddess Brighid or St Brigid, or perhaps both!

The stone head was removed from the owners' possession by Father O' Reilly, the parish priest from 1840–44 and taken to the site of a new church being built at the west end of the parish. There are two versions of the story that explains its fate after the priest took it from its keeper. One story says that the priest arrived at the site of the new church without the head and claimed that on his way there the head suddenly jumped out of the carriage and fell into a lake where it disappeared! However, another version of the tale says that the priest arrived at the site just as dusk was falling and asked for help in unloading the carving. This was done, but the head was never seen again. This suggests that the priest may have placed the head in the foundations of the new church as a way of 'putting to rest' the pagan goddess represented by this carving. There is a precedent for this practice from Guernsey in the Channel Islands where, in 1878, a large ancient goddess statue, today known as *La Gran'mere du Castel*, was discovered buried 30 cms below the floor of the entrance to the chancel of the church. It was undoubtedly placed there as a way of burying the pagan practices that the statue represented. Today the parish church of West Knockbride has a statue of St Brigid in a niche on the front of the building.

Four

The Temple of the Goddess Brighid: the Church of Saint Brigid

The site of the Goddess Brighid's temple at Kildare was situated on top of a clay ridge known as *Drum Cruadh* (rocky backbone), which stands 10–15m (32–49ft) above the surrounding flat land on the western edge of a 2,000 hectare plain known as the *Curragh*. A name that indicates a boat-shaped landscape, i.e. a flat plain surrounded by hills, and described by Gerald of Wales in the 1180s as 'Brigid's pastures', who goes on to say that 'no one has dared to put a plough into them. It is regarded as miraculous that these pastures, even though all the animals of the whole province have eaten the grass right down to the ground, when morning comes have just as much grass as ever'. Archaeology suggests that the *Curragh* has been a sacred landscape since the Bronze Age. The site of the temple of Brighid and the present cathedral are situated at the higher, western end of this ridge and would have been visible from a long way off across this great flat grassland.

While we do not know exactly what the temple of the Goddess Brighid at Kildare looked like, we can suggest it was situated inside a temple precinct, like so many of the Romano-Celtic temples found in Britain and Europe, fenced off with access restricted to the Druidesses who served the goddess. The site was associated with a sacred oak tree or possibly a grove of oaks, the place preferred by Druid's for worship, although the name Kildare is derived from *Cill Dara* which means 'church of the oak tree'. Many Greek and Roman writers mention the importance of the oak to the Druids, for example Pliny the Elder (23–79 AD) wrote: 'While on the subject [of mistletoe] one should not omit the veneration paid in Gaul to this plant and the tree on which it grows, providing it is a hard oak. It is in the groves of these trees that the Druids perform their rites', and 'They select oak for the sake of that particular tree, and will not perform any religious rites without its leaves.'

The form that the temple took is not known for certain but, as the Romans did not conquer Ireland, it would not have taken the shape typical of Celtic temples of the Roman period in Britain. Many of these consisted of a square tower, usually with a roof but occasionally without, with windows high up in the walls if roofed, and a colonnaded portico on each of its four sides. However, underlying many of these Roman period buildings are traces of Iron Age temples, usually of a circular form. For example at Hayling Island in Essex there was a pre-Roman temple consisting of a circular building in the centre of a sacred enclosure.

The Temple of the Goddess Brighid: the Church of Saint Brigid

9 Reconstruction of an Iron Age round house. The temple of the Goddess Brighid and the dwellings of the Druidesses who officiated there probably looked much like this.

Many buildings in the Celtic Iron Age were round with conical thatched roofs, the wattle walls plastered with clay or daub, a mixture of mud, dung and straw to waterproof them and keep out draughts and which, as far as is known, had no openings except for a door. There was often a narrow ditch surrounding it to take away rainwater shed from the roof, and sometimes a porch to protect the entrance. Such structures were described by the writer Strabo (*c*.60 BC–20 AD): 'Their houses are large and circular, built of planks and wickerwork, the roof being a dome of thatch.' There is evidence that decoration in the form of coloured walls, woodwork, plaster and hangings of leather or textiles were used inside such buildings, reflecting the Celts love of colour and decoration. They would also place beautiful gold items and other precious objects inside their temples as offerings to their gods and goddesses, and no matter how valuable these were, no one dared touch them or rob the temples.

It is therefore quite likely that the temple of Brighid would have looked like a typical Celtic round house, with the perpetual fire placed in a central position in the building, the smoke escaping through the thatched roof. Nearby there would have been a settlement of other round huts in which the Druidesses who served the goddess and tended the sacred fire lived, with possibly Druids present as well, each having their own hut as was the case with monks' cells in early Celtic monasteries. This type of arrangement is suggested by the description of Gerald of Wales (*c*.1145–1223) in his *Expurgatio Hibernica*.

He was very impressed by the perpetual fire kept there and described the area where the fire was maintained (page 78), and while his description applied to the Christian monastery, it resembles the layout of many Romano-Celtic temples. While most temples do not seem to have associated dwellings for priests or priestesses (although these may not have been looked for in excavations or lay some distance from the actual temple and outside the excavated areas) some do, and their size suggests there was only one priest or priestess, possibly with a family, or just a couple of priests or priestesses officiating. However, a Romano-British temple excavated in the 1940s on the site of Heathrow Airport, Middlesex, revealed a rectangular timber building with a colonnade on each side, with eleven huts just to the north of this for a small community living by the sacred site.

Strabo (64–24 BC) mentions a community of Druidesses living on an island near the mouth of the River Loire in France. He stated that no man was permitted to land on the island, a practice not unlike that at Kildare where men were forbidden to enter the inner precinct of the temple of Brighid. Pomponius Mela, who wrote his *De Chorographica* in the mid-first century AD, tells of nine virgin Priestesses who lived on the island of Sena off Pointe du Raz on the west coast of Armorica (Brittany) who had a reputation for foretelling the future and giving advice to sailors. An interesting lead tablet found in a necropolis of one 115 graves near Millau in Southern France, dating to 90–110 AD, seems to be a formulae to preserve the dead from the curses of Druidesses and actually mentions the names of nine members of the Order.

Gerald of Wales says that the fire was maintained by nuns but this reflects a practice that pre-dates the Christian monastery on the site when the flame would have been maintained by *ban Druaid* (Druidesses). We have no contemporary description of the temple of Brighid, the Druidesses who served the goddess there or their appearance. We can, however, get some idea of their life and the part they played in Celtic religious and secular society by looking at various authors from the classical Roman and Greek world, descriptions in Celtic literature, and the results of archaeology. The Greek historian Plutarch (c.46–c.120 AD), and the Roman historian Tacitus (56–c.120 AD) both mention a powerful Gaulish woman called Eponina who was probably a Druidess of the cult of the Celtic horse goddess Epona, an example of a priestess taking on a title or name related to the deity she served. Perhaps she was even regarded as embodying the spirit of the goddess on earth, which may also have been the case with the Chief Druidess of the temple of Brighid at Kildare.

Another Druidess, Camma, who lived in Galatia (Turkey), was described as an hereditary priestess. Further evidence for the hereditary nature of this caste is found in the work of the fourth century AD Gaulish writer Ausonius Decimus Magnus who mentions in his *Parentalia*, that he had an aunt called Dryadia (Druidess). This was at a time when the Druid Order had been officially banned for centuries in Europe and Britain, although Druids continued to act as intercessors with the gods at the shrines and temples, and were involved in the healing arts, prophesying and giving advice. They still played an important role in Celtic society but without the political power (except in Ireland) that they once had.

The Celts had a very strong belief in an afterlife, and regarded the gods and goddesses as their ancestors rather than their creators. Kings and chieftains would claim decent from a god or goddess ancestor, so it is possible that the more senior members of the Druid Order would also claim decent from one of the deities they served or revered. They might even claim to be an embodiment of a deity and classical writers mention Druidesses who were regarded as divine by their people. Could this have been the case with the Chief Druidesses of Brighid? Such an appropriate ancestry would give their children, relations or descendants a distinct advantage in being appointed to a senior position in the Druidic priesthood, a process that seems to have continued into Christian times in the case of Kildare (see page 44).

It was not uncommon for Celtic women, including Druidesses, to act as ambassadors, and while women could belong to any of the classes of Druids within the Order, they seem to have commanded particular respect as seers (foretellers of the future). This is clear from the writings of Tacitus describing the Celts:

The Temple of the Goddess Brighid: the Church of Saint Brigid

10 An illustration of the Temple in Jerusalem from the eighth-century *Book of Kells*. This gives a clue to the exterior appearance of St Brigid's Cathedral in the seventh century.

11 Model of Kildare Cathedral as it may have appeared in the seventh century.

> There is, in their opinion, something sacred in the female sex, and even the power of foreseeing future events. Their advice is, therefore, always heard; they are deemed oracular. We have seen in the reign of Vespasian [Emperor 69–79 AD] the famous Veleda revered as a divinity by her countrymen. Before her time, Aurinia and others were held in equal veneration…

However, other references also make it clear that Druids and Druidesses could undertake training for admittance to the Order, so a Druidic lineage was not essential. Posidonius (c.135–50 BC) notes that Gaulish Druidesses were very independent of their husbands, confirming that female Druids could marry, as was the case with the male members of the Order.

Druidesses are mentioned on various inscriptions and dedications from Europe, and there are many references to *ban Druaid* (Druidesses) and *ban Fhlaith* or *ban Filidh* (woman seer) in Celtic literature. The important role of Druidesses is mentioned in the *Rennes Dinnsenchus*, and the names of many Druidesses are mentioned in Irish literature. There are also several accounts of Roman emperors consulting Druidesses because of their powers of foretelling the future. Among the emperors consulting them were Alexander Severus before his military expedition of 235 AD; Lucius Domitius Aurelianus (c.215–275 AD), who asked whether his children would retain the Imperial crown (the answer was no!); Gaius Aurelius Diocletian (284–305 AD) who, before he became Emperor, consulted a Druidess who told him he would eventually attain this position.

In the ninth-century *Tripartite Life of Saint Patrick* the saint warned kings not to accept the advice of Druids or Druidesses. In his Hymn, often called *The Breastplate of Patrick*, he specifically asks God to protect him against, among other things, Druids. In possibly the most famous of the stories from Irish mythology, first written down in the seventh century AD from an older oral tradition, the *Táin Bó Cuailgne* (*The Cattle Raid of Cooley*), Medb, Queen of Connacht, consults a Druidess called Fidelma who had just returned 'from learning verse and vision in Albion' [Britain], confirming Julius Caesar's statement that many Druid's went to Britain, particularly to the Isle of Anglesey off the north Welsh coast, to learn various Druidic arts.

The temple of Brighid was associated with a sacred tree (*bile*), and there were many individual trees, groves and woods in Ireland that were regarded as being special in pre-Christian times. These remained of great importance long after the coming of Christianity. There is strong evidence that many churches were built adjacent to sacred trees, and early accounts of this are found in the eighth century *Book of Armagh* and in the ninth-century *Tripartite Life of Saint Patrick*, both of which mention the *Bile Tortan*, a group of sacred ash trees. Such trees might mark tribal meeting places, or be the spot where tribal leaders were inaugurated, a practice that persisted right up to the sixteenth century in some places, and often indicate sites where the worship of various pagan deities was carried out.

In the *Annals of the Four Masters*, the entry for 995 AD records that the monastic town of Armagh was set on fire after a lightning strike. Not only were the buildings burnt but so was the sacred wood, *fidnemedh*, a name derived from *fid* – wood and *neimed* which indicates a sacred grove. This suggests that this ecclesiastical settlement was located there because it was a pagan site which needed 'Christianising'. Many other monastic sites have names that indicate there was originally a sacred tree or wood located there, e.g. *Doire Choluim Chille* (Columcille's Oakwood), founded by St Columcille in the sixth century, now Derry; *Gleann Uiseann* (sacred glen of yew), now called Killeshin, Co Carlow; *Doire Luráin* (Luran's Oakwood) founded by Bishop Luran; and there are many others.

These indicate the close connection of the early Christian church in Ireland with sacred pagan groves and trees, suggesting a deliberate policy to Christianise these pagan sites by

The Temple of the Goddess Brighid: the Church of Saint Brigid

12 An engraving entitled 'North View of the Cathedral Church at Kildare in Ireland', dated 1 September 1778, from Volume One of *The Virtuosi's Museum* by Paul Sandby. The ruinous state of the building can be clearly seen and in the foreground is the Anglo-Norman chapel whose remains today are known as the Fire House.

attaching them to Christian ritual. This continuity of worship made it easier to convert the pagans from venerating a number of deities to worshipping one Christian God. Kildare seems to have consisted of more than just a sacred oak as suggested by the maintenance there of a perpetual flame in honour of the Goddess Brighid, which required a temple to contain the flame and dwellings for the Druidesses who tended it.

As Brighid was 'conceived' in the first century AD (see chapter two) the temple must date to this period, but was probably established on an existing religious site that contained a notable sacred oak tree (*dáire*). This gave the place its name in Christian times, *cill dara*, 'church of the oak tree', singular, since if it had been established by a sacred grove of oaks the place would have been called *cill doire*, 'church of the oakwood'. One *Life of Saint Brigid*, by Animosus, written *c.* 980 AD, notes the oak tree was still there, although presumably then dead:

> In that place there stood a mighty oak tree, much beloved of Brigid, indeed blessed by her: the trunk survives to this day and none dare cut it with an axe. It possesses a property so great, that any person able to break off a part of it with their hands can hope thereby to win God's aid. Many miracles, by the blessings of Blessed Brigid, have been received through that oak tree.

This had probably been an object of veneration from the time that the site was the cult-centre of the Goddess Brighid and these holy souvenirs were carefully kept as relics. Places, including woods, where saints were said to have worked miracles are also commemorated by place names, such as that concerning St Brigid that is mentioned in Broccán's hymn in her honour.

Brigid: Goddess, Druidess, Saint

13 St Brigid's Cathedral, Kildare. The present building is basically thirteenth-century, but was largely repaired and restored in the nineteenth century. To the west stands the 32m (105ft) tall round tower built *c*.1135. To the north of the cathedral can be seen the low walls marking the building regarded by many people today as the Fire House.

This is recorded in the *Book of Lismore*, a compilation of older material made in the latter half of the fifteenth century:

> A wild boar which was in a certain wood north of Kildare and it used not to allow other pigs to approach it. Brigid with her staff blessed the wood at *Ross na Ferta* [wood of the miracle] in Kildare north of the bell house, so that after it was at peace with them, it was their leader always.

In Christian times, the most impressive building on the site at Kildare was the cathedral, which certainly existed in 639 AD when Áed Dub mac Colmáin was bishop. We are fortunate to have a description of how this looked in the seventh century, although it would have appeared different 150 years earlier when Brigid was the Abbess there. Initially it would have been a re-dedication of the goddess's temple, although such wooden round huts had to be rebuilt every twenty or thirty years. So it is quite likely therefore that at one of the rebuildings its shape was changed to the rectangular form that churches took in western Europe at that time, particularly in Gaul (France). This change of architectural style would have occurred as the fame of St Brigid spread and the church at Kildare become increasingly important to the growing Christian community in the area. A detailed description is found in the *Life of Saint Brigit* by Cogitosus (whose Irish name was Cogitis Maccu Mactheni), who was almost certainly a member of the religious community of the monastery at Kildare and wrote his *Life* about 650 AD:

> Neither should one pass over in silence the miracle wrought in the rebuilding of the church in which those two glorious bodies – namely Archbishop Conlaed and our most splendid virgin Brigid – are laid on the right and left of the ornate altar and rest in shrines adorned with a profusion of gold, silver, gems and precious stones with gold and silver crowns hanging from above and panels with different images presenting a variety of carvings and colours.
>
> And of the old is born a new thing that is a church which, as the number of the faithful of both sexes is growing, is of ample size and built aloft to a menacing height and is decorated with painted panels, and inside there are three spacious chapels which are divided by board walls under the single roof of the greater house. The first of these walls, which is painted with images and covered with wall – hangings, stretches width wise in the east part of the church from one wall to the other.
>
> In it there are two doors, one at either end, and through the door on the right side, one enters the sanctuary to the altar where the archbishop offers the Lord's sacrifice together with his monastic chapter and those appointed to administer the sacred mysteries. Through the other door, on the left side of the aforesaid cross – wall, only the Abbess enters, with her nuns and faithful widows enter to partake of the banquet of the body and blood of Jesus Christ.
>
> The second of these walls divides the floor of the building into two equal parts and stretches from the west wall to the wall running across the church. This church contains many windows and one finely wrought door on the right side through which the priests and the faithful peoples of the male sex enter the church, and a second door on the left side through which the nuns and congregation of faithful women are accustomed to enter.
>
> And so, in one vast church, a large congregation of people of varying status, rank, sex and local origin, with partitions placed between them, prays to the Almighty Master, differing in status, but one in spirit.

So clearly Cogitosus is describing a very impressive and colourful building, spacious and high roofed with many windows, decorated with painted panels and divided into three sections called *oratoria*. Two of these sections made up the nave which was divided down its length by a screen that segregated the men from the women, with separate entrances for each sex into the cathedral. The high alter area (chancel) was divided from the nave area by a screen covered with painted panels and rich embroideries, and separate access to this area from each side of the divided nave so monks and nuns would enter the high alter area separately. There was presumably an opening in the screen running across the width of the church so the congregation could see the altar and the ceremonies. It is clear that the tomb of St Brigid was situated on the north side of the altar and that of Bishop Conlaed on the south side.

Cogitosus says that the church had been rebuilt fairly recently (quite usual with wooden buildings), but the design of the church was an ancient one. The partitioned nave and chancel was a variation of one of the oldest styles of Christian building, the double church with connecting hall. An example has been found at Aquilea in France dating to the time of the Emperor Constantine (*c*.274–337 AD), and similar structures are known from elsewhere in France and Britain. So the inspiration for the early Christian church that replaced the temple of the Goddess Brighid was probably based on designs of buildings in such places as London and Lincoln in early post-Roman Britain, or sites in France.

Inevitably the external appearance of this building must remain conjectural, but there are clues from a variety of sources. Its basic fabric was wood, and this was obviously a traditional

Brigid: Goddess, Druidess, Saint

14 St Brigid's Cathedral, Kildare, Ireland, from the south-east.

material for Irish churches at this period as evidenced by many sources, including the Anglo-Saxon monk Bede (*c*.673–735 AD). He wrote about the Irish bishop, St Finan 'who on the island of Lindisfarne built a church suitable for the episcopal seat, but it was, according to the custom of the Irish, not of stone, but entirely of sawn oak, and he roofed it with reed'. Other early descriptions show some churches were roofed with shingles (oak tiling) still used in some countries today, or *slinn* (slate).

An ancient story recorded in the fifteenth-century *Book of Lismore*, a collection of ancient *Lives* of the saints dating back much earlier, tells how St Brigid acquired the materials to build her first church:

> Brigid went to Bishop Mel, that he might come and mark out her city for her. When they came thereafter to the place in which Kildare stands today, that was the time that Ailill, son of Dunlang, chanced to be coming, with a hundred horse loads of peeled rods, over the midst of Kildare.
>
> Then nuns were sent by Brigid to ask for some of the rods, but their request was refused. The horses straightaway sank to their knees. Stakes and wattles were taken from them, but they would not rise until Ailill had offered the hundred horse loads of building material to Brigid. And therewith was built Saint Brigid's great church in Kildare, and it is Ailill that fed the workers and paid them their wages.

Nothing remains of this cathedral today, but excavations at other sites have shown the size of such early buildings. For example the tenth-century cathedral at Glendalough, Co Waterford was a rectangular building with internal dimensions of 14.75m x 9m (48ft 4in x 29ft. 9in), while the cathedral of Clonmacnoise, Co Offaly, built in 908 AD, had internal dimensions of

18.9m x 8.75m (62ft x 28ft 9in). These suggest that early Irish cathedrals were about 15–18m (50–60ft) long internally, so giving some idea of the probable size of the seventh-century cathedral church of St Brigid at Kildare. Tirechan, a writer in the seventh century, tells how '…Patrick measured out the church of God with sixty of his feet. And Patrick said: "if this church be shortened your kingdom will not long be stable."' This suggests that there was a standard length in the seventh century for important churches and cathedrals that, tradition maintained, was determined by St Patrick himself.

Cogitosus specifically mentions its size, noting particularly its 'menacing height', so how high might it have been? A clue to its general shape can be found in the reliquary containers that take the form of churches, which are rectangular in shape with a hipped roof. This form of building occurs as an illustration of the Temple in Jerusalem in the eighth-century *Book of Kells*. Based on the proportions of this, along with some of the architectural reliquaries and tomb shrines such as that at Clones, Co Monaghan, and assuming its length was 16.7m (55ft), its quite possible that the walls of Kildare Cathedral could have been up to 9m (30ft) tall, suggesting a roof height, floor to roof ridge, of up to 16.7m (55ft). A hipped roof would have been required for a wooden structure of this size and the fact that this is the form taken by reliquaries and shrine tombs, as well as the illustration in the *Book of Kells*, suggests the roof of the Kildare cathedral must have been of this type.

Cogitosus also relates a miracle that concerns one particular feature of the cathedral church:

> Now, when the ancient door of the left hand entrance through which Saint Brigid used to enter the church was hung on its hinges by the craftsmen, it did not fill completely the new entrance of the rebuilt church. A quarter of the doorway was unfilled and a gap was showing and, had a fourth part been added and joined to the height of the door, then it would have been possible to fill up completely the high reconstructed doorway.
>
> And, as the craftsmen were wondering whether to make another new and bigger door which would fill up the entire doorway or attach a piece of board to the old door so that it would after-

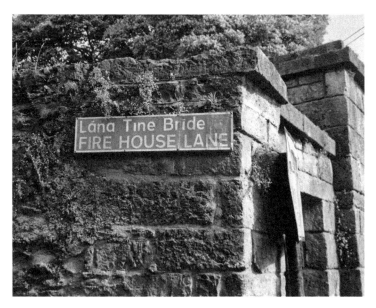

15 Sign on Fire House Lane, Kildare. The Irish reads: 'Lane of the Fire of Brigid'. This leads to Fire Cashal Lane, which is close to where the monastic Fire House once stood that contained a flame kept alight (with one brief interlude) from the time of St Brigid (and probably earlier), until the sixteenth century.

wards be big enough, the Master and leader of all Irish craftsmen offered this wise advice: 'This coming night we must pray to the Lord with faith by Saint Brigid's side so that in the morning she may guide us as to our course of action in this task.'

And so he passed the night praying before the glorious shrine of Saint Brigid and, getting up in the morning after that night, having completed his prayer with faith, he had the old door pushed in and placed on its hinges and it now filled the whole doorway. Neither did it lack anything in fullness nor was it in any respect excessive in size.

And so it was Saint Brigid who extended the door in height so that the whole doorway was filled by it and no gap appeared in it except when the door was pushed back for people to go into the church. And this miracle of God's power was revealed clearly to the eyes of all who see the door and the doorway.

Once again Cogitosus gives some useful information about the church. The 'ancient door' obviously came from an earlier building, and was the one that was used by Brigid as first abbess. Because of this association it was reused in the cathedral described by Cogitosus. It was probably the main or only door of the first building constructed as a Christian church after a rebuilding of the goddess's temple that was converted to a church. This 'ancient door' would date to between 470 AD when Brigid is believed to have established the monastery, i.e. she and her fellow Druids converted to Christianity, and her death about 525 AD. By which time it is probable that the former round temple of the goddess Brighid had been replaced by a rectangular building of a European pattern for churches.

It is also significant that it was placed on the left hand external entrance, that is the side of the site where the nuns' accommodation was situated. This door, because of its direct connection with the saint, would have been regarded as a relic, and so it was appropriate that it was the entrance used by the Abbess successors to Brigid and her nuns. The story also suggests that the doorway of the original church was a lot smaller and simpler than that provided in other wall of the much grander building of his time, which was a 'finely wrought door…' used by the monks, although what the real explanation for the miraculous 'adjustment' of the size of the door was, cannot be determined at this distance in time. From the way Cogitosus describes it, this ancient door would have been one of the 'sights' that pilgrims and visitors to Kildare would have wanted to see, along with the altar which also had a direct and miraculous connection with the saint, and which Cogitosus says was in the cathedral in his day:

Not long afterwards, her parents, in the usual way of the world, wished to betroth her to a man. But Brigid, inspired by heaven decided otherwise: wanting to dedicate herself as a chaste virgin to God. She sought out out the very holy Bishop Mac Caille, of blessed memory. Seeing her heavenly longings, her modesty and her virginal love of chastity, he veiled her saintly head in a white cloth.

She went down on her knees before God and the Bishop, and touched the wooden base (*fundamentum*) that supported the altar. To commemorate her virtue this wood flourishes fresh and green to the present day as if it had not been cut down and striped of its bark, but was still attached to its roots. Even to this day it cures the faithful of afflictions and diseases.

From the description it is clear that this was the simplest and earliest form of altar found in a Christian context. It consisted of a single leg or pedestal that supported a table top. This alter seems to have achieved even more fame after Brigid's time as a medieval note attached to a late eighth-century copy of *Feilire Oengus* (*The Martyrology of Oengus the Culdee* by Oengus mac Oengobann) says: 'Thus was Brigid, with the foot of the alter in her hand, and seven churches were burnt with the foot in one of them, and it was not burnt, being saved by Brigid's grace.'

Another item that pilgrims and visitors would have looked for, particularly if they were sick or injured, was situated near the gate that led into the enclosure that surrounded the cathedral. This was a millstone, almost certainly made of stone from the Red Hills near Kildare which has grits, sandstone and limestone ideal for making them. To this millstone was attached two miraculous stories, apparently widely known in the seventh century, which caused people to believe it possessed healing powers. Cogitosus, after describing the miraculous way that the mill stone was brought with the aid of St Brigid from the top of a high mountain goes on:

> A pagan who lived near the mill surreptitiously sent his grain to this mill employing a simpleton, so the miller, who was doing the milling, did not know that the grain was his.
>
> And when it was thrown and poured between the two millstones nothing could set the mill in motion and thrust it into its revolving orbit and customary round, neither the propulsion and driving power of the strong river, nor the violent force of the waters, nor the efforts of the tradesmen.
>
> And while those who saw this were upset about this and struck with great amazement, they happened to find out that it was the pagan's grain. And so were in no doubt that the millstone on which Saint Brigid performed a divine miracle had refused to grind the heathen's grain into flour.
>
> But the moment they took out the pagan's grain and put their own grain from the lands of the monastery between the same millstones, the revolving motion of the mill stone was immediately restored to its normal way of working without any hindrance.
>
> And after a period of time, it happened that this mill caught fire. And this was no small miracle either. Although the fire consumed the whole mill and the other stone which was joined to the one just mentioned, the fire did not dare to touch or burn the special stone that was blessed by Saint Brigid; instead, it remained alone undamaged without any harm from the great fire which destroyed the mill.
>
> And afterwards, when this miracle had been noted, it was taken all the way to the monastery and placed near the gate of the ornate wall that surrounds the church, where many people gather out of veneration of the blessed Brigid's miracles, and it was given a place of honour in this doorway.
>
> And it drives out diseases and afflictions from the faithful who touch this stone of Brigid, through which she personally performed the miracles recorded above.

In the mid-seventh century the sacred oak from which the place name was derived would have still been flourishing. While its exact location cannot be determined, it would not have been too far from the cathedral, and its branches must have towered above even the exceptionally tall roof of this building. Such sacred trees (*bile*) were objects of veneration and respect even in Christian times, although such beliefs far pre-dated the introduction of the new religion. However, at some point in the next 300 years the tree would die, but still remain an object of veneration for visitors to Kildare.

So popular did the cult of St Brigid become that many pilgrims were attracted to Kildare, the cathedral was richly endowed and facilities grew up around the monastic settlement to cater for visitors, along with houses, shops and inns as Cogitosus clearly describes:

> And who can express in words the exceeding beauty of this church and the countless wonders of that monastic city we are speaking of, if one may call it a city since it is not encircled by a circuit of walls.
>
> And yet, since numberless people live within it and since a city gets its name from the fact that many people congregate there, it is a very great and metropolitan city. In its suburbs, which Saint Brigid had marked out by a clear boundary, no earthly foe or enemy attack is feared; on the contrary, together with all its outlying suburbs it is the safest city of refuge in the whole land of the Irish for all fugitives. In which the treasures of kings are kept and the splendour of its decorated roofs is most excellent.

This plan of a cathedral surrounded by a 'city' with suburbs beyond this is the type of layout that was still found at Armagh in the seventeenth century, but can be found very much earlier. For example, Justiniana Prima, a site in Serbia known today as Caricin Grad, 'the City of the Emperor' was the birthplace of the Emperor Justinian. He rebuilt the cathedral and extended the city as the metropolitan church of Illyria. However, it survived less than 100 years before the area was conquered by the Slavs and the city was abandoned, so it can still be clearly seen how a sixth-century cathedral city was laid out. The ecclesiastical complex lies on top of a hill enclosed within a fortified circuit, while the town is laid out along a main street with numerous churches enclosed within a wall. Outside this, on the slopes of the hill towards the river, lie the suburbs, demarcated by slight banks but not fortifications. This is the model that inspired the monastic cities of the Celtic Christian world, and provided the pattern for Kildare in the seventh century, but with wooden buildings instead of the stone ones of Justiniana Prima.

Cogitosus then goes on to describe the crowds that came from near and far to commemorate the saint, to experience the pageantry, and to have a good time or obtain a cure at the healing mill stone:

> And who can count the different crowds and numberless peoples flocking from all the provinces of Ireland? Some for the abundant feasting, others for the healing of their afflictions, others to watch the pageant of the crowds, others bearing great gifts and offerings – to join in the solemn celebration of the feast of the birthday of Saint Brigid who, freed from care, cast off the burden of the flesh and followed the Lamb of God into the heavenly mansions, having fallen asleep on the first day of the month of February.

Cogitosus's writings also shows the status that Kildare claimed for itself in the seventh century, regarding itself as the 'head of almost all the Irish churches with supremacy over all the monasteries of the Irish, and whose sphere of influence extended over the whole of Ireland reaching from sea to sea'. Although it must be borne in mind that, as an enthusiastic monk of Kildare, he may have exaggerated its status somewhat! The monastery owned tracts of nearby grasslands, the Curragh, and while this was given as gifts by various benefactors, there is a local story that explains how St Brigid acquired the land:

> Brigid went to the King of Leinster to ask him to grant some land to her monastery for grazing sheep and cows, but the king refused her petitions. Eventually, realising the king would not give her the land she wanted, she suggested that he give her as much land as her cloak would cover.
>
> To this the king readily agreed and Brigid removed her cloak and asked four of her nuns to hold a corner each. This they did, then each grasping a corner ran in different directions, the cloak unfolding and expanding as they ran, so that other sisters had to lift the edges of the

cloak, and this continued until it covered a square mile. Whereupon the king grew alarmed that it might cover his whole kingdom, and said she could have the land that the cloak had so far covered. So the sisters stopped running and the monastery had all that land on which to graze their animals.

However, it is possible that this may be based on a real incident. Could it have been that after unsuccessful negotiations with the King to get land gifted to her monastery, Brigid finally said, perhaps in desperation, that she would accept only the land that could be 'covered' by her cloak, to which the King agreed? She, or one of her nuns took her cloak and carried it around an area of land, after which Brigid went to the King, explained what had been done and held him to his agreement. The Celts liked riddles and twists in their stories, and also had a sense of humour. She probably got away with her 'deception' as the cloak had indeed covered the area, even if not in the way the king would have anticipated!

There were two separate communities on the site, one of nuns and one of monks, and there is no doubt that Brigid played a very powerful role. She was head of both communities, and the person who would also have held sway as Governor over the secular settlement that grew up on the monastic land surrounding the main ecclesiastical site, if this existed in her time. However, it must be borne in mind that Cogitosus is describing a settlement that grew up in the century or so after the temple of Brighid became a Christian community. So its quite possible that there was little or no civil settlement in the time of Abbess Brigid beyond the round houses and other necessary facilities required by the community of former Druidesses and perhaps Druids who had once served the goddess and now formed the first Christian community in the area.

From the sixth century Christian religious establishments of monks and nuns were called *Monasteria duplicia*, which is why an Abbot or Abbess of a double monastery presided over a monastery and not an abbey. As a woman Brigid could not hold clerical orders and so would have been unable to ordain clergy or consecrate churches, as is clear from Cogitosus's account. To overcome this situation she appointed a bishop:

> And as by her wise administration she made provision in every detail for the souls of her people according to the rule, as she vigilantly watched over the churches attached to her in many provinces and she reflected that she could not be without a high priest to consecrate churches and confer ecclesiastical orders in them, she sent for Conlaed, a famous man and a hermit endowed with every good disposition through whom God wrought many miracles, and calling him from the wilderness and his life of solitude, she set out to meet him, in order that he might govern the church with her in the office of bishop and that her churches might lack nothing as regards priestly orders.

However, the writer of a ninth century *Life* of St Brigid, the *Bethu Brigte*, overcame this problem of her being unable to ordain priests or consecrate churches by having Brigid accidentally conferred as a bishop when she was 'veiled' by Bishop Mel (see page 83). So it seems that in the first days of the Christian church at Kildare the former Chief Druidess of Brighid, now the first Abbess of the newly converted religious community, held the senior position over both males and females at Kildare monastery. This reflects the Celtic tradition of having women in important positions and this pre-Christian practice continued at the Kildare site. She also seems to have been pre-eminent over other communities further away, some of which were, at least partly, made up of ordained priests which is why she had to appoint Bishop Conlaed (also referred to as Conleth) to deal with this aspect of her community. Another indication of the importance of this church is suggested in Cogitosus's account of Kildare where he mentions, in an account of a miracle, that Conlaed's robes came from abroad:

> After the example of the blessed Job, she never allowed the poor to go away empty handed. On one occasion she generously gave away to the poor the foreign vestments from overseas belonging to his distinguished eminence Bishop Conleth, which he was wont to use on the solemnities of the Lord and on the days of the Apostles when offering the sacred mysteries on the altar and in the sanctuary...

The nuns and monks lived by a Rule, mentioned both by Cogitosus, who used the term *regulariter*, and in the *Life of Saint Kieran of Clonmacnois* where there is a reference to the *regula Sanctae Brigidae*, but no details of her Rule have survived.

From Cogitosus we also know there was a Prior at Kildare, who is mentioned in the miracle of the millstone. He was responsible for organising the 'brothers' engaged in manual labour and those lay workers in the employ of the monastic establishment such as metalworkers and other craftsmen, as well as those working in agricultural activities. He may also have supervised the monks and nuns working in the *scriptoria*, producing the illuminated manuscripts for which Kildare was to become famous. One of these, the *Book of Kildare*, was described by Gerald of Wales in the twelfth century who was very impressed by the work:

Many of the early accounts also say that St Brigid established a school of art for manuscript illumination, carving and metalworking, as well as setting up a great *scriptorium* producing illuminated books, one of which was described by Gerald of Wales:

> This tome contains the concordance of the four Gospels according to Saint Jerome, with different designs on almost every page, all of them in a marvellous variety of colours. ...if you take the time to examine more closely, you may penetrate to the very shrine of art. You will see intricacies, so fine and subtle, so exact and yet so rich in detail, so full of knots and coils, with colours so bright and fresh, that you will not hesitate to declare that you have gazed upon the work, not of men, but of angels.

While most people regard this great book as now lost, a few scholars have identified this as the magnificent work known today as the *Book of Kells*. However, this is generally regarded as being being the work of monks from Iona and brought to Ireland in 814 AD after several Viking raids on the island. A statue in the chapel of St Dredeneaux, near Pontivy, Brittany shows St Brigid with a book in one hand and a quill pen in the other, as does a carving in St Bridget's Church, St Bride's Major, Glamorgan.

There is a reference to the *dairthech* (Oak church) at Kildare in 762 AD, but it is not clear if this is the same building as that described by Cogitosus about 650 AD. Such buildings would have to be regularly repaired and often rebuilt. In 836 AD there is another mention of the *dairthech* when it was blockaded to refuse entry to Forindeán, Abbot of Armagh. By 868 AD however, it was in such a bad state it had to be rebuilt under the patronage of Flanna, wife of the High King Aed Findliath.

The first Viking attack on Ireland occurred in 795 AD with the pillaging of monasteries on the islands of Rathlan (Co Antrim), Inishmurry (Co Sligo) and Lambay (Co Dublin), and later sites on the mainland were attacked. The frequency and areas of such raids increased markedly from the 830s AD and it seems that Kildare may have been attacked in this decade. Fortunately the body of St Brigid had already been removed to Dun (Downpatrick), a place that the monks regarded as being safer.

The Temple of the Goddess Brighid: the Church of Saint Brigid

The cathedral was mentioned once again in 964 AD in a way that emphasises its large size. The first mention of its complete destruction occurs in 1020, when it was destroyed by fire. Up to that date all references to the *dairthech* are in the singular, suggesting that despite repairs and rebuildings, the same basic layout as described by Cogitosus continued. However, after the fire of 1020, there seems to have been a radical change in design, as in 1050 another fire occurred that destroyed both the *dairthech* and the *damliac* (the stone church). So it may be that the wooden church, on or close to the site of the temple of Brighid and later the first Christian church on the site, lay to the north west of the present cathedral which occupies the site of the stone church of c.1020. This seems likely as the 'oak church' would then have been close to the Fire House and the nunnery, which is where activity would have been concentrated when it was a pagan site.

The location of the nunnery to the north west is confirmed not only by recent historical studies and archaeological work, but also by looking at the description of the cathedral church by Cogitosus. He says that the nuns and monks had separate entrances, the nuns entering from the 'left', i.e. west, where their nunnery was situated and the monks from the right which suggests that their accommodation lay to the east of the site, although they worshipped equally in the cathedral at this time. So it appears that during the eleventh century Kildare adopted the pattern of multiple churches on the same site, as was the case at Clonmacnoise near West Meath, Co Offaly and Glendalough in Co Wicklow, providing more places for veneration and making offerings.

It is probable that one of the buildings burnt in 1050 was the church attached to the nunnery, suggesting that a division of the nunnery from the cathedral had occurred at the rebuilding of 1020. It may also be significant that the abbess during the 1020s, Lann ingen meic Selbacháin (abbess 1016–47), was the first to hold this office since Brigid who was not a member of the Fothairt family to which St Brigid belonged. So was more likely to be willing to agree to a major change in the layout and organisation of the ecclesiastical establishment than her predecessors. Gerald of Wales says that the sacred fire was '…surrounded by a circular hedge of withies within which no male enters…', and looking at various examples of Celtic pagan temples the focal point of worship usually stood in a courtyard bounded by a wall or fence. However, other, associated buildings, stood nearby but outside the inner temple precinct, which is likely to have been the case at Kildare in pagan times, and the evidence for this early layout may be discerned in the vicinity of the present cathedral.

Today a sub-rectangular enclosure surrounds the cathedral, possibly on the line of an enclosure surrounding the stone church built between 1020 and 1050. The layout of surrounding streets and a crop mark in a field to the north seems to indicate the line of the very much larger original inner enclosure, that may date back to pagan times. This contained the Fire House where the sacred flame was maintained, and the nunnery which was probably built on the site of the former Druidesses dwellings adjacent to the site of the sacred flame, although the pagan fire temple probably stood in its own enclosure within this larger enclosure.

Cogitosus describes the 'doorway of the ornate *cashel* which surrounds the church', and its been suggested that his use of the word *cashel*, derived from the Latin *castellum*, castle, may have been a stone wall, or earthen wall faced with stone. In the seventh century such a wall may more probably have been of earth and timber, laid in some sort of ornate pattern. So the site was divided into separate enclosures: one fenced off to contain the sacred fire, probably another around the nunnery, one round the church and one around the monks' quarters.

The nuns and monks almost certainly dined separately, so each would have their own refectory, kitchen and other facilities within their own enclosures. All these internal divisions would be contained within the large walled enclosure that encompassed the ecclesiastical site.

Later, when a second church, built of stone, was constructed on the site in addition to the wooden church, this too was situated within its own enclosure. Outside the inner enclosure was a further large area, again indicated by the layout of the present roads, which was presumably that 'which Saint Brigid had marked out by a definite boundary' according to Cogitosus, and in this outer area grew up the 'monastic city' or outer suburbs.

So this was an ecclesiastical site with a secular settlement around its outer precinct, a place of peace under the protection of the monastery, at least in theory. A place of sanctuary for fugitives and regarded as sacrosanct enough for kings to store their wealth there. Although records show that Kildare was burnt (which must often have been accidental as major fires in such settlements were common) or plundered thirty-eight times between 710 AD and 1155. Further evidence for the appearance of this settlement can be found in a document that records that in 909 AD the King of Leinster, Cerball mac Muirecáin, had a fatal accident outside the house of a comb maker as he rode eastwards into Kildare 'along the street of the stone step'. It is notable that houses and workshops flanked by stone kerbs have been excavated at the monastic establishment at Clonmacnoise in Co Offaly and large numbers of dwellings and workshops are a feature of ecclesiastical sites in Ireland in the eleventh and twelfth centuries.

Rebuilding took place after the fire of 1050, and an even grander stone church, mentioned in 1067, constructed. The Abbess's House was mentioned in 1132 and probably also stood within in its own enclosure on the site. About 1135 a 32.6m (105ft) tall round tower (*cloichtech* – 'bell tower') was built to the north west of the cathedral, but there are no further references to church buildings at Kildare until Ralph of Bristol was appointed as Bishop of the See in 1223 and was 'at great expense in repairing and beautifying his cathedral', which was the beginning of the construction of the present building.

In May 1536 the 'Reformation Parliament' declared Henry VIII to be 'the supreme head on earth of the whole church of Ireland', and from this date the days of the monastery at Kildare were numbered. It was eventually dissolved by order of the king in 1540. In 1542, when Henry VIII's Commissioners began removing relics from Irish churches, they recorded that Kildare Cathedral contained 'pictures, relics, images etc.' which they estimated could be sold for £7 4s. The cathedral also acted as the parish church of Kildare and so survived the Reformation. It was not well endowed and, because of a number of factors such as its location and local military campaigns in the seventeenth century, saw its fortunes falling, reviving and falling again. Its connection with the famous saint was not forgotten, and in 1600 the secretary to the Lord Deputy, Fynes Moryson, in describing Kildare town in his four-volume work *An Itinerary*, said that it was 'greatly honoured in the infancy of the church by St Brigit.' The cathedral was described as 'ruinous' in 1611 and 1615, but the choir had been re-roofed by the 1620s, and the chancel was re-roofed in the 1680s.

However, during the eighteenth century the building gradually began to fall into decay, so that by the 1750s the nave, tower, south transept and chapter house of the cathedral lay in ruins, as did the building to the north known today as the 'fire house'. The cathedral's chancel was still roofed and remained in use as the parish church despite its state of dilapidation. In Irish tradition trees on the sites of church ruins were allowed to grow naturally and were never cut down or interfered with, so reached huge proportions, and it was noted that a giant ash grew beside the ruins of Kildare Cathedral.

This situation continued until 1809 when Thomas Trench was appointed as Dean and began the work of repairing the building, a process that continued until the major works of 1875–81 and 1890–96 completed the restoration. The cathedral was officially reopened on 22 September 1896. As part of the ceremony a young oak tree was planted jointly by

The Temple of the Goddess Brighid: the Church of Saint Brigid

the Archbishop of Armagh, the Archbishop of Dublin, Glendalough and Kildare, and the Archbishop of Canterbury. Unfortunately the tree they planted died, but is preserved in the cathedral as a 1.5m (5ft) long verge (a wand carried before the bishop as an emblem of office), carved with acorns, oak leaves and shamrocks and the names of those who planted the tree, along with those of the Dean and his wife. Another oak tree was planted in the grounds in 1985 within a protective railing. Inside the cathedral, St Brigid is depicted on a stained glass in the central tower with a flame on her brow, and an altar in the north transept dedicated to her has a Brigid's cross hanging over it, the Brigid's cross also being reflected on the design of the tile pattern on the floor. Today the cathedral stands as a monument to the Druidess who, over 1,500 years ago, once served the Goddess Brighid, but who became a Christian and still inspires people throughout the world today.

The Fire House at Kildare

Today, what is generally regarded as St Brigid's Fire House where the perpetual flame was once kept, is a rectangular hollow with three steps leading down to a floor surface. It is enclosed by rubble walls measuring 4.44m x 3.48m (14ft 6in x 11ft 2in) that stand an average of 60 cm (2ft) above the present ground surface, and are 1m (3ft 3in) thick. The present walls were constructed in 1988 to show the site of this building more clearly.

Eighteenth-century illustrations show that it had two gabled walls to the east and west that stood between 3–4m high (9–13ft), with a window in each. This structure is shown in a ruinous state in an engraving entitled 'North View of the Cathedral Church at Kildare in Ireland', dated 1 September 1778, in Volume One of *The Virtuosi's Museum* by Paul Sandby. This is the earliest view of the cathedral from the north, and shows the roofless ruins of this building, a single roomed structure, with two gable ends to the east and west containing large Gothic-style windows, although just holes in the engraving with no sign of any tracery surviving. The side walls have substantially collapsed and the north wall has fallen stones strewn around its base, while ivy grows on the walls.

Other illustrations of the cathedral from the north show this building in the same ruinous state. A water-colour, by Gabriel Beranger, shows that the east gable had largely collapsed.

16 Lucky holed stone on the corner of St Brigid's Cathedral, Kildare. The right arm is passed through to touch the left shoulder to 'activate' the luck.

What is notable in these later illustrations is that they all show a small hole, presumably the remains of a window in the west end of the building and not the large window as shown by Sandby, so it appears that a certain amount of artistic licence was used in his engraving. However, other nearby structures have also been described as the Fire House. In Holinshed's *Irish Chronicle*, published in 1577, the Dublin chronicler Richard Stanihurst described how he had visited at Kildare 'a monument lyke a vaute, which to this day they call the firehouse.' On a map of 1757 by John Rocque a 'fire castle' is shown to the north-west of the cathedral churchyard. This is almost certainly the same structure referred to in the sixteenth-century Dissolution documents for the nunnery as a 'small castle or fortlage', suggesting the Fire House was adjacent to this if it was not the actual building.

A further reference by Harris, who republished a work by Ware in 1764, says that 'not far from the round tower is to be seen an old building called the fire-house, where the inextinguishable fire was formerly kept by the nuns of St Brigit', but does not say in what direction from the round tower it was located! However, it was almost certainly the 'Fire Castle' referred to by Rocque. In 1837, when the surveyor John O'Donovan visited Kildare, he showed the site of the Fire House in the position indicated by Rocque to the west of the round tower and outside the churchyard wall, although it seems the remains of this building had been demolished by 1798. St Brigid's fire inspired the poet Thomas Moore (1779–1852) to write:

> The bright flame that burned,
> In Kildare's holy fane [shrine],
> And gleamed through long ages,
> Of darkness and storm.

Documents, old maps and archaeological work suggests that the actual site of the fire temple of Brighid, where the Druidesses once maintained a fire in honour of their goddess and which was later perpetuated by Christian nuns of the Brigidine Order, lay to the north west of the cathedral. Today Fire House Lane leads around the perimeter of the churchyard wall, widening into Fire Cashal Lane near where the Fire House once stood. The building that is today referred to as the Fire House was described by O'Donovan as 'Briget's house level with the ground.' This 'Briget's House' was shown on the first edition of the Ordnance Survey map of Kildare published in 1838, but is not labelled, and it is not until the O.S. map of 1872 that it is shown as 'fire house (site of)'. This attribution may have been influenced by the desire of nineteenth-century antiquarians to identify the famous Fire House which, by that time, had disappeared. Having read the description in Harris that stated it was 'not far from the round tower…', they concluded that the remains of the gabled building north of the cathedral was the Fire House.

What is today generally regarded as the Fire House actually had another function. It may have been either a mortuary chapel or the remains of a small pre-Anglo Norman church or chapel. However, the present 'Fire House', known locally as 'Brigid's House' in the early nineteenth century, is today much visited and the focus for rituals by the Brigidine Sisters and others to commemorate the saint. Candles and even a small fire are lit within it, particularly on St Brigid's Day. This is just as valid as the celebrations held at the original Fire House since the spirit of Brigid permeates the whole site and indeed the whole landscape of Kildare.

Five

Her name and the many Brigids

As a Celtic goddess and later a widely venerated saint, Brigid has long been revered in Ireland, Britain, Europe and America and has always been known by a number of different names in both her pagan and Christian forms. Spellings also vary considerably, depending on whether the name is rendered in Irish, Scottish, English or another language and pronunciation varies not only with the language, but the period in which the name was spoken! In addition some countries use equivalent names that bear little resemblance to 'Brigid' (and its variations). This has, not surprisingly, caused a certain amount of confusion over the years.

Variations of the Name of the Goddess-Saint

Bhríde, Brig-eoit, Brigit, Bridget, Brigid, Bridgid, Brigde, Brighde, Brighid, Bhrighdi, Bhrighit, Bhrighite, Broghidh, Brid, Brig, Bride, Breeda, Breedia, Breedy, Briege, Briga and Brietta (Irish)
Brigh, Bridi, Bridean (Scottish)
Brüd (Scottish)
Briid (Islay, Hebrides)
Breeshey, Vreeshey, Brede, Bahee (Manx – Isle of Man)
Bregit, Breit, Breid, Freit (old Welsh)
Ffraid, Ffred, Fride (Welsh)
Freit (Anglo-Welsh, Herefordshire)
Brigide (English, twelfth-century Belgium)
Bridgett (English, 1703)
Breed (English, 1765)
Brigitae (Latin)
Brigitta, Brigida, Brigte (Latinised forms)
Berc'het, Perhet (Brittany)
Brigette, Britta (French)
Brigitte, Britt (French, Swedish)
Bridget, Birgitta (Swedish)

Brigitta (Dutch)
With the pet forms used as personal names:
Breege, Breda, Bridhe (Irish)
Bridie, Bree (Irish and Scottish)
Biddy, Biddie (Irish and English)

Many people regard the tutelary goddess of a British tribe, the Brigantes, to be the same as the Irish deity, but she is only very distantly related. This British goddess goes under the names:
Briganti (British)
Brigantia (Latinised British)
Brigantiae (Latin)

With her name in Gaul (France) being:
Brigindo, Brigandu (Gallic)

Looking at the way both the classical writers of the Greek and Roman worlds and the Romano-Celtic people equated the Goddess Brigantia with the deities they were familiar with, she can be regarded as being equivalent to, and in many cases interchangeable with:
Minerva (Roman)
Athena (Greek)

The root of Brigid's name, *Brig*, is generally regarded as meaning 'High One', although some interpret it as meaning 'Valour'. It was not unusual in the past to refer to a deity by a title rather than a 'personal name', so both the people of the British Brigantes kingdom and the Irish Celts used the same title for their most important respective deities. In some societies it was felt that the name of a deity was too important for ordinary people to use, so alternatives were used, such as the title of the god or goddess. Another example of 'secret words' were the names of the individual planets, some of which seem to have been linked to Celtic deities, but the planets names were forbidden Druidic words and ordinary people had to refer to them by euphemisms. For example the moon was referred to as 'queen of the night', 'brightness' or 'radiance', and this means that many ancient Irish words for the planets are no longer known.

Even in more recent times, when some pagan deities had been downgraded to fairies in Britain and Ireland, this still applied as many people would only refer to these creatures, which were still held in great awe, as 'the little ones' or 'the host'. In Ireland they were called 'the Gentry' (usually applied to the type of fairy known as a *Leprechaun*) or *Na Daoine Maithe* (the good people), so as not to offend them by using their 'real' name were commonly used phrases. It is possible that the 'real name' of the goddess whom we call Brighid (with its many variations) and whose title was the High One is no longer remembered, perhaps being known originally only to the Druidesses and Druids who once served in her temples. It is possible that every Chief Druidess of the main cult site of the goddess at Kildare in Ireland took the name or title Brighid in both honour of the goddess and as an indicator of their high rank; or perhaps even because they were regarded as an embodiment of the goddess.

The fact that Brighid was a title for a goddess rather than her name is also suggested by the ninth-century work by Bishop Cormac mac Cuilleanáin in his *Sanas Cormaic (Cormac's Glossary)*. He describes Brighid and her attributes and concludes with a statement that suggests all the Irish once used the term Brighid to refer to all goddesses (see page 29). Some writers have claimed that Brigid (or its variations) is derived from a name that means 'fiery arrow', but this is unlikely as the term for fiery arrow in Irish Gaelic, *breosaiget*, does not fit easily with the entomology of the personal name. It has also been suggested that her name may have originated with the Verdic Sanskrit word 'brihati', a title for 'the Divine One', which is possible as most European languages have an Indo-European base.

As a goddess Brighid was known by a number of descriptive titles:

Woman of Healing – *Ban leighis* (Irish)
Woman of Smith-work – *Ban goibnechtae* (Irish)
Woman Poet – *Ban fhile* (Irish)

As a saint Brigid was also given a variety of descriptive titles:

St Brigid of Kildare (general)
St Brigid of Ireland (general)
St Brigid, Patroness of Ireland (general)
St Brigid of the Flame (general)
Queen of the South (Irish)
Christ's Milkmaid (Irish)
Brigid the Holy Woman (Irish)
Blessed Bride –*Bhríde bheannaithe* (Irish)
Radiant flame of gold, noble foster-mother of Christ –*Lasair dhealrach oir, muime chorr Chriosda* (Irish)
Brigid of the Candles –*Bhríde nan Coinnlean* (Irish)
St Brigid of the Isles (Hebridean)
Milkmaid Bride (Scottish)
Golden-haired Bride of the Kine (Scottish)
Victorious Brigid –*Buadach Bhríde* (Scottish)
Mary of the Gaels –*Ban chuideachaidh Moire* (Scottish).
Christ's Foster-mother –*Muime Chroisda* (Scottish).
St Brigid of the Mantle –*Bhríde nam Brat* (Scottish)
The Aid Woman of Mary –*Ban chuideachaidh Moire* (Scottish)
God-mother of the Son of God –*Bana-ghoistidh mhic De* (Scottish)
God-mother of Jesus Christ of the Bindings and Blessings *Bana -ghoistidh Iosda Chriosda nam bann agus nam beannacht* (Scottish)
Melodious Bride of the fair palms –*Bhríde binn nam bas ban* (Scottish)
Bride of Brightness –*Bhríde boillsge* (Scottish)
St Ffraid the Nun –*Sant Ffraid Leian* (Welsh)
Brigid, Worker of Miracles –*Brigitae Thaumaturga* (Latin)

Brigid, Brigit or Bride?

It is believed that the early Irish Christians would not name their children after St Brigid in much the same way as tradition precluded them from naming boys after Jesus, something that still applies in many Christian countries. However, gradually the use of the name Brigid (or one of its many variations) became acceptable and, until the beginning of the twentieth century, almost every Irish family had a child named after the second most important saint in that country. Until fairly recently this was one of the most common Irish girls names. In some parts of the country it was customary to name the first female child in the family Brigid or Bridie. However, its popularity then began to fall due to the increasing use of its derivative form Biddy, which had became more common during the course of the nineteenth century. One English king, Edward V (April until June 1483 when he was murdered) had a daughter named Bridget (died 1517) who was a nun at Dartford in Kent. However, she was a Bridgittine, a member of the Order of St Bridget of Sweden.

There are many variations in both the spelling and pronunciation of the name, and this was not helped by the confusion in the English-speaking areas (rather than Gaelic speaking areas) between the Irish saint and St Bridget of Sweden, 1303–73 (see page 75). This was further compounded during the nineteenth and early twentieth century when people wanting to purchase an image of St Brigid of Kildare were supplied with images of St Bridget of Sweden. The suppliers, many of whom up to the First World War were German, did not seem to realise that there were two saints bearing this name. This caused the Scandinavian spelling to become common, and some children were christened Bridget, although the intention was to name them after the Irish saint. Not only that, the Swedish pronunciation also started to be used, introducing a 'j' sound into the name. The Irish name Brigid should be pronounced with a hard 'g', that is as 'Brigg-id' and not 'Bridjid', in its original form and pronunciation. However, in another one of the ancient forms of the name it was spelt as Brigit, and Latinised as Brigitta (or Brigittae), but can also be found written as Brigid from an early date, Latinised as Brigida (or Brigidae).

The early accounts of St Brigid and the Goddess Brighid were written in the language known as Old Irish which changed around 950 AD to a form known as Middle Irish. In the more modern form of Irish Gaelic, which developed from the fourteenth century, the 'g' became silent, and the name was spelt Brighid (genitive Brighde) and pronounced 'Bree-id'. It is from this pronunciation that the name 'Bride' developed, which is the way it is now pronounced in some parts of England, for example Bride's Mound, Somerset. While there are many variations of the spelling, the generally accepted one in modern English is Brigid, pronounced 'Briggid', but with the increasing interest in reviving Irish Gaelic in recent years, Brigid is being pronounced 'Breed', although in the Irish province of Munster its pronounced as 'Bride' and this is how it appears in written form there. In other written material it is spelt Brighid (which is used to differentiate the goddess from the saint in this work), but the 'h' is a printing convention as there is no 'h' in Irish. An 'h' appearing inside a word represents an aspiration which alters the sound of the way the consonant is pronounced.

When Brigid's name is written in Irish it is very rarely preceded by *Naomh* the Irish word for saint, unlike figures such as Patrick (*Naomh Pádraig*) and other saints. Because of the great love and familiarity the Irish have with Brigid they do not feel it necessary to have to explain her status as 'everyone' knows who she is. St Brigid is generally called *Sant Ffraid* in Welsh, *sant* meaning saint while Bhríde became Ffraid as the 'B' transmuted into 'v' sound, so it would be pronounced Vraid. However, 'vr' is not a natural start to a word, and the 'v' sound in Welsh is written as 'f', although her name usually starts with a double 'ff', but still pronounced 'v'. Further confusion has been caused by the fact that there were a number of saints that had the same name (see page 73).

Her name and the many Brigids

17 Mór-Ríoghain (Morrigan), the mother of the Goddess Brighid, was a war goddess. She possessed a herd of Otherworld cows and is commemorated in the Irish landscape by ritual cooking sites and two hills known as *da chich na Mór-Ríoghain*, the breasts of Mór-Ríoghain. She is also sometimes visualised as three goddesses with individual names and attributes.

Her parents: their names

Ancient Irish stories give the names of the parents of both the goddess and the saint and, as these are in Irish Gaelic, their pronunciation is not always clear to non-Gaelic speakers, and a guide to the way they should be pronounced is given below. The first column gives the standard Irish spelling of the name, the second column gives the usual English spelling and the third column gives a guide to the Irish pronunciation.

The Parents of the Goddess Brighid

	Her Father:	
Daghdha	Dagda	Daye

	Her Mother:	
Mór-Ríoghain	Morrigan	Moer-ee-en

The Parents of Saint Brigid

	Her Father:	
Dubhthach	Dubtach	Duffach

	Her Mother:	
Broicsech	Brocsech	Brocksheh

Saint Brigid – some basic questions

Saint Brigid's parents

The Goddess Brighid had a traditional ancestry and a family that was recorded in the medieval period in a rather muddled way (see page 231–32), but there has also been a lot of confusion about the ancestry of St Brigid, with some scholars even doubting her existence. The earliest surviving *Life* of the saint, written by Cogitosus, *c.* 650 AD, describes her origins in the following terms:

> Now Saint Brigid ... was born in Ireland to noble Christian parents stemming from the good and most accomplished tribe of Eochtech. Her father was Dubhtach, her mother Broicsech.

The Eochtech Fúath nAirt were ancestors of the Fothairt, a group of families, all nobles, who occupied land in the Kildare area and various other parts of Leinster. Brigid was a member of the Fothairt family, and another early reference confirming this is found in an eighth century history of the Forthairt (see page 44).

Cogitosus and many later writers say that her parents were Christians, but at this early period the compilers of the *Lives* of the saints were not so much concerned with historical accuracy as showing the sanctity of their saint. So to have a saint the daughter of pagans, which they probably were, especially as she herself was a senior Druidess before conversion, would not have been regarded as 'suitable', so she was allocated a more respectable genealogy in most *Lives*, but for an exception see page 39.

All St Brigid's histories says she was the daughter of a minor king or local chieftain called Dubhtach. He is believed to be the son of Demri, who was eleventh in descent from Fedlimidh Rechtmar, known as 'the Law Giver', a notable Irish King of the second century AD (110 – 119 AD according to the *Annals of the Four Masters*). This genealogy for St Brigid is given in full by Geoffrey Keeting in *The History of Ireland*, *c.* 1630:

> Brigid, daughter of Dubhthach Donn,
> Son of Dreimhne, son of Breasal Borr,
> Son of Dian, son of Connla, son of Art,
> Son of Cairbre Nia, son of Cormac,
> Son of Aonghus Mor, of high dignity,
> Son of Eochaidh Fionn, Fúath nAirt,
> Son of Feidhlimidh Reachtmhar the noble,
> Son of Tuathal Teachtmar, the excellent,
> of the race of Eireamhon.

Her mother, Broicsech, is usually described as the bondmaid of Dubhthach, which many writers equate with slave (*cumal*). However, in Irish law, a bondmaid who gave birth to a child by her master assumed the status of a second lawful wife, although she had considerably less rights than the primary wife so was a *dormun*, in effect a concubine. However, the child of such a union enjoyed the same legal rights as a child born to the first wife. Broicsech was said to have been the daughter of Dallbronach, son of Aedh Meamhair, of the Dail Conchobhar in South Bregia, a petty kingdom now covered by part of counties Dublin and Kildare.

According to most of the *Lives*, Dubhthach's first wife was jealous of Broicsech and forced her husband to sell her, and such tensions between principle wives and *dormun* were probably common. However, since the child had rights and status, Dubhthach only sold the pregnant Broicsech on condition that when the child was born it should be returned to him. This ancient Irish system of bondage continued even into the Christian era and was considered quite acceptable. It was only abolished when a decree was passed at the Council of Armagh following the Anglo-Norman invasion in the twelfth century which forbade both bondage and slavery.

The Birthplace Question

As a saint, tradition places Brigid's birthplace at one of two places in Ireland, but the claim she was born in Scotland by John Major (*Historia Majoris Britanniae, tam Angliae, quam Scotiae*) and a few other early historians can be discounted. However, in all likelihood, bearing in mind the landholdings of the family to which she belonged, she was almost certainly born in the vicinity of the place now called Kildare (*Cill Dara*), although it would not have been called this at the time of her birth, only acquiring this name after she established a church (*cill*) there.

18 St Brigid shrine on the spot where tradition says her father, Dubhtach, lived and she built her first church. Standing by the shrine is Peter Gorry who maintains both this and her 'birthplace shrine' and has vowed to supply them with fresh flowers for the rest of his life in thanks for the help of the saint in curing a medical condition.

Brigid: Goddess, Druidess, Saint

19 An old engraving of the ruins of Foughart church, Co Lough. One tradition said that the church was built on the site of the house where Brigid was brought up, and locals used to point out a stone on which they claimed the saint was laid after her birth.

There is a local tradition that St Brigid was born near Umeras, Co Kildare, which is about 8km (5 miles) north west of Kildare. On a crossroads on the road from Monasterevin to Rathangan (R414), is a shrine with a statue of St Brigid, bearing the inscription, in both English and Irish: 'Authentic tradition states that near this place Brigid, Mary of the Gael, was born. Erected by the Lackagh Committee, 1982'. This area lay in the territory of the Luigne, one of five main tribal groups that held the area in the late fourth and early fifth century. About 8km (5 miles) east of this shrine, in the townland of Knocknagalla on the outskirts of Kildare, there is another Brigid shrine which bears the inscription: 'Dubhthach, St Brigid's father lived here and here she built her first church. Erected by the Lackagh Memorial Committee, 1982.' Not far away is an ancient mound that shows it was the site of a dwelling of an important individual, although whether it really belonged to St Brigid's father cannot be known.

For centuries many writers have claimed that St Brigid's birthplace was at Foughart, near Dundalk, Co Lough. There a ruined church dedicated to the saint is said to have been built on the site of the house in which St Brigid was born, a belief that is widespread today, although in the mid-nineteenth century the tradition said the church was built by the saint 'although they knew not that here was her birthplace'. During the early nineteenth century the locals would point out a stone in the graveyard on which, so they said, St Brigid was first placed 'when she came into the world.' Another local tradition claimed that St Brigid herself built the church and stayed there for two and a half years before leaving to establish the monastery at Kildare.

In the seventeenth century the small village of Foughart was called *Fochart Brighde* (Fochart of Brigid), in honour of it being her supposed birthplace. Near the church is a large ancient mound, 12m (40ft) in diameter and surrounded by a deep ditch. It is the site of a fortified dwelling which, tradition says, was the house of Dubhthach. About 0.62km (one mile) away is a modern shrine and place of pilgrimage (see page 224). This has been developed in a wooded valley with a stream and holy well and is said to have been the place where St Brigid, as a young girl, tended her father's animals. However, Foughart is unlikely to have been the place where Brigid was born, and this belief seems to have arisen because of the name of the village is similar to the name of the family to which St Brigid belonged, the Fothairt.

Her date of birth?

All authorities agree that she was born around the middle of the fifth century and died in the sixth century but there have been many 'exact dates' given for these events by a variety of writers and historians over the last 1000 years. For example the earliest date suggested for her birth is 436 AD, the *Annals of Dublin* give 439 AD, the *Annals of Roscrea* give 449 AD, while Bishop Usher determined that she was born in 453 AD. The Welsh *Annals Cambria* give 454 AD, the *Annals of Inisfallen* and the *Annals of Senat Mac Magnus* both give 456 AD; while Henry of Marlborough puts her birth as late as 468 AD. The only common factor that all the historians and martyrologies agree on, is that she was born on 1 February!

Her date of death?

The date of her death shows a similar wide variation to her date of birth. The *Annals of Boyle* gives the date 504 AD; *Annals of Innisfallen* 514 AD; Gerald of Wales 518 AD; *Annales Cambriae* 521 AD; Henry of Marlborough and many other early sources put it at 523 AD, *Martyrology of Donegal* 525 AD; *Annals of the Four Masters* 525 AD; *Annals of Ulster* 527 AD; *The English Martyrology* 540 AD; while one early *Life of Saint Brigid* gives the date as 548 AD, possibly a copyists error. However, one thing they all agree on is that she died on the 1 February – her birthday!

According to an eighth-century *Life of Brigid* by St Coelan, when St Brigid was dying she was attended by St Ninnidh, afterwards known as 'Ninnidh of the Clean Hand', as he had his right hand, which he used to give her the last rites, encased in a metal covering to prevent it being defiled. Her fame and deeds lived on far beyond her death, providing inspiration for many throughout the world, and in 1962 St Brigid was formally made a patron of Ireland.

The other Saint Brigids

There was more than one St Brigid, the original being Brigid of Kildare. She was so influential in the Celtic and later the Catholic church, that it is not unexpected that many females were named after her or took her name when they became nuns. In the early Celtic world a person automatically became a saint if they founded a nunnery, monastery or church, so it is not surprising that other St Brigids are found. The *Martyrology of Tallaght*, written *c.*800 AD by Oengus the Culdee, which appears in the twelfth-century *Book of Leinster*, lists seven St Brigids, each with a different Saints Day. G.O Simms (1910–91) Archbishop of Armagh, estimated that there may have been as many as twenty-four St Brigids in addition to Brigid of Kildare. Geoffrey Keating in his *General History of Ireland*, published *c.*1630, says that there were fourteen holy women bearing the name Brigid, in addition to Brigid of Kildare, and he lists twelve of them:

St Brigid, daughter of Aodh
St Brigid, daughter of Colla
St Brigid, daughter of Damhar
St Brigid, daughter of Dioma
St Brigid, daughter of Eachtar Ard
St Brigid, daughter of Eanna
St Brigid, daughter of Fiadhnat
St Brigid, daughter of Luinge
St Brigid, daughter of Mianach
St Brigid, daughter of Moman
St Brigid of Seanbhoth
St Brigid of Inis Brighde

Other Saints named Brigid

Saint Briga – Saints Day: 21 January
Also known as St Brigid of Kilbride (formerly Kilbriga), Ireland, she was a holy virgin who lived and died in the sixth century. Local tradition says she was visited by St Brigid of Kildare several times, and on one occasion when St Brigid of Kildare arrived, according to the custom regarding important guests, her feet were washed. After the water was removed, it cured another nun whose feet were crippled. St Briga was once much venerated in the Diocese of Lismore, Co Waterford, her cult being approved by the Catholic church.

Saint Britta – Saints Day: 28 January
There is a place in France that claims to be the location of the bodies of two virgin martyrs, one of whom was called Britta, a form of the name Brigid, but who seem to be the same two sister saints that are also buried in Picardy, France! St Gregory of Tours explained how St Euphronius, Bishop of Tours 556–73 AD, was told about a mysterious light seen by many people over a bramble-covered hill at Touraine. Some of them had also seen a vision of two young girls pleading for a chapel to be built on the hill as two holy virgins, presumably the maidens who had appeared, were buried there.

Euphronius visited the spot where the light had been seen and the two girls appeared to him, informing him that they were called Britta and Maura. They related how they had been solitaries at Ariacum (now called Sainte-Maure), and had died in the fifth century. They asked Euphronius to grant their wish of having a chapel built, and so he instigated a hunt on the hill for their remains, which were eventually discovered, and the relics placed in a chapel that Euphronius built on the spot. A cult, approved by the Catholic church, grew up in the area venerating the two saints, and their Day is still kept at Tours on 28 January.

Saint Brigid of Opaco – Saints Day: 1 February
This holy virgin lived in the ninth century, and tradition says she was the sister of St Andrew, Abbot of St Donatus Monastery at Fiesole in Tuscany, Italy. Andrew had left Ireland many years before to live in that holy community, and as he lay on his death bed in the monastery

he thought of his little sister Brigid whom he had left behind so many years before, and heartily wished he could see her one more time before he went to his maker.

Brigit, now an old and pious woman, was eating a meal of herbs and small fishes when suddenly an angel transported her across the seas to the bedside of her dying brother. She did not recognise the emaciated old man who gazed back at her, and she was scared of the monks in their unfamiliar garb who stood around him. Brigid thought this must be a vision and that she was still in Ireland, but Andrew assured her that she was really there and that God had ordained that she would remain in Italy living in great austerity and penance.

After her brothers death Brigid went to live in a cave that can still be seen today below the church in the village which bears her name, Santa Brigida a Opaco, setting an example of mortification, self-sacrifice and devotion, until her death at almost 100 years old. She was buried where she died and a church erected at the spot that still bears her name. Her cult was approved by the Catholic church which celebrates her feast day on the same date as St Brigid of Kildare.

Saint Brigid of Picardy – Saints Day: 13 July

There was a cult of two sister saints, Brigid and Maura, in Picardy, France. According to local legends they were two fifth-century British princesses from the kingdom of Northumbria who went on a pilgrimage to Rome. On their return journey they were passing through Gaul (France), when they were attacked and killed by pagan outlaws at Balagny-sur-Thérain. They were buried there and venerated as martyrs, which created a local cult honouring them, and various miracles were attributed to them. However, these seem to be the same two sister saints which tradition maintains are also buried at Touraine, not the only example of a saint's body being claimed by two places!

St Bathildis, the Queen of King Clovis II, tried to move the remains of these two martyrs to her monastery at Chelles in the mid-seventh century, but was prevented from taking them all the way 'by divine intervention', and so was forced to leave their remains at Nogent-les-Vierges (Oise). On 13 July 1185 the relics of saints Brigid and Maura were solemnly re-interred in a richly decorated shrine. King Louis IX, was very devoted to them, and was a great benefactor of their shrine and church to which he made regular pilgrimages. Brigid and Maura were generally credited for the cessation of the terrible plague at Beauvais, which gained them even more adherents.

Saint Bridget of Sweden – Saints Day: 23 July

The most prominent 'other' saint of the same name who has sometimes been confused with Brigid of Kildare, is St Bridget of Sweden, also known as St Birgitta. She was born in 1303, her father being Birger Persson, Governor of Upland, while her mother, Ingeborg Bengtsdotter, died soon after Bridget's birth on 14 June. She was then brought up by her aunt at Aspenas. At the age of sixteen she married Ulf Gudmarsson, Prince of Nercia, had eight children and, in 1335, was appointed as Principal Lady-in-Waiting to Blanche of Namur, Queen of Sweden.

In 1340 she made a pilgrimage to the shrine of St Olaf at Trondheim in Norway, and on her return to Sweden renounced her title and resigned her position at court. Later she went on another pilgrimage, this time accompanied by her husband, to Santiago de Compostela in Spain. Her husband became ill on the return journey, but recovered as Bridget knew would happen, having been told of it in a vision by St Denis. However, soon afterwards Ulf died at the Cistercian monastery of Alvasta and Bridget spent the next four years (1343–46) living there as a penitent. She experienced many visions and revelations, all of which were recorded by the Prior of the monastery who assured her that they were authentic. Following one of these revelations she denounced the king and queen for their frivolous life styles and later divided her estates and wealth among her eight children.

In 1346 she founded the Order of St Saviour, also called the Bridgettines. Despite her criticism of the royal family, she gained the support and patronage of King Magnus and was able to build a monastery at Vadstena on Lake Vattern the same year. She was inspired by the Brigidine establishment at Kildare to establish a double monastery with sixty nuns and twenty-five monks. They lived in separate enclosures but shared the same church, and seem to have followed the Rule of St Augustine with certain amendments. All surplus income was given to the poor, and luxury was forbidden, although the inmates could have as many books to study as they wanted. In 1346 Bridget travelled to Rome to gain papal approval for her Order, but this was not granted until 1370 and she never returned to Sweden. She spent the rest of her life in Italy, visiting many of the most famous churches or on pilgrimage visiting the holy places in Palestine. She continued to experience visions and revelations and lived an ascetic life, caring for the poor and sick, despite suffering severe bouts of illness in her later years.

She died in Rome, aged seventy-one, on 23 July 1373 with a son and daughter present at her sackcloth bed. She was declared a saint on 8 October 1391 for her virtue not her visions, and her relics were translated to her abbey at Vadstena on the same day. She is now the patron saint of Sweden. Initially 8 October was allocated as her Day, but later her life was celebrated on 23 July, the date of her death, as it still is today. Her Order once had seventy houses, but this has dwindled today to twelve nunneries world wide. One, now located at Chudleigh in Devon, England, has an unbroken existence since it was founded at Syon in Middlesex in 1415 and endowed by King Henry V.

Saint Brigid of Cill Muine – Saints Day: 12 November.

St Brigid of *Cill Muine*, also called Brignat, was an Irish nun, a contemporary of St Brigid of Kildare, who later went to Wales, sailing there from Ireland on a piece of turf. She may have been one of the original Welsh St Brigids (St Ffraid) (see page 152).

Six

Goddess of fire and fertility, Saint of hearth and midwifery

The fire connection

While some early peoples worshipped fire itself, others revered the materials used in making it. From this simple belief many religions evolved deities who were associated with fire, or even a specific aspect of fire such as the Egyptian goddess Sekhet who symbolised the destructive heat of fire and the Indian goddess Devi who dwelt in natural fires found in oil fields. The Lithuanian god, Dinstipan, was solely responsible for directing smoke up chimneys!

There were deities associated with the domestic fire and hearth, the best known being the goddess Vesta of the Romans and her Greek equivalent Hestia, who were deities of altar and the hearth rather than fire itself. Despite its importance to home life neither goddess had any public festivities associated with them. In many societies fire was regarded as a gift of the gods to mankind, and the sun too was seen as a deity by some peoples, and this worship of the sun may have been the prime reason for the construction of many prehistoric stone circles. As religions evolved and became more sophisticated fire still continued to play a central part in sacred symbolism and ritual practices, and often fire rituals and beliefs of a semi-religious nature continued in a secular context until quite recently in some places.

The difficulty of creating fire is reflected in the importance of the perpetual fire in many religions all over the world, from the Temple of Vesta in Rome and the Temple of Hestia at Delphi, both in the Mediterranean area, to the Asian altars of Zoroaster. In many cases great importance was attached to keeping the ritual fire pure and uncontaminated, and since those chosen to tend such sacred fires had such an important role in religious practices, they too were expected to be pure and of high moral virtue.

There is no direct evidence connecting the Goddess Brighid with fire, but she almost certainly had some connection with the element. Once again it is the ninth-century *Sanas Cormac* (*Cormac's Glossery*), of Bishop Cormac mac Cuilleanáin that gives a clue to the characteristics of the goddess when he described Brighid as a 'woman of smith-work' (*ban goibnechtae*). This suggests she was a patron deity of metalworkers who, of course, use fire, rather than being a goddess of fire itself. Most of the evidence for Brighid's fire connection is indirect as it concerns customs that were later associated with St Brigid, last Chief Druidess of the goddess and first Abbess of Kildare. These have to be extrapolated back to practices that may have originated with the goddess. One of the most compelling pieces of evidence for the Goddess Brighid being commemorated by a perpetual fire was the practice of the nuns at the cathedral church of Kildare who maintained a perpetual fire in a shrine to which men were not admitted, as described in *Expurgatio Hibernica* written by Gerald of Wales in 1185:

> In Kill-dara of Leinster, which the glorious Brigid made illustrious, there are many miracles worthy of being remembered. Foremost among which is the fire of Brigid which they call inextinguishable; not that it cannot be extinguished, but because the nuns and holy women so carefully and diligently cherish and nurse the fire with a supply of fuel that during many centuries from the virgin's own day it has ever remained alight and the ashes have never accumulated, although in so long a time so vast a pile of wood hath here been consumed.
>
> Whereas in the time of Brigid twenty nuns here served the Lord, she herself being the twentieth, there have been only nineteen from the time of her glorious departure and they have not added to that number, but as each nun in her turn tends the fire for one night, when the twentieth night comes the last maiden having placed the wood ready, saith, 'Brigid, tend that fire of thine, for this is thy night'. And the fire being so left, in the morning they find it still alight and the fuel consumed in the usual way.
>
> That fire is surrounded by a circular hedge of willow withies within which no male enters, and if one should presume to enter, as some rash men have attempted, he does not escape divine vengeance. Only women are allowed to blow the fire, and then not with the breath of their mouth, but only with bellows or winnowing forks.
>
> At Kill-dara an archer of the household of Earl Richard crossed over the hedge and blew upon Brigid's fire. He jumped back immediately, and went mad. Whomsoever he met, he blew upon his face and said 'See! That is how I blew out Brigid's fire'. And he ran through all the houses of the whole town, and whenever he saw a fire he blew upon it using the same words. Eventually he was caught and bound by his companions, but asked to be brought to the nearest water.
>
> As soon as he was brought there his mouth was so parched that he drank so much that, while still in their hands, he burst in the middle and died. Another who, upon crossing over to the fire, had put one leg over the fence, was pulled back and restrained by his companions. Netherthless, the leg that had crossed the fence perished immediately with its foot. Ever afterwards, because of this, he was lame and an imbecile.

This would seem to be a continuation of the way the Goddess Brighid was venerated at her cult centre before Christianity took hold in Ireland. It is interesting that the *Book of Lismore* records that Druids could make a fence over which 'whoever passes dies'. A perpetual fire would have been maintained in honour of this goddess, which may perhaps suggest she was particularly associated with fire, although it is symbolic of purity among other things. It would have been tended by Druidesses (*ban Druaid*) and, if the later practices were based on what was happening in pre-Christian times, then males would have been excluded from the focal centre of the temple, the Fire House. It was sometimes the case in pagan religions that the chief priest or priestess of a temple or shrine would identify themselves with the deity they venerated, sometimes taking the deities name as part of their title. Bearing in mind that *Brig* means 'High', the Chief Druidess of the goddess's temple could well have coupled her name with that of this goddess.

This 'perpetual fire' seems to indicate the continuation of a pre-Christian practice and was apparently maintained until 1220 when Bishop Henry of Dublin ordered it to be extinguished. While this may have been because he regarded it as superstitious, it must also be borne in mind that at that time the church did accept a great many superstitious practices. It had been actively, but not too successfully, trying to discourage the veneration of stones, springs and wells for some centuries, but this still remained prevalent in his time. So veneration of a flame by nuns, along with the exclusion of males from the shrine, may have been regarded by the bishop as something too pagan to be allowed to continue in a Christian context. However, the flame was almost certainly re-lit some time later and maintained until the monastery was 'dissolved' in the sixteenth century when it was finally extinguished by Archbishop George Browne.

The custom of making Brigid's crosses (see page 96), once widespread in Ireland, may also have a pre-Christian origin since the most common style of cross seems to be symbolic of the sun; in ancient times, emblems for fire and sun seem to have been interchangeable. Part of the celebration of the festival of St Brigid, especially in Ireland, involved the preparation of food for a special meal that required the use of a cooking fire or oven, although for practical rather than symbolic reasons (see page 128). In some places in Ireland, Scotland and the Isle of Man a special bed was made next to the fire for St Brigid (see pages 102, 136 and 162).

There are many traditions, most strongly seen in Scotland and particularly in the Hebridean islands, connecting St Brigid with fire particularly that of the hearth. These arrived there with settlers from Ireland or with St Columcille and his followers. These Scottish customs were recorded by Alexander Carmichael between the 1850s and 1900. Carmichael describes some elaborate rituals involved with the safe preservation of the domestic fire through the night to the next morning, and it is hard to appreciate today how difficult it used to be to make fire.

It was important to preserve the domestic fire from one day to the next, and some people claimed that their family fire had been alight for decades or even centuries, something that actually had a basis in fact. However, keeping a fire going for a long time also represented a fire hazard, particularly at night in wooden houses or stone houses with wooden beams, wooden dividing walls and, quite often, thatched roofs, so the practice of 'keeping the fire in' had to be done carefully to avoid disaster.

The household fire in the past formed the focal centre of the dwelling physically, as it was the place where cooking was done and where warmth could be got when coming in from the cold. Socially it was where the family and friends gathered to talk and, at times, to hear stories, sing songs and hold celebrations. It was vital to the life of the household. In many of the houses in Ireland and Scotland, right up till the end of the nineteenth century, the fire was situated in the centre of the main room as it had been for over 2000 years.

In Ireland it was the custom, last thing at night, to rake up the ash of the hearth over the glowing embers, an operation referred to as *coigilt na tine* (raking the fire), which preserved the *an mháthair* (the mother fire), until it was persuaded back into life the following morning. Hearth prayers (*guí tinteáin*), and hearth blessings (*beannacht tinteáin*) were said in Ireland, and a similar practice prevailed in Scotland. The exception to this was when a person was sick, since the people believed that the fire was symbolic of human life and so it was not allowed to die down during the time of sickness in case the ailing person further weakened or died.

It is thanks to the recording of traditions of the Highlands of Scotland and the Hebridean Islands in the nineteenth century by Carmichael that we know some of the rituals connected with the practice of conserving the fire overnight there. As wood is scarce in these areas peat is the more common fuel and much easier to 'keep in' than fires of other materials. This process was called *smaladh* in Scottish Gaelic (which means little hill or mound), *smooring* in Scottish dialect, and perhaps should be described in English as smothering or subduing.

There was a set ritual involved in this operation which seems generally to have been done by the woman of the house. The embers were evenly spread on the hearth and formed into a circle. This was then divided into three equal sections with a small mound left in the middle. Three pieces of peat were laid between each section, with each piece touching the central mound. The first peat was laid down in the name of the God of Life, the second in the name of the God of Peace, and the third in the name of the God of Grace. This was then covered over with sufficient ashes to subdue the fire but not extinguish it, which was done in the name of the Three of Light. Once this was completed it was referred to as *Tula nan Tri* (the Mound of the Three). The woman then closed her eyes, stretched out her hand and softly intoned one of the *Beannachadh Smaladh* (Smooring Blessings) praying for the house and the family to be kept safe during the night. For example:

> *Smooring the Hearth (*Smaladh An Tula*)*
> I will smoor the hearth
> As Bride the Fostermother would smoor.
> The Fostermother's holy name
> Be on the hearth, be on the herd,
> Be on the household all.

Similar smooring invocations were also said in Ireland, such as this example from Connacht:

> As I save this fire tonight
> Even so may Christ save me.
> On the top of the house let Mary,
> Let Brigid in its middle be,
> Let eight of the mightiest angels
> Round the throne of the Trinity
> Protect this house and its people
> Till the dawn of the day shall be.

Carmichael also quotes a Scottish blessing that invokes the *ainghle geallaidh* (the angels of promise) and it is interesting to note that in Scottish Gaelic the word *ainghle* has the dual meaning of 'angel' and 'fire', while from Ireland some hearth prayers also mention angels (*ainghle*). For example:

> I save this fire, as noble Christ saves,
> Mary on the top of the house and Bride
>   in its centre,
> The eight strongest angels in Heaven
>   preserving this house and keeping its
>   people safe.

In the morning the Hebridean housewife would encourage the fire back into life, and while doing this would recite a *Beannachadh Beothachaidh* (a Blessing of the Kindling), a number of which mention St Bride, such as:

> I will raise the hearth-fire as Mary would.
> The encirclement of Bride and of Mary
> On the fire, and on the floor,
> And on the household all.

Who are they on the bare floor?
John and Peter and Paul.
Who are they by my bed?
The lovely Bride and her Fosterling.

Who are those watching over my sleep?
The fair loving Mary and her Lamb.
Who is that near me?
The King of the sun, He himself it is.
Who is that at the back of my head?
The Son of Life without beginning, without time.

While a similar invocation from Ireland goes:

Brigid, excellent woman,
Sudden flame,
May the fiery bright sun
Take us to the lasting Kingdom.

I will smoor the hearth
As Brigid the fostermother
   would smoor,
The fostermother's Holy name
Be on the hearth, be on the herd,
Be on the household all.

The ringing of its busy bent anvils,
The sound of songs from poets tongues,
   the heart of its men or clean contest
   its beauty of its women at high assembly.

The use of ashes, particularly for foretelling future events, was an ancient and widespread practice and this applied to the ashes from the hearth on the Eve of St Bride's Day in Scotland (see page 136). Ashes from fires were also associated with the fertility of crops and were spread on fields to increase the yield of crops, which worked as the ash is rich in potassium. Ash from the sacred fires lit on the old pagan Quarter Days, *Imbolc* (1 February), *Beltane* (1 May), *Lughnasadh* (1 August) and *Samhain* (31 October) was ritually spread on the fields to promote fertility, while the ash from the burning of old Brigid's crosses were also deposited on fields for the same purpose (see page 107).

From the *Life of Saint Brigid*, written by Cogitosus *c.* 650 AD, comes two miracle stories. One involves the sun while the other concerns fire in which a loom was burnt and which seems to have given rise to Irish 'end of work customs'. For example a spade was placed briefly and symbolically in the fire to indicate it was no longer needed once the potato crop had been gathered in.

A similar custom also applied to spinning and after a long period of performing this activity part of the spinning wheel was put in the fire when the work was completed. It was quickly rescued by the woman of the house before it was damaged. She was then expected to prepare a feast for the other workers, which was consumed to the accompaniment of singing, dancing and story telling. This curious practice, known as *féil searra* (stretching of the limbs) or *clabhsúr* (closure), may possibly have originally been derived from a story recorded by Cogitosus.

Miracle of the wood of the loom that was burnt and later restored

To these miracles also ought to be added her glorious and most celebrated sojourn as a guest of a certain faithful woman. Saint Brigid, as she was making a happy journey according to God's will in the wide expanse of Mag Breg, came to the woman's house when the day had turned to dusk and stayed the night with her.

She received her with open arms and a warm welcome, giving thanks to the Almighty for the happy arrival of the most revered Brigid as if she were Christ. Since, on account of her reduced circumstances, she had no fuel to make a fire with or food to feed such a guest, she threw into the fire as fuel the wood of the loom on which she used to do the weaving. On this pile of wood, she laid her cow's calf, which she had killed, and cooked it with eagerness.

When the supper in God's honour was over, the night was passed in the customary vigils. And when they got up the morning after, our hostess, who had lost her cow's calf, found with her cow another calf of the same kind, which she loved as much as the previous one. She beheld the wood of the loom which had likewise been restored for her in place of the other in the same form and quantity as the previous lot had been. Hence, she suffered no loss at all as a result of having welcomed and regaled St Brigid.

And so, St Brigid, after a happy turn of events and having wrought a wonderful miracle, bade farewell to the house and its occupants and peacefully continued her auspicious journey.

The miracle of the garment hung on a sunbeam, recorded by Cogitosus

> As Brigid was grazing her sheep in the course of her work as a shepherdess on a level grassy plain, she was drenched by a very heavy downpour of rain and returned to the house with her clothes wet.
>
> There was a ray of sunshine coming into the house through an opening, as a result, her eyes were dazzled and she took the sunbeam for a slanting tree growing there. So, she put her rain soaked clothes on it and the clothes hung on the filmy sunbeam as it were a big solid tree. And the occupants of the house and the neighbours, dumbfounded by this extraordinary miracle, began to extol this incomparable lady with fitting praise.

In the early Irish huts a large branch or small tree, stripped of its smaller branches and bark, was often incorporated into the wattle and daub wall inside the building to act as hooks from which clothing and other items were hung. This is what Cogitosus meant when he referred to St Brigid mistaking a sunbeam for a 'tree'.

A later variation of the story makes it even more miraculous:

> So great was Brigid's fame, especially in the giving of charity to all who needed it that Brendon, himself later to be a saint, visited her at Kildare. When he arrived she was out working in the fields and, as it had been a showery day, she was wearing a rain cloak. Upon seeing Brendon so unexpectedly in the distance, she flung off her rain cloak without bothering to hang it up and to Brendon's astonishment it caught on a sunbeam and hung there until it was dry!

There is a twelfth-century manuscript copy of the *Life of Saint Brigid*, written in Irish Gaelic in the Bodlian Library, Oxford that gives an account of the many miracles attributed to St Brigid and among these are three that show a connection with the saint and fire:

Goddess of fire and fertility, Saint of hearth and midwifery

20 Reconstruction of an ancient Irish hut showing a tree used as a hanger. See the story of St Brigid hanging her clothes on a sunbeam (page 82).

The Miracle of the phantom fire

> One day Broicsech, the mother of Brigid, went to milk and leaves no one in the house except the Holy Maiden who was asleep. They saw the house burning behind them. The people run towards it, thinking they would find one beam against another. The house is found intact and the maiden asleep and her face like [word indecipherable]. And Brigid is revered there as long as it shall exist.

Saint Brigid and the fiery column

> The Druid and the female slave and her child were at Loch Mask, and the Druid's mother's brother was there also, the latter was a Christian. In the middle of the night the Druid was watching the stars and saw a fiery column arising out of the house in which were the slave and maiden. He wakes his mother's brother and he saw it too, and the latter said she was a holy maiden. 'It is true', said he, 'if I were to relate to you all her deeds.'

Saint Brigid conferred with the order of a bishop

> There the bishop being intoxicated with the Grace of God did not recognise what he was reading from his book, and consecrated Brigid with the orders of a bishop. 'This virgin' said Mel, 'alone in Ireland will have the episcopal ordination.' While she was being consecrated a fiery column ascended from her head.

St Brigid also has a fire connection in the story of her life which is included in the *Lives of the Saints* from *The Book of Lismore*:

> On a certain day the bondmaid went to milk her kine, and left the girl [Brigid] sleeping in the house. Certain neighbours beheld the house, wherein the girl lay, ablaze, so that one flame was made thereof from earth to heaven. When they came to rescue the house, the fire appeared not, but they said that the girl was full of the grace of the Holy Spirit.

It was not uncommon for a miracle story about one saint to also be credited to another or even to more than one. This seems to be the case with St Brigid as a story originally concerning another saint was attributed to her. There was a story about St Latiaran of Co Cork who, because of her holiness, innocence and goodness, was able to carry burning coals in her apron without her clothes burning, a similar story also being credited to Eiche, the sister of St Mel.

However, in Ardagh, Co Longford, there is a tradition that it was St Brigid who carried the burning coal in her dress from one place to another, not only remaining unhurt herself, but with no damage to the dress in which she carried it. St Brigid has completely overshadowed St Mel, the patron saint of the parish, and it is St Brigid's Day (1 February) that is now celebrated rather than Mel's (6 February). Today the distance that St Brigid, not Eiche, walked with the coal is pointed out, and a well, dedicated to St Brigid, is said to mark the spot where she eventually dropped the hot coals.

Some writers have suggested that the Goddess Brighid was originally a sun deity, although there is no clear contemporary evidence for this with the possible exception of the Brigid's crosses representing the sun. As sun and fire were interchangeable, they more likely represent fire with which Brighid probably did have a connection, but she may have had an indirect sun association because of her possible connection with the fertility of crops. This theory of a sun connection is supposedly reinforced by verses in various ancient hymns that couple St Brigid with the sun, but early Christianity often described Christ as the sun. Old hymns and prayers to many other saints also compare them with the sun. A few examples of St Brigid being connected or compared with the sun:

A hymn in the *Leabhur Iomaun*, c. 580 AD, by St Ultan mac Concubar, Bishop of Ardbraccan in Co Meath, contains the verse:

> She vigil spent; as the bright sun on high
> Her radiance warm'd the earth, and fill'd the sky.

There is an invocation known as *Brigid's Arrow*:

> Most Holy Brigid,
> Excellent woman,
> Bright arrow,
> Sudden flame.
> May your bright fiery Sun
> take us swiftly to your lasting Kingdom.

An early prayer to Brigid runs:

> Brigid, ever excellent woman,
> Golden sparkling flame,
> Lead us to the eternal kingdom,
> The dazzling resplendent sun.

In the fifteenth-century *Book of Lismore* it says that a Druid predicted that Broicsech, St Brigid's mother, will: '…bring forth a daughter conspicuous, radiant, who will shine like a sun among the stars of heaven.'

There are two traditional Irish stories that explain why St Brigid's Day on 1 February comes before Candlemas on the 2 February, the Feast of the Purification of the Virgin Mary, which also has flames as its central theme:

> When Mary and the Christ Child were fleeing into Egypt from King Herod, they found the road blocked by a party of soldiers, and Mary did not know what to do. Just then Brigid came along and offered to distract the soldiers from the fugitives. She did this by placing on her head a bright head-dress with lighted candles attached to it. She approached the soldiers, who were so amazed and intrigued by this strange sight, that they did not notice Mary and Jesus slip by. In thanks for the help she rendered Mary told Brigid that henceforth the feast of this Irish saint would come one day before her own.

Another story to explain why St Brigid's Day comes before that of the Virgin Mary goes:

> When Mary took the infant Jesus to the temple for the first time she found great crowds before it. However, Mary, being very diffident, dared not push her way through the crowds, so Brigid, who happened to be there, said she would help Mary by distracting the crowds. She then got some candles and placed them in her head-dress, causing the crowd to press round Brigid to see the strange sight of a young girl wearing a crown of lighted candles, so enabling Mary to slip into the temple. In thanks for this kindness Mary decreed that the feast day of Saint Brigid should be celebrated on the day before that of the Purification and the Candles.

A third story to account for the date of the saint's feast day from Scotland says:

> That as the night was dark when the Virgin Mary went to the temple for purification, Brigid walked before her carrying a lighted candle in each hand and although a very strong wind was blowing they did not go out. For this kindness Mary said that Brigid's feast day should come one day before her own, which is why the saint is sometimes referred to in Scottish Gaelic as *Bhríde nan Coinnlean* – 'Bride of the Candles'.

The fertility connection

There are suggestions that fertility aspects were connected with the Goddess Brighid and these are reflected in the miracles ascribed to St Brigid. They can also be seen in many of the beliefs and customs once carried out in honour of the saint, although some probably have a pre-Christian origin. St Brigid's Day (1 February) coincides with the start of the ancient, pre-Christian celebration of *Imbolc*. This is generally translated as 'in the belly', which may refer to either the pregnancy of animals (or even humans) or the production of milk (lactation), or both.

However, some writers have suggested that *imbolc* is derived from the Gaelic words *imb-fholc*, which means 'washing oneself', and claimed that it refers to ritual and actual cleansing that was carried out at this time. This day also marked the first day of spring and many customs carried out at this time seem designed to ensure or encourage fertility. In the Hebridean

Brigid: Goddess, Druidess, Saint

21 Stained-glass window in the church at Llansanffraid-Cwmdeuddwr, Powys, showing St Brigid's head wreathed in flames.

islands it was customary to clean the house in preparation for celebrating St Brides Day on 1 February, and the custom of 'spring cleaning' is still widely known if not so widely observed today.

It was a significant time for humans as well, since during winter fresh milk would have been scarce and, in Ireland at least, it was considered a great delicacy to drink the first sheep's milk. This has a higher fat content than cows' milk and so would have provided much needed energy after the privations of winter. There was a strong belief in Ireland, that persisted into recent times, that from St Brigid's Day milk production in cows increased, as 'the milk had gone up into the cows' horns from Christmas until after the Feast of St Brigid.'

The meals that were specially prepared to celebrate the St Brigid's Day feature dairy produce, particularly butter and often this was carefully saved up in the weeks proceeding St Brigid's Day because of the scarcity of milk for making it (see page 131). This day was also regarded in Celtic countries as marking the first day of spring. It is a time when the worst of winter is left behind and new life begins to appear in both the plant and animal world. Farmers prayed for good weather to enable the spring ploughing to begin on time and for strong grass growth to nourish and fatten their animals, while fishermen hoped for calm weather to begin the fishing season. Many traditional customs linked to St Brigid's Eve or Day were carried out to bring blessings on crops, animals and the household in general.

There is strong verbal and written imagery reflecting this time of new life, some of the most vivid of which come from Scotland, where St Brigid is generally referred to as St Bride. One tale from the Scottish Highlands clearly visualises Bride, as a goddess rather than a saint in this case, as the 'spirit' of spring. It appears in *Scottish Customs* by Sheila Livingstone. It was said that Bride spent the winter imprisoned inside the mountain known as Ben Nevis by the *Cailleach* or Hag, who represented the spirit of winter, destroying everything and bringing darkness to the land.

However, at *Imbolc* Bride was rescued from the mountain prison by her brother Angus (Oengus mac Oc), an ever youthful god who rode a white horse. On her release great storms were raised by the *Cailleach*, but eventually Bride chased away the old Hag who had held sway over the land since Hallowe'en (*Samhain* in the pagan calendar, 31 October), and so spring returned to the land once again.

Alexander Carmichael, in *Carmina Gadelica* also records graphic imagery from the Hebrides: 'Bride is said to preside over the different seasons of the year and to bestow their functions upon them according to their respective needs. Some call January *am mhíos marbh* (the dead month), some December: while some apply the terms *na tri miosa marbh* (the three dead months), *an raithe marbh* (the dead quarter) and *raithe marbh na bliadhna* (the dead quarter of the year) to the winter months when winter is asleep'. Bride, with her white wand, was said to breathe life into the mouth of the dead Winter and to bring him to open his eyes to the tears and the smiles, the sighs and the laughter of spring. Cold was said to tremble for its safety on Bride's Day and to flee for its life on St Patrick's Day (17 March), i.e. the cold begins to decrease from St Bride's Day and disappears from St Patrick's Day, summed up by the saying:

> Bride put her finger in the river
> on the Feast Day of Bride,
> and away went the hatching mother of the cold.
> And she bathed her palms in the river
> on the Feast Day of Patrick,
> and away went the conception of the mother of cold.

Weather sayings applying to St Brigid's Day are found in Ireland, Scotland and the Isle of Man. One ancient Scottish hymn begins: 'The day of Bride, the birthday of Spring…' On the Hebridean island of Uist the flocks of sheep were dedicated to Bride on her Day, and it was a time when, so tradition claimed, birds began nesting as is clear from the saying:

> On the Feast Day of beautiful Bride,
> the flocks are counted on the moor.
> The raven goes to prepare the nest,
> and again goes the rook.

Other Scottish verses also indicate St Bride's connection with spring:

> Nest at Bride, egg at Shrove, chick at Easter,
> If the raven has not he has death.

In a society that was heavily dependent on animal husbandry and agriculture and reliant on human power to carry out many day to day activities, the time of conception and birth of babies was significant. Ideally conception should take place around May, when *Beltane*, with its own fertility associations, is celebrated. If successful the future mother would be at her least mobile during the quietest part of the year when activities were generally confined to the house rather than the fields. She would give birth at the beginning of February, a time when nourishing animal milk for the mother to drink was beginning to be available again, and other food crops would come to maturity in the following months, so the baby had the best part of the year to gain strength before winter came once again.

The Goddess Brighid's association with fire may also be significant in having a fertility connection, as in Ireland she was patron goddess of smiths and in Scotland was closely connected to the domestic fire. Ashes from fires were valuable in enhancing the fertility of crops, both ritually and in fact. Wood ash spread on land does increase its fertility by adding nutriments to it, the main one being potassium. Many places in Ireland once began planting crops on St Brigid's Day, although this varied around the country. In the far south of Ireland not only ploughing traditionally began on this day but they began sowing wheat and potatoes immediately after St Brigid's Day. However, in the far north of Donegal the farmers saw little connection between St Brigid's Day and the planting of crops, with the exception of sallow (willow) used for basket making, as this was regarded as the luckiest time to sow it.

Today, 1 February, the day traditionally said to mark the beginning of spring in the Celtic world, seems very early, but this tradition dates far back into antiquity. The climate has always been subject to change, sometimes slowly, at other times much faster, with many complex factors accounting for this. There is much debate as to the exact cause of climate change at the present time. By looking at both past climate change and human development it may be possible to suggest when this date was first chosen to mark the beginning of spring in the west European Celtic world. The last 10,000 years is divided into three climatic periods determined by the prevailing climate and predominant vegetation:

Boreal Period, c.7500 BC–5500 BC
The climate was warmer and drier than today, being very much Mediterranean in character, with pine trees giving way to hazel, elm, lime and oak over much of Britain and Ireland. The people living there were hunter gathers.

Atlantic Period, c.5500 BC–3500 BC
The climate gradually became wetter but was still warm with a Mediterranean range of temperatures. Oak and alder forests proliferated, with elm and limes common, and hazel dominant as understory shrub. The Mesolithic hunter-gatherer society began to give way to the settled Neolithic communities that relied on agriculture from around 5,000 BC.

The Neolithic farmers began to clear the trees and the forests of Dartmoor, the chalk downland's and other areas of Britain, leading to almost complete deforestation in some places.

Sub-Atlantic Period, c.3500 BC–recently
Around 500 BC the climatic conditions deteriorated from the warm and wet of the Atlantic Period to one that was much colder and wetter. This was the climate that prevailed until the present noticeable changes. Within this were warm and cold cycles, but these were of relatively short duration.

It was when farming began to take hold that the specific marking of the seasons assumed a vital importance to these new farmers, with the development of customs to mark the start of these particular periods. One of the most important times would have been the start of spring, when the sowing of crops began. Bearing in mind the prevailing climatic conditions when agriculture developed in Britain and Ireland, it would suggest that it was during the Atlantic Period, with its prevailing warmer climate, that the traditional date of 1 February was chosen for *imbolc* (although this is a Celtic word and it would have been called something else when devised in the Neolithic period). Therefore, the tradition of choosing 1 February to mark the beginning of spring appears to date back around 7000 years.

Whether or not sowing actually began, St Brigid was often called upon to bless the coming crop. In parts of Ireland on St Brigid's Eve the family rosary was offered to the saint to bring a blessing on the crops for the coming year, after which they put a small sheaf of oats and a potato on the doorstep so the saint could bless it as she 'travelled' the country. The next morning they took the oats and potato indoors and hung them on a rafter until it was time to sow the crops. When that time came the sheaf of oats would be rubbed between the hands and mixed with the seed to be sown, while the 'blessed' potato was cut into small pieces which were put in with the seed potatoes. While this was being done the farmer called upon the help of the saint to keep the crops disease free. In some parts of Scotland it was the custom to feed the last ear of corn from the last sheaf of the previous harvest to the livestock on St Bride's Day, an act to ensure the passing on of fertility from one year to the next, another indicator of the close connection of St Brigid and fertility.

This fertility aspect applied not only to crops and animals but humans as well, particularly in Ireland, Scotland and the Hebridean Islands, where the blessing of the saint was also invoked to protect the members of the household. The subject of fertility was of vital importance to the Celtic people whose economy and, in many cases, individual livelihood and survival, were mainly based on agricultural activities until quite recently. Further evidence for the pre-Christian origin of the fertility aspects of St Brigid is suggested by the fact that from the earliest descriptions of her as a saint she is described as a virgin, on the face of it an unlikely characteristic for a person so heavily involved with fertility!

However, although the Chief Druidess who became the first abbess of Kildare could have been married, as Druids, male and female, were entitled to do, although not obligatory, the fertility aspects that passed to the saint probably came more from the ceremonies she would have presided over rather than her marital status. As Chief Druidess she would been involved in ceremonies to invoke the help of the goddess in the blessing of crops, animals and humans. Carrying out actions to encourage fertility of the land, livestock and people, was a role she would have continued as an early Christian abbess on her conversion from a Druidess.

The Book of Leinster says:

> Fainche, daughter of Dallbronach, Brigid's maternal aunt, was a wife of Neman's. Now Fainche had long been barren, so ... Brigid fasted a three days' fast in the church at Kildare, and an angel came and said to her: 'O holy Brigid, bless thine aunts womb, and she will bring forth a distinguished son'.

Fainche had a son, Colmán, and went on to have a further three Conall, Eogan and Cairpre.

St Brigid in Scotland, seems to have been particularly closely connected with childbirth in various ways to judge by the customs that were carried out. Once again this seems a strange connection for a virgin saint, but the Scots devised stories to account for this, some of which date back many centuries. They had a curious, widespread, and illogical belief that she was the midwife to the Virgin Mary and assisted with the birth of the Christ child. For this reason St Brigid became the patron saint of midwives and newborn babies. In the Scottish islands St Bride was often described as *ban – chuideachaidh Moire*, 'the aid women of Mary' (i.e. midwife), while in Ireland she was even equated to Mary herself, often being referred to as the 'Mary of the Gael.' The Hebrideans had a story to explain how St Bride gave help to the Holy Family and was involved in the holy birth:

> Bride, it is said, was the daughter of poor, pious parents and the serving maid in the inn at Bethlehem. A great drought occurred in the land and the master of the hostelry went away with his cart to get water from afar, leaving Bride a bowl of water and a bannock of bread to sustain her till his return. The man left orders that Bride was not to give food or drink to anyone, as he had left only enough for herself; and she was not to give shelter to anyone until he returned.
>
> However, as Bride was working in the house two strangers came to the door. The man was old, with brown hair and a grey beard, and the woman was young and beautiful, with an oval face, straight nose, blue eyes, red lips, small ears, and golden brown hair which fell below her waist. The strangers asked Bride for a place to rest, for they were foot sore and weary, for food to satisfy their hunger and for water to quench their thirst. Bride dare not give them shelter, but she did give them her bannock and water from her bowl. The couple ate and drank outside the door, and having thanked Bride the strangers went their way, with Bride gazing wistfully and sorrowfully after them.
>
> Bride saw how weary was the young woman and her heart was heavy that she did not have the power to give them shade from the heat of the sun and cover from the cold of the dew. When Bride went back into the inn as dusk began to fall she saw, to her amazement, that the bannock of bread was whole, and the bowl of water was as full as it had been before! She could not believe it! The food and the water that she had given the strangers and had herself seen them consume were as they had been before.
>
> When she recovered from her wonderment Bride went out to look for the pair who had gone on their way, but she couldn't see them. Just then she saw a brilliant golden light shining from the stable door, and realising it was not a *dreag a bhais*, [a meteor of death], she went into the stable and there she saw the Virgin Mother about to give birth. Bride assisted the mother in the birth of the child, who was Jesus Christ, the son of God who had come to earth and the strangers were Joseph and Mary. When the child was born Bride put three drops of water from the nearby spring of pure water on his forehead, in the name of God, in name of Jesus and in the name of the Holy Spirit.
>
> As the innkeeper was returning home he ascended the hill on which the inn stood and suddenly heard the sound of music. This seemed to flow past his inn like a flowing stream, while above his stable he saw a bright star. From these signs he knew that the expected Messiah had come and that Christ was born, for it had been foretold to the people that Jesus Christ, the Son of God, would be born in Bethlehem, the town of David.
>
> And the man rejoiced and was exceedingly joyful at the fulfilment of the long awaited prophecy and so he went to the stable and worshipped the new Christ, who lay in a cradle made from the manger of the horses. It is because of this that Bride is called *Ban-chuideachaidh Moire*, 'the aid-woman of Mary', while others call her *Muime Chriosda*, 'foster-mother of Christ' while some call her *Bana-ghoistidh Mhic De*, 'the god-mother of the Son of God, and others still *Bana-ghoistidh Iosda Chriosda nam bann agus nam beannacht*, 'God-mother of Jesus Christ of the bindings and blessings'. Christ himself is called *Dalta Bhríde*, 'the foster-son of Bride', and *Dalta Bhríde bith nam beannacht*, 'the foster-son of Bride of the Blessings', and even *Dalton Bhríde*, 'little fosterling of Bride'.

Another, probably more recent, alternative Scottish story explains how St Bride ended up in the Middle East:

> Bride was set adrift in a basket as a helpless infant, and eventually floated ashore in a place where Druids dwelt. They found her and brought her up, and she remained happily with them until one day she was summoned by a white dove who guided her through a grove of hawthorn trees to a desert. She crossed this and arrived at Bethlehem where she assisted at the birth of the Divine Child and on his forehead she placed three drops of water.

One of the ways for a woman in Scotland to encourage or increase her fertility if she was failing to conceive was to rock a cradle, either empty or with a doll laid in it, while singing:

> O Bride, Brideag, come with the wand
> To this wintry land;
> And breathe with the breath of Spring so bland,
> Bride, Bride, little Bride.

In Scottish Gaelic the term *Bean ghlun* was sometimes used, literally meaning 'Woman on her knees', as a euphemism for a midwife, presumably describing the position she took when she was assisting at a birth. In the Highlands and Hebrides, however, it was at one time traditional for women to give birth kneeling on one knee. This is said to be the origin for the phrase, once common in Scotland *Is i Bhríde mhin chaidh air a glun* – 'It was Bride fair who went on her knee', referring to her role as midwife rather than being on her knees to pray. St Bride was often invoked when a woman went into labour. When this occurred either the woman acting as midwife or her assistant went to the door of the house, stood on the doorstep with her hands on the door jambs and entreated St Bride to enter by calling out softly:

> Bride! Bride! Come in,
> Thy welcome is truly made,
> Give thou relief to the woman,
> And give the conception to the Trinity.

According to Alexander Carmichael the mother-to-be might also appeal directly to St Bride:

> There came to me assistance,
> Mary fair and Bride;
> As Anna bore Mary,
> Mary bore Christ,
> As Eile bore John the Baptist
> Without flaw in him,
> Aid thou me in mine unbearing,
> Aid me, O Bride!
>
> As Christ was conceived of Mary
> Full perfect on every hand,
> Assist thou me, foster-mother,

> The conception to bring from the bone;
> And as thou did'st aid the Virgin of joy,
> Without gold, without corn, without kine,
> Aid thou me, great is my sickness,
> Aid me, O Bride!

The time of birth was regarded as dangerous and, if the labour were prolonged, it was sometimes believed that this might be caused by a witch who had somehow managed to get hold of some of the mother-to-be's hair or a paring of her nails. To counter such a spell, all the women present would gather round the bed and chant an anti-witchcraft spell, an example of which is given in *Guy Mannering* by Sir Walter Scott, published in 1829:

> Trefoil, vervain, St John's-wort, dill,
> Hinders witches of their will;
> Weel is them, that weel may
> Fast upon Saint Andrew's Day.
> Saint Bride and her brat [cloak],
> Saint Colme and his cat,
> Saint Michael and his spear,
> Keep the house frae reif and wear.

If the birth went smoothly they knew that the saint had indeed been present and the family had her blessing. However, if things had not gone so easily, the family felt that they had somehow offended the saint and she had not entered the house. One of the oldest recorded Irish customs was the bathing of a newborn child in milk, which was noted by Benedictus Abbas in *Gesta Regis Henrici Secundi* in 1171: 'The infant was thrice dipt in milk; which was then thrown into the drains or some other unclean place.' This was probably based on an old story concerning St Brigid's birth, a version of which is given in the fifteenth-century *Book of Lismore*, a compilation of earlier material:

> But on the morrow, when the bondmaid went at sunrise with a vessel full of milk in her hand, she put one of her two footsteps over the threshold of the house, the other foot being inside, then she brought forth the daughter, even Saint Brigid. The maidservants washed Saint Brigid with the milk that was still in her mother's hand. Now that was in accord with Brigid's merit, even with the brightness and sheen of her chastity.

It was common until recently in Ireland, Scotland and England that women went to church for a short service of thanksgiving after the birth, a custom known as the 'Churching of Women'. This has almost completely died out, but consisted of a psalm, some prayers and a communion celebration to offer thanks for the safe delivery 'from the great pain and peril of childbirth'. More recent services dropped the reference to the common and ancient belief that a woman who had borne a child was 'unclean'. However, for centuries 'Churching' was regarded by most people as a ritual purification, and a woman was regarded as unlucky until she had been 'Churched', and in many places she was discouraged from entering other peoples houses, speaking to or touching another pregnant woman, or even crossing a road or river in case she spread her bad luck far and wide.

This service has its origins in Jewish custom mentioned in *Leviticus* (12:2–7) that, forty days after the birth of a male child and eighty days after a female, the mother would complete her ritual cleansing and could then go to the temple to offer a sacrifice of thanksgiving. According to ancient stories from Ireland, St Brigid also assisted the Virgin Mary in this task. There are several variations of the story, the following example from Co Galway being fairly typical:

> The Blessed Virgin was about to be 'Churched' and, as she was going to the church, she met Saint Brigid. Our Blessed Lady was very shy in going to the altar rails before the whole congregation and she told Brigid how she felt. 'Never mind', says Brigid, 'I'll manage that part all right'. She got a harrow and put it on her head turning the points upwards. They went into the church and no sooner had Brigid entered than every point of the harrow turned into a lighted candle. The whole congregation turned their eyes on Brigid and her crown of lighted candles, and the Blessed Virgin preceded her to the altar rails and not an eye was turned on her until the ceremony was over. The Blessed Virgin was so delighted with Brigid that she gave her her day before her own, and that is the reason that Saint Brigid's Day is before the Feast of the Purification.

Another Irish story explaining how St Brigid helped the Virgin Mary goes:

> As Mary went up to the temple for purification she was preceded by Saint Brigid carrying a lighted taper in each hand, but despite the wind being very strong on the temple mount and the tapers being unprotected, they did not blow out or even so much as flicker.

Because of this Candlemas (2 February) is sometimes referred to as *La Fheill Bhríde nan Coinnle*, 'the Feast Day of Bride of the Candles'. The many associations of St Brigid with fertility and nature has inspired a modern Irish invocation to her (author unknown):

> Dear Saint Brigid of the kine,
> Bless this little field of mine,
> The pastures and the shady trees,
> Bless the butter and the cheese,
> Bless the cows with coats of silk,
> And the brimming pails of milk.
> Bless the hedgerows and I pray,
> Bless the seed beneath the clay,
> Bless the hay and bless the grass,
> Bless the seasons as they pass,
> And Heaven's blessings will prevail,
> Brigid, Mary of the Gael.

Seven

Celebrating the Goddess-Saint in Ireland

Throughout Ireland, Britain and Europe, many customs and traditions associated with St Brigid continued to be practised until quite recently. A few clearly had their origins in pre-Christian times when the Goddess Brighid was commemorated with activities related to her aspects as a fire and fertility deity. In Ireland, Scotland and a few other countries some of these customs still continue, mostly to honour the saint, but a few still commemorate the goddess, while some of the Christian celebrations have a distinctly pagan aspect to them.

One of the four main pagan festivals celebrated by the Druids, who divided the year into four quarters, was *Imbolc*, which means 'in the belly', i.e. when sheep are pregnant and begin lactating. The start of this period, 1 February, was regarded as marking the beginning of first day of spring. Another name for *Imbolc* was *Óimelc* meaning 'ewe milk', a term also used in ancient Ireland as in his ninth-century *Sanas Cormac (Cormac's Glossary)*, Cormac mac Cuilleanáin defines *Óimelc: is aimsir anisn tic as cairach. melg…i. as arinni mblegar* ('it is the time the sheep's milk comes. Milk. i.e. milk that is milked'). In Welsh this day was called *Gwyl Fair*.

As the pagan Celtic day ran from sunset on one day to sunset on the next, *Imbolc* was celebrated from the evening of 31 January to the evening of 1 February (sunset being approximately 4.45 pm on that day). This ancient method was noted by Julius Caesar in his *De Bello Gallica*, VI, 18 (written in 53 BC):

> …they count periods of time not by the number of days but by the number of nights; and in reckoning birthdays and the new moon and new year their unit of reckoning is the night followed by the day.

However, the Christian church calculated days as running from midnight to midnight, so when celebrating St Brigid, her Eve (*Oíche Fhéile Bhríde* in Irish) started from midnight on 30 January, i.e. 31 January, while St Brigid's Day (*Fhéile Bhríde*) was celebrated from midnight on 31 January, i.e. 1 February.

In ancient Ireland *imbolc* was a day special to the Goddess Brighid, and associated with the lighting of bonfires in honour of this deity who is connected with fire. It was one of the most important of the Celtic fire festivals. It also marked the beginning of spring and the fact that from this time onwards the days can be seen to be becoming noticeably longer. Although in practice this has been happening since the winter solstice around 21 December, it does not become obvious until some time after this date.

When the church wanted to celebrate the life of St Brigid, it was decided that her principal Feast Day should be the same as that on which the goddess was celebrated, 1 February. Most church authorities agreed that the saint was both born and died on this day. Other special days as far as St Brigid was concerned were 24 March, which commemorated the discovery by Bishop Malachy in 1185 of the bodies of saints Patrick, Columba and Brigid at Downpatrick and 9 June when the translation, that is the re-internment of her body, along with those of the other two saints, took place (see page 221). Although some authorities said this should be celebrated on 10 June.

On 6 August 1854 Pope Pius IX granted an *Indult*, that is a licence authorising an Act that the common law of the church does not sanction, which said that 1 February was 'to be observed, as a double of the second class, thoughout all Ireland.' In the united dioceses of Kildare and Leighlin the Feast of St Brigid was observed as 'a double of the first class, with an octave, commencing on 1 February, and terminating on the eighth day of the month.' This means that in the saint's 'own' diocese the services and celebration of St Brigid were to last for a period of eight days. All authorities note 1 February as St Brigid's principal Feast Day with the exception of the *Annals of Roscrea*, as in this it states that she was born on a Wednesday 'on the eighth moon of February', in 449 AD, which suggests her Feast Day should be held on 8 February. However, this is so out of line with all other sources it is unlikely that any church actually accepted this.

In Ireland the church services included the reading of nine lessons, and church services in other countries took a similar form. In Britain the *Breviarium Chlorisopotensis* records that the service also had nine lessons, while in Cologne, Germany, the church dedicated to her marked her Feast as a 'double'. In other German churches dedicated to her in Utrecht, Treves, Mentz, Herbipolis, Constance, Strasburg and many other places she was celebrated by a straightforward service of thanksgiving. She was also celebrated in France and an office of St Brigid had been published in Paris as early as 1622. Divine services to St Brigid were once also common in England, Scotland, Germany and other countries where Irish monks and priests had influence. In Irish Gaelic the festival of St Brigid gave its name to the month of February (*Mí na Fhéile Bhríde*) but most of the associated customs were performed on St Brigid's Eve and St Brigid's Day, as was the case in Scotland and the Isle of Man.

Outside the confines of the church St Brigid was commemorated with a great variety of customs, not all involving religious aspects and special meals were prepared and consumed to honour her. The most numerous examples of customs and beliefs associated with St Brigid come from Ireland. In January 1942 the Irish Folklore Commission issued a questionnaire about 'The Feast of Saint Brigid' which asked what people did on the eve and day of the saint. The results of this questionnaire, covering these traditions in most of Ireland, filled 2,435 pages of manuscript! The range and number of customs and beliefs concerning St Brigid is almost certainly greater than those attached to any other saint.

Brigid's crosses (Crois Bhríde)

The making of specially woven rush or straw crosses, the *Crois Bhríde* (Brigid's cross) or *bogha Bhríde* (Brigid's Bow), is an Irish custom that formed an important part of the St Brigid's Eve (*Oíche Fhéile Bhríde*) celebrations, and the crosses were put up in the house and also some-

Brigid: Goddess, Druidess, Saint

times in outbuildings on St Brigid's Day (*Fhéile Bhríde*). In a few places, such as Portglenone, Co Antrim, it was traditional to make them on the night of 1 February and put them up on the next day, Candlemas. The making of crosses was once very widespread in Ireland and although much less common, it is still carried out, often in schools and pubs, in some places today, while knowledge of the custom, even if not practised, is still known to many people.

Brigid's crosses, in various materials, can be bought in Irish craft centres, 'tourist' shops, at attractions and even at Irish Airports. They are also available from some 'New Age' shops in England but it is a distinctively Irish custom. There are a great many patterns of cross, ranging from simple four-armed examples to those of very elaborate construction and including a design of a three-armed 'triskele' type. In 1942 the Irish Folklore Commission received several hundred examples of Brigid's crosses from correspondents all over the country in conjunction with the questionnaire they sent out regarding the customs surrounding St Brigid's Day. It seems, however, that no one particular type of cross was held in greater veneration than any other. Many of these are now among the collection of Brigid's crosses held by the National Museum of Ireland, and housed at Castlebar Museum, Co Mayo.

There are many elaborations on the basic design, but whether these were of symbolic significance or due to the artistic flair and inspiration of the maker is hard to say. It was a matter of pride to be able to make very elaborate crosses. Among the museum's collection is one that incorporates a total of fourteen diamond crosses. In this case it should be borne in mind that there was an old tradition that the more frequent the repetition, then the greater and more certain would be the blessing that follows, which may account for the existence of multiple-diamond crosses.

The most common design (*crosóg*) more resembles a swastika than the simple Latin cross that is generally associated with Christian practices, despite the fact that the latter is a lot easier to make. The swastika, its name derived from the Sanskrit word *Svastika*, which means 'well being' or 'being fortunate', is a symbol connected with auspicious things in India. While the Brigid's crosses of this type do not have the right angled bends at the ends of the arms

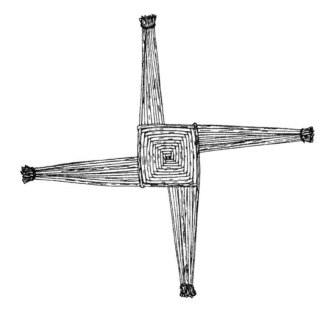

22 'Swastika'-shaped Brigid's cross, the most common type made today. Traditionally constructed of reeds, they can also be made from straw or other material.

Celebrating the Goddess-Saint in Ireland

of a true swastika, it has been suggested that this is what they are derived from, although the swastika was designed independently in many different cultures. It is found on a Palaeolithic carving on mammoth ivory from the Ukraine, dating to about 10000 BC., but occurs in ancient Greece, southern Europe, China and Japan. It also appears on the oldest coins of India, Persia and Asia Minor, as well as those of Troy of about 1000 BC.

Most scholars agree that it represents the sun and/or fire, the two often being interchangeable symbolically. Its shape also indicates movement, i.e. turning; while Christians referred to it as the *Crux dissimulata* – 'disguised cross.' There is an inscribed stone of the early Christian period located near St Brigid's Well at Cliffoney in Co Sligo which bears a Latin cross, but the upper arm incorporates a swastika symbol, perhaps suggesting an association of this with the saint in this instance, although this is the only example known.

As there is a close tie symbolically between fire and the sun, the practice of making sun symbols from rushes or straw at *Imbolc* in honour of the Goddess Brighid, one of whose associations was with fire, meant that such tokens would form a powerful talisman. Brighid was connected with fire as she was the patron goddess of smiths (see pages 29 and 31), and may also have been associated with the domestic fire if this characteristic was 'transferred' to St Brigid as suggested by Scottish and Irish hearth customs (see pages 79–81), and she may have had an indirect sun association (see page 30).

Since a very similar method of veneration continued after the Goddess Brighid's temple was converted to a Christian establishment it seems that the making of these crosses may pre-date Brighid's 'conversion'. It would therefore be derived from a practice associated with the goddess, perhaps a representation of the sun emerging from the darkness of winter. It can also be seen as representing the continuing cycle of the year and the four seasons in the sunwise rotation of the offset arms of the cross (if made in the traditional way), which may also symbolise the movement of the sun with its life-giving energy.

In none of the many ancient stories about St Brigid and her miracles is there any direct connection with an incident that might account for the making of these crosses, and the

23 Three-armed 'triskele'-patterned Brigid's cross. This type was usually put up in stables, cow houses and outbuildings.

stories that explain them all seem to be of a later date. The ordinary people would have continued to practice their customs at the relevant time, particularly those connected with the agricultural year. Many of these the Christian church found difficult to suppress, so often allowed people to continue their traditional activities but provided a Christian 'reason' for carrying them out.

Further evidence for the pre-Christian origin and significance of the crosses can be seen in the way they were used till recent times to ensure the fertility of crops, animals and even humans (fertility being an attribute of the Goddess Brighid that passed to the saint), and as a method of protecting from illness and curing ailments in both humans and animals (the goddess was also a healer). It was also believed that a cross fixed to the house would protect it from fire (an element that was closely associated with the goddess).

Today they are just said to be 'lucky', a great simplification of the complex multi-functionality of the crosses in the past. While we can only guess that the reason that the crosses were made in the time of the goddess was as a fertility and/or sun or fire emblem, we can see a definite fertility connection in the way they were used in Christian times. While the reason for making them in pagan times has been forgotten, there is a Christian explanation for their origin and, not unexpectedly, there is more than one story to account for their origin:

Saint Brigid and the chieftain
One day Saint Brigid was passing by a hut and heard a moaning sound coming from it. She entered it to find a dying man, a tribal chieftain, lying on a bed. Brigid realised he was a pagan and went over to him and began to explain about the Lord God and his Son, but he would not listen to her. Undaunted she continued to try and win him over to the Lord, but all her efforts were in vain as he refused to listen.

Finally, she went outside, pulled up some rushes and quickly wove them into a cross. She returned to the hut and held the cross up before him and explained to him what it represented and the sacrifice that Christ had made, whereupon the chieftain was moved to sorrow and embraced the Christian religion. He then made his confession and Saint Brigid administered the last rites and so his soul was received into heaven.

Saint Brigid and the sick friend
On a visit to a friend who was very sick with a deadly and contagious disease Saint Brigid, while ministering to him, reached down and picked up some straw from the floor. She was an expert weaver, and wove this into a braided cross and hung it on the rafters above his bed.

Soon after this her friend fully recovered, and it seems all the household was protected from the disease because of her blessing and their possession of this token. So today the crosses are made to commemorate this event, bring St Brigid's blessing on the household, and to show it is a good Christian family.

Saint Brigid and her father's conversion
Brigid had been a Christian for many years, but had never been able to convince her pagan father, Dubhthach to embrace this new religion. Eventually she had to nurse her father as he was on his deathbed, and sitting there with little to do she picked up some rushes from the floor and wove them into a cross. Her father asked what she was doing and Brigid explained the significance of the cross and took the opportunity to once again explain Christianity, and at last her father was convinced and became a Christian just before he died. So crosses are

still made today to commemorate the work of Brigid in continuing Saint Patrick's work of converting Ireland.

Saint Brigid and the sneezing death

From Co Waterford comes this curious story (recorded in the late nineteenth or early twentieth century) to explain the reason for the crosses:

> A Druid lived in the hills of Comeragh a long time ago and he used to light a great fire on Saint Brigid's Eve, but whoever saw this fire would sneeze till they died. Saint Brigid was going by and saw that the people were closing their doors, covering their windows and staying indoors although it was early in the afternoon.
>
> She inquired the reason for this strange behaviour, and on learning its cause she told each person to each make a small cross of rushes or willows and to hold it up between them and the fire, and to say three times 'God and Mary with us!'
>
> If their sneezing persisted they then to say 'God and Mary and Patrick with us!' three times. If even that did not prove effective then they were to say 'God and Mary and Patrick and Brigid, Mary of the Irish be with us!'

Presumably this would then work! This story seems to be related to, or more probably derived from, an ancient legend concerning St Patrick, who extinguished a 'baleful (harmful) nocturnal light which killed all who saw it.' The Patrick legend also explains the custom, still common, of saying a blessing after a person sneezes, as the baleful light induced a fatal attack of sneezing until St Patrick introduced the blessing and broke the power of this harmful light.

Another reason why the crosses may have continued to be made in Christian times is that in the early centuries they were regarded as relics of the saint herself. Relics could be actual bits of the body of the saint or items associated with them or which had come into contact with an actual relic of the saint, e.g. a piece of cloth that had touched a saint's bone or even their shrine. There is an example of a sixth-century churchman, Caillín, who later became a saint, who during his life was well known for his collection of relics, but who also had the ability to produce battle tokens (*Cathach*) in return for payment from the Conmaicne sept, as is clear from his *Life* in the *Book of Fenagh* : 'Caillín also ordained from himself a *Cathach* for the Conmaicne, to break battles before them on condition of this tribute being kept up, to wit, a hazel cross to be cut, and its top through its middle – that is the *Cathach*'. It seems that the formula for making this item was regarded as a 'relic' as it was handed down personally from St Cailli. Since the idea of making the Brigid's crosses was also said to have been handed down from the saint herself, they too may have been regarded as 'relics' in early Christian times.

Rushes (*luachair*) have traditionally been used for making the crosses and in other customs to commemorate her. Among the reliquaries in the National Museum of Ireland, Dublin, is a small silver reliquary box 2 x 2cm which was found at Straidcayle, Co Antrim. From its style the box was made either in England or Germany in the twelfth century and, when opened, was found to contain a wreath of plaited rush wrapped in linen. While it cannot be definitely connected to a St Brigid custom or practice, the rush circlet does suggest a possible association. The earliest written mention of Brigid's crosses is found in an account by George Story, a chaplain, who was travelling from Loughbrickland to Newry, Co Down, in 1689 in the wake of King William III's army:

Brigid: Goddess, Druidess, Saint

24 Diamond-patterned Brigid's crosses were usually made of straw on a wooden cross base, but are sometimes made of rushes or rushes and straw together. They are made so they appear to have no beginning and no end.

> I went abroad into the Countrey, where I found all the Houses deserted for several miles; most of them that I observed, had Crosses on the inside above the Doors, upon the Thatch, some made of Wood; and others of straw or rushes, finely wrought, some Houses had more, some less…

While this description of them only dates to the seventeenth century, it is not too surprising that they are not mentioned earlier as the practices, particularly 'superstitious' ones, of the lower class or peasants, were of no interest to many people. There are huge areas about the daily life of ordinary people in the distant past of which historians know very little, something that particularly applies to the Irish. The Brigid's cross and the power it was believed to have in protecting the house from fire is expressed in a poem, *The Irish Hudibras*, written by William Moffet and first published in 1728:

> An' if perhaps you do admire,
> That this great house did ne'er take fire,
> Where sparks, as thick as stars in sky,
> About the house did often fly,
> And reach'd the sapless wither'd thatch,
> Which dry like spunge the fire would catch,
> And where no chimney was erected,
> Where sparks and flames might be directed,
> St Bridget's cross hung over door,
> Which did the house from fire secure,
> As Gillo thought, O powerful charm
> To keep a house from taking harm:
> And tho' the dogs and servants slept,
> By Bridget's care the house was kept.

Crosses were also put up to provide general protection for the house as is clear from a description of crosses in Co Limerick by the Revd James Hall in *A Tour Through Ireland*, 1813:

Celebrating the Goddess-Saint in Ireland

Here, and in general in the interior of Ireland, they have crosses stuck up in the roofs of their houses, to keep away mischief. Some of these are of straw beautifully plaited; and after being blessed by the priest, they are fixed on the inside of the roof…

The reason for making the crosses

While the crosses were generally made to bring St Brigid's blessing on the house, give it her protection, and for luck, there were more subtle reasons for this practice, that is, to encourage fertility. In some areas they were made from unthreshed straw or the maker would insert some grains of corn into a rush cross. In other places the cross was fixed to the roof beams with a wooden peg on which a potato was impaled. Later, to ensure fertility and the saint's blessing on the crop, some of the grain from the cross was added to the first seed sown that year, while the potato was planted as the first seed potato of the coming crop for the same reason. The crosses were usually fixed to the house and outbuildings, but might be taken down to encourage fertility.

For example, sometimes a Brigid's cross was put into a basket of seed potatoes and taken out into the fields at planting time to bring the blessing of St Brigid on the crop. Other ways of using them to ensure fertility and the blessing of the land was to crumble or burn the crosses (or the previous year's crosses), and then bury the dust or ashes, or alternatively scatter them over the land. This is an ancient practice that is related to the Corn Dolly or Kern Baby, a humanoid shaped figure made from the last sheaf of corn to be cut and treated with reverence in various harvest customs. Its origins go back to the pagan festival of *Lughnasa* (1 August) when the gathering of the harvest was celebrated and thanks given to the gods.

The Corn Dolly, embodying the 'spirit' of the land, was carefully kept until the following spring when, in many areas, it was either ploughed back into the soil, or was burnt and the ashes scattered in the fields, practices to ensure the continuity of fertility from one year to another. This method of ensuring the continuation of fertility was once widespread in Ireland, Britain and parts of Europe. In the parish of Killeenadeema, Co Galway they used to plait a

25 A multiple-diamond-patterned Brigid's cross. An old tradition said that frequent repetition would bring greater blessings, which may be the reason for making this type of cross.

sheaf into the shape of a cross and hang it up under the rafters until it was time to sow the seed. Then it was taken down and threshed, and the corn from the sheaf was mixed with the seed that was about to be sown as a way of invoking St Brigid's help for a good, healthy crop.

Even the material left over from making the crosses was believed to have special properties. In a few places the straw left over from making the crosses was placed on the floor to make a rough bed, the *Leaba Bhríde* (Brigid's Bed), to welcome St Brigid to the house and encourage her to spend the night. This practice was more common in Scotland than Ireland. In some households rush-lights were made from the remaining rushes and lit in honour of the saint, while in Co Antrim a small ring of rushes was made and hung on the spinning wheel so St Brigid would bestow her blessing on the work it produced in the coming year.

In the parish of Kilfiddane, Co Clare, they kept the remainder of the straw from which the crosses were made and mixed it with the oats when they were being sown in the spring as it was believed that this would prevent wireworms affecting the crop. In Co Tyrone they used the material left over from making the crosses to make a small *riddle*, a sort of sieve, used for sowing the first few handfuls of corn, and to make a little basket for carrying out the first seed potatoes to the field. The rushes or straw might also be used to make up the *Bhrideog*, the image of the saint (see pages 111–119).

The leftover material could also be used to protect the farmer's animals. In Co Mayo each animal on the farm was provided with a 'golden collar', usually called 'tieings' (*buathrach*), woven from the straw to wear on St Brigid's Eve to bring blessing or luck to the animal and protect it from disease and evil in the coming twelve months. Fidgety cows were soothed by them and difficult horses could be calmed. This material might also be used as 'protective' bedding for horses or calves (Co Tyrone and Co Donegal), and in Donegal the cows were struck with a small switch made from the straw or rushes to protect the animals, presumably by driving out any evil. In Co Derry and Co Kilkenny a portion of this material was put into each outbuilding to protect it and the animals or used to make the Brigid's Bed (see page 79).

The crosses were also believed to be able to protect the house from natural dangers as is clear from an account by a correspondent to the 1942 questionnaire of the Irish Folklore Commission:

> Some people went to great pains in making the cross and I have seen some very large and artistic ones made, and it certainly looked a nice emblem when it was stuck or secured to the ceiling, or rather the scraw [sic], usually above the kitchen window, in line with the kitchen bed.
>
> The cross was supposed to be a safeguard against storm or the blowing away of the roof of the house during the coming twelve months. It was supposed this danger of the winter's storm or its crisis had passed on St Briget's eve, when provision was made for protection against the next winter by the exhibiting of St Briget's cross in the manner described.

There has been some debate as to whether the practice of placing crosses in stables, byres and other outbuildings as well as in the house was universal or only confined to certain regions, and whether different patterns of cross were used in these different locations. The results of the 1942 Irish Folklore Commission questionnaire shows that in some places (Counties Antrim, Derry and Donegal) the type of cross made for the house, the 'swastika' style, was not used in byres where the three armed triskele type was used instead, while another pattern, an open interlaced cross, was used in stables.

However, placing them in outbuildings may never have been a widespread practice as there were many regional variations concerning them. When the 1942 survey was carried out

Celebrating the Goddess-Saint in Ireland

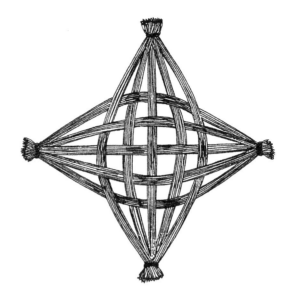

26 The interleaved pattern of Brigid's cross. In Ulster it was used only in stables, but in some other places was put up in the house.

correspondents in many areas had no memory of crosses being put in outbuildings, although they were still put up in houses in those places. It is possible that this may be because this aspect of the custom had changed many decades earlier. While traditionally the crosses were made on St Brigid's Eve or occasionally on St Brigid's Day, they were sometimes made at other times for presentation to friends. The making and giving of the crosses was believed to not only honour St Brigid, but bring a blessing on the maker, increasing their wealth, and strengthening the friendship between the donor and the recipient.

Collecting the material for the crosses

Traditionally the rushes from which the cross was made were picked, not cut, on 31 January, St Brigid's Eve. In some areas the material for the crosses would be taken into the house before sunset on St Brigid's Eve, in others it was after sunset. In Counties Galway, Mayo and Wexford the person who gathered the material for the crosses had to do so in secret without telling other members of the family. One gatherer described how the rushes were cut without speaking a word to his companions and they all prayed until they reached home. It was usual for the rushes or straw, once collected, to be hidden in an outbuilding or somewhere outside the house until brought in with due ceremony. However, in Co Donegal, some people made their crosses from straw which came from the last sheaf cut at the previous harvest and carefully stored. Rush crosses sometimes incorporated weathered rushes from the previous year with the fresh rushes giving contrasting colours, although eventually this coloured effect disappeared.

Bringing in the material

There were regional differences in both the way that the material for the crosses were brought into the house, and the amount of ceremony involved in its 'entry'. In some places a special meal preceded the entry of the material and their construction, often referred to as Brigid's Tea or Brigid's Supper. In other places the material formed part of the meal's preparation or was used symbolically during its consumption (see page 132). Traditionally, in some parts of Ireland, for example Donegal, the crosses were woven on St Brigid's Eve after eating a special

Brigid: Goddess, Druidess, Saint

meal. The straw that was used to make them formed part of the ritual associated with the preparation of the dish known as *Brúitín* or 'Poundies', mashed potato and onion.

In some areas the rushes or straw were brought into the house by a girl chosen to represent St Brigid (or sometimes by the eldest daughter, the father or the mother), and they had to ask permission to enter. In other places the bundle of rushes or straw were brought into the house by the daughter named Brigid, not too difficult as many Irish households used to have a daughter named after the saint. The person chosen to bring in the rushes, whether a female representing the saint or the head of the house, had to demand entry, and those inside called out a traditional reply. While there were minor variations in this part of the celebration, an account from Co Mayo shows the way that the entry of the rushes or straw was carried out.

When the Brigid supper was ready on the table, the head of the household said he was going out to bring in St Brigid, as she too must be present at the feast. He then went outside, collected the rushes and went round to the back door where he knelt and then in a loud voice cried out to those who were inside '*Téar ar do ghlúna, oscail do shúile, agus lig Bhríde isteach*' ('Go on your knees, open your eyes and let Brigid in'). Those within the house cried out: '*Sé beatha, Sé beatha, Sé beatha*' ('She is welcome, She is welcome, She is welcome').

This was repeated twice more and at the end of the third response the man got up, picked up the bundle of rushes and went round to the open front door. Those in the house now called out '*Sé beatha*' ('She is welcome') until the man entered the house when they finished the response by saying '*Mush! Sé beatha agus a sláinte*' ('Low!, It is life and it is health'). The bundle of rushes was then laid against the edge of the table or placed under it, and the family sat down to supper, proceeded by a short prayer. After supper a thanksgiving prayer was said and they made crosses from the rushes. Quite often a garment or cloth that was later to become a Brigid's Cloak (*brat Bhríde*) was bound round the sheaf (see page 125), or a number of garments, one from each member of the family, was wrapped around it. In some regions the *Brúitin* was mashed on straw, or the food, in its dish, was placed on a bed of straw (see page 132).

The people of Innishmore, Co Galway, one of the Aran islands, traditionally made only a plain Latin cross of plaited straw, and obtained the material for it from the *Bhrideog*, the image of St Brigid made from straw and cloth (see page 111). The *Bhrideog* was taken by young women from house to house on Brigid's Eve, saying a prayer as they were invited to enter each building. Once inside, each member of the family took some of the straw from the *Bhrideog* which they later used to make their cross. Straw left by the *bhrideog* party was also used to make crosses on Lettermore Island, Co Galway and elsewhere.

Brigid's crosses and the Brigid's Eve Supper

In some places the crosses seem to have been used to 'bless' the Brigid Eve meal. The crosses were placed on the table or even had food placed on them, as was the case in Co Derry in the 1830s, where bread and butter was actually eaten off the crosses, before they were placed over the door of the house after being blessed by a priest. In early nineteenth-century Co Limerick it was the custom to hang up a small bag containing three slices of bacon along with the Brigid's cross over the main door on St Brigid's Eve to ensure a year of plenty and the multiplication of stock and crops. The use of the three pieces of bacon suggest this practice was inspired by a miracle story concerning the saint, although the Goddess Brighid was also closely connected with food production and prosperity:

> A guest came to her fathers house and so she began boiling a flitch of bacon, cut into five pieces, for the meal. However, a hungry dog was attracted by the smell and Brigid, in her compassion,

gave it a fifth of the bacon, but the animal was not satisfied by this and so she gave it a further fifth. This was observed by the guest who told her father on his return what she had done. Her father then went to the pot and counted the pieces, and there were five. The guest said he had seen her give away two pieces so there could not possibly be five remaining, but her father calmly replied that this was just another example of a miracle that Brigid had performed!

A practice, unique to Co Leitrim, also involved food with the crosses, as well as an image of the moon and stars. There a young girl was chosen to represent the saint at the St Brigid's Eve feast and any food the 'saint' left over was divided among the guests who took it home. When they had made their Brigid's crosses of green rushes and bits of wood, they inserted a small piece of the left over 'saint's' food into it, and then constructed another item, unique to this county. They took a piece of board about 30–45 cm long and stuck on it, with the viscous exudation from a partly boiled potato, an image of the sun, moon and stars made from peeled green rushes, and to the moon and each star they fixed a morsel of the 'saint's' food.

When finished the cross and board were placed over the main door and the Rosary recited in honour of St Brigid. Prayers were said to get her blessing to keep the household free from sickness, sin and scandal for the coming year. The cross and board would be replaced on the next St Brigid's Day. While the crosses are symbolic reminders of a story concerning the saint making a rush cross to demonstrate the power of the Christian cross or the power of Christianity over paganism, they are also regarded as a lucky talisman. They also recall St Brigid's close association with weaving.

Curiously, while an essential part of celebrating the festival of St Brigid was the weaving of the crosses, all other forms of weaving were strictly forbidden. The prohibition on wheel turning not only applied to the spinning wheel, but all others as well, e.g. on wagons, bicycles, mill wheels, wheeled farm machinery and so on. In Co Kerry and Co Cork dressmakers used to refuse to operate their sewing machines, and men would walk long distances rather than using their bikes. In a few places ploughing and smithing were also forbidden on this day. This is summed up in the traditional saying '*Bíonn Lá 'le Bhríde ina shaoire ar chasaibh*' ('St Brigid's Day is a holiday from spinning').

The making of the crosses

Brigid's crosses can be found in a number of designs and sizes. There are many local traditions concerning their method of construction, with variations not only in districts but in individual families. Sometimes the rushes were blessed when initially picked, sometimes after woven into crosses. Traditionally they were made working in a sunwise direction, i.e. from left to right, perhaps another indicator of their origin as a symbol of the sun and their pre-Christian origins.

In some localities the making of Brigid's crosses involved a division of labour between both the sexes and the generations within the household. For example, after the Brigid supper the younger members of the household would begin making the crosses, while the male head of the household, if a farmer, began making 'tieings' (*buarach*) a neck band, from the first of the rushes to be used as lucky charms for the lambs or calves that would shortly be born. The tieings were usually placed behind the family crib until required. When a lamb or calf was born one of the 'tieings' was put round its neck to bring it luck and the blessing of St Brigid. It was quite common to use a tieing made from or incorporating left-over cross material to lead an animal to the fair, but care had to be

Brigid: Goddess, Druidess, Saint

27 A wheel-patterned Brigid's cross. Made in parts of Counties Clare, Tipperary, Limerick, Cork and Kerry, it was made of straw, hay or hazel rods.

taken to bring home the rope and not give it to the animals new owner. Once these had been made, the man would begin making his own cross.

In some places only women made the crosses, while in others they would only be made by the member of the household called Brigid, a very common name. However, in many cases, crosses were made by all members of the family, including quite young children, one being made for each room in the house. In Co Leitrim each member of the family made a cross on St Brigid's Eve, the largest, made by a parent, was later fixed over the door, while other adults would make crosses to hang up in the stables and byres, and those made by the children were hung over their beds. Generally when the family had made their crosses they would be left in a pile until the morning of Brigid's Day when they would be fixed in position.

In Co Tyrone left-over rushes were sometimes used by young men and women to determine who their future spouses would be. After making the crosses they would take two or three rushes and shape them into the form of spinning wheels and ladders, the young women putting a ladder under their pillow and the young men a spinning wheel under theirs. That night the sleeper would, hopefully, have a dream in which they would see their future spouse climbing a ladder or sitting at the spinning wheel!

Until quite recently in Co Donegal, where cross making is still carried out, there was a tradition that any rushes left over from making them were boiled up in a pot of water and a note or photograph of a loved one would be thrown into the pot by anyone with a grievance or bad thoughts who wanted the help of the saint. On Cruet Island, Co Donegal it was the custom to place a bundle of rushes or straw left over from the making of the crosses in the fishermen's boats to provide protection at sea. In some coastal communities it was traditional to weave a little ribbon from the left over straw and carry it when out fishing.

The blessing of the crosses

Sometimes a blessing was performed before the cross was put up. This might be a formal blessing by a priest in the church, although it has been suggested that the priestly blessing only applied to crosses put up in the house. However, the possibility of a formal church blessing was variable, as in some parishes the priest would only make himself available to bless them at a particular time. In one parish in Co Donegal in the 1950s the priest condemned the weaving of Brigid's crosses as superstition.

In Co Armagh the crosses made on St Brigid's Eve were taken to church on Candlemas (2 February) for blessing before being put up in the house or distributed among friends, a practice that had almost ceased by 1945. Alternatively, an informal blessing done in the house either by simply reciting a prayer, or sprinkling Holy water on the crosses while saying a prayer. Occasionally a woven Brigid's cross is used at weddings, and at one, held in July 2001, on the island of Eoghan, Co Donegal, the priest blessed the couple with words modified from an old Gaelic saying:

> God's blessing upon you from
> Samhain to the Feast of St Brigid,
> From the Feast of St Brigid to Beltane,
> From Beltane to Lammas,
> From Lammas to Samhain,
> And may Almighty God bless you,
> Father, Son and the Holy Ghost.

After this the couple was presented with a woven Brigid's cross.

Fixing the crosses

Generally the crosses made on St Brigid's Eve were put up on the saints day (1 February). In some areas (Counties Clare, Offaly, Sligo and Tipperary) a prayer was said as the cross was being fixed up in the house or outbuilding: 'May the blessing of God the Father, Son and Holy Ghost be upon this cross, the place in which it hangs and every person that looks upon it.' While in Co Sligo they said: '*Ibhfad urine gach orc is gach orchid, Grace is i gclocha, Is i mean fawna farrago*' ('May we be safe from every evil and harm on land and the deep sea') as the cross was put up.

In Co Leitrim the head of the household blessed the house with the cross before its final fixing by taking the cross and going round the outside of the building, holding it against every window and door, saying at each '*Naomh Bhríde, saor sinn ó gach tinneas, gorta agus tóiteán*' ('St Brigid save us from all fever, famine and fire'), all attributes covered by the goddess as well as the saint. He then entered the house and fixed the cross over the kitchen door. The usual place to fix crosses, in both the house and outbuildings, was to the internal rafters usually over the door. However, a change began with the replacement of unceilinged thatched houses with modern dwellings and the modernising of old houses. This led to the crosses being fixed to the wall, often in bedrooms, but sometimes over the fireplace. This change in building style clearly led to a decline in the making of the crosses from the 1930s.

Replacing the crosses

In some places (e.g. Co Kildare) they were taken down the following St Brigid's Day and replaced with a fresh one. The old ones were disposed of, usually by burning or sometimes by rubbing them between the palms of the hand until they turned to dust. The dust or ashes from the crosses was then often spread on the land to ensure its fertility. In some places the crosses from the previous year were mixed with the first batch of grain sown in the spring.

Brigid: Goddess, Druidess, Saint

However, it was a widespread practice to leave them in place and put up a new one each year and it was claimed that you could calculate the age of a house by counting the crosses. This was not an infallible method however, as some people were said to have left the crosses up for seven years before starting a fresh series, while in parts of Co Kerry, newly married couples began putting up a fresh collection of crosses, so indicating how long they had been together. When the house passed to the next generation the incoming couple would remove all the old ones and begin their own collection. Julius Rodenburg in *A Pilgrimage Through Ireland*, 1860, referring to his time in Co Kerry, noted that:

> The lath-work of the roof, which stood uncovered over the cabin, was coated with a greasy deposit of many years' peat smoke, and in this black rusty mass I saw here and there a small Maltese cross made of barley straw. 'That is the cruse' said Bridget, 'in honour of my patron saint; on Saint Bridget's night… we weave it and place it on the rooftree. It protects the house from fire.

Using the crosses for other purposes

Crosses and the material from which they were made was used to ensure the fertility of animals, crops and the land, but their use in fertility rituals could also apply to humans. For example, in Co Mayo when a couple got married the mother, or mother-in-law, would make a straw cross and singe each of the ends. This was then placed under the mattress of the couple's bed to ensure they would produce a family.

One of the reasons for fixing the crosses to houses and stables was to protect their occupants, both human and animal, from illness, but since St Brigid was connected with healing, as was the Goddess Brighid, it is not unexpected that the crosses and the material from which they were made should also have been used to cure or alleviate a variety of ailments. For example, rushes left over from the making of crosses in Co Donegal were burned under the noses of sick cows to cure a variety of problems. In Co Down the cross from the byre was

28 A Latin-patterned Brigid's cross, which could be made from rushes, straw or occasionally simply two slivers of wood nailed together. This type, usually made from grass or grass and hay, was made on the Aran Islands.

Celebrating the Goddess-Saint in Ireland

taken down and used to cure milk fever in cows, bless a cow that was milking poorly, and could even, so it was said in Co Mayo, cure cows who had a slipped shoulder.

In Co Kilkenny a Brigid's cross was placed on the sick person, while in Co Cork a cure was carried out by placing the cross under the mattress. The material left over from the making of the crosses was also used in cases of illness. In parts of Co Donegal people would tie some of this around their heads or other parts of the body subject to pains on St Brigid's Eve so the saint could cure the person when her spirit 'walked abroad' that night. The sick person would burn the straw the next morning to ensure a rapid cure, a form of transference magic where the illness was passed to the straw and 'destroyed' when it was burnt. In Co Donegal ashes left over from the burning of old crosses were carefully retained as they were believed to have curative powers.

Designs of the crosses

While rushes are the traditional material from which to make Brigid's crosses, many other materials have been used in the past, including straw, sedge, bent grass, hay, wood, willow, goose quills, tin, wire, cardboard and leather. Brigid's crosses made of cloth were sometimes worn by children to celebrate St Brigid's Day and jewellers have made them in silver and gold as pendants or brooches. The crosses also vary in size, pattern and method of construction, and have been divided into seven main types by John C. O'Sullivan based on the collection of crosses in the National Museum of Ireland: Swastika, Diamond, Three armed triskele, Wheel, Interlaced, Latin, Greek and a selection of Miscellaneous types.

A Brigid's cross was used as a logo by the Irish radio and television service, Radio Telefís Éirinn (RTE), for some years, which gave widespread exposure to this ancient emblem and the making of the crosses. Although less common than it once was, this tradition is still maintained. Many Irish schools encourage their pupils to carry out Brigid projects and make traditional Brigid crosses. In the Teach Ruer, a pub in Gortahork, Co Donegal, they still spread rushes on the floor on St Brigid's Eve and a dish of mashed potatoes, *Brúitín*, is served to all those present. Then everybody makes a Brigid cross, with a prize of a bottle of whiskey to the person who makes the most elaborate one. At the end of the night everybody takes their cross home and hangs it up. Locals say this has been going on in the pub for as long as anyone can remember.

Many Irish local papers carry stories about the making of Brigid's crosses and how to make them during the last week in January, so helping to keep the tradition alive. A mass Brigid's cross-making session occurred at the Goddess Conference at Glastonbury, Somerset in 2004, with over 200 people making one each! Today Brigid's crosses can be found made of rushes or straw in some shops in Ireland, but are more widely available made in resin as wall plaques, fridge magnets and on cards. They are also made from a variety of materials such as compressed turf, pottery, of Connemara marble on a key ring, and as items of jewellery such as a brooch, lapel pin and tie clips in gold, silver and alloy. In Somerset a modern diamond pattern variant of the Brigid Cross is sometimes seen, made from a wooden cross with coloured wools spanning the arms. While some people call this a 'God's Eye' others refer to it as a 'Brigid's Cross' and examples are sometimes hung up near a sacred spring and well at Glastonbury.

How to make a Brigid's cross

Preparation

Cut and gather some fresh rushes (*Scipus lucustris* or *Phragmites australis*), or Soft Rush (*Juncus effusus*) or Common Rush (*Juncus communis*) and sort out so that they are approximately the

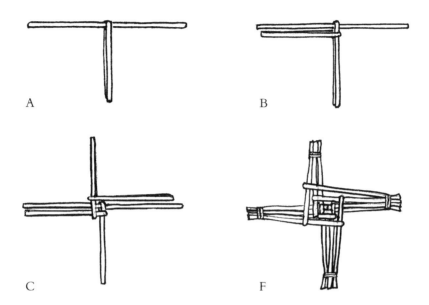

29 How to make a Brigid's cross.

same length and thickness. As an alternative you can also use straw, drinking straws or pipe cleaners.

To make the cross
A. Take two rushes, fold one in half and place it over the first to form a T shape.
B. Fold another rush in half and place it over the rush already folded. Turn the cross so that you are holding the most recently added piece.
C. Take another rush, fold it and place it over the last rush you added. Turn the cross again so you are holding the most recently added piece of rush. This will create the four-armed cross shape.
D. Repeat step C, turning the cross each time. Each additional rush added holds the previous one in place. Continue to do this.
E. You will soon see a pattern forming in the centre of your cross. Care must be taken not to overlap the rushes.
F. When you feel you have added sufficient rushes you will need to tie off each arm of the cross. Begin with the arm to which you last added a rush. Once this is tied it will lock the rest of the cross together. Then tie the other three arms. Trim the end of each arm so they look neat and they are all of the same length.

You have now made a Brigid's cross!

The Brigid's cross, shield and veil custom

A custom, that was once apparently confined to the Louth-Armagh border and died out sometime in the nineteenth century, was described by the Revd L. Murrey in the *Louth*

Celebrating the Goddess-Saint in Ireland

Archaeological Journal in 1914, and by T.G.F. Patterson in *The Ulster Journal of Archaeology* in 1945. It was ridiculed by some people, 'even by some of those whose duty it was to guard the morality and piety of the people', which seems to have been the reason for its eventual demise.

On *Oíche Fhéile Bhríde* (the Eve of St Brigid's Day), it was traditional for people in that area to prepare a number of *Crois Bhríde* (Cross of St Brigid), *Sgiath Bhríde* (the Shield of St Brigid) and *Crothán Bhríde* (the Veil of St Brigid). These seem to have been miniature representations of the items, made from the soft rush (*Juncus effusus*) found in wet meadows that is strong and easily plaited. The different areas would compete with each other to produce the neatest and most 'ingenuously wrought' shields and crosses.

The people of each district would assemble in the evening with the purest and most modest young girl in the area being chosen to represent St Brigid (*An bhrídeog*), and it was considered a bonus if she was actually named Brigid. She put on the veil and, taking the cross in her right hand and the shield in her left, she went to each house, accompanied by a group of people who earnestly prayed to God for the Holy Spirit to fill the family they were visiting, that they might keep his commandments according to the example set by St Brigid.

When they reached a house the *bhrídeog* asked those inside: 'Are you resolved, with God's assistance, to obey His laws and those of His church, and to lead blameless lives like the great St Brigid?' The answer was invariably 'Yes' upon which the 'saint' presented the household with a cross saying: 'Take the sword with which the great St Brigid fought against her enemies, the world, the flesh and the devil, and remember to bear the crosses of this life with true Christian fortitude after the example of the great St Brigid.'

She then presented one of the shields, and said: 'Take ye this shield, the shield of faith; remember the many victories gained by St Brigid under its protecting influence, and bravely follow her example.' Then presenting the veil, she asked: 'Will you follow the rules of virtue, piety and general good conduct laid down for your guidance by St Brigid?' The housewife then called on her daughters and the female domestics to 'Be modest, chaste, and virtuous according to the example which the saint, whose festival we celebrate, has left for your imitation.'

Then the cross was held up, and the whole assembly were called on to remember their redemption, at which point everybody would kneel and join in a prayer asking for God's help to keep the household pious and virtuous in the coming year. The whole assembly then moved on to the next house where the ceremony was repeated. The crosses, shields, and veils were fixed over the doors as a reminder to all the members of the household of the promises they had made on St Brigid's Eve.

The image of Brigid – the Bhrideog

Another of the customs carried out on *Oíche Fhéile Bhríde* (Eve of St Brigid's Day) was the making of an image of St Brigid which was carried around all the houses in the neighbourhood to announce the coming of the saint. It was firmly believed that on the eve of her festival she, or at least her spirit, travelled around the country, accompanied by a white cow, bestowing her blessing on tokens left out for her to touch so creating the *Brat Bhríde* (Brigid's Cloak, see page 124) and *Ribín Bhríde* (Brigid's ribbon, see page 126).

This image was generally fairly crude, but still treated with great reverence, and was referred to as the *Bhrideog* (little Brigid) and sometimes, more recently, as 'Miss Biddy'. It was carried

Brigid: Goddess, Druidess, Saint

30 A traditional Irish *Bhrideog* (little Brigid), an image of St Brigid made of straw. This example comes from Ballycreggan, Kiltoom, Athlone, Co Roscommon.

from house to house by young people who were also referred to as *bhrideog*, 'little Brigid' (plural – *bhrideoga*) in a custom that involved the whole community. This may have its origins in medieval processions where images of saints were carried around the district on the saint's day, a practice still common in some parts of Europe.

Generally the *Bhrideog* was carried by children and young people up to the age of twenty, and while in some places it done by a mixture of boys and girls, in others the procession consisted only of girls, all wearing their best dresses. In Co Kerry, however, only young men took part. In English the boys taking part were called Biddies or Biddie Boys and very occasionally adults also participated. This custom was once widespread throughout Ireland and followed the same basic pattern, although with regional variations. While most accounts are from the first half of the twentieth century, there are a few descriptions from an earlier period when the custom seems to have been more elaborate, with each participant doing a 'turn' to entertain the household they visited. This is described in a ninteenth-century account in which both boys and girls took part.

There were often three or more groups taking the *Bhrideog* round the neighbourhood depending on the number of houses and density of population, but preparations started some days earlier. During the last week in January the young people, in this case up to their late 'teens, gathered at a house in the village, using either the kitchen or a barn, to rehearse what they would do. It was usual for the participants to carry round the *Bhrideog* wearing a disguise, often with boys wearing girls' clothes and girls being dressed as boys, although other accounts refer to costumes made entirely of straw. Some of the *bhrideoga* wore multi-coloured clothes, with masks and elaborate hats of straw designed to hide their faces. More recently they wore brightly coloured clothes decorated with ribbons or sashes, along with fancy hats and face masks.

One early account of the *Bhrideog* describes how it was made from a peeled turnip, with the eyes, nose and mouth cut out and coloured with soot to represent her head. This was put on a stick and straw or rushes tied to it to form the body. Everyone who was to take part in the procession prepared some form of entertainment to perform on entering the houses. This might be a song, music played on the flute or violin (although later an accordion was more common), poems or rhymes. In addition prayers and invocations to the saint, in Gaelic, were said asking for her blessing on the house and the household for the coming year.

Early on the evening of 31 January, St Brigid's Eve, the party met up and began their rounds, all wearing their disguises. Each group of *bhrideoga* had a certain traditional area in which they operated. To visit houses in another groups territory was not only considered unmannerly and unfair, but could lead to long-running feuds. Quite often, when the group visited a house, they might be asked 'What Biddies are you?', and the leader, sometimes referred to as the Captain in more recent times, would raise his mask to assure the householder that they were the local group.

The person who carried the *Bhrideog* was the first to enter the house and the group was almost always given a warm welcome as it was regarded as not only inhospitable not to invite them in, but unlucky. After the performance and some prayers the household were then asked to contribute something to the *bhrideoga*. The early accounts say that this was a gift of bread, butter, potatoes or other items of food, particularly eggs, and bags or baskets were carried in which to put it. This food was used later for a feast after they had finished their rounds.

According to some accounts, a pin was given, especially by poorer households, which the housewife pushed into the *Bhrideog* or pinned to the costume of the *bhrideog* who carried the image. However, the symbolism of the pin as a gift or offering is an ancient one, since the pin was regarded as an emblem of childbirth and fertility, and was to some extent connected with the curing of illness. Pins were quite valuable items in the past and housewives would look after them carefully, and they were often deposited in springs and wells as offerings. This value gave rise to the proverb: 'See a pin and pick it up, and luck will follow all the day', so the older accounts of a gift of a pin may possibly have an underlying significance. A gift was invariably given as it was regarded as unlucky not to do so as St Brigid was so good at unselfishly giving charity herself.

A detailed account of the *Bhrideog* custom appeared in the Dublin newspaper *An Claideam Soluis* (*The Sword of Light*) dated 6 August 1910. The article, in Gaelic, *Old Customs of the Irish: The Bhrideog*, regrets the passing of many of the old traditions and the writer (Cáit Ni Dhonnchadha) described her experience of the custom in the second half of the nineteenth century and which, in her area, was carried out only by boys:

> A group of boys, about nine or ten altogether, arranged between themselves that they would go out on the Eve of St Brigid's Day carrying the *Bhrideog*. This *Bhrideog* was made in many different ways, but the most common method was to get a wooden cross, and push it through a white shirt, with rags or straw put inside to fatten it up. A head was put on top of it and a mask or perhaps a little cap in placed on it. Strips of cloth, ribbons, sashes or things of that sort were put on it to decorate it. I leave it on oath that there was no more prestigious work than dressing the *Bhrideog*.

The boys were often working hard for a fortnight beforehand thinking about how they would make it, buying (or stealing) goods for it, sweet talking the women of the households to do the sewing and collecting money between themselves to buy their masks. But making the *Bhrideog* wasn't even half the work.

Every person of the group of friends had to dress himself up similarly, get a mask and an old cloak (the longer the better) and a neat stout fine stave that would not bend or break. Each of them had to transform himself as much as he could in both his voice and appearance. They had every single item ready for the night, and immediately after nightfall, out they went walking.

The *Bhrideog* would be given to one person who was in charge of it throughout the night. They would go round from house to house in the neighbourhood, and it is many a mile that they would walk from then on until midnight. But what would they do when they arrived at a house? Yes, it was customary with each group that they had rehearsed something beforehand – a song or two perhaps, some dancing, or some funny entertainment. Then the leader of the *bhrideoga* went up to where the man of the house was and he had to give the leader something – a shilling or a sixpence or whatever thing he was willing. When there was an end to this event, the money was divided among the *bhrideoga*.

The article goes on to describe how she saw the custom from a child's eye point of view:

It is funny the thing that happened to me the last night I saw this *Bhrideog*. This is quite a few years ago when I was young. I remember the night very well and a lot of what occurred. Sure! For a week before St Brigid's Day no single thing was worrying me except how it would go for me on the night of the *Bhrideoga*, would I or any of the young children, be afraid of them? Would I find it in me to remain down in the kitchen while the members of the *bhrideoga* were inside, watching what was going on?

It was usual for us, however, when we heard the crowd approaching the house, to run up the stairs and listen to all their talk and all their acting. But on this particular night, however, we were all sitting in front of the fire, and whatever got into me I could not say, but I said boldly that I would bet that I would not be afraid of the *bhrideoga*, and so would stay downstairs while they were in the house. 'Yes, wisha' said my father, 'I will bet that you will regret it by the end of the time. I bet that you won't have that much courage as you have now when they are finished with you.'

It wasn't long until we heard the noise outside coming towards us. The youngsters scrambled themselves up the stairs, but I sat myself on a stool in a corner of the fireplace. It wasn't long before supernatural scrapings came from the door! It was opened for them. In they came with a leap, one after the other. The leader of the *bhrideoga* well out in front. He wore a large, devilish old hat, high on his head, a ginger beard the length of two hands hung from his face, a flannel jacket on him, a straw rope around him as a belt, and an old woman's skirt down to his feet. The company was about ten, and everyone of them in clothes funnier than the other.

These devil's were not long inside when they noticed me sitting in the corner. Straight in my direction came the leader of the *bhrideoga* and shoved his face in mine. 'Yeah!' (God!), I let out a high, near death scream, and I covered my face with my handkerchief to keep that frightening sight out of my view. I would have given up a lot of the wealth of life at that moment to be upstairs together with the other children, I was that scared.

It wasn't long until two of them came out and stood facing each other in the middle of the floor and recited *Aighneas an Pheacaigh leis an mBás* (*The Argument of the Sinner with Death*) for us. When they was finished another one of them started to sing *An fear óg singil* (*The Young Single Man*). During all this time my face was covered, and I would not move a foot for all the wealth in the world. However, I had that much interest in the song that I took my head out from behind the handkerchief to look around, but my 'friend', the leader of the *bhrideoga*, pulled away the hanky and brandished it over me, and shook his beard at me, and as soon as I hid behind the hanky he pulled it away.

Anyway, they received some money and they gathered themselves together to go to some other house. The rest of the family had their fill of laughter about me and my antics in front of the *bhrideoga*, and it was a long time before they let me forget it.

More recent accounts show that the ceremony performed at each house had became simpler, individual performances were dropped and the traditional contribution expected had changed to a gift of money, although food was still given by some households. A begging verse was said at each house to elicit a gift, for example in Co Kerry they said:

> Something for poor Biddy!
> Her clothes are torn.
> Her shoes are worn.
> Something for poor Biddy!

While in Co Claire they cried:

> Here is Brigid, dressed in white.
> Give her a penny for her night.
> She is deaf, she is dumb,
> She cannot talk without a tongue.

The money was used by the older *bhrideoga* to provide food and drink for a celebration, called the Biddy Ball, held at a local house. Other accounts indicate that the money collected was sometimes divided among the *bhrideoga*, while in some cases they sold the food they had been given and shared out the money raised. In ninteenth-century Co Kerry the pennies collected bought candles which were lit on St Brigid's Day in honour of the saint. It is interesting to note that in one of the early accounts of this tradition it seems that girls taking part in this custom was frowned on by the church and, if the procession encountered the priest, he would stop them and take their disguises off to see if there were any girls taking part. If he discovered any the priest would severely reprimand them and order them to go home, only allowing the boys to continue their rounds. However, this view seems to have been the exception and probably concerned one individual priest or at most a very small number.

Occasionally the *bhrideoga* consisted of girls no older than fourteen, and they did their rounds on the morning of St Brigid's Day rather than her eve. They were just as warmly greeted, with the woman of the house addressing the image of St Brigid in the following terms: '*Móire is dachad ar maidin duit, is a Chriosta tóig, tá an bhlian caite agus tánn tú tagaithe arís inár dtreo*' ('More than forty good mornings to you young Christian, this year is spent and you have come to us once again'). She would then take the *Bhrideog* in her arms and kiss it.

Around 1900, on Inisheer, one of the Aran Islands, there was a poor woman who, on St Brigid's Eve, used to go to all the houses begging for food and carrying a large *Bhrideog* made of straw, dressed in an immaculate white dress with a picture of the saint pinned to its breast. The woman would say the following verse:

> Here comes Brigid dressed in white,
> Give her something for the night,

> She is deaf, she is dumb,
> For God's sake, give her some.

After this she was usually given some food 'in honour of the saint', for which she blessed all the people in the household, some of whom used to take a little straw from the image with which to make their Brigid's crosses. The tradition long survived the old women, with small groups of girls carrying round a *Bhrideog* dressed in white Conformation robes and conferring a blessing on each house that gave a gift, usually money, part of which they spent on sweets with the rest donated to the church.

The construction of the image varied greatly. In many cases the basis of this was a churn dash, a long wooden rod which was used in the churn used for butter making, since most farms and many households made their own butter in the nineteenth and early twentieth century. Padding was tied around this rod to make a humanoid shape which was then clothed, often in a dress loaned by a village girl.

St Brigid was very much associated with cattle and the dairy, and churn dashes used as the basis of a *Bhrideog* were said to give increased yields of butter after they were returned to their proper use. One of the gifts that the saint was said to have possessed was the ability to increase the yield of food and drink. Something often featured in her miracle stories, but which was almost certainly also an attribute of the Goddess Brighid. The use of churn dashes as a basis of the *Bhrideog* was popular as it could be stood upright on the floor.

Sometimes a broom or just some sticks were used as the basis of the *Bhrideog*, while the face might consist of a mask or a white cloth painted with a face. In some places it was the custom to attach a Brigid's cross to the figure which was then referred to as the *rionnag Bhrideog* (the star of little Bride). This suggested knowledge of the story, more prevalent in the Scottish Islands, that St Brigid was present at the birth of the infant Christ (see page 90). In some places, especially in more recent times, a child's doll was dressed up and used as the *Bhrideog*, being carried round by young unmarried girls. When welcomed into a house, they announced they were bringing the blessings of the saint and gave one of a number of Brigid's crosses they carried to the head of the household.

As time went on the custom became diluted, as has happened to the effigy aspect of the English Guy Fawkes celebrations over the last thirty years. The *bhrideoga* would consist of a few children who would simply carry a doll round houses in the neighbourhood, collecting a few pennies from each. In west Co Galway in the period 1949–56 the *bhrideoga* 'music' had been simplified into a tune being performed using a comb and a piece of paper!

An interesting account of the custom as it was carried out in North Galway, in the 1960s was given by Paul Connaughton in *The Tuam Herald* (January 27, 2005):

> For readers who have never seen *bhrideogs*, we were dressed up so that we would be totally unrecognisable. I must say it was not very artistically done but we used all types of camouflage from pulp bags to outrageous dresses to nightwear, and don't talk to me about the range of masks! Masks were certainly not as elaborate as they are now and with breathing so difficult in an ill fitting mask you would be fit to faint when the dancing got under way in the kitchen. Another big problem was when the eyelet of the mask did not synchronise with your eyes and you simply went blind. Try managing to get out of a small house in the company of maybe ten or fifteen others with the head of the house after you with the handle of a broom as a result of something said or done that he obviously didn't like! More about that later.

Celebrating the Goddess-Saint in Ireland

31 *Bhrideog* (little Brigid) at the Ninth Annual Goddess Conference, Glastonbury, Somerset, 2004. The theme that year was 'Celebrating Bridie and the Maiden Goddess'.

Our group normally worked the parishes of Newbridge, Mountbellew, Moylough & Kilkerrin. We even got so enthusiastic one year we went out two nights — talk about milking the system. I remember one woman remarking rather caustically that we were poor timekeepers as the *Bhrideogs* night was always on the night before 1 February, she said. We, however, boldly informed her we would be quitting before twelve midnight and we were still all right for St Bridget's Night.

Our only method of transport was the humble bicycle and very humble the said bicycles were, believe you me. I remember one late night coming through the village of Moylough, at least fifteen lads and girls, and not a stitch of light to any of us. Were it not for the few smokers who had cigarettes lighting in their mouths, we may as well have been a herd of deer so difficult were we to recognise.

To our horror didn't we meet the late Sergeant Garvin who was on duty in the village. A nice man, but even he was taken aback at the poor 'rolling stock', we had no lights, little by way of brakes, several mudguards missing, chains coming off and the worst affliction of all, a missing 'gear wheel' which basically meant that irrespective of how fast you pedalled you got nowhere. One of our *bhrideogs* who had the problem that night could not explain to the Sergeant why he was well over ten minutes after the main bunch. How could he tell the Sergeant that he had tried the oldest cure known since bicycles were invented – urinate on the small 'gear wheel' to get it going.

The Sergeant asked us to remove the masks and he knew most of us immediately. Worse was to come. Some of us were in *Macra na Feirme* at the time and had sort of notions of ourselves. The activities of our local *Macra na Feirme* branch in Mountbellew used to be published in the local newspapers. We all got a caution never to go on the road again without a light and certainly never to appear in Moylough again without lights and brakes. Just as we were scampering off with our tails between our legs the Sergeant whispered to me: 'I thought *Macra* was an educational organisation'. No need to say any more.

That same night was eventful in another way. We decided to visit Slieve in the parish of Moylough as we always got a good reception in all the houses. When we called to the late Batt Cosgrove's house we noticed his wife, who was an excellent bread maker, had left a newly made cake out to cool on the window ledge. The dancers surrounded Batt as a diversion and, if I am not mistaken, it was Tom Quigley of Springlawn who swiped the cake. We had a good laugh at the head of the road as we devoured that lovely cake to our utter satisfaction, and of course the news spread around the village the next day of how we pulled a fast one on Batt.

However, we repeated the call a year later and to our great glee wasn't another cake in the very same place on the window. Batt himself was down in bed with a bit of a flu, and to create another diversion didn't Jimmy Leahy and Mick Gavin, both of Springlawn, go down to the bed and lie down beside Batt. Naturally enough his wife grabbed the tongs to get them out of the bedroom and that was my chance to swipe the cake. Again we devoured it, as we were over four hours cycling, dancing and generally acting the maggot all over the parish.

But, we noticed there was a bad taste in the last few mouthfuls of the cake and, sure enough, a few hours later as we wearily cycled home we were more often off the bicycles than we were on them. In fact we were more often inside a ditch than on the road. Batt had the final laugh. He put soap through the cake and, believe me, a soapy cake is the best laxative I have ever seen. We had stomach pains for days afterwards.

Another year we called to the village of Annaghmore, and to the late Mikie Healy's house. Mikie and his family were always delighted to see us and after we did part of a Siege of Ennis in the kitchen and got 'something for the Brideogs' we started to file out of the house into a pitch-dark night. Mikie always had two beautiful box hedges about two to three feet high each side of the path going all the way to the gate, and he was very proud of his hedges. When I was about three yards from the door I was tripped by Tom Quigley and pushed by Mattie Donnellan in over the box fence, only to have to endure Paddy Gordon coming down on top of me after he got the works too.

Everybody got inspired then, and the worst day at Beecher's Brook at the English Grand National would not compare with the pandemonium and the shrieks as men and women were put flying all over the garden. Some of the women to this day would swear that male *bhrideogs* were trying to 'knock a squeeze' out of them under the box hedges that night. It was amazing what *bhrideogs* had on their mind on that holy night. You should see the clamour to get up off the ground and not let your face be seen when Mikie switched on the light.

I remember a youthful Declan Kelly in Abbeylands providing the music on the accordion all night long and in many houses, including the late Jack Naughton in Mountbellew, who would always take down the fiddle and play a tune to himself. On another night we called to the late Paddy and Peggy Finnerty's house in Gurteen. Again, always a good reception, but Peggy, who was the local domestic economy teacher, had one of what was then the most modern kitchens in the parish. There was a tiered tripod for pans, saucepans and other utensils standing up at one corner of the kitchen. We were halfway through the 'bundle of fun' which comprised of at least ten big *bhrideogs* but someone released the grip on one of the biggest men which had the effect of catapulting him right into the tripod. The sight of a big *bhrideog* on his back with pots and pans dropping down on the top of his head was a sight to behold.

On another famous occasion we covered the parishes of Newbridge and Kilkerrin, but as we approached the village of Kilkerrin we were hungry, tired and would do anything to get a bit of grub. Younger readers will understand there was no takeaways and no fish and chip shops in local villages that time, and Kilkerrin was a quiet lonely village while its inhabitants prepared for bed. So delirious with the hunger were we that we knocked on Maureen Feeney's grocery shop and begged her to let us in, and to her eternal credit she did.

Well, I never saw such gluttons. All the poor woman had was red lemonade in big bottles and Geary's square penny biscuits. We sat down in a circle on the shop floor outside the counter. We consumed every drop of lemonade, and if any customer called the next morning for a penny biscuit they were out of luck. We paid our dues with gratitude and then proceeded to divide what was left equally between all the *bhrideogs*. We didn't have much for our troubles but we sure had a lot of fun. It is great to see Maureen hale and hearty and still running the shop.

I remember another night when we joined forces with Pakie Kenny and his Rushestown gang and we toured all around Cappagh, Buggauns, Kilclough and Lissavruggy. Some of us, for some unknown reason, made the cardinal error of calling twice to the one house. I can see to this day the face of late Mick Gavin in Cappagh opening the door to a gang that was there a few hours earlier. There wasn't exactly a great welcome this time round and, pointing his finger in my direction, he shouted: 'That long string of misery was here earlier tonight.'

It might not appear to a younger generation that so much fun and banter could be got from cycling around dark country roads on a cold winter's night dressed up in the most outrageous attire, but for us it was the social night of our year. If St Bridget never did anything else other than give us *Bhrideog* Night, she will be forever remain a favourite saint of mine.

Today, in many areas, groups of *bhrideoga* dressed in weird costumes go visiting pubs and other places that people gather to collect money for various charitable causes. Unfortunately, in some places the custom has degenerated into groups going from house to house without a *Bhrideog* simply asking for gifts, and even doing this on two or three nights following St Brigid's Eve, something that makes them very unpopular.

In some places (Counties Clare, Cork, Galway, Kildare, Mayo, Offaly, Roscommon, Sligo, Waterford and Westmeath) a girl, dressed in white representing St Brigid and carrying a beautifully woven Brigid's cross of the local style, led a procession from house to house. One account suggests that the 'saint' was carried around the houses by the other *bhrideoga*, where she would hand out a small Brigid's cross to each household. A similar tradition of making an image of St Brigid was also carried out in Scotland (see page 135), while in Somerset an ancient custom used a doll to mark the saint's supposed death in that county (see page 159).

Brigid's girdle (*crios Bhríde*)

In some parts of Ireland another part of the traditional activities of *Fhéile Bhríde* (St Brigid's Day) was passing through the *crios Bhríde* (Brigid's girdle or belt). This was a woven straw rope (*súgán*) between 1.2m (4ft) and 3.6m (12ft) long formed into a continuous circle large enough for a grown man to pass through and, according to region, had either a single cross, three crosses or four crosses tied to it at the top. The *crios* was taken from house to house by the village youths, and members of the household would climb through it to gain the blessing of the saint. As in other aspects of St Brigid's Day customs there were regional variations in both the way the girdles were constructed and how they were used.

Brigid: Goddess, Druidess, Saint

32 A Brigid's girdle (*Crios Bhríde*) made of straw. Usually carried round by boys to each house on St Brigid's Eve, each member of the household climbed through it to obtain the blessing of the saint.

Making the girdles

The *crios Bhríde* seems to have always been made by the boys who were to take it around the village or district and, in some places, the boys would be busy making it for a week before *Oíche Fhéile Bhríde* (Eve of St Brigid's Day). Because of its size and complexity the final finishing touches were put to it on St Brigid's Eve, although in other districts tradition dictated it had all to be made on the Eve, especially if made from reeds. The majority of girdles in more recent times were made of plaited straw, but there were exceptions to this. In Leenaun, Co Mayo it was made entirely from rushes, while on Innishmore (*Inish Mor*), one of the Aran Islands off the west coast of Ireland, they made either a plaited grass or a plaited grass/hay mixture *crios* because the barren island did not support much in the way of cereal crops or rushes.

In an account published in 1889, it was stated that straw would only be used if rushes were not available or in short supply. In which case the straw girdle had three green rushes woven into it, which suggests that originally it was traditional to make the *crios Bhríde* from rushes which, like those gathered for the Brigid's crosses, were plucked on the eve of the saint's day. There were variations in the method of making them. Some people simply twisted the straw into a rope, but more usually they were woven or plaited. In Knock, Co Mayo, they used straw which was first woven on some sort of device called a 'rough weave', making a rope 5–6m (16–20ft) in length, which was then doubled over and the two sections woven together. After this the two ends were joined together to form a large loop to which three crosses were attached.

The crosses

Either one, three or four crosses were fitted to the girdle, although an account from Rosmuc, Connemara, Co Galway, suggests that their *crios Bhríde* had two crosses. While most were made of plaited straw, the straw girdle from Rosmuc had crosses made of rushes and there is mention of twig crosses used at Carna, Co Galway, while a mixed grass/hay *crios* from the Aran Islands had a single cross of wood. Usually the crosses are only about 15 cms (6in) long, although one account from Innishmore mentions a cross that was about 60 cms (2ft) high and 30 cms (12in) wide, with ribbons or strips of cloth sewn onto the girdle. In Knock, Co Mayo, the crosses were fitted to the girdle 'in the name of the Father, Son and Holy Ghost.'

However, the most common number of crosses used seems to have been four and the various verses sung to accompany the girdle ceremony usually mention the 'The girdle of the four crosses'. If three of the crosses represent the Father, Son and Holy Spirit could the fourth represent St Brigid herself? An exception to the verses referring to four crosses is found in one from from the parish of Killannon, Co Meath that mentions 'The girdle of the three crosses.'

The crios Bhríde *ceremony: why was it done?*

The reason why the passing through the girdle ceremony was carried out was said to be because it 'puts luck on the house and on the people of the house throughout the year'. 'Luck' really meant the blessing of St Brigid. The reason for this custom may be connected to a story related in the *Life of St Brigid* attributed to Animosus (Anmchadh, a tenth century Bishop of Kildare):

> A certain woman came to Brigid seeking alms and the saint offered to give her either her cloak or a calf that had recently been presented to her. The woman, however, explained that such gifts would be useless to her as robbers whom she would meet in her travels would take them from her. The saint then told her she had heard that illness was rife in her native district and gave her the belt she was wearing, telling her that the water in which it was dipped would cure the invalids and that she would receive food and clothing for curing them.

Variations of the story are found in other *Lives* such as those in the fifteenth-century *Book of Lismore*. The belts of many saints were used to perform miracles and appear in stories about many of them. A number of reliquaries contain belts or pieces of belts reputed to have belonged to saints and which were believed to have curative powers. So the custom of Brigid's Girdle may be a remnant of the veneration of saints' belts as relics. This tale may also account for the custom of the Brigid's ribbon (*ribín Bhríde*, see page 126) and Brigid's cloak (*brat Bhríde,* see page 124).

Tradition dictated that to obtain the saint's blessing the person had to pass through the girdle rather than just touching it. Passing 'through' holed stones and other items has a long tradition, and was done for a variety of reasons, but in many cases was believed to effect a cure. The Men-an-Tol, a holed stone near Morvah, Cornwall, was used to cure rickets in children by passing them through it three times and there are many other 'curing stones' known.

There was a millstone at Kildare Cathedral in the seventh century that was believed to cure illness since it was associated with a miracle performed by St Brigid (see page 57), and presumably it had a central hole in it. A holed stone on the north west corner of Kildare Cathedral today brings 'luck' (the blessing of St Brigid?) to people who put their arm through it and touch their opposite shoulder. Another tradition linking St Brigid with effecting a

Carrying out the ceremony

It was usually boys, ranging in age from about eight to thirteen years, who carried the *crios Bhríde* around the houses on St Brigid's Eve, although there is evidence from a few places that adults accompanied them. Occasionally the *crios* seems to have been taken around by boys at the same time that the girls took round the *Bhrideog*, the image of Bride (see page 111). In a few cases the boys carried both a *crios Bhríde* and a *Bhrideog*. It appears that generally each girdle was taken around by one pair of boys, but there could be four or five pairs going round each town or village, each with their own *crios*, although how they 'divided' the area between them is unclear!

They carried the *crios* carefully folded up until they reached the house and opened it up on entering and, as in so many of the customs connected with St Brigid, there were regional variations in the details of the ceremony. An account from Innishmore, gives some details of exactly how the 'passing through' was carried out there:

> When the houses were visited each of the parents used to gather up the girdle. It would then be about two feet long. The girdle and cross were then taken in the right hand and with the face of the cross inwards it was moved around the body three times – being changed from the right hand to the left as necessary to complete the circle, saying at each pass: 'In the name of the Father and of the Son and of the Holy Ghost.' It was not folded around the waist as a belt would be.
>
> Next the parents went through the girdle with the right foot first, the cross in the right hand, and the girdle and cross passed over the head saying: 'In the name of the Father and of the Son and of the Holy Ghost.' This was repeated three times.

An account of the ceremony performed in the nineteenth century shows some variation in the details:

> ...the master of the house, holding it doubled up in his right hand, makes the sign of the cross with it in the name of the Trinity and passes it three times from right to left around his body. Then holding it out at arms length in his right hand he lets one end drop so as to form a circle, through which he passes three times, putting the right foot through first each time. He then doubles up the rope and again passes it three times around his body as he did initially. He is followed in turn by every member of the family. In some cases the girdle is simply laid on the floor in a circle and everyone passes through it by lifting up one side and steeping under it and then raising the other side to step out again.

In Spiddal, Co Galway the boys, on entering the house, said:

> Here is the belt inside, the belt of the four crosses,
> Arise, O woman of the house and travel through the belt,
> And whatever state you are in tonight,
> May you all be seven times better from tonight.

After this the woman of the house passed through the girdle three times and blessed herself with the crosses that were on it, after which other members of the household passed through it. The method of passing through the *crios* was also laid down by both tradition and modesty, although there were regional variations. The man would place his right foot inside the girdle, and then his right arm and shoulder, then his head, followed by the left shoulder, left arm and left foot, usually doing this three times. To protect the women's modesty the practice was often slightly different as the *crios* was lowered over their head and shoulders, and lowered down the body to the floor where they stepped out of it, right foot first. This was repeated twice more.

The *crios* Verses

There are a number of variations on the verse that the boys recited when they had entered the house with their *crios*, most of which invoke St Brigid, but some also include mention of Christ, the Virgin Mary, St Patrick, the Apostles, or the Holy Ghost. For example a verse from Carraroe, Co Galway mentions the Virgin Mary:

> The Girdle of Brigid is my girdle,
> The girdle of the four crosses,
> Brigid who went out,
> Mary who waited inside,
> However you are today,
> May you be seven times better from today.

Among the acceptable 'gifts' given to the boys who brought the *crios* to each house was an egg and one version of the girdle verse from Oranmore, Co Galway specifically mentions this. It is also unusual in that the housewife was encouraged to put her child through the girdle rather than passing through it herself:

> Girdle, Brigid's Girdle, my girdle,
> A penny from the pocket of the women of the house,
> Or an egg from the backside of the yellow hen,
> That is over there in the bottom of the house.
> Arise, O women of the house,
> And put your child under the Girdle.
> As good as he is going through it,
> He will be seven times better coming out.

In a few places it seems that the boys taking round the girdle also took with them a bundle of straw and would leave three stems at each house, from which a little cross was made. There are accounts from a few places that indicate that a special, very large girdle was made through which animals were driven so they too could be blessed by St Brigid. At Rosmuck, Connemara, this large *crios* was placed on or over the barn door before the animals left the building. In the parish of Ardclinis, Co Antrim, they made a smaller version of the *crios Bhríde* on St Brigid's Eve, which they used to encourage the fertility and prosperity of the crops. A few handfuls of the first seed of the year to be sown was passed through this ring, as were the first five or six seed potatoes if this crop was being sown, to bring a blessing on them and the forthcoming crop.

At the end of the ceremony the boys were rewarded with either a few pennies or, if it was a very poor household, a gift of eggs, since almost everyone kept a hen or two, and the boys carried a small basket in which to carry any eggs given. They took the girdle from house to house repeating the ceremony and at the end returned to the house where the *crios* was made. It was not generally kept after the end of St Brigid's Day, although an account from Innishmore suggests that there the *crios* was hung up under the roof. Money collected by the boys was divided up between the pair and was usually used to buy sweets.

It is clear that the majority of households welcomed the visit by the bearers of the *crios Bhríde*, and old people were particularly pleased to take part. However, it seems the occasional household was either not so welcoming or did not give a sufficient reward in the form of pennies, eggs or potatoes. This is indicated by the existence of a verse that incorporates a threat from Moycullen, Co Galway that was chanted at anyone who would not pass through the *crios*:

> The Girdle of Brigid is my Girdle,
> The Girdle of the four crosses,
> The person who will not go through the Girdle,
> It will be seven thousand times worse he will
> be a year from tonight.

From Rosmuck, Connemara, comes the threatening 'curse rhyme':

> The Girdle of Brigid is my Girdle,
> The Girdle of the four crosses,
> The person who will not give me a penny,
> May the Devil break his foot!

While also from Connemara comes the even more worrying:

> This is the Girdle of Brigid for you,
> The Girdle of the four crosses,
> The person who will not go through the Girdle,
> It's unlikely that he will be alive a year from today!

The custom is still carried out today in a few places on the Aran Islands, and in Co Galway and features in a poem *A Brigid's Girdle for Adele* by Seamus Heaney:

> Now it's St. Brigid's Day and the first snowdrop
> In county Wicklow, and this is a Brigid's girdle   I'm plaiting for you,
> an airy, fairy loop   (Like one of those old Crinolines they'd trindle)
> Twisted straw that's lifted in a circle,
> To handsel and to heal, a rite of spring
> As strange and lightsome and traditional
> As the motions you go through going through the thing.

Brigid's cloak – the *brat Bhríde*

Another custom associated with the celebration of St Brigid's Day was the 'making' of the *brat Bhríde* – the cloak or mantle of Brigid. This was a custom that mainly involved the women and young girls of the household, with the exception of those *brats* that played a part in the bringing in of the rushes to make the Brigid's crosses. In its most basic form a piece of unwashed cloth, handkerchief, blanket, towel, stockings, socks, or any other material, was taken out of the house during daylight on the Eve of St Brigid's Day (*Oíche Fhéile Bhríde*) and left on the window sill, wall, low roof or a nearby bush till either late that night or the following morning when it was taken back into the house. In the south of Ireland it was tied to the door handle so it would be touched by the saint when she entered the house.

In some households shawls or other items of clothing were put out, including waistcoats which were usually worn by men, although in a few places there was a tradition that a women would wear a waistcoat when giving birth. Belts, braces and ties were also left out to gain this special protection, being worn when the owner was engaged in particularly hazardous jobs. This was based on the belief that the saint wandered abroad on the Eve of her festival accompanied by a white cow. Seeing the cloth, she would touch it, so endowing it with her blessing and her power of healing, thereby creating a *brat Bhríde*.

Rather than leaving a cloth or garment out overnight the *brat Bhríde* could also be 'made' by using it in conjunction with the bundle of rushes or sheaf of straw used for making the Brigid's crosses. In Co Mayo it was the male head of the house who procured the item for this type of *brat Bhríde*, selecting an article of clothing usually worn by the member of the household who had the most dangerous occupation. In that area the head of the household was often a fisherman and so regarded as the person most in need of the saint's protection in the coming year, so his coat, waistcoat or muffler was usually used. There are many accounts of fishermen who claimed to have been saved from drowning because they were wearing a *brat Bhríde* or had one in their boat. The man would go out of the house, collect the sheaf of straw or bundle of rushes and wrap the chosen garment, and often an additional piece of specially chosen new cloth around it in such a way as to make it resemble a human body. After this it was concealed in a convenient place until brought in with due ceremony (see page 104).

This garment was then regarded as being blessed by St Brigid, while the additional piece of cloth was cut into strips for each female member of the household to use for cures, and was said to be particularly good for headaches. A variation of this in a few places saw a garment from each member of the household being either placed in a basket with the sheaf, or even just put in a basket and placed near the sheaf. However, in Co Donegal the basket of clothes was placed outside overnight, and was brought back in when, it was believed, each item had been blessed by St Brigid and so was a *brat Bhríde*. The uses of the *brat Bhríde* were very varied. For example, in Co Cork the cloth was torn into strips and a piece given to every female in the household so each would have the protection of the saint wherever they went. In Co Kerry they sewed a piece of the *brat* into the clothes of young girls, not only to protect them against any misfortune befalling them, but also to preserve their virginity by placing them under the special protection of this virgin saint.

In some places the power of the *brat* was believed to last only a year and a new one had to be left out for the saint's blessing annually. In other cases the possessor of it was convinced it remained potent for many years. The belief in Co Kerry that the *brat* was an anti-fertility token is unusual, and in most places the reverse is true, since the *brat* was regarded as a powerful way of ensuring fertility and curing illness in both animals and humans. This is clearly shown by an account from Donegal recorded in the 1940s:

> There was a poor old woman going around this place long ago and she had a shawl which was a *Bratach Bhríde* of fourteen years' standing and any request she made in the name of the shawl, she

was granted it. She went into one of the houses here once. There was a cow tethered at the lower end of the house, about to calve, but the calf wasn't coming and appeared unlikely do so and the cow seemed doomed.

The poor woman inquired — 'Are you not doing any good?' 'They said they weren't. 'Well, go down again,' says she, 'and try her once more.' The men went down and tackled the cow again. And the old woman shook the shawl over the cow and went down on her two knees there and began to pray to Brigid the Holy Woman. It wasn't ten minutes till the cow was all right.

There was another woman in this place, long married, unable to have a family and nor looking like she was ever going to have one either. This old woman came in to visit her one day. She told the poor woman how things were and how she would like to have a family. The poor woman removed her shawl and shook it three times over her in the name of God and Mary and Brigit the Holy Woman. She had her family after that.

There are accounts of midwives, often referred to as 'handy women' in Ireland, using the *brat Bhríde* to help with illnesses connected with childbirth. In Co Kerry a woman sick during pregnancy had a *brat Bhríde* placed on her head which, she said, had given her 'relief'. This form of treatment applied not only to humans but to cows as well. In this case the *brat Bhríde* was placed on the hindquarters of the cow if it had problems following calving. On at least one farm they laid a *brat Bhríde* over the back of the cow after calving even if all had gone well to ensure the animal would have good luck in the coming year, that it would give an abundance of milk, and that its calf would thrive. In some places the *brat Bhríde* was used if a mare would not bond with its foal or a ewe with its lamb, being rubbed on the baby animal or placed around it, after which the mother would accept it.

There was a strong belief in fairies in Ireland, and one of the fears was of a human child being abducted and a 'changeling' (*iarlais*) being left in its place, so to prevent this a *brat Bhríde* was laid over the cot to give it the protection of St Brigid. There was a traditional prayer invoking Brigid's *brat*:

A Bhríde scar os mo chionn,
Do bhrat fionn dom anacal.

Oh, Brigid spread above my head
Your mantle to protect me.

Brigid's ribbon – the *ribín Bhríde*

One traditional story says that St Brigid wove the first cloth in Ireland and worked into it white healing threads which were said to have kept their healing power for centuries. In many places in Ireland it was customary to put a piece of silk ribbon, red being the preferred colour, outside the house on the Eve of St Brigid's Day, much in the same way as articles of clothing or cloth left out on the saints eve would be endowed with St Brigid's blessing when they became known as the *brat Bhríde* (Brigid's cloak).

It was believed that St Brigid, when travelling around the country on the Eve of her Day, would see and touch the ribbon, so endowing it with her blessing and conferring on it some of her healing power. After this it was referred to as the *ribín Bhríde*. An account of the practice published in 1895 describes this custom:

On St Brigid's Eve a silk ribbon is placed on the window sill [outside] during the night, in honour of our saint. This ribbon is said to lengthen during the night, and is ever after preserved as a remedy against headache. For this purpose *ribín Bhrigide* is used as follows – First it is rubbed or drawn around the patient's head three times, saying each time the invocation, 'in the name of the Father, and of the Son, and of the Holy Ghost, Amen', after which it is knotted around the head. The *ribín Bhrigide* was good for curing other ailments connected with the head such as toothache, earache and sore throats.

In Co Donegal one member of the household was chosen to bring in the *ribín* on the following morning. They would go outside, pick it up and then carry it around the house three times before knocking at the door, upon which a special welcome prayer was recited and the *ribín* carrier admitted. Once endowed with the blessing the *ribín Bhríde* would be carefully stored until it was required for use. Some people said that the older it was the more healing power it had, while others maintained it achieved its greatest power after seven years.

The belief that it would lengthen once blessed was used in Counties Clare, Limerick and Tipperary to foretell the future. Before being put out overnight its length was carefully recorded by holding it against a whitewashed wall and making a mark on the wall showing its length. When brought in the next morning it was compared to the mark made the day before and if its length had increased it was seen as a sign that the members of the household would have a long life, that they would get a good return from their crops and stock, and would be free from accident, illness and misfortune in general. In most cases the ribbon was indeed found to have lengthened, which may have been due to its being blessed by St Brigid, although it is more likely to have been the damp night air that caused its expansion.

Other Irish Brigid customs and beliefs

The time of marriage and baptism
It was once common in Ireland for couples to try to marry before St Brigid's Day so they could move into their house on 1 February, a day regarded as particularly lucky. Any female children born on the saint's Eve or Day were invariably named Brigid (with variations of spelling). Even female children born in the two weeks before or after her Day were often baptised with her name unless a sister had already been called this, in which case it might be given as a second name.

Rush strewing
In some households a bundle of rushes or layer of fresh straw was laid on the threshold of the main door of the house on which it was hoped St Brigid would kneel to bless the building or, even better, on which she might wipe her feet before entering the house!

Charcoal blessings
In Co Meath there is an account of a cross being marked with a charred stick on the wall of an outhouse on *Fhéile Bhríde* (St Brigid's Day), while in Cork, Kerry and Limerick a cross was made with a charred stick on the forearm or forehead of each member of the household, a custom that was also carried out on St Patrick's Day.

Frost gathering

It was said in some parts of Ireland that hoar frost gathered on St Brigid's Day would be a certain cure for headaches.

The hiring of servants

It was usual for farmers who did not have enough help from within their own families to take on additional helpers for the busiest part of the year, employing young men and women who were paid a wage and had their food provided. Traditionally the time for employing such casual workers was from 1 February, St Brigid's Day, to 31 October, *Samhain* (Hallowe'en).

Traditional Brigid's Day food and customs

> She who put beam in moon and sun,
> She who put food in earth and herd,
> She who put fish in stream and sea,
> Hasten the butter up to me.
> Pray Brigid, see my children yonder,
> Waiting for buttered buns,
> White and yellow.
>
> Traditional Celtic Butter Prayer

In Celtic society, both past and present, hospitality has almost been regarded as a duty and feasting, along with impromptu parties and singing sessions are common. In addition many of the ancient festivals, whose origins lay in pagan times but were still celebrated when Christianity took precedence, were associated with special meals and, in many cases, with particular kinds of food. Today such festivals are celebrated on one day, but before the use of the Christian calendar, the day was calculated differently (see page 94) and the festivities extended over two days as we now count them.

The food associated with *Imbolc*, the pagan celebration that marked the beginning of spring (1 February), the time when the Goddess Brighid was celebrated, featured dairy products and this tradition continued after the day became associated with St Brigid. Many of the miracles attributed to her involved the multiplication of existing stocks or improving them, such as changing water into ale. Animosus (Anmchadh), writing in the tenth century, says that she 'changed stinging nettles into butter, and the bark of trees into the richest and sweetest of fat [lard].' Since many of her miracles featured dairy products and ale, these influenced the type of food and drink consumed to commemorate her.

Irish customs connected with celebrating the saint took place on both the Eve of St Brigid's Day (*Oíche Fhéile Bhríde*), 31 January, and on St Brigid's Day (*Fhéile Bhríde*), but her eve required a special meal and certain foods associated with it. This particularly applied to the way her festival was marked in Ireland where ritual also played an important part in both the preparation of some of the dishes and in the consumption of the meal. In Ireland it was also a time for stocktaking, when the housewife checked her store of flour, salted bacon, potatoes and other foods to see how well they were lasting and estimate how the remaining stocks should be used till they could be replaced. The farmer would do the same, checking his stores of hay and other animal fodder.

Celebrating the Goddess-Saint in Ireland

33 The tower of St Bride's Church at the end of St Bride's Avenue off Fleet Street, London, about 1830. This inspired the cook William Rich to make the first tiered wedding cake in the eighteenth century.

The kitchen and main, and often only, living room was carefully cleaned in preparation for receiving St Brigid on her Eve, and a prayer was often said to bless the kitchen. One of which was said to have been written by St Brigid herself:

> My kitchen,
> The kitchen of the White God,
> A kitchen which my King hath blessed,
> A kitchen stocked with butter.
> Mary's Son, my friend, come thou
> To bless my kitchen.
> The Prince of the World to the border,
> May He bring abundance with him.

An account from Co Mayo shows that the whole house was cleaned and tidied. Some households even went to the trouble of re-limewashing the walls. When darkness began to fall the family built up the fire and, if it was an Irish long house, bedded down the animals that lived at one end. Everything possible was done to make the house warm and cosy for the coming celebration and to welcome the saint's presence into the dwelling.

It was a widespread belief that St Brigid, accompanied by her favourite white cow, travelled about the country on her eve when she bestowed her blessing on her people, their livestock and crops. Many of the customs were to encourage the saint to give her blessing to that particular house. A common way of encouraging the saint, or attracting her attention, was to leave a piece of cake or some bread and butter outside on the window sill on her eve, often accompanied by a small sheaf of hay for her cow, although in Co Donegal it was customary to leave out a bowl of porridge for her. In Co Mayo it was traditional to stick the handle of the churn dash into the ground and place the cake on its flat top.

In some parts of Ireland the bread left out was a specially baked oatmeal loaf in the form of a cross, soda bread being the usual everyday loaf. There are also accounts of pieces of meat, portions of butter, salt and fresh water being left for the saint. Some people regarded these as acquiring the power to cure ailments and so they were sometimes carefully preserved for this purpose. Leaving food and fodder for the saint and her cow was a tradition also practised in Scotland and the Hebrides. An Irish variation of this was the custom of throwing a sheaf of oats, a cake or a loaf at the doorstep on St Brigid's Eve, an act that was believed to drive away hunger from the the house and ensure the family would have a good supply of food in the coming year.

In some places it was accepted that the food put out would be taken away by a poor person from the district or a tramp, echoing the saint's reputation for giving charity to the poor. It might also be taken by the young people who carried around the *Bhrideog*, the image of St Brigid, and added to their traditional feast (see page 115), although in other areas it was taken back inside the house next day and ritually consumed by the members of the family. It was the custom for the more prosperous farmers in the district to give gifts of butter and buttermilk to their poorer neighbours since it was regarded as essential that butter was present at the meal because of the saint's close association with cattle and the dairy. Some generous farmers even killed a sheep and gave pieces of meat to the poor, so important was the feasting aspect of the celebration regarded.

One of the verses sometimes recited when the custom of the *crios Bhríde* (Brigid's girdle) was carried out (see page 119) emphasised the important connection of butter with this day:

Celebrating the Goddess-Saint in Ireland

Oíche Shamhna gan bia,
Oíche Nollag Mór gan aran,
Oíche Fhéile Bhríde gan im,
Is e an gearan tinn.

Samhain Eve without food,
Christmas Eve without bread,
St Brigid's Eve without butter,
That is a sorry complaint.

Many households felt the St Brigid's Eve supper was almost as important as that of Christmas and mutton and fowl were also sometimes served at the meal. An effort was made to do something special in the poorest households, even if it was just larger portions than normal of the usual meal if they were unable to prepare any of the traditional foods. However, an effort was always made and the housewife in many homes would have been preparing for the festival for a week or ten days before by 'gathering a drop', that is, putting aside a little milk each day for churning into butter on St Brigid's Eve. This was necessary as milk was scarce at that time of year and butter was regarded as an absolutely essential part of the supper. In some cases the meal was eaten with prayers and solemnity, in others with music and dancing, as described in Colonel Vallency's *Essay on the Antiquity of the Irish Language* published in 1781:

> On St Briget's eve every farmer's wife in Ireland makes a cake called *bairin-brac*, the neighbours are invited, the madder of ale and the pipe go round, and the evening concludes with mirth and festivity.

This traditional cake, the *bairm breac* (speckled bread), was probably ploughed into the first furrow cut by the plough in pagan times as an offering to the deity of fertility or the earth goddess, a practice common among the pagan Saxons of Britain. Christians, however, tended to consume such celebratory foods. In the Scottish Highlands and the Hebridean Islands the first cake baked on the first day of spring which, as in Ireland, was regarded as being 1 February, was called the *Bonnach Bhríde*, Bride's Cake.

In many places the meal only commenced when the person representing St Brigid, carrying rushes or straw from which crosses would later be made, asked for admittance to the house. A description of how this was carried out in Carrickmore, Co Tyrone about 1900 appeared in *Ulster Folklife* in 1957:

> One of the girls of the family chosen to represent the saint left the house, collected the rushes that had been picked previously and then knocked on the front door three times, saying in Gaelic, each time she knocked, 'Go down on your knees, do homage, and let Blessed Brigid enter the house'.
>
> After she had said this for the third time, those inside the house responded: 'Oh, come in, you are a hundred times welcome'. Then the door was opened and she entered, placing the rushes under the table on which the Brigid's Eve supper had already been laid. The 'saint' was given a place of honour, and grace was said by the mother and father: 'Bless us, O God, bless our food and our drink; it is Thou who hast redeemed us at great price; deliver us from all evil!'

The family then began the supper in honour of the saint and when it was finished both parents recited a thanksgiving prayer:

Love to Thee, O Lord, glory and thanksgiving for this meal and every meal which Thou hast ever given us! O Redeemer who has given us life, grant us eternal life in Heaven! May we be seven times better off at the end of the year, in the greatest graces and the smallest sins! Health of soul and body in people and their cattle safe from accident, from the year's hardship, from fever and diseases; and particularly we pray that Thou will put nothing in our path which will deprive our souls of eternal glory!

If the Eternal Father sees anyone among us here present in doubt, may He counsel and guide them for our good so that our souls may attain eternal glory; that He may keep the temptations of the Devil and of the ugly besmirched world from our hearts and our minds and keep us in a state of grace so that our souls may attain eternal glory. Amen.

After this the family began making their Brigid's crosses. An account from Co Derry in the 1830s describes a custom involving bread and the rushes that played such an important part in the celebration of St Brigid:

> A *strone* or large cake of oatbread is made in the shape of a cross. The rushes are thrown on the floor, and the *strone* placed on the rushes. All kneel round the rushes and the bread, and at the end of each short prayer a piece is taken off the *strone* by each person and eaten.
>
> When all is eaten the crosses are made, and when blessed by the priest or sprinkled with holy water, are placed over the door, usually by the third day after the Feast in the hope that the family may have a plentiful supply of bread until that time twelve months, and in honour of St Brigid.

It was common to put the rushes or straw, from which crosses would be made after the supper, under the table or leaning against it. In counties Donegal and Sligo the traditional dish of *Colcannon* (mashed potato mixed with cabbage and onions or leeks), was placed on the material to be used for the crosses when mashing the potatoes which formed the basis of this food. A similar practice was carried out in the preparation of Poundies, sometimes called *Brúitín*, another potato based dish. On St Brigid's Eve a potful of potatoes for the Poundies was brought into the house and peeled. In some places the children would select the potatoes which were used in this dish. As night fell the potatoes were put on to boil and when they were cooked a sheaf of straw was brought into the house and spread on the floor.

The pot was then drained of water and placed on the straw 'bed', the *Leaba Bhríde* (Bride's Bed). The beetle (a wooden masher) which was used to mash the potatoes was used by each member of the household in turn from the youngest to the oldest to involve everyone in the custom. The woman of the house took first turn saying 'Thanks be to Brigid for what she sends us' as she began to pound the potatoes. The man of the house took the next turn and so on till all the family had all made their contribution and given thanks. Onions were then added to the mash and a large dish was filled up for the men of the household and placed on the table.

A depression was made in the centre of the mashed potato and a large lump of butter placed in it which quickly melted. Each man took up a spoon, a short prayer of thanks was said and all began eating. The women and the younger children ate the Poundies straight from the pot after a lump of butter was added to it. Sometimes the dish was placed on a 'bed' of straw on the table for serving and in some places it was placed on the Brigid's crosses to ensure the saints blessing. Sometimes the crosses were just put on the table, but in some places slices of bread were laid on them.

It was not unusual for an extra place to be laid for St Brigid at the supper table and in Co Donegal, they always provided an extra spoon for the saint. However, in Co Leitrim they went one step further by inviting a guest, who symbolised the saint, to the Brigid's

Celebrating the Goddess-Saint in Ireland

Eve feast at a house selected to represent the hospitality of all the households in the district. This practice, however, seems to have been confined only to that one county. All the people in the neighbourhood agreed the house for this ceremony and a young girl was chosen to represent the saint. At the appointed time all the neighbours gathered in the house, after which the 'saint' arrived and knocked on the door. All those inside cried out '*Cé tá ann?*' ('Who is there?'), whereupon the girl answered '*Bhríde bheannaithe*' ('Blessed Bride'). The door was immediately opened allowing the 'saint' to enter when she was then shown to the best seat at the table.

The feast then commenced and any food left over by the 'saint' was then divided among the guests who took this token back home with them. They later used it in conjunction with the making of their Brigid's crosses and the custom, apparently unique to that county, of making a moon and star board (see page 105). It was the custom in Co Antrim and a few other northern counties to set up a table in the kitchen with food as for a guest and, after the time for the saint's visit had gone by, the first poor person to pass was invited in to partake of the meal. Animals were not forgotten as it was also common to provide some special food and treats for the cows and horses on St Brigid's Day.

In Wales St Brigid, where she is known as St Ffraid, was not particularly celebrated on her Day and there are no specific foods or recipes associated with her. St Brigid was well known for her hospitality, so it is appropriate that at Llansantffraid (place of the church of St Ffraid) in Powys it was traditional to make a large pie for visitors at festive seasons. This consisted of a boned roast goose stuffed with boiled tongue, the whole encased in pastry lined with mincemeat and eaten cold.

Indirectly St Brigid was responsible for the design of the modern wedding cake! William Rich (1755–1811) walked from Gloucester to London to seek fame and fortune and was apprenticed to a master cook called Pritchard at 3 Ludgate Hill. William was a gifted cook who married his master's daughter and eventually acquired the business and improved its reputation still further. One day he was asked to make a special wedding cake and was looking for inspiration. Seeing the steeple of St Bride's Church, which could be clearly seen from Ludgate Hill, William decided to make a wedding cake based on the soaring tiers of the steeple of St Bride's Church, the tallest, at 71.3m (234ft), that Christopher Wren had ever built.

His tiered wedding cake was infused with the finest French brandy 'even when it cost a guinea a bottle' and this novel tiered style of wedding cake quickly became fashionable. The idea rapidly spread and the multi-tiered wedding cake remains popular to this day. There is a photograph, dated 1954, of a cake in the form of a replica of the steeple of St Bride's Church in London being cut by the Revd Cyril Armitage, and it is believed to have been an old tradition, but the next 'steeple cake' was not made until 2005 when one was made for Guild Sunday on 6 February, and none have been made since. One wedding at St Bride's is of particular interest as Elenor and Ananias Dare were married there just before they sailed for the new American colony where their child was baptised Virginia on 18 August 1585, the first child of English descent born in America.

Eight

Celebrating the Goddess-Saint in Scotland and the Hebrides

St Brigid, almost always referred to as St Bride in the Highlands of Scotland and the Hebridean Islands, was greatly revered there. The Hebrides are even said by some people to be named after her, as is one of the islands, *Eilean Bhríde* (Bride's Island), one of a group known as the Treshnish Islands 9.6km (6 miles) west of Mull. It is said that there were more chapels and churches dedicated to St Bride than any other saint in the Hebrides and the west and south of Scotland, areas subjected to Irish influence and settlement. Fairs in honour of St Bride were once held at Forres, Grampian and at Inverness. So strong was the veneration of St Bride in the Highlands that St Bride's Day (*La Feill Bhríde*) displaced that of the Virgin Mary celebrated on Candlemas (2 February), and some Candlemas customs became attached to St Bride's Day, sometimes referred to as *La Feill Bhríde nan Coinnlean* (the Feast Day of Bride of the Candles). There was even a long standing tradition that St Bride had been born in Caithness, while a legend from Arran said she 'played with her brother, St Patrick, on the banks of the River Clyde and gathered cockles with him near Dumbarton'. It was also claimed she had died and was buried at Abernethy, Perthshire.

In the past a great many customs probably existed in connection with celebrating *La Feill Bhríde*. Many of these traditions and even the stories and invocations to the saint, however, were actively discouraged and ridiculed, particularly during the nineteenth century, by clerics, school teachers and others who saw such practices and beliefs as 'primitive' with no place in the modern world. This led to the loss of much folklore connected with this saint, along with many other traditions and beliefs. Fortunately, Alexander Carmichael realised in the 1850s that much valuable folklore, stories and customs were on the verge of disappearing, and so made a great effort to record what remained before they disappeared completely. He published the first of several volumes of his *Carmina Gadelica*, in 1900.

St Bride's Day was important and celebrated by the people of the Highlands and Islands with much merrymaking at one time, but it become more muted as time went on, as is indicated by an account dating from the nineteenth century:

> You need not ask me about these things, they were long done before my day… but I heard that they kept *Latha Feill Bhríde* (St Bride's Day) until perhaps seventy or eighty years ago in some places. They had big feasts and a lot of merrymaking and each had on his best suit of clothes, and mind you

no one thought of doing work that day. Oh, no, and then the women would take a big cloth, such as a bedcover, and they would take hold of it by each corner and then they would dance, and now and then call out together; and at other times only one would call '*Briedean, Briedean, thig an nall 's dean do leabaidh* ' (Bridean, Bridean, come over and make your bed), and at times a little bundle of straw would be thrown out in a corner as if it was meant that the saint was to come and sleep on it.

As in Ireland, one of the most important aspects was the taking round of an image of the saint from house to house, called in Scotland and the Islands the *Bhrideag* (little Bride). One of the earliest published accounts of this custom in the Hebrides is found in *A Description of the Western Islands of Scotland* by Martin Martin, 1716, where he describes the practice on Islay, although he says it took place on Candlemas Day:

> Another ancient custom observed on the second of February which the papists there yet retain is this: the mistress and servants of each family take a sheaf of oats and dress it up in women's apparel, put it in a large basket, and lay a wooden club by it, and this they call Briid's-bed and then the mistress and servants cry three times; 'Briid is come, Briid is welcome.'

More recent accounts supply further details. On the Eve of St Bride's Day the young girls would get a sheaf of corn and shape it into a humanoid figure as a representation of the saint. This was dressed in clothes and then decorated with pretty shells, sparkling quartz crystals, greenery and whatever flowers were available, usually primroses and snowdrops. The mild climate of the Hebrides meant that a few other flowers remained in flower during the winter and so in these islands the *Bhrideag* presented a colourful image.

A particularly pretty or bright shell or crystal was selected by the girls to be fixed over the heart of the *Bhrideag*. This *reul-iuil Bhríde,* 'the star of knowledge of Bride', represented the star that appeared over the stable door in Bethlehem which guided the saint to the birthplace of Christ according to a traditional Scottish story. In places where shells and quartz crystals could not be easily obtained the girls would decorate their *Bhrideag* with intricately interlaced straw, perhaps a variation of the Irish Brigid's crosses.

Once the figure was constructed the young girls would take the *Bhrideag* from house to house. The girls who formed the *banal Bhríde*, 'the maiden band of Bride', were dressed in white, the colour that tradition said St Bride wore. They all had their hair let down and loose to symbolise purity and youth. They visited each house where they were always warmly welcomed, each person in the household curtsying or bowing to the *Bhrideag* before presenting a small gift in the form of a shell, crystal, flower or bit of greenery to add to the decoration of the image. The housewife would offer a special gift to the *banal Bhríde*, either a Bride bannock (*bonnach Bhríde*) a special cake baked on the first day of spring, a Bride cheese (*cabag Bhríde*) or a lump of Bride butter (*rolag Bhríde*).

Having completed their round of the district, the girls would take the *Bhrideag* and all the food they had collected back to a house where they held the *féis Bhríde* (the Bride feast). The windows were barred, the door locked and the *Bhrideag* put in a place of honour at the table. Before long the young men of the community came to the house where the girls had prepared the feast and formerly and humbly asked permission to enter, which was granted after some ritual and light hearted negotiations. When the young men entered the house they bowed to the *Bhrideag.*

Then the feasting began, accompanied by dancing, singing and general frolics which lasted until dawn began to break, which signalled that it was time for all the participants to form a circle to sing the hymn *Bhríde bhoidheach muime chorr Chriosda* ('Beauteous Bride, choice

Brigid: Goddess, Druidess, Saint

foster-mother of Christ'). All the remaining food was gathered up and tradition dictated that the participants ate only sparingly of the gifts. The remnants of the feast (*fuidheal na feis*) were then distributed among the poorer women of the district, reflecting the help the saint would give to the poor.

However, while taking round the *Bhrideog* was a custom confined to the younger girls, older women also had a tradition of making a figure of the saint, the *dealbh Bhríde* ('image of Bride'). This practice began on St Bride's Eve with the construction of an oblong basket in the form of a cradle, the *leaba Bhríde* (the bed of Bride), which was decorated with flowers and greenery. However, in a few places this part of the custom was less elaborate with the bed simply consisting of corn or straw placed on the floor near the front door of the house with some blankets laid on top. In some places the bed was made of birch twigs and when ready the woman would call out '*Bhríde, Bhríde, thig astigh, tha do leaba deante*' ('Bride, Bride, come in, your bed is ready').

An eighteenth-century account of the practice by John Ramsey of Ochtertyre says that one or more candles were left burning near St Bride's bed all night. They then took a carefully selected sheaf of corn, and formed it into a humanoid figure to represent the saint. Sometimes this was made from the last sheaf cut during the previous harvest, as is clear from a description by the Revd James Macdonald in *Religion and Myth* (1893). He explains how the Scots of the South Western Highlands left a single small stand of straw at the corner of a field. The youngest girl among the reapers was delegated to cut this while the others all stood round in a circle, and this special small sheaf was later used to make a *Bhrideog* which was placed in a basket representing Bride's bed.

The image was clothed in women's clothing and embellished with ribbons, pretty shells, coloured stones and crystals, along with any flowers they could gather such as primroses, snowdrops, dandelions and daisies. Once the *Bhrideog* and *leaba Bhríde* was completed one of the women of the household went to the door of the house, opened it and stood on the step with her hands on the door jambs and called quietly into the darkness '*Tha leaba Bhríde deiseal*' ('Bride's bed is ready'). Then a second woman, who was standing behind the first, called out '*Thigeadh Bhríde steach, is e beatha Bhríde*' ('Let Bride come in, Blessed be is Bride.')

The first woman again addressed the saint, '*A Bhríde! Bhríde thig a steach, tha do leaba deanta. Gleidh an t each dh'an Triana*' ('Bride! Bride, come thou in, thy bed is made. Preserve the house for the Trinity'.) The women then carried the image of the saint to the bed they had prepared, and placed the *Bhrideog* in it with great reverence. Beside the *dealbh Bhríde* a white wooden rod with the bark peeled off, was placed, known as the *slatag Bhríde* (the little rod of Bride), *slachdan Bhríde* (the little wand of Bride) or *barrag Bhríde* (the birch of Bride), it was traditionally made of a 'sacred wood' such as birch or white willow, care being taken not to use an 'unlucky' wood.

This rod had to be straight as it was said to represent justice, while its colour, white, was chosen to symbolise peace and purity. Rods or wands of sacred wood were presented to the Irish kings at their coronation and a similar one was given to the Lord of the Isles when he was installed. The Druid's used wands of yew (*slachdan Druidheachd*) for divination and the reason for placing it by St Bride's bed also seems to have been for foretelling the future.

The next stage was for the woman to carefully level the ashes on the hearth, although in a few cases the ashes were carefully placed on a board and a roll of cloth placed around it to prevent it being disturbed by draughts. Once this had been prepared the family went to bed and, on rising, eagerly looked at the ashes to see if there were any marks there that might have been made by St Bride's rod. If there were the whole household was pleased, but were overjoyed if they thought they could see the *lorg Bhríde*, the 'footprint of Bride'. This was regarded as a sign that the saint had been present during the night and a sure indicator that there would be an increase in the family, their flocks and their crops in the coming year.

However, if they could see no marks in the ash they were very dejected as it was regarded as a sign that they had offended the saint in some way, so that their appeals to her would go unanswered. If this was the case the family had to make an offering to her. At one time this was usually a cockerel or occasionally a pullet, which was buried alive near the junction of three streams, and the family was obliged to burn incense on the hearth that night before they went to bed.

Chickens seem to be particularly associated with St Bride in the Highlands and the Hebrides and at one time St Bride's Day was called *Lá Cath Choileach* ('the Day of Cockfighting'), and the boys took cocks in to school and pitted them against each other. At the end of the contest the most successful cock was called *Lá choileach buadha* (the victor cock), and its owner was given the title 'King of the School' for the coming year. A losing bird was called *fuidse* (craven) and all the defeated, injured and dead cocks were kept by the school master as a perk of the job.

One of the most curious aspects of the traditional observances of St Bride's Day in Scotland was that involving the serpent which, Scottish tradition maintained, 'came forth' from its hole or hollow in the hills on this day, suggesting winter had ended and spring had begun. However, on St Bride's Day a hymn was sung to propitiate it and there are several versions which vary according to region. Carmichael says that only the first verses survived which he recorded, and these are given below:

1. Early on Bride's morn
   The serpent shall come from the hole,
   I will not molest the serpent,
   Nor will the serpent molest me.

2. The serpent will come from the hole
   On the brown Day of Bride,
   Though there should be three feet of snow
   On the flat surface of the ground.

3. The Feast Day of the Bride,
   The daughter of Ivor shall come from the knoll,
   I will not touch the daughter of Ivor,
   Nor shall she harm me.

4. On the Feast Day of Bride,
   The head will come off the 'caiteanach',
   The daughter of Ivor will come from the knoll
   With tuneful whistling.

5. On the Day of Bride of the white hills
   The noble queen will come from the knoll,
   I will not molest the noble queen,
   Nor will the noble queen molest me.

Some writers have suggested that these songs had their origin in Ireland and although snakes have never existed in that country, the serpent is often mentioned in Irish mythology.

A ram-headed serpent is a common motif in Celtic art from Britain, Ireland and Europe, but snakes were common in Scotland. The term 'Daughter of Ivor' was one of several terms used in the Hebrides for a serpent as it was regarded as unlucky to refer to a snake, particularly the adder, by its correct name

By the mid-nineteenth century snakes in the Hebrides were small and rare, but earlier had been more numerous and considerably larger. One snake killed at Bailemonaidh on Islay at the beginning of the nineteenth century measured 2.74m (9ft) in length and 45cm (18in) in diameter! These verses seem to relate to a strange custom that was described to Carmichael by an eyewitness, suggesting that it was a practice that had apparently died out by the mid-nineteenth century when he began recording Hebridean traditions.

This incident took place at Uignis on the island of Skye. The woman concerned was Mrs Macleod, the widow of Major Macleod of Stein, who was in her eighties. One morning at breakfast, someone commented that this was the Day of Bride. 'The Day of Bride' repeated Mrs Macleod meditatively and with a bow of apology rose from the table. All present watched what she did with curiosity when she went to the fireplace, picked up the tongs and a bit of peat and took them out to the doorstep.

She then removed her stocking, put the peat into it and began pounding it with the tongs. As she hit the peat filled stocking she said: '*An diugh La Bhríde, Thig an righinn as an tom, Cha bhean mise ris an righinn, Cha bhean an righinn rium.*' ('This is the Day of Bride, The queen will come from the mound, I will not touch the queen, Nor will the queen touch me.') Then, having thoroughly beaten the peat and put her stocking back on, she returned to the table, apologising to those present for not remembering that it was Bride's Day earlier.

Carmichael suggested this indicated traces of vestigial serpent-worship, but the beating of the 'serpent' does not look like veneration of the snake but rather its defeat, perhaps inspired by the biblical reference in the Book of Psalms 91:13 'Thou shalt tread upon the lion and adder; the young lion and the dragon shall thou trample under feet.'

The line in version four, 'The head will come off the caiteanach' is more understandable as, in Celtic belief, winter was visualised as an old hag called *Cailleach*. Each year she was 'defeated' by spring, the season with which the Goddess Brighid was associated, which suggests a pre-Christian origin for this verse, indicating that spring would defeat winter. The phrase 'the head will come off the caiteanach' also brings to mind the habit of the ancient Celts of cutting off the heads of their enemies. A more recent Hebridean visualisation saw the coming spring as St Bride, with her white wand, breathing life into the mouth of the dead winter, a period which was called *an raithe marbh*, 'the dead quarter' (see page 87).

The advent of spring also saw the start of the fishing season in Ireland and Scotland. On the Hebridean island of Barra in the nineteenth century the fishermen cast lots for the local fishing banks. They knew their exact boundaries and fishing qualities, and once it was determined how many boats wanted to go long-line fishing that year they divided the banks into the appropriate number. All the fishermen went to church on St Bride's Day where the blessings of the saint were recited, and many examples of her virtues given. The priest then reminded the congregation that as God made the land and everything upon it, he also made the sea and everything within it.

He then urged the fishermen to avoid disputes and quarrels over their fishing, to bear in mind the dangers of the deep and the precariousness of life, and to remember the poor, widows and orphans. Following the service the fishermen come out of the church and cast lots for the fishing banks, after which they dispersed to their houses, happy at getting a good place on the bank or down-cast if they had been unlucky. To mark their claim the fishermen would set a line and buoy with a few hooks attached. When the fishing season began these first hooks were taken up.

It was a widespread belief in Britain and Ireland that the fairies would sometimes take away a human baby leaving a magic image, known as a stock that looked just like the child that had been removed. However, it was said these stocks would later sicken or show some deformity or other problem. For this reason many babies that had some form of medical or mental problem were regarded as fairy changelings, and it was believed that by ill treating them or 'exposing them', the fairies would be compelled to bring back the human child. One of the more curious and tragic beliefs connected to St Bride in Scotland was that a fairy changeling placed on the beach below the high water mark on St Bride's Eve would be taken away and the original human child restored. This is likely to have led to the drowning of a number of 'disabled' babies.

Until the Reformation in the sixteenth century it was not unusual for pageants and plays to be performed on Candlemas Day (2 February), but these were regulated by magistrates and town councils to prevent such occasions getting too boisterous and out of hand. The first such Ordinance concerning these dramas is dated 1442. Among the figures in this play were an emperor, three kings, the Virgin Mary, bishops, angels, Moses and other religious and secular figures, including St Bride. The drama was obviously a mixture of scriptural and legendary stories, and such pageants, usually including St Bride among the 'cast', were performed at intervals until their disappearance.

While *La Feil Bhríde* (St Bride's Day) was of great importance in Scotland and particularly the Hebridean islands, St Bride was not forgotten during the rest of the year. The three great Irish saints Brigid, Patrick, and Columcille (Columba) were all appealed to in invocations and prayers covering many aspects of everyday life and work. St Patrick plays a much smaller role in Scots' Gaelic beliefs than in Ireland. St Columcille is associated with cattle in the Highlands and Islands along with St Bride who is connected with the protection of herds, but who, in addition, is also appealed to in prayers for healing the sick and midwifery.

So strong was the veneration of St Bride that she seems to have rivalled the Virgin Mary in the affection of the people, who even regarded them as contemporaries. According to Highland and Hebridean tradition, she assisted Mary with the birth of the Christ child (see pages 90–91), earning St Bride the title *ban chuideachaidh Moire*, 'the aid woman of Mary', i.e. her midwife. One of the ways of looking into the future was to look through a circle made by the fingers of the hand at the person to whom the prediction concerned. In the Hebrides they said that the first people to use this method was Joseph and Mary when looking at the Christ child, and St Bride was the first person to learn its use. She was present as the midwife to Mary, and looked at Jesus through a circle made by the Virgin herself. This is why St Bride was invoked in a method of foretelling the future using a technique known as Hand Tube Divination, referred to as *Frith Bhríde*, the Augury of Bride.

Many sayings, prayers, incantations and even magic spells were said that mentioned St Bride along with the other saints regarded as most important in that region. For example the Hebridean islanders went to bed early and rose early, a practice embodied in the saying:

> Supper and light the Night of Saint Bride,
> Sleep and light the Night of Saint Patrick.

When preparing to sleep St Bride's protection was requested by the Hebrideans. The first verse of one *Coisrig Cadail* (Sleep Consecration), of which a number were recorded by Carmichael, goes:

Brigid: Goddess, Druidess, Saint

34 A painting, tempera on canvas, by John Duncan, 1913, entitled 'St Bride'. It shows the saint being carried by angels to the Holy Land, illustrating the Scottish tradition that St Brigid assisted Mary with the birth of Jesus. (National Gallery of Scotland)

> I lie down tonight
> With fair Mary and with her Son,
> With pure-white Michael,
> And with Bride beneath her mantle.

St Bride was also called on to protect the house at night, and this prayer, recorded in a manuscript dated 1684, was said by people just before they went to bed:

> Who sains the house the night,
> They that sains it like night.
> Saint Bryde and her brate,
> Saint Colme and his hat,
> Saint Michael and his spear,
> Keep this house from the weir;
> From running thief,
> From burning thief;
> And from a'ill Rea

That be the gate can gae;
And from an ill wight
That be the gate can light…

There was a nineteenth-century saying that suggested the most appropriate material for bedding according to the season:

For the cold of *Calluinn* (winter) thickest woollen homespun is right.
For the cold of Saint Bride's Day mixed stuff will do.

For a people involved in agriculture and fishing the weather played an important part in their lives, particularly in the Highlands and Islands where the weather was often poor and sometimes even extreme. Not surprisingly there were many weather sayings, some of which involved St Bride. For example:

As far as the wind shall enter the door
On the Feast Day of Bride,
The snow shall enter the door
On the Feast Day of Patrick.

In the Hebrides they said:

Good weather on the Feast Day of Bride
indicates a famine or foray.

They also said that another indicator of the approaching spring was that on St Bride's Day the lark sings with a clearer voice. While St Bride's Eve and Day were of great importance to the Scots and the Hebridean islanders, she was invoked during many everyday activities all through the year. Many of these invocations were concerned with the household fire (see page 80), but others were said when the Scots were carrying out their daily occupations to achieve what was, in many cases, an uncertain and precarious living. There were also the more general invocations that asked the saint for her protection from the many dangers that might be encountered:

The Blessing of Bride (Beannachadh Bhríde)
Bride daughter of Dugall the Brown,
Son of Aodh, son of Art, son of Conn,
Son of Criara, son of Cairbre, son of Cas,
Son of Cormac, son of Cartach, son of Conn.

Each day and each night
That I say the Descent of Bride,
I shall not be slain,
I shall not be sworded,
I shall not be put in cell,
I shall not be hewn,
I shall not be riven,
I shall not be anguished,

I shall not be wounded,
I shall not be ravaged,
I shall not be blinded,
I shall not be made naked,
I shall not be left bare,
Nor will Christ leave me forgotten.
Nor fire shall burn me,
Nor sun shall burn me,
Nor moon shall blanch me.
Nor water shall drown me,
Nor flood shall drown me,
Nor brine shall drown me.
Nor seed of fairy host shall lift me,
Nor seed of airy host shall lift me,
Nor earthly being destroy me.

When a girl went out at night to draw water from the well she would often croon a song called a *Rune of the Well* in the firm belief that calling on St Mary and St Bride would protect her from coming to any harm:

Rune of the Well (Rann Tobair)
The shelter of Mary Mother
Be nigh my hands and my feet
To go out to the well
And to bring me safely home,
And to bring me safely home.

May warrior Michael aid me,
May Bride calm preserve me,
May sweet Brianag give me light,
And Mary pure be near me,
And Mary pure be near me.

In times of fear, danger or when a person was distressed they would make the *caim*, an invocation which put a protective 'circle' around them. To do this the person stretched out their right hand with the forefinger extended, turned round sunwise to make a circle with the tip of the forefinger while saying the Encompassing Invocation. This practice, once common in the West of Scotland, was carried out by both educated and illiterate Protestants and Catholics.

Encompassing (Caim)
The holy Apostles' guarding,
The gentle martyrs' guarding,
The nine angels' guarding,
Be cherishing me, be aiding me.
The quiet Bride's guarding,
The gentle Mary's guarding,
The warrior Michael's guarding,

Be shielding me, be aiding me.
The God of the elements' guarding,
The loving Christ's guarding,
The Holy Spirit's guarding,
Be cherishing me, be aiding me.

As in Wales, the blessing of St Bride was called on from those undertaking a journey or pilgrimage:

The Pilgrim's Safeguarding (Comaraig Nan Deor)
I am placing my soul and my body
Under thy guarding this night, O Bride,
O calm Fostermother of the Christ without sin,
O calm Fostermother of the Christ of wounds.

I am placing my soul and my body
Under thy guarding this night,
O Mary, O tender Mother of the Christ of the poor,
O tender Mother of the Christ of tears.
I am placing my soul and my body
Under Thy guarding this night, O Christ,
O Thou Son of the tears, of the wounds, of the piercings,
May Thy cross this night be shielding me.

I am placing my soul and my body
Under Thy guarding this night, O God,
O Thou Father of help to the poor feeble pilgrims,
Protector of earth and of heaven,
Protector of earth and of heaven.

St Bride was associated with the fertility of crops in Scotland as she was in Ireland. Many customs, some dating back to pagan times, were carried out in the Hebrides to assist the sowing of seed. For example, three days before it was sown the farmer sprinkled it with clean cold water in the name of the Father, Son and Holy Spirit – a Christian blessing, but which was said while walking sunwise around the seed to invoke the blessing of the sun – a pagan aspect of the custom. Once the corn was ready for reaping, further customs were observed in the Hebrides.

On the day of reaping the whole family would put on their best clothes and go to the field to acknowledge the 'God of the harvest', the Christian God but with traces of an older pagan deity underlying the ceremonies. When they reached the cornfield the head of the household would carefully place his hat on the ground, take up his sickle, face the sun and ritually cut a handful of corn. Then he would wave the corn three times round his head in a sunwise direction and began to sing the *Beannachadh Buana* (Reaping Blessing), with the whole family joining in to give thanks for the corn, bread, food, their flocks, wool, clothing, health and strength. A a number of saints are mentioned in such invocations including St Bride:

God, bless Thou Thyself my reaping,
Each ridge, and plain, and field,

> Each sickle curved, shapely, hard,
> Each ear and handful in the sheaf,
> Each ear and handful in the sheaf.
> Bless each maiden and youth,
> Each woman and tender youngling,
> Safeguard them beneath Thy shield of strength,
> And guard them in the house of the saints.
> Guard them in the house of the saints.
> Encompass each goat, sheep and lamb,
> Each cow and horse, and store,
> Surround Thou the flocks and herds,
> And tend them to a kindly fold,
> Tend them to a kindly fold.
>
> For the sake of Michael head of hosts,
> Of Mary fair-skinned branch of grace,
> Of Bride smooth-white of ringleted locks,
> Of Columba of the graves and tombs,
> Columba of the graves and tombs.

Following the reaping, workers carried out a custom to foretell certain events in the coming year. They would throw their sickles up into the air and look carefully to see how each struck the earth and lay on the ground. From these observations they were able to determine which among them was to remain single, who was to be married, who might fall sick and which might die before the next reaping season! Once the corn was gathered in it was ground into flour, and for centuries the people of the islands used a hand quern to grind the corn between two circular stones after it had been winnowed. As the women turned the quern it made a noise, and a song echoing the rhythm of the turning often accompanied the process, which acted as a blessing on the flour. This song spoke of the hope that the family would have meat, barley, mead, wine, milk, honey and music, while the last verse asked for the blessing of St Bride and the other saints on the family:

> The calm fair Bride will be with us,
> The gentle Mary mother will be with us.
> Michael the chief of glancing glaves,
> And the King of Kings
> And Jesus Christ,
> And the Spirit of peace and of grace will be with us,
> Of grace will be with us.

Many jobs were accompanied by songs, and since cattle were of great importance and many households kept a cow for their own use, there existed many milking songs, often called a Milking Croon (*Cronan Bleoghain*). These lilts were sung in unison with the movement of the hands as they were milked, and the cows would often not give milk unless they were sung to, and occasionally a cow would withhold her milk until her own favourite song was sung to her! It was said that one of the essential requirements for a milkmaid was the possession of a good singing voice. Carmichael gives a number of examples of these milking croons, all of which have a verse that mentions St Bride, for example:

> Come, beloved Colum of the fold,
> Come, great Bride of the flocks,
> Come, fair Mary from the cloud,
> And propitiate to me the cow of my love,
> Ho my heifer, ho heifer of my love.

St Bride was also invoked in Herding Blessings (*Beannachadh Buachailleachd*). These were sung as the farmers sent their herds and flocks out to the pastures or on to the hills and glens in the morning. There were variations in the different districts, and not unexpectedly St Bride appears in most of them, for example:

> I will place this flock before me,
> As was ordained of the King of the Word,
> Bride to keep them, to watch them, to tend them,
> On ben, on glen, on plain,
> Bride to keep them, to watch them, to tend them,
> On ben, on glen, on plain.

St Bride was invoked in general charms (*Sian seilbh*) for the protection of the farmers stock, and even when taking leave of his animals he would commend them to the safe keeping of St Bride and St Mary, waving to them as he said:

> The herding of Bride to the kine,
> Whole and well may you return.
> The prosperity of Mary Mother be yours,
> Active and full may you return.

Dairy products were of great economic importance in the Hebrides and Highlands, and with St Bride's connection with butter making she is naturally mentioned in one verse of a Churning Invocation (*Eoir A' Mhuidhe*) said as the housewife turned the churn:

> Come, thou Bride, handmaid calm.
> Hasten the butter on the cream;
> Seest thou impatient Peter wonder
> Waiting the buttered bannock white and yellow.

The inhabitants of the Hebrides were dependent on seaweed to manure their land as the island soils are mostly peaty or sandy containing little lime and other minerals. Seaweed is detached from the seabed during winter storms and driven ashore where it can be gathered, but a prolonged calm period was serious as with little seaweed available there was the possibility of such poor crops that famine might ensue. So when the seaweed was driven ashore the people sang hymns of praise:

> Come and come is seaweed,
> Come and come is red sea-ware,
> Come is yellow weed, come is tangle,
> Come is food which the wave enwraps.

Come is warrior Michael of fruitage,
Come is womanly Bride of gentleness,
Come is the mild Mother Mary,
And come is glorious Connan of guidance.

During the winter many women of the Highlands and Hebridean islands were engaged in processing and working the wool from its raw state to the finished product. Like so many other activities, this was accompanied by songs that acted both as blessings on the task and its product and helped pass the time. As St Bride was associated with sheep her name occurs in one Warping Chant (*Duan Deilbh*):

Thrums nor odds of thread
My hand never kept, nor shall keep,
Every colour in the bow of the shower
Has gone through my fingers beneath the cross,
White and black, red and madder,
Green, dark grey, and scarlet,
Blue, and roan, and colour of the sheep,
And never a particle of cloth was wanting.

I beseech calm Bride the generous,
I beseech mild Mary the loving,
I beseech Christ Jesu the humane,
That I may not die without them,
That I may not die without them.

St Bride was connected with healing, as was the Goddess Brighid, and in the Hebrides the saint often appeared in charms to cure a wide variety ailments, and Carmichael records many of these which invoke St Bride and other saints, and a few examples are given below.

*Charm of the Sprain (*Eolas An T – Sniamh*)*
Bride went out
In the morning early,
With a pair of horses;
One broke his leg,
With much ado,
That was apart,
She put bone to bone,
She put flesh to flesh,
She put sinew to sinew,
She put vein to vein;
As she healed that
May I heal this.

St Bride was also invoked in the case of eye problems such as getting a mote in the eye, including charms to counteract the 'evil eye' (*cronachdain Suil*). The connection with eye problems is probably due the story of St Bride plucking out and replacing her eye as given in various sources, such as this from the *Book of Lismore* :

> Not long therafter came a certain man of good family unto Dubhthach to ask for his daughter [in marriage]. Dubhthach and his sons were willing, but Brigid refused. A brother said to her "Idle is the pure eye in thy head, not to be on a mattress beside a husband." Brigid replied "The son of the Virgin knows it is not right if it bring harm upon us." Then she put her finger under the eye and plucked it out of her head, so it lay on her cheek. When Dubhthach and her brothers saw that, they promised she should never be forced to take a husband, except a husband whom she should like. Then Brigid put her palm to her eye, and it was immediately healed.

*Charm for the Eye (*Obi ri Shúil*)*
Bride, be by mine eye,
Mary, be my support,
Glorious King, be by my knee,
Loving Christ, be by my body.
The mote that is in the eye
Place, O King of Life,
Place, O Christ of Love,
Place, O Spirit Holy,
Place upon my palm.

May the King of Life be giving rest,
May the Christ of Love be giving repose,
May the Spirit Holy be giving strength
May the eye be at peace.

*Spell to Counteract the Evil Eye (*Cronachdain Suil*)*
The spell the great white Mary sent
To Bride the lovely fair,
For sea, for land, for water, and for withering glance,
For teeth of wolf, for testicle of wolf.

Who so laid on thee the eye,
May it oppress himself,
May it oppress his house,
May it oppress his flocks.

Let me subdue the eye,
Let me avert the eye,
The three complete tongues of fullness,
In the arteries of the heart,
In the vitals of the navel.
From the bosom of Father,
From the bosom of Son,
From the bosom of Holy Spirit.

Brigid: Goddess, Druidess, Saint

Sandy Skipper's wife, who lived at Latheronwheel, Caithness in the nineteenth century, was a well known healer and one of her incantations, a Thread Charm, features St Bride. To bring about the healing and avert the evil eye, three threads, red, white and black were tied around the affected part.

> *Charm of the Thread* (Eolas an t-snaithein)
> An eye will see you
> Tongue will speak of you,
> Heart will think of you,
> The Man of Heaven blesses you.
> The Father, Son and holy Ghost.
> Four caused your hurt -
> Man and wife,
> Young man and maiden.
> Who is to frustrate that?
>
> The three persons of the most Holy Trinity,
> The Father, Son and Holy Ghost.
>
> I call the Virgin Mary and St Bride to witness
> That if your hurt was caused by man,
> Through ill-will,
> Or the evil eye,
> Or a wicked heart,
> That you [persons name] may be whole,
> While I entwine this about you.

Another Hebridean incantation, *The Charm Sent of Mary*, refers to St Bride as a nun and describes the landscape of Ireland where she lived, keeping sheep on the huge, flat grassland known as the Curragh and referred to in Highland and Hebridean terms as 'the floor of the glen'. This particular incantation was probably inspired by the wording of *Saint Broccán's Hymn*, written in the seventh century which, in the first stanza, says: 'She was not a milkmaid of a mountain side; she wrought in the midst of a plain.' The second stanza mentions one of her many miracles: 'When the first dairying was sent with the first butter in a hamper, it kept not from bounty to her guests, their attachment was not diminished', which refers to the miracle of 'The Churned Butter Given Away to the Poor' as recorded in the fifteenth century *Book of Lismore*:

> Afterwards Brigid went to visit her mother who was in bondage and sick. She was at a mountain-dairy of a Druid with twelve cows and making butter. Now Brigid helped her mother, tidied the dairy and then divided the butter into twelve portions in honour of the Lord's twelve apostles with a thirteenth portion in honour of Christ which was greater than every other portion and was given to the poor and to the guests.
>
> For Brigid used to say that Christ was in the person of every faithful guest. The Druid and his wife went to the dairy, carrying with them a great hamper eighteen hands high to be filled with butter. Brigid made them welcome, and the Druid's wife asked Brigid how much butter she had made. However, she had made only one and a half churns of butter and went into the dairy and

said a prayer to God, asking for his blessing and abundance. She then brought the half churn of butter and placed it before the Druid's wife.

The Druid's wife said mockingly 'So, you think this will fill our large hamper!' To this Brigid replied 'Fill your hamper and God will add what is required.' So the Druid's wife put the butter in the hamper, and Brigid took the churn back into the dairy. She brought it back half filled again, and kept on repeating this, singing all the time. So much butter was produced that not only did she fill the Druid's great hamper, but if all the hampers of the men of Munster had been brought to the dairy there would have been enough butter to fill them all!

The Druid and his wife were amazed at this miracle and he said to Brigid 'I offer you the butter and the cows that you have milked, and you shall not be my servant but free to serve God.' To this Brigid replied 'Keep your cows, but grant my mother her freedom.' The Druid immediately freed her mother, but still gave the cows to Brigid, which she later gave to the poor and needy. As a result of witnessing this miracle the Druid became a Christian.

The Charm Sent of Mary (Ean Eoir a Chuir Moire)
The charm sent of Mary Virgin,
To the nun who was dwelling
On the floor of the glen,
On the cold high moors —
On the floor of the glen,
On the cold high moors.
She put spell to saliva,
To increase her butter,
To decrease her milk,
To make plentiful her food —
To increase her butter,
To decrease her milk,
To make plentiful her food.

What also survived in the Hebrides was a belief, that is very ancient and widespread, that sometimes men and women could change into other forms, usually animals and could also make themselves invisible, an ability of particular use to hunters and travellers. One transformation charm invokes both St Bride and St Mary:

Transmutation charm (Fath fith)
Fath fith
Will I make on thee,
By Mary of the augury,
By Bride of the corslet,
From sheep, from ram,
From goat, from buck,
From fox, from wolf,
From sow, from boar,
From dog, from cat,
From hipped-bear,
From wilderness-dog,
From watchful 'scan',

Brigid: Goddess, Druidess, Saint

> From cow, from horse,
> From bull, from heifer,
> From daughter, from son,
> From the birds of the air,
> From the creeping things of the earth,
> From the fishes of the sea,
> From the imps of the storm.

Charms of Protection (*Seun Sabhalaidh*) were also used to protect all parts of the body from harm, specifically mentioning protection from weapons, the sea, assault and the wiles of women! One of which begins 'The charm placed of Bride about her Foster-son …'. With domestic animals of such importance to people in the Highlands and Islands it is not surprising that there are charms to protect the creatures. With St Bride's close association with cattle and sheep it is not unexpected that St Bride's Charm was used to protect them from a variety of dangers:

> *Saint Bride's Charm* (Sian Bhríde)
> The charm put by Bride the beneficent,
> On her goats, on her sheep, on her kine,
> On her horses, on her chargers, on her herds,
> Early and late going home, and from home.
>
> To keep them from rocks and ridges,
> From the heels and the horns of one another,
> From the birds of the Red Rock,
> And from Luath of the Feinne.
>
> From the blue peregrine hawk of Creag Duilion,
> From the brindled eagle of Ben-Ard,
> From the swift hawk of Tordun,
> From the surly raven of Bard's Creag.
> From the fox of the wiles,
> From the wolf of the Mam,
> From the foul-smelling fumart,
> And from the restless great-hipped bear.
> [Line lost]
> [Line lost]
> From every hoofed of four feet,
> And from every hatched of two wings.

St Bride was taken as their patron by the Scottish family of Douglas, who invoked her help when going into battle which was noted in *The Bruce; or The Metrical History of Robert I, King of the Scots* by John Barbour, written in Scottish dialect:

> The folk upon the Sonounday
> Held to Saynet Bridis Kyrk thair way;
> And thai that in the Castell war

> Ischyt owt, both les and mar,
> And went thair palmys for to ber.

The association of St Bride with the Scottish islands inspired Michael Wilkson to write a song for the piano in 1931 called *St Bride of the Isles*, and mentions some of the traditional beliefs about the saint that were well known in the Hebrides:

> Saint Bride of the Isles, who wanders far,
> with a Dove, a Thrush, and a Golden Star.
> To nurse in her arms Saint Mary's Child,
> Saint Mary she gives, with her eyes so mild,
> to Bridget her Mantle with Stars shining bright.
> To calm the wild Elements by their Light.

Nine

Celebrating the Goddess-Saint elsewhere in the United Kingdom and the Isle of Man

Wales

In Wales St Brigid is often referred to as St Ffraid, and sometimes as *Sant Ffraid Leian* (St Ffraid the Nun). However, there seem to be few St Brigid Eve (*Nos Wyl Sant Ffraid*) or St Brigid's Day (*Dydd Gwyl Sant Ffraid*) customs connected with her compared to those in Ireland, Scotland or even the Isle of Man. However, there has been considerable confusion about St Brigid in Wales for many centuries and while the original Welsh St Brigids are not the same as St Brigid of Kildare, so popular was the latter that they merged in people's minds many centuries ago, and now seem almost inseparable!

In the *Martyrology of Donegal* there is listed a 'Brigid of Cill-Muine or Menevia', commemorated on 12 November. The *Gorman Martyrology* records saints 'Brig and Duthracht, from Cill-Muine were they', so quite clearly there was, at one point, a clear distinction between St Brigid of Kildare and St Brigid of *Cill Muine*, a name that indicates she came from a church of teaching, i.e. a monastic school.

St Brigid of *Cill Muine* has been identified as a nun called Brignat, a disciple of St Darerca (also known as St Moninna), who was, according to Concubran's *Life of Saint Darerca*, a contemporary of St Brigid of Kildare. She eventually established a community of virgins beside a lake at a place called *Cenn Trágha* (landing place of the boats) about 6.5km (4 miles) from Brigid's monastery at Kildare in Ireland. At one stage Darerca sent Brignat to Britain for some time to learn the Rule of a monastic community there:

> Among other servants of God a certain maiden, Brignat by name, is said to have resided with a holy virgin; who observing in her the marks of developing holiness, is reputed to have sent her from the monastery at Rosnatensi to the Isle of Britain, in order to undertake the discipline of monastic life.

Brignat was careful to comply with the instructions of her Abbess and she stayed in the British monastery, presumably a *Monasteria duplicia*, an establishment of both monks and nuns, where she was given a little cell in which she read the psalms and other necessary books and learnt about the discipline of the Order. Later, 'to fulfil holy obedience, with God's help, she made the return voyage safely just as she had made the outward passage and came back to her abbey.' One day, after Brignat had returned to Darerca's abbey, she had an encounter with angels, which is described in another version of the *Life of Darerca*.

This tells how Brignat made her way to the cell where St Darerca used to pray and converse with angels. As she approached Darerca's cell, she saw two swans flying from it, and sensing that these were more than just birds, she fell to the ground trembling at this marvellous vision. When Brignat told Darerca what she had seen, Darerca explained that Brignat had been privileged to witness angels, which others could not see. She went on to tell Brignat that the time was now right for her to establish her own religious community, but from then on she would be blind because she had seen angels. The lack of sight, however, would mean that she would better be able to contemplate God without distraction.

Brignat, later went to Wales and her arrival there is described in the *Life of Saint Darerca* (or possibly another Brignat/Brigid as Concubran got rather muddled in his work and may not have meant Brigid of *Cill Muine*). Darerca, Brignat and two other nuns, Luge and Athea, sailed from Ireland on a piece of detached ground and 'landed at the castle named '*Daganno*' [Deganwy Castle near Conwy], beside the shore of the great sea.' Brignat and Luge remained there while the other two went elsewhere as missionaries. The piece of ground on which they had arrived afterwards became fixed to the coast, and on it the Chapel of St Ffraid was built. This building was about 0.4km (a quarter of a mile) west from the present church at Llansantffraid Glan Conwy. By the 1740s part of the chapel had been lost to the sea, and today it has completely disappeared.

There is another, more recent but traditional, Welsh story that accounts for St Brigid's arrival in Conwy:

> Ffraid was an Irish princess whose father was determined that she would marry a man whom he had chosen for her. However, Ffraid was equally determined that she was going to become a nun and devote her life to God. Her father decided to punish her by making her work as a dairy maid looking after his cows and she was so good at this that the cows in her charge gave milk three times a day instead of the more usual twice.
>
> As this lowly role did not seem to be encouraging her to marry the man her father had chosen, he decided to use force to make her marry. However, Ffraid heard about this plan and managed to escape from the palace, but was soon being pursued by the king's soldiers. She continued to flee before them until eventually she reached the coast and could go no further.
>
> As the soldiers began to close in on her she knelt and prayed to God for help, and suddenly the piece of ground on which she was kneeling separated from the land and floated out to sea, and so she escaped her pursuers. She continued sailing along on her rocky raft until eventually it bumped into a bank of the River Conwy in Wales, where it attached itself to the land, forming a mound which became known locally as *Y Twyrch Ffraid* (Ffraid's Turf). Ffraid then went to a nearby village on the banks of the river and found that the people were in great distress and starving because of a famine.
>
> She decided that God had brought her to that place to help the people so, gathering an armful of rushes from the shore, she strode out into the river and threw them upon the water where they immediately turned into a shoal of small silver fish called by the English Sparling, and by the Welsh *Brwyniaid*, from the Welsh word *brwyn* which means 'rush'. The people of the village then cast their nets on the river and caught a great catch of *Brwyniaid* so ending their famine.
>
> Ffraid decided to settle in the village and built a chapel on the mound. This building survived until the eighteenth century and in honour of the saint and her saving of the people, the village

Brigid: Goddess, Druidess, Saint

took the name Llansffraid Glan Conwy, the name by which it is still known today and they held a special fair, the *Ffair Ffraid* (Ffraid's Fair) to mark her special day on 1 February.

Sparling (*Osmerus eperlanus*) come into fresh water to breed and are traditionally said to enter the River Conwy on 1 February to mark St Ffraid's Day, but their name, *Brwyniaid* (rush), is said to have been given to them because they taste of rushes! In the nineteenth century sparlings were also called locally *pysgod Sant Ffraid* (St Ffraid's fish).

There is another alternative traditional story that claims St Ffraid sailed across from Ireland on a large lump of turf, but in this case landed at Trearddur Bay on Holy Island (*Ynys Gybi*). This lies just across from the island of Anglesey off the North Wales coast, once the main centre of Druid worship in Britain. There she built *Capel Sant Ffraid* (St Ffraid's Chapel) which stood on a mound by a sandy beach called *Tywyn y Capel*, but this stone building too was washed away by the sea. In 1997 the east side of the burial ground, the only part of the site surviving, was excavated and showed that a typical coastal Christian community was settled there between the sixth and sixteenth century. Yet another tradition has St Ffraid landing in the estuary of the Dovey near Talybont, Powys, where a brook called Ffraid runs into the Eleri, a tributary of the Dovey.

The tradition of floating across on a turf is obviously related to an Irish belief that claimed that many stones associated with saints were capable of floating and some saints even used rafts and boats made of stone. For example St Boec sailed from Wexford to Brittany on a stone, which returned to Ireland after he landed.

To mark the millennium in 2000, St Ffraid's Cross was erected at Holyhead to commemorate both this event and mark St Ffraid's traditional landing on Holy Island, said to have occurred about 500 AD. Dedicated by the Bishop of Bangor on 10 September, it was designed by Dr J. K. Roberts and sculpted by Meic Watts with stone from an Anglesey quarry. It incorporates a carved rush Brigid's cross on its head and on the column is the hand of St Ffraid holding a flame. On its base it bears an inscription in Welsh, English, Latin and Irish:

Y Santes Ffraid Gweddia Drosom ni
Saint Bride Pray for Us
Sancta Brigida Ora Pro Nobis
A Bhrid Bheannaithe Guidh Orainn

Today, however, some people claim that her turf raft can still be seen in the form of a mound that is situated near the newer church of *Capel Sant Ffraid*. This lies on the left as you cross the Four Mile Bridge from Anglesey to Holy Island, some 2km (1.2 miles) from Trearddur Bay, while the east window of St Ffraid's Church, Trearddur depicts the saint crossing the sea.

There is yet another ancient tradition that Brignat spent time at St David's monastery in Wales, which is why the large bay on which the cathedral city of St David's stands is called St Bride's Bay, after St Brigid of *Cill Muine*. This belief was based on the way Brignat was described as 'Brigid of Cill-Muine or Menevia' a place-name given in both Gaelic and Latin. Gerald of Wales, who visited the city of St David's in 1188, said that the Romans called it Menevia, information which he got from the *Antonine Itinerary*, compiled in the early third century AD. This is a road book giving distances between many places in the Roman Empire, and records a place on the south west coast of Wales described as 'Menapia from which one could sail for Ireland', but Gerald either misread it or his copy was inaccurate.

Celebrating the Goddess-Saint elsewhere in the United Kingdom and the Isle of Man

35 St Ffraid's Millennium Cross, Holy Island, Anglesey, Wales. Dedicated by the Bishop of Bangor on 10 September 2000, it commemorates the tradition that St Brigid landed there about 500 AD after sailing across from Ireland on a piece of turf.

Just to confuse the issue, the Romans called the Isle of Man *Monavia* and there is a tradition, recorded by Hector Boece in his *Historia Scotorum*, that St Brigid was 'veiled', that is made a nun, on the Isle of Man by St Maughold, Bishop of Sodor. However, although the bishopric of Sodor was based on the Isle of Man it also covered Anglesey and Holy Island. The Latin name for Anglesey was *Mon*, so could this be a muddled tradition that the novice Brignat/Brigid was made a full nun some time after she arrived on Holy Island by the Bishop of Sodor?

In medieval Wales St Brigid of Kildare, whose legends got intermingled with the St Brigid of *Cill Muine*, the Brigid who landed on Holy Island and the Brigid of Conwy, became a very popular figure, generally regarded as fourth in importance after St Mary the Virgin, St Michael the Archangel and St David, the patron saint of Wales. There are at least seventeen churches dedicated to St Ffraid, although some use the alternative name Brigid, Bridget or Bride, along with four chapels, a convent and several holy wells. She is to be found in place names such as Llansantffraid (*Llan* means place of the church), St Bride's Bay, Pembrokeshire, which has, on its south side, St Bride's Haven, near which stands St Bridget's Church built in 1868, which has a stained-glass window showing St Brigid. This replaces an earlier chapel dedicated to St Brigid, the earliest mention of which occurs in 1291 when the '*Ecclesia de Sanct Brigida*' was required to pay a tithe of £1 6s. 8d to the crown. From the churchyard can be seen an imposing residence, once the home of Lord Kensington, who also held an Irish title, called St Bride's Castle.

Not all of the churches dedicated to St Ffraid feature images of the saint. Among those that do is the Church of Ffriad and Non at Llanon, Ceredigion which features a stained-glass window that shows St Brigid holding a bowl of milk standing next to St Non with her young son, St David, the patron saint of Wales. At St Bride's Church, Llansantffraed-Cwmdeuddwr, Powys the stained-glass window shows St Brigid holding a book and wearing a crown of flames. St Brigid's Church, St Bride's Netherwent, Monmouthshire has a 1950s stained-glass window showing the saint holding a crosier, in front of an altar with flames upon it. An inscription within the window identifies her as 'St Brigid of Kildare', but above her fly three birds, which are presumably supposed to be swans, but look more like geese, an image inspired by an incident from the life of St Brigid of *Cill Muine*. One of the most impressive images of the saint is to be found in St Bridget's Church, Skenfrith, Monmouthshire where a carved figure of St Brigid is found on a lectern dated 1909.

St Brigid does not seem to have assumed the importance in Wales, as far as customs are concerned, as she did in other western Celtic countries. The most important February festival in Wales was Candlemas (*Nos Gwyl Fair*), although it should be remembered that in the popular mind there was a close connection between St Brigid and St Mary. St Brigid's Day on 1 February is also Candlemas Eve, Candlemas being a celebration of the Purification of the Virgin Mary.

St Ffraid is frequently mentioned in medieval Welsh literature and in an anonymous poem in the twelfth century *Black Book of Carmarthen* she is invoked by travellers: '*Sanffraid suynade in imdeith*' ('St Ffraid bless us on our journey'). It is recorded in the fifteenth century that Lewis Glyn Cothi swore an oath at one of her shrines. There are also references in the medieval period to several people who bore the title of *Gwas Sant Ffraid* (the tonsured servant of St Ffraid), suggesting they were priests or monks of a Welsh Brigidine Order.

The story of St Brigid of Kildare's life was well known in Wales, and a Welsh poet, Iorwerth Fynglwyd (Edward Greybeard), 1480–1527, wrote a poem about St Ffraid describing her life and various miracles, but his poem also incorporates legends of other St Brigids, such as Brigid of *Cill Muine* and Bridget of Sweden. He says she was a beautiful Irish nun who was the

36 St Brigid carved on a lectern dated 1909 in St Bridget's Church, Skenfrith, Monmouthshire.

daughter of Duke Dipdacws (Dubhthach) and performed various miracles: she got honey from a stone to give to the poor; she gave her distaff (a spindle on which wool is wound for spinning), to a man whose coulter (the part of a seed drill that makes the furrow for the seed), had broken; she converted butter that had turned into ashes back to butter again; and although she gave to a poor man all the cheese in the steward's store, not one cheese was missed by him.

She knew the fifteen prayers; when it rained she threw her white winnowing sheet on the sunbeams; when her father wanted her to marry a man she refused and one of her eyes fell out of its socket, although she later put it back and it worked as well as ever; she sailed on a turf from Ireland and landed on the bank of the Dovey River in Wales; she created fish in Gwynydd by throwing pieces of rush among the water cress; she went to Rome to St Peter's; and Jesus established her festival on Candlemas Eve which was observed with as much solemnity as was Sunday.

Another, longer version of the poem, almost certainly by a different writer, adds the information that she sailed over to Wales with her fellow nuns on green turfs landing at Porth y Capel near Holyhead where she built a chapel on a little mound. She then went to Glan Conwy and founded Llansantffraid, and there turned rushes into fish in the River Conwy. The veneration of St Brigid of Kildare was very popular and started to spread widely from the seventh century, so its not surprising that that many women bore this name, or adopted it when they joined a religious order.

Welsh Saint Ffraid customs

While St Brigid (or a mixture of Brigids!), was venerated in Wales from an early period, there seem to have been few customs connected with the saint, or perhaps few of the traditions that were once carried out to commemorate her were recorded. Over a period of time it appears that any veneration of St Ffraid on her Eve (*Nos Gwyl Sant Ffraid*) or her Day (*Dydd Gwyl Sant Ffraid*), 1 February, was superseded by the importance of Candlemas Day on 2 February. This may have been because the Virgin Mary was equated with St Brigid, as was the case in other western Celtic countries.

Blessing of the cow stalls

In a few places Welsh people on St Ffraid's Day would bring blessings on their cattle by hanging rushes up in the cow stalls.

Saint Ffraid's Ale

In Wales there was a custom called *Cwrw Sant Ffraid* (St Ffraid's Ale). This was mentioned in *The Red Book of Asaph* and in the works of various poets such as Dafydd ap Gwilwm and Iolo Goch. It also occurs in a rhyming calendar written in 1609, which passed into the language as a common saying:

> *Digwyl San Ffraid ydoedd fenaid,*
> *I bydd parod pawb ai wyrod.*

> St Ffraid's Day it was, my soul,
> everyone be ready with his drink.

In the ancient Welsh *Wishes of Brigid*, sometimes called *The Quiver of Divine Desires*, which is preserved in an eighth century manuscript in Brussels, ale also features, and ale customs

Brigid: Goddess, Druidess, Saint

37 St Brigid on a stained-glass window in St Bride's Church, St Bride's Netherwent, Monmouthshire. She holds a bishop's crosier and stands before a fire alter with three birds flying in the background. The window is in memory of Susan Florence Reece who died in 1954.

38 Stone carving of St Ffraid in St Bridget's Church, St Bride's Major, Glamorgan, showing St Ffraid with a book under one arm and a quill pen in the other.

are connected with St Brigid of Kildare. Many of the miracle stories about her concern the production of this drink, something for which she was renowned:

> I should like a great lake of ale
> For the King of Kings.
> I should like the angels of Heaven
> To be drinking it through time eternal.
> I should like excellent meats of belief and pure piety.
> I should like flails of penance at my house.
> I should like the men of Heaven at my own house;
> I should like barrels of peace at their disposal;
> I should like vessels of charity for distribution;
> I should like for them cellars of mercy.
> I should like cheerfulness to be in their drinking.
> I should like Jesus to be there among them.
> I should like the three Mary's of illustrious renown
> to be with us.
> I should like the people of Heaven, the poor, to be
> gathered around us from all parts.
> I should like to be a rent payer to the Lord
> So that should I suffer distress
> He would bestow a good blessing upon me.

England

While *Imbolc* was celebrated in a few places in England by traditionalists and has always been special to witches both past and present, there seems to be little that connects 1 February with the beginning of spring or St Brigid in England as a general rule. A Somerset custom to celebrate St Brigid was transferred to May Day Eve, 30 April, which is also the start of the ancient pagan festival of *Beltane*, regarded as marking the first day of summer. May Day has always been strongly associated with fertility and was once a time when sacrifices were offered for the protection of cattle. So it may not be a coincidence that since fertility and cows are 'interests' common to both the Goddess Brighid and St Brigid then *Beltane*/May Day eve would be regarded as an appropriate date to celebrate Brigid in either personification.

There are a number of precedents for celebrating St Brigid on a day other than her own, even in Ireland. For example in the nineteenth century the saint was celebrated with singing and other activities on the first Sunday in August at St Brigid's Well about 1.6km (1 mile) from the ruined church of Ballinakill, Co Galway. Traditionally this was regarded as the first day of the harvest. Today the National Pilgrimage, when a relic of St Brigid is carried in procession between Kilcurry and Foughart, Co Louth, occurs on the second Sunday in July.

A Somerset custom: the Biddy bed
According to accounts from the early years of the twentieth century, noted by the folklorist Ruth Tongue, St Brigid was celebrated on May Day Eve on farms in the Brendon Hills of West Somerset by making a 'Biddy's bed' of Mountain Pansies (*Viola lutea*) and laying a

Brigid: Goddess, Druidess, Saint

doll on it, which was then covered in petals of the Periwinkle (*Vinca minor*). In Somerset the common country name for pansies was 'Biddy's eyes', Biddy being a form of the name Brigid/Bride. It seems that before the bed was made the children would go round to neighbours to collect gifts of flowers or pennies, singing the verse:

> Ring the bell Biddy's dead,
> Please give us a flower (or penny) for Biddy's bed.

This is reminiscent of the custom of preparing a bed of rushes or straw for the saint on the Eve of her Feast Day (31 January) which was once common in the Scottish Highlands, the Hebrides and the Isle of Man, all Celtic countries. It was believed that the 'spirit' of St Brigid travelled around the countryside, accompanied by her favourite cow, blessing crops, animals and houses. To demonstrate both their hospitality and veneration of the saint, households would make a bed of rushes or straw so the saint could rest in the house if desired, so guaranteeing that she would bless the family.

While the Somerset tradition of making a Biddy bed was probably related to the practices of these countries it seems to have evolved to represent something other than a bed for this saintly traveller. It may be a modification of another custom, since in some parts of England it was once traditional for the May Doll to appear. These was usually an ordinary female doll seated in the centre of a garland of flowers which, in many places, was carried from house to house by small girls to solicit little gifts.

39 A 'Biddy's bed' of Mountain pansies (Biddy's eyes) and Periwinkle petals. An ancient Somerset custom that may reflect the twelfth-century belief that St Brigid died in that county and was buried at Glastonbury Abbey.

Various explanations were put forward regarding the May Doll. Some researchers have said it was a representation of the Virgin Mary and others that it was Flora the goddess of flowers or just represented the 'May Queen'. However, at Bampton in Oxfordshire a folklore researcher was told that their May Doll was 'a goddess', while another person there explained more precisely that it was 'Minerva'. On the face of it this is a strange answer, but the ancient British goddess Brigantia was equated by the Romans with their goddess Minerva and some people in more recent times have equated the Goddess Brighid with Brigantia. So could the Bampton doll also be related to Brighid/St Brigid?

However, Somerset seems to be the only English county where the doll was specifically identified as Biddy (Brigid/Bride) and the flowers used to make the bed had to be 'Biddy's eyes'. The Somerset image was also not carried around the houses like May Dolls. Making a Biddy bed in Somerset was probably due to the strong connections the county had with Ireland from at least the ninth century with many Irish monks based at Glastonbury Abbey, where St Brigid was certainly venerated from the tenth century, and probably much earlier.

The verse 'Ring the bell Biddy's dead' implies that the custom dates back to at least the twelfth century when there seems to have been a local belief that St Brigid died while staying at Glastonbury and was even buried in the abbey there. This suggests that the 'bed' in Somerset might symbolise a death bed or coffin, so the custom would commemorate the supposed death of the saint in the county. In early Christian usage the word for 'grave' was *lectus* in Latin and *lecht* in Old Irish, a word that also means 'bed'.

While Glastonbury is about 58km (36 miles) from where this custom was recorded, the tradition of making a Biddy bed may be the remnant of an old practice that was once widespread in the county. The Brendon Hills, an area which adjoins Exmoor was, until recently, very isolated where a custom may well have survived longer than in other parts of Somerset. It is not far from where Irish travellers once landed on their pilgrimage to St Brigid's Chapel at Beckery.

The Biddy bed custom in Somerset was performed on May Day Eve rather than on St Brigid's Feast Day Eve (31 January) or her Feast Day, which is regarded as the traditional first day of spring in Ireland. While this may be because the people of Somerset believed St Brigid died on 30 April/1 May, it is possible that this date was chosen because in England this was the time more usually associated with fertility than 1 of February, and St Brigid is strongly linked with fertility. May Day Eve (30 April) is also the start of the ancient pagan festival of *Beltane*, regarded as marking the first day of Summer.

It was once a time when sacrifices were offered for the protection of cattle, and fertility and cows are 'interests' common to both the Goddess Brighid and St Brigid. So something to mark St Brigid's 'death' in Somerset would have been just as appropriate at a time when people's thoughts were also on fertility! The custom of making a Biddy bed seems to have died out by the end of the First World War.

A Somerset custom: Imbolc at the Chalice Well

For many years a ceremony has been held at the Chalice Well, Glastonbury on 1 February to celebrate the pagan festival of *Imbolc*, a time that is particularly associated with the Goddess Brighid and which, in Celtic tradition, marks the first day of spring. This date is also St Brigid's Day, so she can be honoured in either form by participants. The well is decorated with greenery and snowdrops and celebrants place lighted candles and often other small tokens around the well. Songs are sung extolling the virtues of Brighid/Brigid. After this a procession goes from Chalice Well Gardens to Bride's Mound at Beckery where further ceremonies are carried out in honour of Brighid as goddess. Both these events are well attended and provide

Brigid: Goddess, Druidess, Saint

a moving experience that keeps alive the connection of Somerset with this highly regarded goddess-saint.

A Somerset custom: The Festival of Brighid at her Temple

The Goddess Temple at Glastonbury, celebrates Brighid and her season from 1 February (*Imbolc*), by holding ceremonies to honour her and decorating the temple for about six weeks with white material, images, objects and plants connected with Brighid. This creates a beautiful tribute to this widely venerated and highly regarded deity. Once a month the temple is a Flame Keeper for the Flame of Avalon, dedicated to the Goddess Brighid of the Triple Flame.

Other English Brigid customs and beliefs: bathing in the Brigid's Day dew

It was once a widespread belief that dew collected from plants and applied to the face improves the features and many ladies, including the diarist Samuel Pepys's wife, used to do this. In the English Midlands it was said that dew gathered and used on St Brigid's Day would not only help the complexion but would also make old people young!

Saint Brigid's Summer

In Scotland and the Hebridean Islands St Brigid is connected with sayings regarding the weather expected in the forthcoming spring, but in England she is associated with a weather phenomenon at the other end of the year. Often there are a few fine days towards the middle

40 A ceremony at Chalice Well, Glastonbury, Somerset, to celebrate *Imbolc*, 1 February, a day particularly associated with both the Goddess Brighid and St Brigid. It also marks the first day of spring in the Celtic calendar.

41 The Goddess Temple, Glastonbury, Somerset, decorated for *Imbolc* (1 February) 2008, when the Goddess Brighid is celebrated.

of October and, depending on the day this begins, it is sometimes referred to as a particular saint's 'summer', a practice that was once widespread in both England and Europe.

In England a spell of warm weather beginning on 8 October was known as 'St Brigid's Summer', but if it began on the 11th it was known as 'St Martin's Summer'. If this phenomena was as late as the 18th it was called 'St Luke's Little Summer' as this coincides with his saints day. However, in the case of St Brigid and St Martin the date does not coincide with their festival days, so the association of a saint with a late warm spell does not necessarily correspond with their feast day!

The Isle of Man

The Isle of Man, situated in the Irish sea midway between England, Ireland and Scotland has had a turbulent history. Its people were mostly Celtic with close ties to other Celtic countries particularly Ireland and Scotland, with a Scandinavian influence from later Viking settlement there and considerable English input when the island came under their rule.

Initially almost everyone spoke Manx, the local Celtic language, but this began to decline quite rapidly from the nineteenth century, which also led to the loss of many of its traditional customs. In 1874 there were 12,000 Manx speakers, but by 1901 there were only 4,419 out of a population of 54,752. By the 1920s the language was hardly heard and by 1961 only 160 speakers of the language remained. The last person for whom Manx was the mother tongue was Ned Maddrell, who died aged ninety-seven, in 1974. C.I. Patton, who wrote *Manx Calendar Customs* in 1939, says in his introduction that most of the customs he describes were 'now extinct'.

As in other parts of the Celtic world, the Manx divided their year into four quarters with the spring season, *Arragh*, beginning on 1 February and lasting till the end of April, while February was *Mee-hoshee yn niarragh* (the first month of spring). In the past St Brigid, known as St Breeshey in Manx Gaelic, was strongly venerated, with seven ancient *keeill* (small churches or chapels), a nunnery and a parish church dedicated to her, along with the parish of Kirkbride. Of 135 sacred 'wells' (most more accurately described as springs) known on the Isle of Man, only three are dedicated to St Brigid and these are to be found in the parishes of Braddan, Bride and Patrick.

Local tradition says that St Brigid came to the Isle of Man to take the veil (i.e. become a nun) from St Maughold, the fourth bishop of the island (see also page 155). She was said to have founded and lived for a while at the nunnery of St Brigid which lies on the banks of a river near Douglas, the capital town. In ancient times the Prioress there was said to have held courts in her own name since she was accorded the rank of Baroness of the Isle of Man, possessing authority equal to that of a Baron. The nunnery was 'dissolved' in 1540 during the Reformation and later a house was built on the site, only the twelfth century chapel surviving. This was used as a stable until the late nineteenth century when it was re-opened for worship, which continued until 1998 when the new owner stopped the Anglo-catholic services being held there.

The Kirk Bride Fair was held on 1 February until the change in calendar in 1752 after which it was held on 12 February, and continued to be an annual event until the late nineteenth century. As in Ireland, St Brigid's Day (*Laal'l Breeshey*) was celebrated on 1 February, although many people seem to have preferred to keep to the pre-1752 calendar and so commemorated her on the 12 February.

As in Ireland, one of the main customs recorded was to offer the hospitality of the house to the saint who was believed to travel around the district on the eve of her Day, at least in spirit. A bundle of green rushes was collected during the day, and that evening a member of the family stood at the open front door holding them and called out a plea for St Brigid to enter:

Breeshey, Breeshey, tar gys my hie, tar gys thie ayms noght. Foshil-jee yn dorrys da Breeshey, as lhig da Breeshey cheet stiagh.

Brigid, Brigid, come to my house, come to my house tonight. Open the door to Brigid, and let Brigid come in.

Once the invitation was issued the rushes were either spread on the floor or put down in a pile as a token bed for the saint. C. Roeder, writing in *Yn Lioar Manninagh* (*Journal of the Isle of Man Natural History & Antiquarian Society*) says:

I have heard from my mother that … on St Briget's Eve the old farmers' wives used to sweep out the barn, and put a bed and a chair and a table in, and light a large mould candle that would burn all night, and set bread and cheese on the table with a quart jug of good Manx ale, all in the hope that Breeshey would pay them a visit; and used to say at the open door before going to bed '*Quoierbee y thie hig – oo, huggey tar gys y thie ainyn*' ('Whosoever house you come to, come to ours tonight').

A young man, a relation of my mother, was once coming home at a late hour and passed one of these lighted barns. He went in to have a look, and ate as much as he could of the bread and cheese, and finished the ale; and then went and rolled himself well in the bed, and shut the barn door again. The farmer's wife, as soon as she got up in the morning, went to the barn to see if Breeshey had been, and when she saw the bread and cheese and the ale removed, and someone had lain in the bed, her joy knew no bounds. She was all day going about telling the neighbours that Breeshey had paid her a visit and she would be alright, blessed with peace and plenty for the year.

Other customs such as the carrying round of an image of St Brigid, the making of Brigid's crosses, girdle, cloak and other traditions performed in Ireland and, to a lesser extent, in Scotland, don't seem to have been recorded on the Isle of Man. They may well have existed in the eighteenth century or earlier and accounts of them have simply not survived. An indicator that such customs were once carried out there is suggested by a description of a Brigid supper being held on the island around 1810–20 by Edward Farquhar, at which time this was probably a widespread tradition.

As in other parts of the western Celtic world, there was weather lore regarding St Brigid's Day:

Laa'l Breeshey bane,
Dy chooilley yeeig lane.
Dy ghoo ny dy vane.

Brigid's Feast Day white,
Every ditch full
Of black or white.

Which means that if it snows on St Brigid's Day, there will be a wet or snowy spring. While another says:

Choud as hig shell ny-gah-ghreinney stiagh Laa'l Breeshey,
Hig sniaghtey roish Laa Boay Idyn.

As long as the sunbeam comes in on St Brigid's Feast Day,
The snow comes before May.

That is, if it is mild on St Brigid's Day it will be a cold spring. There is also the saying:

My nee yn ushag gherrym er laa Breeshey,
nee e keayney roish Laa Patrick.

If the bird crow on St Brigid's Day,
She will cry before St Patrick's Day.

As the importance of celebrating St Brigid's Day declined many of the weather sayings were transferred to St Mary's Feast Day of the Candle (Laa'l Moirrey-ny-gianle) on 2 February, more commonly known as Candlemas. This is not unexpected as St Brigid and the Virgin Mary are closely entwined, Brigid often being referred to as the 'Mary of the Gael', while one tradition claimed that St Brigid was the midwife of Mary (see page 90).

Ten

Saint Brigid visits Glastonbury!

There is a long-standing belief St Brigid visited Glastonbury, Somerset, in 488 AD, first noted in *The History of the King's of Britain* by William of Malmesbury, *c.* 1120:

> Wherefore the report is extremely prevalent that both Saint Indract and Saint Brigid, no mean inhabitants of Ireland, formerly came over to this spot. Whether Brigid returned home or died at Glastonbury is not sufficiently ascertained, though she left here some of her ornaments; that is to say, her necklace, bag, and implements for embroidering, which are yet shown in memory of her sanctity, and are efficacious in curing divers diseases.

This mentions the local belief that St Brigid had died in Somerset and was buried in Glastonbury Abbey. William of Malmesbury later wrote a history of the Abbey, *An Enquiry into the Antiquity of the Church at Glastonbury*, between 1129 and 1139, making extensive use of books and manuscripts in the abbey library, but there are slight variations of the story in different editions of William's work, the earlier one reading:

> Hence the custom developed among the Irish of visiting that place to kiss the relics of their patron. Whence the well-known story that Saint Indract and the Blessed Brigid, prominent citizens of that land, once frequented the place. They say that after Saint Brigid, who had come there in 488 AD, had tarried for some time on the island called *Beokery* she returned home but left behind certain of her ornaments, namely a bag, necklace, a small bell and weaving implements, which are still preserved in memory of her.

A later edition of his work does not record her date of arrival at Beckery, either because William or the copyist may have considered it irrelevant, or because there were doubts about it, so the account concerning St Brigid's visit is shorter:

> Hence Irish pilgrims came to the spot. Saint Brigid dwelt long in the island of *Beokery* and, returning home, left memorials of wonder-working power.

St Brigid's stay at Beckery was also related by John of Glastonbury who wrote his work *The Chronicle of the Ancient Church at Glastonbury* between 1340 and 1342:

> Saint Brigid flourished at about the end of the life of the greater Saint Patrick, with whom we dealt earlier. She survived him, as Gildas writes, by sixty years, and she came to Glastonbury about 488 AD. Saint Columcille was born four years before the death of Saint Brigid and came to Glastonbury later, about 504 AD. These saints, indeed, frequented the spot, along with some of the Irish nobility, because of the venerable relics of their patron Patrick.
>
> Saint Brigid made a stay of several years on an island near Glastonbury, called *Bekery* or Little Ireland, where there was an oratory consecrated in honour of Saint Mary Magdalen. She left there certain signs of her presence – her wallet, collar, bell, and weaving implements, which are exhibited and honoured there because of her holy memory – and she returned to Ireland, where, not much later, she rested in the Lord and was buried in the city of Down. The chapel on that island is now dedicated in honour of Saint Brigid; on its south side there is an opening through which, according to the belief of the common folk, anyone who passes will receive forgiveness of all his sins.

Saint Brigid's chapel at Beckery and the Irish connection

The chapel in which St Brigid is said to have stayed was dedicated to St Mary Magdalen, although one local tradition says St Brigid founded it. This was on the Isle of Beckery, a raised area which lies 2km (just over a mile) south west of Glastonbury Abbey which owned it, the chapel being on the highest part of a low lying island that was once almost surrounded by reed marshes. While in old documents it is always referred to as an 'island' it was really a ridge about 1.07km (two thirds of a mile) long. It would never have been a true island, although it must have seemed very like one, particularly in winter with the water of the marshes at a high level. The ridge remained intact until the construction of a sewage works in the twentieth century that levelled its eastern end, leaving only the western part, where the chapel of St Brigid is located, intact. This remaining ridge is now seen as a distinct elongated hillock, referred to today as 'Bride's Mound'.

John of Glastonbury said that there was a 'monastery of holy virgins' located on nearby Wearyall Hill, the first mention of a women's community in the area which, local tradition claims, was also founded by St Brigid. In the ninth-century *Sanas Cormac* (*Cormac's Glossery*) of Bishop Cormac mac Cuilleanáin he refers to 'Glasimpere of the Gadhaels' (Glastonbury of the Gaels, i.e. the Irish Celts), suggesting the strong link between Glastonbury and Ireland even at that time. There were many connections between Glastonbury and Ireland with Irish monks based at the abbey, and the abbey owning land in Ireland from the twelfth century.

In the thirteenth century Archbishop Henry de Loudres of Dublin and other Anglo-Norman bishops in Ireland granted indulgences to Irish pilgrims to Glastonbury. This meant their sins were completely absolved and they could avoid going to purgatory if they undertook a pilgrimage to this holy site. By this period Glastonbury had long been on the Irish pilgrim's route to the continent. Travellers crossed from Wexford or Waterford in Ireland to Bridgwater Bay in the Bristol Channel, up the River Parrett to Bridgwater then along the pilgrim's route to Glastonbury. This took them along an ancient road running eastwards on top of a long ridge on the Polden Hills (today the route of the A39).

From Bridgwater to Beckery is 21km (13½ miles), so easily walked, although the River Brue disgorges into the River Parrett at Highbridge. So it is probable that some pilgrims may have transferred to a smaller boat there and travelled up the Brue to St Brigid's Chapel at Beckery since this river passes within a few metres of the island. The Brue was improved and modified by successive abbots to connect Glastonbury with the coast and keep it navigable.

After the pilgrims had visited Beckery Chapel and Glastonbury Abbey, they would have continued their pilgrimage via Salisbury, Winchester and Canterbury, then across the Channel to visit holy sites on the continent. Somerset had three churches besides Beckery that have St Brigid as their patron saint. There was a St Bride's Church at Bridgwater, although known only from the existence of a medieval Deed of Gift to the church, a St Bridget's Church at Brean on the coast north of Bridgwater and another St Bridget's Church at Chelvey south west of Bristol.

Nechtán, King of the Scottish Picts had, by 625 AD, become patron of a group of nuns from St Brigid's monastery at Kildare in Ireland who had gone to his kingdom to establish a daughter house at Abernethy (near Perth). So it may be that the nunnery on Wearyall Hill was not actually founded by St Brigid but followed the Rule of her Order as a daughter house and could well be another early foundation, particularly considering the close ties that the abbey had with Ireland.

A calendar in the *Bosworth Psalter* written, it is believed, for the use of the church at Canterbury when St Dunstan was Archbishop of Canterbury (960–980 AD), is derived from the *Calendar of Glastonbury*. Among the saints especially commemorated in this is St Brigid. This shows her special veneration at Glastonbury perhaps as early as the 930s AD when Dunstan would have become familiar with this *Calendar* as a monk at Glastonbury. He was Abbot of Glastonbury from 943–957 AD, although even before his ordination he was educated there by Irish monks. This suggests that the nunnery and the possible connection with St Brigid and her Order at Kildare could also date to this time or even earlier since the *Glastonbury Calendar* may well have considerably pre-dated Dunstan's time at the abbey. A calendar in the *Leofric Missal*, a little later in date, is also copied from the *Glastonbury Calendar* and confirms the importance of St Brigid at Glastonbury.

John of Tynmouth (c.1295–c.1348) compiled a collection of abridged lives of the saints of Britain and Ireland, *Saint's of England, Wales, Scotland and Ireland*, based on early works he examined during his travels. Since there is no evidence that he ever visited Ireland he seems to have got the details of the life of St Brigid from manuscripts in the library of Glastonbury Abbey as he makes clear in the introduction to his work: 'However, what is further expounded below, I have drawn from books in the library at Glastonbury.'

Beckery Island in documents and the origin of the name

The earliest mention of Beckery (*de insula Bekeri*) is in a charter dated 670 AD, in which Beckery was given by Cenwealh, seventh King of the West Saxons (reigned 643–672 AD), to Abbot Beorthwald of Glastonbury Abbey (abbot c.670–80 AD). However, according to John of Glastonbury writing in 1342, Beckery was owned by the abbey much earlier than this but lost it to the Saxons when they occupied this part of the country in the late sixth century:

> Cenwalh, the seventh king of the West Saxons after Cerdic, was converted to Christ through Birinus, apostle of the West Saxons, in the twentyninth year of his reign; and he gave Abbot Beorhtwald the land which is called Meare, and two small islands, to wit Westhay and Godney, on either side

of the pool, along with the fishery, marshes, woods, and the very wide moors that belong to those lands. This was in 670 AD. The same king also gave three other islands, namely Marchey, Nyland, and *Bekeri*. One ought to know that that land and the aforesaid islands, and many other holdings, had previously been in the possession of the church of Glastonbury in the time of the Britons. But when the pagan Saxons came, and the Britons fled beyond the Severn into Wales and the country's western corner called Cornwall, the invaders seized everything; afterwards, when they had been converted to the faith of Christ, they restored it all with the greatest good will.

The name Beckery is derived from the word *beocera*, a mixture of Anglo-Saxon and Latin. *Beo* means bee in Anglo-Saxon while *cera* is Latin for wax. It would have been referred to as an *Eo*, 'island' in Anglo-Saxon or *insula*, 'island' in Latin, so its original name was Beeswax Island. Some authors have suggested Beckery is derived from the Anglo-Saxon name for beekeeper, but this is unlikely as the Anglo-Saxon term for a Beemaster (they did not use the term beekeeper) was *Beo-ceorlaf* (pronounced Beo chorlaf). Many beehives must have been located there when the site was occupied as a monastery, and Beeswax Island suggests this may at one time have been the main place that produced wax for making beeswax candles for use on altars. They would also use the honey to make mead, a honey-based drink that was consumed at the abbey on days when beer was not served.

The charter of 670 AD was almost certainly a later copy of the original document since the name of the island is given as *Bekeri*. As the charter records the gift of the island to the abbey there would have been no beekeeping monks there at that time, but the clerk copying the charter at some later date used the name of the island by which it was known in his day, rather than any earlier name that might have been given on the original document. A charter of King Ine of Wessex (reigned 688–726 AD) mentions Beckery, confirming that it was 'Cenwealh, who granted Mere, *Beokeri*, Godney, Marchey and Nyland, at the insistence of Archbishop Theodore…'. A further charter of King Edgar (959–975 AD) mentions the chapel: 'I also confirm and corroborate, what has hitherto been scrupulously observed by all my ancestors, that neither the Bishop of Wells nor his servants shall have any power at all over this monastery or its parochial churches, namely Street, Moorlinch, Butleigh, Shapwick and Zoy, or over their chapels, or even those located on the islands, namely *Bekeria*, which is called *parua Ybernia* [Little Ireland]…'

William of Malmesbury lists among the main places owned by the abbey '…the island of *Beokery* with it's appurtenances', while a Papal Charter of 1168 refers to Beckery as the 'first' of the islands in Glastonbury Abbey's estates, indicating that it was regarded as the most important of the several 'islands' owned by the monastery. John of Glastonbury says in 1342 that 'This island [Glastonbury] also has islands lying near it and bordering it, of which these are the names: *Bekery*, which is called little Ireland, where Saint Brigid once sojourned…'.

'Little Ireland' is a name still sometimes used today and it has been suggested that this is a misinterpretation of Beckery that assumed its original name was *Becc-Eire* or *Beag-Eire*, which means in Irish Gaelic 'Little Ireland'. The term 'little Ireland' does not appear until it is used in a charter of King Edgar, by which time a small monastery had been established at Beckery for some time, probably from shortly after the abbey reacquired the island in 670 AD. Many of the monks at Glastonbury in the tenth century would have spoken Anglo-Saxon English (Old English), Latin and, in many cases Irish Gaelic so would have known that the name of the island was a Latinised form of the Saxon words for Beeswax Island (*Beocera Eo*). It seems that 'Little Ireland' was being used as a descriptive name, suggesting a particularly strong Irish connection with Beckery at an early date.

At some point during the 1260s Glastonbury Abbey lost possession of Beckery Island. John of Glastonbury describes how 'Abbot John of Glastonbury (abbot 1274–91) acquired possession of the appropriated churches of Doulting and East Brent, purchased Milton, Shapwick, Wytheges, Ashcott, Greinton, and *Bekeri*, and restored them to the monastery's use.' Adam of Domerham records in his *History and Activities at Glastonbury* that the chapel at Beckery was 'repaired' (rebuilt) by Abbot John, presumably once it came under monastic control again. During the time that Beckery was no longer owned by the abbey the chapel of St Brigid still seems to have continued as a place of worship, and it is likely that the priests in residence continued to occupy their dwelling adjoining the chapel.

Today local tradition maintains that the area of Bride's mound was once known as the 'Women's Quarter' because a community of nuns lived at Beckery from the time of the visit of St Brigid, where they maintained a perpetual flame following the practice of her church at Kildare. This may be a memory of a flame, perhaps a special candle or oil lamp, tended at the chapel by nuns from the nearby nunnery of St Peter on Wearyall Hill as there is no evidence of a female community actually living at Beckery.

The custom of keeping a flame continually burning was not unusual. For example Henry de Blois, Abbot of Glastonbury 1126–71, made a grant to the abbey 'to pay for a candle to burn night and day before the image of the Blessed Virgin Mary in the Old Church at Glastonbury'. Alternatively, as St Brigid is often referred to as 'St Brigid of the Flame' because of the perpetual fire which was maintained at Kildare in her honour, this story has simply been attached to the tale of her visit to Beckery Chapel.

The disappearance of the chapel

When approaching Glastonbury from the west using one of the ancient roads, the chapel at Beckery would have been the first place that a traveller would have encountered, as was the case with those arriving by water up the River Brue from Bridgwater. It was said that pilgrims would spend the night in prayer there before going on to Glastonbury Abbey. In addition to the chapel, another reason to visit Beckery was the existence of a spring called St Brigid's Well. When Glastonbury Abbey was 'Dissolved' by order of Henry VIII in 1539 Beckery Chapel and the Priest's House seem to have been abandoned.

In the seventeenth century the ridge was called 'Bride's Hay', i.e. Brigid's Island, a name still used in the nineteenth century when it was written 'Brideshay', and the ruins of the chapel survived until the late eighteenth century. When John Collinson published his *History and Antiquities of the County of Somerset* in 1791 he said that: 'In the Isle of Beokery was likewise a chapel to the honour of St Briget, now entirely in ruins'. These had completely disappeared by the mid-nineteenth century, but local people still called the part of the ridge where the chapel once stood 'Bride's Hill', and surrounding fields 'The Bride's', so preserving the tradition of St Brigid's chapel and the saint's visit, a name by which the fields are still sometimes referred to even today.

The excavation of Saint Brigid's chapel

While the ruins were no longer visible by the mid-nineteenth century, the foundations showed as parch marks in the grass. The site was excavated by John Morland in 1887–88 who uncovered the remains of two chapels, one within the other, with walls built of the local Blue lias stone. Just north of the chapel, he found another structure which he interpreted as a priests house.

Morland found numerous fragments of fourteenth-century decorated floor tiles and there were also plain black and white tiles which were presumably used as a border to the decorated ones. Other finds included fragments of heavier plain floor tiles, thin stone probably from the roofing, ridge tiles, nails, a few pieces of slate probably from Devon, some lead, fragments of green marble and shards of glass. Two silver coins of Edward I (1272–1307) or Edward II (1307–27), a 'Nuremberg token' and six skeletons were also found.

The 1967–68 excavations

A more extensive excavation was carried out by Philip Rahtz in 1967–68, funded by the Chalice Well Trust. Prehistoric flint tools and Roman pottery indicated at least occasional occupation or visitation, but the earliest buildings discovered were in the form of post holes and timber slots, the most important being a chapel, approximately 6m (19ft 6in) long, orientated west–east. Inside this was a stone-lined grave for a person of exceptional importance, perhaps the original founder of the chapel, a male 25 to 35 years old at the time of death. Radiocarbon dating of the skeleton gave a date range of 730 AD ± 80 years (650–810 AD) indicating that the skeleton could have been interred as early as the seventh century or as late as the ninth century.

The wooden chapel could well pre-date the skeleton if the interment was of a later benefactor or other significant person, and so the wooden chapel just might have been there at the time that St Brigid is reputed to have visited the site. To the north west of this chapel was another wooden building with a further one north east. It is difficult to say exactly what this early chapel looked like as this level had been considerably disturbed by later foundations, ditches and burials, and was probably altered or rebuilt at least once. Wattle and daub were found during excavation, but seems to have come from another structure on the site that burnt down. It is probable that the chapel, being an important building, was constructed of wooden planks fixed on a framework as was the case in some other early Saxon buildings such as those excavated at West Stow in Suffolk (fifth to seventh century AD) and Charlton in Hampshire (seventh to eighth century). Planked Saxon buildings are depicted on the Bayeux Tapestry made 1066–77.

A small religious community was initially settled at Beckery and dating evidence indicates that this was founded in the early Saxon period, perhaps shortly after 670 AD when Beckery was given back to the abbey by King Cenwealh. The last British Abbot of Glastonbury, Bregored, was succeeded in 668 AD by Abbot Beorthwald, an Anglo-Saxon who introduced Roman practices in place of those of the Celtic church. He changed the Celtic tonsure that the monks wore to that approved by the Roman church, used the 'new' date of Easter and established the Benedictine Order at the abbey, which suggests the Beckery community had a planned and organised monastery, rather than the unsystematic layout typical of Celtic monasteries in Western Europe.

While the abbey and Beckery Chapel now came under 'Roman rule' they never ceased to be a centre for Celtic pilgrimage. Both played an important part in reconciling the Celtic and Roman church, particularly after Beckery Chapel was re-dedicated to St Brigid, one of the most venerated of the Celtic saints and second in importance in Ireland to St Patrick. This early chapel was surrounded by a cemetery containing the remains of sixty-three individuals. Most of the skeletons were fully extended with arms beside the body, but six were buried face downwards which, in pagan Celtic and Saxon society, usually indicates the individuals were criminals or evil doers, but is unusual in a Christian context. The reason for these at Beckery is difficult to explain.

Brigid: Goddess, Druidess, Saint

All but three burials were male, the exception being one adult female, either a domestic servant, someone who had specifically requested burial at Beckery or a traveller who had died there. The other two, both children, one aged six or seven, the other eight or nine, may have been noviciates (who were admitted very early in the first monasteries), children of the woman, or orphans related to a monk. However, this was clearly a male community and, if a perpetual flame was maintained by nuns as local tradition asserts, they would have come from the nunnery on Wearyall Hill, about 0.8km (half a mile) away, presumably taking it in turns to tend it each day following the tradition of St Brigid's church at Kildare. This would have been particularly important to them if they believed the chapel and their nunnery were founded by St Brigid or it was a daughter house of the Kildare monastery.

Around 1000 AD the cemetery went out of use and the wooden chapel was replaced by a building 11m (36ft) long and 6m (19ft) wide. Only the lower few courses of Blue lias stone remained, but it was constructed with stone walls rendered in white stucco, with ashler stone around the doors and windows. The roof was probably made of thatch, turf or shingles since no roofing material was found. The nave of this second chapel encloses the site of the wooden chapel, and while it has been suggested that it may actually have been built around it so enclosing the earlier structure, it is more likely that the wooden chapel was dismantled prior to the construction of the larger building. The chancel was probably built at the same time as the nave, although the excavators thought it just might have been a later addition. The stones

42 Reconstruction drawing by Judith Dobie of the Anglo-Saxon chapel at Beckery built *c.*1000 AD. Just to the north of the stone-built chapel was a wooden building to house a priest (ordained monk) or priests to minister to the needs of pilgrims.

that were visible were not obviously bonded into the nave wall. It is therefore possible that this building was a single-celled chapel to which a chancel was later added.

There were two doors in the stone chapel, one into the nave and one into the chancel, both on the north side of the building. On the south side there was a separate outer wall lying parallel to the chapel wall and 1m (3ft 3in) from it. This may have been roofed and had a staircase to an upper gallery, perhaps a chamber where the relics of St Brigid were kept. Pilgrims would have been admitted to pray in the presence of the relics which, William of Malmesbury in his work of *c.*1120 says, were 'efficacious in curing divers diseases.' When this second chapel was built it seems the monastic community on the site was discontinued and the chapel was probably attended by one or two priests (ordained monks) living in a timber building 8 x 6m (26 x 20ft) to the north of it.

About 1290, the chapel was totally rebuilt and enlarged for the third time in a typical medieval pattern, and was now 15m (50ft) long and 7m (22ft) wide with the foundations of the earlier chapels underlying the interior of this new structure. It was mainly of dressed Blue lias laid in mortared courses, with buttresses on the corners, Doulting stone facings and probably a roof of Cornish slate capped with ceramic ridge crests. A single entrance seems to have been located in the north wall of the nave. Internally it had a floor of decorated glazed tiles and white plaster on the walls.

John of Glastonbury, when describing this chapel in 1340–42, says that 'on its south side there is an opening through which, according to the belief of the common folk, anyone who

43 The foundations of Beckery chapel, Glastonbury, Somerset, from the north-west during excavation in 1967–68 by Philip Rahtz which revealed successive rebuildings. The earliest structure can be seen as a few surviving post holes and slots (with a grave), while the second chapel, a stone two-roomed structure, can be seen 'inside' the walls of the Norman period chapel. Originally dedicated to Mary Magdalene, the dedication of the chapel was later altered to St Brigid.

passes will receive forgiveness of all his sins.' In the medieval period diseases were often seen as a punishment by God for committing a sin, so if St Brigid's relics were able to be visited after passing through this 'opening' (presumably a narrow door into a upper story reliquary room or sub-chapel), the supplicant would have been able to pray in the presence of the relics and even, apparently, kiss them. They would have appealed for the blessing and help of St Brigid who was so strongly associated with healing.

To the north of the chapel the priest's house, also now rebuilt in stone, had a blue slate roof and was 11m (36ft) by 6.5m (21ft), with two main rooms and two smaller ones, a chimney stack and a *garde robe* (toilet). It was separated from the chapel by a substantial fence, and could have doubled as accommodation for a few better off pilgrims. By this period the chapel seems to have been visited by even larger numbers of pilgrims, which explains the rebuilding and improvements that were carried out to reflect its importance, and perhaps the need to provide accommodation for some of them. This would confirm the tradition of pilgrims spending the night in prayer at the chapel of St Brigid before going on to abbey. The chapel continued in use until the Dissolution of Glastonbury Abbey in 1539 when the chapel and priest's house were abandoned.

Selected Finds from the Excavations

Finds from the 1960s excavation suggest that the wooden chapel and the later stone buildings contained elaborate fittings and objects. See Rahtz and Hirst *Beckery Chapel Glastonbury, 1967 – 68*. Among the bronze finds were a hinge with traces of gilding from a book cover, a piece of gilded edging either from the spine of a book or a book marker, the edging from a box and a small bell shaped object, probably from the cover of a book. However, they are all difficult to date accurately. Closer dating can be obtained for a stud (15mm in diameter) with 'marigold decoration' surrounded by a circle of dots, that is typical of the early Christian period, so would have come from the wooden chapel. A strap end or part of a book marker (42mm long) has an incised animal that looks like an otter on a spotted background, not unlike a design on the Kells crozier in the British Museum, and dates to the late Saxon period, so may have been used on an item from either the wooden chapel or the second chapel.

Some of the bronze finds may be of particular relevance because of their possible St Brigid connection. These include fragments of what appears to be a grill or relic holder and a sheath-like object (49mm long) which bears an incised fish with spots on its side resembling a salmon, significant because of the association of this fish with sacred wells dedicated to St Brigid. Another item of interest is a fastening for a book cover which consists of a round disc (18mm in diameter) with a hook (22mm long) coming from it. On the disc are incised two crosses, one of which consists of four arms radiating outward from a circle, while the other, overlying this, has its four arms indicated by a double line that join each other at their inner end. The four arms do not meet each other at the centre and appear to be offset in relation to each other so it resembles a Brigid's cross (*Crois Bhríde* in Irish Gaelic), a design uniquely associated with this saint. If this is a Brigid's cross it is the earliest representation of the symbol discovered, and it is tempting to speculate that it came from a book, possibly brought from Ireland, dealing with the life and miracles of St Brigid, kept in the chapel.

Did Saint Brigid visit Somerset?

The local tradition that St Brigid herself visited Beckery is as strong today as it was when it was first noted by William of Malmesbury over 880 years ago. So is it possible to discover the

Saint Brigid visits Glastonbury!

44 A fastening for a book cover (18mm in diameter) with a hook (22mm long) coming from it. The incised design seems to resemble a Brigid's cross. This was found during the excavation of Beckery Chapel in 1967–68.

origin of this story? Brigid was a real person, a member of the Fothairt family, who founded a Christian religious establishment at Kildare in Ireland around 470 AD and whose generally accepted date of death was around 525 AD. However, the written stories told about her visit and stay in Somerset only date to *c*.1120 in the earliest surviving source by William of Malmesbury, and a date for her visit, 488 AD, is first recorded by William around 1129–35, but he does not state on what authority he arrived at this except 'they say…'.

This suggests he got it from a local tradition at Glastonbury Abbey, which at that point had kept the relics 'left by the saint' for over 120 years in Beckery Chapel. Later historians such as John of Glastonbury writing 1340–42 accepted this date as fact. There is no supporting evidence for St Brigid's visit to Somerset in early Irish sources, and the first Irish writers to mention her visit to Glastonbury were Archbishop James Ussher (1581–1656) in his *Chronological Index* and John Colgan in his *Trias Thaumaturga* (*The Miracle Works of Saints*), 1647.

Brigid, as Abbess of the important double monastery at Kildare, would have travelled widely in Ireland, but it seems very unlikely that she actually visited Somerset. What is certain is that about 400 years after St Brigid's death there was extensive contact between Somerset and Ireland, with many Irish monks both visiting Glastonbury and being based there. To answer the question as to whether St Brigid actually visited Glastonbury we need to look at when and why the Beckery Chapel was rededicated to her.

John of Glastonbury says that the original dedication of the Beckery chapel was to Mary Magdalen, but had been changed by his time (1340) to that of St Brigid in honour of her visit. However, it is difficult to be absolutely sure when this rededication occurred and it could just possibly have applied to the original wooden chapel on the site. In a charter of 971 AD the 'island' is described as '…*Bekeria*, which is called little Ireland…' which sounds almost like a nickname for the place, perhaps suggesting that most or all the monks there were Irish. Irish monks had certainly been at Glastonbury Abbey for at least fifty years by that time. It also raises the possibility that the chapel was, at that date, already dedicated to this important Irish saint and being visited by many Irish pilgrims, truly making Beckery a 'little Ireland'!

However, the most likely time for the rededication of the Beckery chapel was when the wooden one was replaced by the larger stone building around 1000 AD. The replacement of the first wooden chapel by a more impressive stone structure could indicate the increasing importance of the site due to the story attached to it that the saint had spent time there. However, John of Glastonbury says that when St Brigid left Beckery for Ireland, she left behind certain articles which were '…honoured there because of her holy memory', and it is quite likely that the stone chapel was built to house these relics.

This would certainly have been an appropriate time to re-dedicate the chapel to the saint whose relics were displayed there. These items remained as objects of veneration in Beckery Chapel until Glastonbury Abbey was 'Dissolved' in 1539, when they disappeared. The relics consisted of a bag, a collar (or necklace according to William of Malmesbury), a hand bell and weaving implements. Not all the St Brigid relics were kept at Beckery Chapel, as among the many relics in Glastonbury Abbey noted by John of Glastonbury in 1340 was a 'rib of the virgin St Brigid'.

The small monastery at Beckery probably consisted mostly of Irish monks and they would have been keen to see a chapel to an Irish saint there, particularly as Beckery was the first place that Irish and other pilgrims encountered when approaching Glastonbury from the west. It would therefore have been the most appropriate place for the abbey to build a chapel to St Brigid, turning the monastic site into what was primarily a place of pilgrimage to venerate St Brigid's relics, with only a few priests (ordained monks) at the site to attend the chapel and minister to the needs of the pilgrims.

There is no doubt that this second chapel at Beckery did house relics of St Brigid which must have come from Ireland. In view of the importance placed on saints relics in the early medieval period, and particularly of items that were reputed to have belonged to this second most important Irish saint, they would have been brought to St Brigid's Chapel by a senior member of the church. The most obvious person to have brought the relics to Beckery from Ireland was the Abbess of St Brigid's monastery at Kildare.

The Abbess of Kildare from 979–1016, the period when the second chapel at Beckery was built, was Eithne ingen Uí Suairt, whose official title was *Comarba Brigte* (successor to Brigid). She was also a member of the Ui Chulduib branch of the Fothairt, the family to which St Brigid herself belonged. Following the visit and gift of St Brigid's relics to the chapel by Abbess Eithne, the account of this event became distorted over time, so that the 'successor of Brigid' and an actual descendent of St Brigid, who was also a Brigidine nun, became confused with the saint herself. So by the time that William of Malmesbury was writing around 1120 the monks and local people had convinced themselves that St Brigid had actually visited Beckery.

The discovery of a relic from Brigid's chapel

In 1913 one of the relics from Beckery Chapel, Brigid's Bell, was rediscovered when an elderly Glastonbury man of 'slender means', Robert Kitch, lost an old friend who lived at a small farmhouse on the Somerset Levels, not far from Glastonbury. The deceased's effects were

Saint Brigid visits Glastonbury!

45 Sister Mary Minehan, a Brigadine nun, holds the perpetual flame brought from their house at Kildare in Ireland to Bride's Mound, Beckery, where a special garden was built to honour the Goddess Brighid/St Brigid in 1994. The circular, arched entrance represents the opening in the south side of Beckery Chapel described by John of Glastonbury in the fourteenth century (see page 173).

46 *Imbolc* ceremony (1 February) at Bride's Mound, Beckery. Offerings of water, fruit and cake are being made to the Goddess Brighid.

sold at auction and on the day of the sale Kitch attended to buy a memento. Unfortunately everything was beyond his means except for an old oak box, which he bought, unopened, for five shillings (25p).

On returning home he opened it and found some sewing tools and a roll of worn linen enclosing something. On unwrapping the linen he found it was protecting an ancient hand bell, in perfect condition but minus its clapper. The bell was obviously made for a woman as Robert could not get his fingers through the loop handle at the top. This find was reported in the local paper at the time and the bell was later examined by specialists at the British Museum and the Dublin Museum. They came to the same conclusion: 'that it is a most ancient Celtic bell, and there is no reason why it should not be St Bride's'. This view was confirmed after it was later examined by Dr R. A. S. Macalister, Professor of Celtic Archaeology at Dublin and President of the Royal Society of Antiquaries of Ireland.

By the early 1920s the bell was in the possession of Amy Hiscock of Chilkwell Street, Glastonbury who inherited it from her Great Uncle, Robert Kitch. Unfortunately the sewing tools and linen wrapping were thrown away, but it is interesting to speculate whether these may have been some of the other relics once kept at Beckery Chapel. However, it is more likely that someone knew that Brigid's Bell was associated with weaving implements and so added more recent sewing tools to 'complete' the relic. If this was the case, it would confirm that this was the relic bell from Beckery.

Amy Hiscock lent the bell to Alice Buckton, playwright, writer and the owner of the Chalice Well Gardens and adjoining buildings where she opened the Chalice Well Training College for Women and the Chalice Well Hostel. Alice used the bell in ceremonies and events at the well and it was regularly shown to visitors staying at the hostel. The bell, being held by Alice Buckton standing by the Chalice Well, appears in a short ten-minute film which was made to promote her longer film, 'The Glastonbury Pageant', made in 1922 by the Steadfast Film Company (now in the national film archive of the British Film Institute). In this she is shown striking the bell with a small hammer, but at some time after this the clapper was replaced. This repair is mentioned by Lionel Lewis, Vicar of St John Baptist Church, Glastonbury in 1925 who described the bell as giving out a sound 'most true and musical.'

In the last few minutes of the film there is a parade through Glastonbury with figures representing various personages, among whom is a nun carrying a crozier. She was a real nun of the Order of St John, two of whom were based in Glastonbury at that time. The nun carrying the crozier was, in real life, an Abbess of that Order. It is unusual for an Abbess to carry a crozier, except in the case of St Brigid who was also consecrated as a bishop according to tradition (see page 83), so she is almost certainly played the part of St Brigid in the film.

It was planned to present a live performance of the Pageant in 1923 based on the film of the previous year, but there are no indications from the local newspaper that this was ever performed. St Brigid, holding this bell, is shown on a stained-glass window made by Danielli in 1924 in the east window of St George's chapel in St John Baptist Church, Glastonbury. In 1944 Alice Buckton, in failing health, moved to Wells to stay with some friends, and took the bell with her. Its owner, Amy Hiscock requested its return several times, but even personal calls from members of Amy's family failed to secure the bell or even any firm information regarding its exact whereabouts. When Alice died in December that year all traces of the bell disappeared.

Another local story to account for the bell's disappearance says that Alice Buckton, fearing an invasion by the Germans, either hid the bell by burying it somewhere in the vicinity of the Chalice Well, or gave it to someone for safekeeping, but died before she disclosed who then had it. Every effort was made by Alice's executors to trace it and the Revd Lionel Lewis offered a large reward for its return, but to no avail. It has now been lost again, although locals hope that one day it will again be rediscovered.

Bride's Well at Beckery

Memories of St Brigid's connection with the site persisted even after the ruins of the chapel disappeared. Pilgrims still visited St Bride's Well at Beckery, although more recently it did not present a very impressive appearance. It was described at the beginning of the twentieth century as resembling a muddy pond into which water from nearby fields used to drain through a sluice. A hawthorn tree grew next to it on which visitors tied pieces of cloth, prayers on scraps of paper and other offerings, a practice found at holy healing wells all over Britain and Ireland, suggesting that this was still regarded as a healing well.

When the then owner of the land, James Mapstone, investigated the well, he found deposited in it many small items, including a twopenny coin of George III, two ornaments of Mother of Pearl, two glass bracelets and a solitaire jewel. In the 1920s the neglected St Brigid's Well was still fairly obvious, being described as a shallow basin surrounded by bramble bushes and marked by an inscribed stone. The exact original site of the well is now unclear. An inscribed stone, located close to the River Brue and a few hundred metres from the mound, marked the site of the original well at one time as it bears the inscription:

> Saint Bride
> ✝
> This stone marks the
> traditional spot of
> Saint Bride's Well.

However, this was not the stone that marked the well in the 1920s, as this, the St Bride's Memorial Stone as it was called at the time, was dedicated 'to the Glory of God' on 12 September 1952 by the Revd J.A. Batcup, Vicar of St Benedict's Church, Glastonbury. During the dedication ceremony, reference was made to St Brigid's work with women, 'whom she helped live dedicated lives', and part of the ancient *Hymn of St Patrick* was sung.

This stone was originally situated in the middle of the field in line with the trackway known as Hulk Moor Drove but in 1957 the bank of the River Brue was widened, the old sluice covered in, the 'bush' (presumably the hawthorn tree) removed, and the memorial Stone moved to its present position. Today this is known locally as Bride's Stone, and flowers and other small tokens and offerings can be found placed at its base.

Alice Buckton, owner of the Chalice Well Gardens, wrote to a friend around 1920 to explain the route of a pilgrimage she had devised around Glastonbury and the prayers and rituals to be carried out at various points. This included St Brigid's Well and gives an insight into the rituals carried out there, at least by some people, at that time and their romantic views of ancient Glastonbury. From the instructions it is clear that she thought the pilgrimage was ideally done by a man and woman together, although the directions address the pilgrim as 'he'. From Cradle Bridge, which carries Porchester Drove Road across the River Brue north of Beckery, the pilgrim walked south along the bank of the river until:

47 The Bride's Stone (originally called the St Bride's Memorial Stone), which marked the site of St Bride's Well at Beckery in 1952 and was moved to its present location in 1957.

...he reaches a Willow Tree standing alone. This is the Tree of Sorrow. Baring his head, he prays for all pilgrims who have been unable to follow the path and have failed to carry out the conditions laid down by their predecessors. Here he leaves all sorrow and with a heart full of joy approaches the Thorn Tree.

All prayers and ritual at 'Brides Well' must be undertaken by the woman if present. This 'Thorn Tree' was consecrated in the name of Joy and in the service of Bride, at the destruction of the thorn which sprang from the staff of Joseph of Arimathea upon Weary-all-hill by a puritan. After that the Watchers no longer met there as of yore so a common thorn was chosen by them, in an isolated spot, where seekers could come to learn the truths taught there and to carry out their instructions. The pilgrim should encircle the tree clockwise 3 times, stopping N, S, E and W to offer up a prayer.

At the first turn he prays for the Watchers, that workers may be found to carry out their work in the world. At the second turn he prays for the World, that it may be prepared to receive new illumination from on High. At the third turn he prays for any personal wish or ideal. Then, at the mouth of the well, he throws into the water a piece of glass, or some fish symbol, as a sign of the Astral Plane through which he has been passing. He should leave at this spot also a symbol of his work as an offering to the Bride and dedicate his life to the work of the Holy Spirit.

Then he seeks inspiration upon the stone in front of the thorn. He can write any great desire upon a piece of paper and cast it upon the waters of the Brue Stream, praying that it may meet its fulfilment. Upon the thorn he ties a piece of ribbon (this can also be done at the Cradle Bridge and at the Willow Tree) as a sign to pilgrims that he has passed that way and left his prayers.

Here in the far ages the Goddess of Wisdom hid her treasures and knowledge which can only be found by true seekers and pilgrims. Here the teachings of the Bride were given out and taught by a Druid centre and then a Christian church. It was to a nunnery on Beckery that King Arthur went to be taught. There the vision of the Virgin Mary appeared to him which was afterward symbolically represented on his shield. When dying it was to the mystic isle of Avalon he passed.

Saint Brigid visits Glastonbury!

From earliest times the spot has been dedicated to the Bride and her teachings, to whom the pilgrim offers his life and work.

This part has also been named 'the Salmon's Back', due to the belief that under the ground are remains of the greatest fish idol in the world. It was believed that in the mouth of the salmon, the 'Salmon of knowledge', treasures were to be found. Passing over the hill the pilgrim reaches an orchard, through which he passes. He picks an apple and partakes of it, giving it first to the woman if she is present, thus symbolising the partaking of the apple in the Garden of Eden. The pilgrim then joins the road and proceeds to Northover Bridge.

The pilgrimage then took the participants via various points in Glastonbury, to the top of the Tor and then down its eastern slope where they joined the road. Taking a footpath they entered Paradise Lane, but at the crossroads they separated:

The women take the turn to the right which leads into Maidencroft Lane. A little along this lane, to the left, there is a path across a field. There the woman pilgrim sought the divine wisdom of the Mother and the Bride, praying for guidance for her work in the world. Joining her fellow pilgrims where the path meets the road they pass together down Wick Hollow, stopping at St John's Church for a few moments to pray. Before reaching the station there lies, to the right of the road, a field called Paradise. The pilgrim enters here with a heart full of joy and thanksgiving that he was permitted to complete his pilgrimage and prays earnestly that the act of service and dedication may be accepted. [This marked the end of the pilgrimage].

Saint Bride's Well and the glass bowl

in 1885 Dr John Goodchild, a doctor with a fascination for antiquarian and mystical studies bought, for £3, a shallow glass bowl with a silver leaf pattern and blue, green and amber floral designs at Bordighera, Italy. After bringing it back to England, it was examined by the British Museum who suggested it was 'ancient', but could not identify where it was made. Goodchild was interested in both Christian and Eastern philosophy, and pre-Christian Celtic beliefs. His studies led him to believe in a supreme female deity, and that her cult had become attached to the figure of Brigid.

In 1897 Dr Goodchild had a psychic experience in which a voice indicated that this vessel had been carried by Jesus, but was not the grail. It instructed him to take the Cup, as he now called it, to Bride's Hill at Glastonbury where it would end up in the care of a woman. In September 1898, following another psychic instruction he placed the Cup beneath a stone in the murky waters of the well. Its fate was now out of his hands, but every year from 1899 to 1906 Goodchild visited the well. In September 1906 two sisters, Janet and Christine Allen from Bristol went to see Goodchild and told him they has discovered a 'primitive looking' blue glass bowl in the water, but sensing its sanctity had replaced it. Goodchild was overjoyed as at last it had been discovered by a woman, or women in this case.

A small but enthusiastic group of young women was formed known as the Guardians of the Cup, and on 1 October the Cup was recovered and taken to 16 Royal York Crescent, Bristol. There the group created a small chapel in an upper room, later known as the Bristol Oratory. The public were invited to meditate and receive healing, and there were several reports of cures. A service was designed with emphasis on the female aspect of Celtic Christianity represented by the Cup. The Guardians presided over services, baptisms and marriages, with the Cup being used to hold consecrated wine. St Brigid played a major part in their services as is clear from an extract from one of their service books:

> O Thou Most Holy Virgin Bride, we greet thee.
> Thou who comest robed in the Greater Glory of
>   the Holy Spirit.
> Bride of Supreme Wisdom, Beauty and Truth.
> Come on Wings of Deliverance, bearing the shield of a Dove.
> Whisper to the saddened heart of My Humanity that Redemption draweth nigh.
>
> Come, O Bride, My Servant to Wait.
> Come, O Bride, My Servant to Wait.
> Come, O Bride, My Servant to Wait.
> Come in the Light of the Shining Moon,
> Come in the Dawning of the Day,
> Hear our cry, O Bride.
> Most Glorious Virgin of Supreme Loveliness
>   be gracious unto thy servants.

By 1909 the Bristol Oratory was so popular that the Guardians had to introduce an appointment system. Over time the Guardians became less involved as they married or were occupied by other activities, and in 1913 one of the Guardians took the Cup to Letchworth, Hertfordshire when she moved there. Later it was kept by Wellesley Tudor Pole, who eventually placed it in the custody of the Chalice Well Trust. For the full story of the Cup see *The Avalonians* by Patrick Benham.

A pageant play: *The Coming of Bride*

In 1914 a pageant, *The Coming of Bride*, was written by Alice Beckon and students at the Chalice Well Training College for Women. Probably inspired by the discovery of Brigid's bell (see pages 176–79), it was performed on 6 August that year. The incidents featured in the play were drawn from some of the ancient *Lives* of St Brigid and the miracles associated with her, along with beliefs from both Ireland and Scotland. In addition it included traditional tales about St Brigid and her supposed visit to Glastonbury taken from the writings of William of Malmesbury and John of Glastonbury. This was performed in the Glastonbury Assembly Rooms, at the Chalice Well and at Crispin Hall in Street in the Summer of 1914.

The Brigid landscape swan

Kathy Jones in *The Goddess in Glastonbury*, has suggested that the outline of a swan in flight can be discerned in the shape of the 15.2m contour of the peninsula on which Glastonbury is located. This she feels is significant since many people connect swans with St Brigid who has a long connection with Glastonbury. However, the tale of St Brigid and the swans applies to St Brigid of *Cill Muine* (also known as Brignat), who was a contemporary of St Brigid of Kildare, a story that is related in *The Life of Darerca* (see page 153).

The Somerset churches of Saint Brigid

While Beckery Chapel at Glastonbury and the church dedicated to St Bride at Bridgwater have now disappeared, two others remain as standing structures today. The possession of four churches dedicated to this saint is unusual among the English counties and is yet another indicator of the importance of St Brigid in Somerset. A field at Cannington, not far from Bridgwater, is called Bride's Field and although the origin of the name is uncertain, it may at one time have belonged to the church of St Bride at Bridgwater. The church rented out this type of land, given by benefactors, and used the income for charitable purposes.

Saint Bridget's Church at Brean

While the spelling used at Brean Church looks 'foreign', being the same spelling as St Bridget of Sweden, this was the way it was recorded in the tenth century. The church is located on the coast of Bridgwater Bay, north of the mouth of the River Parrett that goes up to Bridgwater, so is close to the Irish pilgrims' route to Glastonbury. The church is said to have been founded by Irish monks in the sixth century, although the present building dates from the thirteenth century. A charter dated 693 AD records that Ine, King of the West Saxons, granted an estate called Brent to Glastonbury Abbey. This seems to have included Brean within its bounds, although there is no later reference to the abbey's possession of Brean manor.

A stained-glass window at the west end showing St Brigid holding a sheaf of corn, was installed in 1984 in memory of church warden, Dennis Arthur Willis, 1919–84. Outside the church, water flows in a channel round two sides of the building from a spring whose water has been found to originate in the Brecon Beacons of Wales, flowing right under the Bristol Channel and emerging at this point. This is probably the reason why the church was built at this spot as many springs and wells in Ireland were dedicated to St Brigid. It is quite likely that the Irish monks who established Brean Church located it by a spring which they also dedicated to the saint, which would have endowed it with healing properties, although no such tradition is remembered there today.

In the light of the connection of St Brigid with cows and curing their illnesses, a curious discovery was made in a church account book dated 1771 in which was found 'infallible receipts', one of which was for curing sick cows: 'Take an ounce of dragon's blood [the juice of the Snapdragon plant, *Antirrhinum majus*], an ounce of Bol Avmeniack [Sal Ammoniac – Ammonium chloride], half an ounce of Irish slait and one handful of nettles boiled in milk.'

Saint Bridget's Church at Chelvey

The church at Chelvey, called Calviche in the *Domesday Survey*, 1086, also uses the 'foreign' spelling of the saint's name, but it too has St Brigid of Kildare as its patron saint, although its difficult to estimate when the church was founded.

The Temple of Brighid

In 1999 a group of people who venerated the Goddess began the quest to create a permanent temple in Glastonbury. They began raising funds and drawing awareness to the project by building a series of temporary Goddess temples. In 2001 a suitable building was located and the following year saw the opening of the Goddess Temple at 2–4 High Street, Glastonbury,

Brigid: Goddess, Druidess, Saint

48 St Bridget's Church, Brean, Somerset. Founded in the sixth century by Irish monks near a spring, the present building dates to the thirteenth century.

49 Stained-glass window showing St Brigid of Kildare holding a sheaf of corn, St Bridget's Church, Brean, Somerset.

the first public Goddess Temple in Europe for over 1000 years. A further step forward occurred in 2003 when the temple was formally registered as a Place of Worship.

The temple, open to all, is dedicated to the Goddess Brighid, but also honours those goddesses particularly associated with Avalon (Glastonbury). Ceremonies are held to honour the Goddess and it is a place to celebrate and explore the many aspects of the divine feminine, and to allow people to worship and explore their relationship to the Goddess. Ceremonies for individuals such as hand-fasting, rite of passage and healing also take place, along with teaching sessions. The eight festivals of the year are celebrated and the temple is decorated appropriately for each.

Saint Brigid images in Somerset

Glastonbury Abbey
Further evidence that the story of St Brigid's visit to Glastonbury was known locally in medieval times can be found in a carving, now very weathered, of the saint milking her cow on the Norman doorway, c.1220, on the south side of the Lady Chapel at Glastonbury Abbey.

The tower of Saint Michael's Church, Glastonbury Tor
Another carving of St Brigid, in much better condition than that at the Abbey, can be seen on the west side of St Michael's Tower on Glastonbury Tor. This is the only remaining part of a church built after an earthquake had destroyed an earlier church on the hill on 11 September 1275. The tower was built by Abbot Adam of Sodbury (1323–34), and in the carving St Brigid is shown sitting on a stool milking a cow, with what appears to be a long flowing cloak beneath her.

Saint John Baptist Church, Glastonbury
A stained-glass window in the chapel of St George made by Danielli in 1924 shows St Brigid holding the bell that was kept at Beckery Chapel until 1539 and rediscovered in 1913.

Saint Mary's Catholic Church, Glastonbury
St Brigid, with a cow and holding a milking stool, is shown on a tapestry hanging behind the altar in St Mary's Church, Glastonbury. This was designed by Brother Louis Barlow of the Benedictine Abbey at Prinknash, Gloucestershire. It took five weavers, using an upright loom, a total of 240 working hours to complete, between February and June 1965.

Wells Cathedral
While the cathedral at Wells does not have a chapel dedicated to St Brigid, she appears on a stained-glass window to be found in the Corpus Christi Chapel which shows St Brigid with a bundle of rushes in her right arm and holding up a lighted oil lamp with her left. Above her head is a Celtic cross flanked by two angels. This window was installed as a memorial to Douglas Hamilton McLean in 1901.

In the Chapter House, started c.1250 and completed in 1306, one of the stone bosses has a carving known as 'the Green Lady of Wells'. This is an unusual way of depicting a Green Woman, the female equivalent of the more common Green Man, as she is shown as a woman with long flowing hair, wearing a round necked dress with long sleeves and holding a bunch

Brigid: Goddess, Druidess, Saint

50 A very weathered carving of St Brigid milking a cow, c.1220, above the doorway of the Lady Chapel, Glastonbury Abbey, Somerset.

51 Carving of St Brigid milking a cow on St Michael's Tower on top of Glastonbury Tor, Somerset, built by Abbot Adam, 1323–34.

Saint Brigid visits Glastonbury!

52 St Brigid holding a milking stall, on a tapestry made in 1965 and designed by Brother Louis Barlow of the Benedictine Abbey at Prinknash, Gloucestershire. This hangs behind the altar in St Mary's Church, Glastonbury, Somerset.

53 The Green Lady of Wells. A plaque, made by Black Dog of Wells Pottery, based on a carved roof boss in the Chapter House of Wells Cathedral which was built c.1250–1306. This carving probably depicts St Brigid who was highly venerated locally.

of plants. These have heads that look like reed mace (the false bulrush, *Typha latifolia*) rather than flowers and she is surrounded by hawthorn leaves. The hawthorn tree is also known as the May tree as it flowers in that month and, bearing in mind that in Somerset a custom celebrating St Brigid was transferred to the Eve of May Day (see pages 159–161), it could be that St Brigid was associated locally with this tree.

The masons working at Wells would also have been employed at Glastonbury Abbey from time to time, so would have been very aware of the veneration of St Brigid and the local stories told about her there. It is obvious from many examples all over the country that the sculptors of figures in churches had a sense of humour. So it may have appealed to the carver of this particular boss to depict St Brigid, a female who had been ordained as a bishop according to tradition, overlooking the deliberations of the senior cathedral officials in the Chapter House! This particular boss can be seen by sitting in the canon's stall labelled 'Comba XI' and looking straight up.

The Taunton Bridewell

The former prison at Taunton, built in 1754, was known as the Bridewell until it was enlarged in 1843 when its name was changed to The County Gaol. It was named the Bridewell after one of the most famous English prisons, the Bridewell in Fleet Street, London. Originally built by Edward IV, it remained a Royal residence until 1553 when it was converted into a combination of prison, hospital and school. Later it become a prison only, and was one of the 'sights' of London where people would go to see the public flogging of prostitutes. Somerset was not unique in having a jail called the Bridewell as many prisons in eighteenth- and nineteenth-century England and Ireland were known by this name.

Eleven

Saint Brigid's Holy wells

In the past water was linked with life and some form of water veneration seems to go back as far as the Neolithic period, 5000–1500 BC. This applied to rivers, streams, lakes, bogs, springs and wells, and many items were deposited in water as an offering to the gods, a practice that has never ceased to judge by the coins to be seen in 'water features' in shopping centres today! The Celts deposited many valuable items such a weapons, shields, chariot fittings, cauldrons, iron bars and jewellery in water, which they believed was inhabited by its own deity. Water was also seen as a way of connecting to and communicating with the Otherworld, their 'heaven' and the dwelling place of their deities. While we do not know whether formal rituals were carried out when making such offerings, human nature suggests that such depositions to the deities would be accompanied by certain actions and invocations.

The veneration of water was so strong that even with the coming of Christianity to Britain and later Ireland it proved impossible to suppress, despite some initial attempts to do so. A Cannon issued by the Second Council of Arles in 452 AD stated:

> If in the territory of a bishop, infidels light torches or venerate trees, springs or stones, and he neglects to abolish this usage, he must know he is guilty of sacrilege.

In practice it proved impossible to ban the veneration of water and, in the case of Ireland, also that of trees and stones. The church absorbed the wells into Christian practices by dedicating them to a saint in place of the pagan deity with which most, if not all, were associated. They allowed or adapted 'water rituals' but now of a Christian nature, and wells were now used for baptisms and formal washing by priests. In some places pilgrimages to wells still continues while in Derbyshire and a few other places in England Well Dressing, the elaborate decorating of the well, still continues, a practice that probably pre-dates Christianity.

There are a great many wells and springs that have been or are regarded as holy, and an estimate of the numbers suggests that there are around 3000 in Ireland, 1000 on the Scottish mainland (not counting those on the islands), 1200 in Wales and at least 2000 in England, making a total in excess of 7,000. Many of these are believed to be able to cure particular ailments and have stories and traditions associated with them, so only a few examples with

Brigid: Goddess, Druidess, Saint

a St Brigid connection will be included to give an idea of the general beliefs about such water sources.

One of the Goddess Brighid's attributes was healing and so many curing wells in Ireland are dedicated to Brigid, although this is as a saint rather than a goddess. It is impossible to say how many, if any, of these wells were connected to the Goddess Brighid between her conception in the first century AD and the time from the sixth century when her characteristics and stories began to 'blend' with stories about the last Chief Druidess of her shrine at Kildare, who was to become the first Abbess of the Christian monastery that was established there.

As an Irish saint there are more wells dedicated to St Brigid in Ireland than in Britain and, due to the nature of Irish religion and conservatism in their customs, they have more pilgrimages to St Brigid's wells. However, wells dedicated to St Brigid are found elsewhere and also visited by pilgrims and those seeking help. In Scotland, because of the local tradition that St Brigid was the midwife to Mary (see page 90), women having problems conceiving visited wells dedicated either to St Brigid or St Mary, while a few wells dedicated to St Brigid are also found in Wales, England and the Isle of Man.

The Brigid Wells of Ireland

There is little doubt that the majority of Irish holy wells were once dedicated to a pagan deity or were seen as a way of communicating with the Otherworld. Christians changed such veneration by dedicating wells to various saints and using them for baptisms and the ceremonial washing of church vessels. The pre-Christian tradition of leaving votive offerings and using the wells for curing illnesses persisted and still does today. Rituals performed at many wells of a Christian nature probably replaced pagan rituals once carried out when asking the help of a goddess, as it was generally female deities that were connected with water. An example of such a 'conversion' can be clearly seen at a famous healing well in Co Donegal, where the *Tobar Bhríde* (Brigid's Well), was formerly known as *Tobar Aibheog* (Aibheog's Well), Aibheog being the name of a Celtic goddess.

While wells all over Ireland are dedicated to a variety of saints, some very local, others are of national importance (Patrick and Columcille) and some to universal saints (Mary, John etc.). A great many are dedicated to St Brigid and a few examples will give an indication of the veneration that wells were and are held in Ireland, along with the traditions and beliefs connected with them. There are many stories which tell how a well or spring appeared due to the action of a saint such as striking their staff against the ground and St Brigid is no exception. An early story says that one day the saint and Bishop Erc sat down at a church with their attendants who were very tired after their journeying.

A youth in their company said that whoever gave them food would confer great charity upon them, to which St Brigid replied that food and drink would be provided but he must be patient. Before long, alms, in the form of food, arrived to refresh the weary travellers. They gave thanks to God, but had only received edibles and still wanted for drink. So to allay their thirst St Brigid told some of the attendants to dig the earth at a certain spot. On obeying her order, a spring of clear water issued from the ground, and ever afterwards was known as St Brigid's Well.

St Brigid is also credited with causing a well to appear where she dropped some burning coals, in a tale that is also told about another saint (see page 84). According to some stories

St Brigid also made a well appear when she needed to replace her eye that she plucked out on one occasion. Wells were often regarded as good for curing a specific ailment so it might be necessary to travel some distance to visit the 'right well' to cure a particular illness. Between them wells were able to help with almost all medical problems: backache, bowel illnesses, eye ailments, fertility difficulties, gynaecological problems, headaches, mental illness, rheumatism, rickets, scurvy, skin diseases, sprains, stomach illnesses, toothache, wounds, general non-specific illness, and giving protection against getting diseases in humans; while some wells were regarded as specific for curing and protecting animals.

Wells (*tobar*) range from elaborate examples covered by Well Houses to a pond surrounded by a wall or fence. Wells are usually supplied by springs and springs themselves were also regarded as sacred in pre-Christian times, but were still venerated after the arrival of the new religion. The term 'well' also covers springs, deep areas in some rivers and streams, and even standing water in hollowed out rocks, either formed naturally or artificially. The Irish even regarded water found in tree stumps, if associated with a saint, as suitable for curing various ailments.

The pilgrimages and rituals at the many wells dedicated to St Brigid throughout Ireland on *Fhéile Bhríde* (St Brigid's Day) were regarded as the most significant of such events. These were attended by large numbers of people despite occurring on 1 February, a cold time of year. This day corresponds to the first day of spring in Irish tradition. It is also the pagan festival of *imbolc*, which also marked the beginning of this season, although a few wells dedicated to St Brigid were ritually visited on other dates, often in August.

St Brigid is regarded as the most important Irish saint after St Patrick, and a large number of churches are named after her. Parishes have taken her as their patron saint, so there were lots of pilgrimages held to honour her and invoke her help and blessing. One of the most famous was that which is still held at Foughart near Dundalk, Co Louth, where, so tradition claims, St Brigid was born, her father lived and where she plucked out her eye to discourage a suitor. Today a National Pilgrimage, started in 1934, is still held on the second Sunday in July and attended by large numbers of people. An open-air Mass is celebrated in her honour at a shrine there and people with eye ailments tie rags to a hedge or bathe their eyes in small, naturally formed pools in rocks that line a stream, and which are associated with legends about the saint.

While there are many wells and springs dedicated to St Brigid all over Ireland, two of the most famous, which are still very much visited today, are at Kildare, Co Kildare where St Brigid established a monastery and a cathedral is dedicated to her. One, known as St Brigid's Wayside Well has an arc shaped stone bearing the inscription: '*A Naomh Bhríde, Muire na nGail, guí orainn.*' ('St Brigid, Mary of the Gael, pray for us.'). The other, with its prayer stones, was laid out in its present form in the 1950s. A number of stones mark the places where prayers may be said, each symbolising an aspect of the saint. The first stone represents Brigid – A Woman of the Land, the second: Brigid – The Peacemaker, the third: Brigid – The Friend of the Poor, the fourth: Brigid – The Hearthwoman and the fifth: Brigid – Woman of Contemplation.

The water from the well flows through an arch bearing a carved Brigid's cross and through two stones known as St Brigid's shoes. Local tradition says that she used to graze her cows here and make butter by the stream. Stepping stones cross the stream to the shrine which originally had a small statue of St Brigid, which was replaced by a larger glass fronted statue of the saint dressed in a black habit with a white wimple that was donated by the Brigidine Community at Goresbridge, Co Kilkenny, where it had stood for many years within the Convent School. More recently this was replaced by a bronze statue of St Brigid that shows her more as she would have been dressed in the fifth century. Other notable wells include St Brigid's Well at Dunleer, another in the Parish of Marlerstown, both in Co Louth and

54 St Brigid's Well, Kildare, Ireland, showing the two stones known locally as St Brigid's shoes (*bróga coilteach Bhríde*). On the reverse side of the arch is a carved Brigid's cross.

55 Bronze statue of St Brigid holding a flame, St Brigid's Well, Kildare, Ireland.

St Brigid's Well at Cliffony, Co Sligo. The St Brigid's Well at Castlemagner, Co Cork was so famous for its healing powers that huge crowds used to turn up on 1 February, as was the case with another St Brigid's Well at Buttevant Parish, Co Cork that lay in a townland called Mountbrigid.

There was a St Brigid's well in the grounds of the former palace of the Archbishop of Armagh, which may have been the one mentioned in the *Book of Lismore*, a collection of earlier manuscripts compiled in the fifteenth century: when St Brigid was visiting Armagh, two people passed her carrying a tub of water, and asked the saint for a blessing, which she gave. They accidentally dropped the tub which rolled over and over as far as Lough Lapham, but it was not broken or any water spilled. St Patrick, who happened to be there, said part of the water should remain in Armagh and that 'every disease and every ailment in the land was cured.' Pilgrimages to the well were once common and the water was said to be a cure for eye troubles as well as being very good to drink. In the mid-nineteenth century Archbishop Beresford had some of this water shipped to London so he could drink it when staying there, presumably not trusting the local water! He is also said to have given a sample to Queen Victoria who was very pleased with its special taste and requested that more should be sent to her.

So popular was St Brigid that she sometimes displaced the local saint. For example, in the seventeenth century the parish of Killore, Co Westmeath, was dedicated to St Aodh, but he had been displaced by St Brigid by the 1840s. Pilgrims went to St Brigid's Well near an ancient church called *Teampall Brighide* (Brigid's Temple), and today Aodh is no longer remembered there. A similar situation is found in Outeragh Parish, Co Leitrim, where St Finnabhar has been eclipsed by St Brigid. Pilgrimages to wells in the earlier period, involved more than just rituals and prayers. After these there was a more secular celebration, often involving huge numbers of people which might go on for two or more days and was one of the few holidays for many poorer people. These pilgrimages were often referred to as *patterns*, a word derived from Patrons Day, a celebration of the saint to whom the well was dedicated who was often also the patron saint of the parish. Such events were also called 'rounds', a reference to the ritual walk that often took place. There are a number of descriptions of these *patterns* dating from the eighteenth and nineteenth century, usually by people who disapproved of the practice, although their descriptions are valuable as they give a good idea of what went on.

In the *Parochial Survey* compiled in the nineteenth century there are a number of references to *patterns* such as the one from St Peter's Parish, Athlone, Co Westmeath. This records that three such pilgrimages were held in the district, at Brideswell, St John's Well at Lecarrow and another at Clonmacnoise, the latter of which attracted 3000–4000 people from all over Ireland and lasted for two days. The entry notes the 'drunken quarrels and obscenities' that always took place, before continuing: 'At these places are always erected booths or tents as in fairs for selling whiskey, beer and ale, at which pipers and fiddlers do not fail to attend, and the remainder of the day and night (after their religious performances are over and the priest withdrawn) is spent in singing, dancing and drinking to such excess that it seems more like orgies of Baccus than the memory of a pious saint'.

Church Wells

The majority of Irish wells are located close to ancient churches, either existing buildings, ruins or the site of an old graveyard. This is probably because the well or spring was already a focus of worship by pagans before Christianity arrived in the district, and building a church (*cill*) near it ensured there was a continuity of worship as the local people were used to coming there. The water could be used for Christian practices and so rituals remained focussed on it. Many churches were abandoned and fell into ruin in the eighteenth century, but so strong was the tradition of carrying out well veneration that the practice of visiting, either singly or as part of a ceremony and leaving votive offerings continued.

Many of these *patterns* and pilgrimages survived almost to the present day, despite the Reformation of the sixteenth century and the opposition of the reformed Irish Church which regarded such customs as superstitious and idolatrous. The civil authorities also tried to ban such practices as they were often boisterous affairs which ended in drunkenness and fighting, making the authorities fearful of civil disorder. In 1704 the Irish Parliament passed an Act to ban *patterns* with a penalty of a whipping or a fine of ten shillings. The seventeenth and eighteenth centuries were a difficult time for many Irish priests. It was usual during a *pattern* for a collection to be taken from among the pilgrims, since at this period many priests were in hiding and could not collect their tithes, so relied on the generosity and aid of sympathetic people. In more recent times the police often tried to put a stop those *patterns* they regarded as 'riotous assemblies'.

Trees at Holy wells

Many pilgrimages to holy wells also involved the ritual walking round of nearby trees, and sacred trees (*bile*) play a large part in Irish stories and mythology. The tradition of venerating them dates back to pre-Christian times, and the inclusion of trees in pilgrimage rituals to wells suggests its origin lies in pre-Christian practices that were so deeply embedded in Irish culture it had to be accepted by the church and many early churches were built near sacred trees.

In Ireland it was believed that holy wells could be 'insulted' by someone not treating them with due reverence such as bathing animals in them, washing clothes, urinating in them or other unsuitable activities, after which they would often lose their curative powers. They could also be 'attacked' by filling them in or blocking them off, although there are many accounts of wells then reappearing nearby, and often the action rebounded on the person who damaged the well.

It was also possible to 'insult' a sacred tree associated with a well. An example of this concerns a nineteenth century Protestant policeman from Co Galway. He insulted Biddy's Tree, near St Brigid's well in the Parish of Buttevant, Co Cork, by swinging on its branches to amuse himself, much to the horror of the locals. When he got back to the police station at Churchtown he was struck down with violent pains in all his limbs and died six months later.

Stones at wells

As with water and trees, certain stones were also venerated in pre-Christian times and they played a part in a number of holy well pilgrimage rituals. In many instances this simply took the form of adding a stone to a cairn built by earlier pilgrims, and a number of writers describe the pile of stones to be seen adjacent to wells. A particularly notable example was once to be seen at St Bride's Well in Kildare Parish, Kildare. At many sites single stones or piles of small stones are used to mark the stations (*leachta*) where prayers were said during the pilgrimage. Sometimes nearby stones are incised with a cross that was touched or kissed by the pilgrims, while others were crudely carved with a face to represent the saint.

Often the votive offering deposited at the well, rather than just an addition to a cairn, was a carefully selected stone. While this may be because the pilgrim was too poor to afford anything else, the more probable explanation is that it was seen as something 'special' and so a worthy offering, and leaving the stone was also felt to be the equivalent of leaving a part of the pilgrim in the presence of the saint. Frequently these stones are of white quartz, a stone that has been regarded as special since prehistoric times and is found on ancient sites and built into monuments 5000 or 6000 years old. One notable Irish example is the impressive prehistoric structure known as *Brú na Bóinne* at Newgrange in Co Meath, which tradition says was built by the Daghdha, father of the Goddess Brighid. While quartz is a very common stone it looks very special and jewel-like, particularly when pure and freshly broken. It was considered a suitable offering to either a water goddess or a saint.

St Adamnan's seventh-century *Life of Saint Columcille* clearly shows that beliefs about the powers of white stones continued into Christian times:

> In the same district mentioned above, he [St Columcille] took a white stone from the river and blessed it so that it should effect some cures. Contrary to nature that stone, when immersed in water, floated like an apple.

There are traditions from many parts of Ireland and Britain of stones that returned to their original spot when moved and an example of such a 'homing stone' involves St Brigid's Church, Loch Hyne, Co Cork. There, about 45m (150ft) north east of the church is a piece of a broken pillar about 45 x 38cm (18 x 15in) bearing an incised cross. One day this was carried away by a fisherman who took it to his house, but the very next morning he found it had gone and it was discovered back in its original spot. The fisherman drowned shortly after and everyone knew he had been punished for daring to remove this holy stone.

Fish and Holy wells

Many fish occur in tales from Irish mythology, the most famous being the Salmon of Knowledge that lived in the Well of Segais in the Land of Promise which spread knowledge into all Ireland's waters. There are a number of Irish saints, including St Patrick, who have stories connected with them that involve fish, but there is no doubt that some holy wells did house sacred or magic fish (*Iasc sianta*), usually either salmon or trout, another sacred fish in Celtic tales. Fish in wells are also known from Wales and England. One early reference to fish in Irish wells is found in the *Annals of the Four Masters* which, under the year 1061, records:

> An army was led by Aedh an Gha-bhearnaigh O'Connor to Kincora and he demolished the fortress and destroyed the enclosing wall of the well and ate its two salmons, and also burned Killaloe.

There are also many references from all over Ireland in the nineteenth and twentieth century to fish living in holy wells. In most instances, the fish were real, although some of the appearances seem to be of a supernatural nature suggesting that imagination played a part in some of the stories. One account tells of two trout that used lived in the Brideswell, Co Roscommon until about 1820. On one occasion one of the fish was caught by a man who took it to his home and put it on a gridiron to cook, but it flew off the fire and returned to the well, where it lived for several more years. People who saw it said that the marks of the gridiron could clearly be seen on its side. Later a boy caught and ate one of the trout 'to his own destruction' and later another child caught the other one.

Sacred 'fish' were recorded as living in St Brigid's Well at Birchfield, Co Clare but, in this case, it was a large eel and eleven small ones, but they were only seen on certain occasions. If pilgrims who had gone there for a cure saw them they were instantly cured. Its water was remarkably clear and cool but water taken from the well for cooking was said to be useless as it did not boil. This belief about the water from a number of holy wells was common, suggesting a widespread taboo against using them for mundane purposes.

On one occasion this well actually disappeared when it was insulted. On the day that a large dinner party was being thrown by the O'Briens, the owners of the land on which the well stood, a servant took water from this well to boil potatoes. The dinner was late and when O'Brien inquired why was told the potatoes would not boil no matter how hard the fire was stirred and blown upon. O'Brien, realising what had probably happened, examined the contents of the pot and found an eel in it, upon which he angrily berated the servant. He then ordered the water and eel to be taken back to the well but the eel was found to have disappeared! However, it later returned to the well.

This well was very famous for its healing properties and visited by many pilgrims who were given free access to it across the O'Briens lawn. One member of the family became seriously ill while in London, and several doctors who examined him despaired of his life. When some water from St Brigid's Well was brought to him and he drank it, he immediately recovered. So to commemorate this miraculous cure he built a beautiful wall around the well, planted the enclosure and even built some stables nearby where pilgrims could leave their horses.

Another account of a fish in a well dedicated to St Brigid at Burren, Co Galway, was recorded by Lady Gregory around 1900 in *A Book of Saints and Wonders*. It also shows the touching faith that the saint could help those who were ill:

> I brought my little girl that was not four years old to Saint Brigit's well on the cliffs, where she was ailing and pining away. I brought her as far as the doctors in Gort and they could do nothing for her and then I promised to go to Saint Brigit's well, and from the time I made that promise she got better. And I saw the little fish when I brought her there; and she grew to be as strong a girl as ever went to America. I made a promise to go to the well every year after that, and so I do, of a Garlic Sunday, that is the last Sunday in July. And I brought a bottle of water from it last year and it is as cold as amber yet.

While another account of the fish in the well at Burren was also recorded by Lady Gregory:

> But if Brigit belonged to the east, it is not in the west she is forgotten, and the people of Burren and of Corcomruadh and Kinvara go every year to her blessed well that is near the sea, praying and remembering her. And in that well there is a little fish that is seen every seven years, and whoever sees that fish is cured of every disease. And there is a woman living yet that is poor and old and that saw that blessed fish, and this is the way she tells the story: 'I had a pearl in my eye one time, and I went to Saint Brigit's well on the cliffs. Scores of people there were in it, looking for cures, and some got them and some did not get them. And I went down the four steps to the well

and I was looking into it, and I saw a little fish no longer than your finger coming from a stone under the water.

Three spots it had on the one side and three on the other side, red spots and a little green with the red, and it was very civil coming hither to me and very pleasant wagging its tail. And it stopped and looked up at me and gave three wags of its back, and walked off again and went under the stone. And I said to a woman what was near me that I saw the little fish, and she began to call out and to say there were many coming with cars and with horses for a month past and none of them saw it at all. And she proved me, asking had it spots, and I said it had, three on the one side and three on the other side. That is it she said. And within three days I had the sight of my eye again. It was surely Saint Brigit I saw that time; who else would it be? And you would know by the look of it that it was no common fish. Very civil it was, and nice and loughy, and no one else saw it at all. Did I say more prayers than the rest? Not a prayer. I was young in those days. I suppose she took a liking to me, maybe because of my name being Brigit the same as her own.'

The miraculous changing of water

From the *Lives of the Saints* in the *Book of Lismore* comes a story about St Brigid turning well water into ale (*coirmm*):

Then Dubhthach and Brigid go to their country in the province of Offaly. And her nurse was along with Brigid and illness seized her nurse as she was wending her way. So Brigid and another girl were sent to ask a drink of ale for her from a certain man named Baethchu, who was making a mighty feast. He refused Brigid. Then Brigid went to a certain well, and filled her vessel thereat and blessed the water, so that it turned into the taste of ale and she gave it to her nurse, who straightaway became whole. As to the feast at which she was refused, when they go to drink it, not a drop thereof was found.

Pilgrimages to Brigid's Wells in Ireland

On the saint's day to which the parish or well was dedicated there was a pilgrimage (*pattern*) to the well. This was generally regarded as a holy day of obligation to hear Mass and no work was done. This was a practice that lasted in some places into the twentieth century and it was regarded as very unlucky to work on that day. There are many stories concerning people who broke this taboo, as for example the one concerning a stone mason in Co Leitrim who worked on *Fhéile Bhríde* (St Brigid's Day) despite dire warnings about what might happen. However, he ignored all these and nobody was surprised when the unfortunate man was blinded by a splinter of stone. *Patterns* held during the summer were always well attended and some attracted over 5000 people, with the celebration extending over two or three days. Once the rituals had been completed the participants would join in the rest of the days activities, eating, drinking, boxing, wrestling, dancing, courting, getting drunk and fighting!

A letter from Father Charles O'Connor to his brother, written in the nineteenth century, is of interest as it shows the reasons given by ordinary people at that time for undertaking the *patterns*, although it must be borne in mind that he was not really in favour of such practices:

> I have often enquired of your tenants what they themselves thought of their pilgrimages to the wells of Kil-Archt, Tobar Brighde, Tobar Muire near Elphin, and Moore near Castlereagh where multitudes assembled annually to celebrate what they in broken English termed Patterns (Patron's Days): and when I pressed a very old man, Owen Hester, to state what possible advantage he expected to derive from the singular custom of frequenting in particular such wells as were contiguous to an old blasted oak or an upright unhewn stone and what the meaning was of the yet more singular custom of sticking rags on the branches of such trees and spitting on them – his answer, and the answer of the oldest men, was that their ancestors always did it: that it was a preservative against *Geasu-Draoidacht,* i.e. the sorceries of the druids: that their cattle were preserved by it, from infections and disorders: that the *daoini maethe,* i.e. the fairies, were kept in good humour by it: and so thoroughly persuaded were they of the sanctity of those pagan practices that they would travel bareheaded and barefooted from ten to twenty miles for the purpose of crawling on their knees round these wells, and upright stones and oak trees westward as the sun travels, some three times, some six, some nine, and so on, in uneven numbers until their voluntary penance's were completely fulfilled.

Many pilgrimages in the past involved the pilgrims walking barefoot or often crawling around the well or the various station markers at which prayers were said at the cost of great pain to themselves, with bleeding knees and feet being endured even by women and children. Even today the pilgrimage to the top of Croagh Patrick, Co Mayo is undertaken by some people in bare feet. Vigils and fasting were also part of some pilgrimages, although a number of these only seem to date to the nineteenth century, but may be a revival of a much older tradition. Such pilgrimages, although less numerous than formerly, still attract thousands of people in some cases.

Today the Catholic church encourages the visiting and praying at holy wells, and has recently reorganised some pilgrimages which has increased their popularity, especially if the date is moved to a more attractive one by changing its timing from early or late in the year to the summer. Many private visits are still made to wells today to pray, ask the blessing of the saint and to pray for their help with curing illnesses. It was traditional to collect water from a well dedicated to St Brigid on her Day and sprinkle it on the house, farm buildings, livestock and fields to invoke the blessing of the saint. Such practices still continue today and many wells have small offerings and tokens left nearby. At Doon Well, Co Donegal, a nearby bush is festooned with such items as handkerchiefs, glasses, asthma ventilators and even menstrual tampons.

The pilgrimage to Saint Brigid's Well, Outeragh, County Leitrim

A pilgrimage was always held on 1 February in honour of St Brigid in the parish of Outeragh in Co Leitrim, as she was the patron saint of the parish. Pilgrims, either singly or in small groups, walked towards the old graveyard in the early morning darkness reciting the rosary. As they entered the graveyard the pilgrims first walked three times around an ancient ash tree which stood near the site of the medieval parish church in a sunwise direction, saying a few *Our Father's* and *Hail Mary's* while doing so. The pilgrims then walked to another tree and repeated the ritual before returning to the site of the old church where they knelt by an ancient stone that was said to be a carving of St Brigid's head. They then said a few *Our Father's* and *Hail Mary's* before depositing a few coins on the ground by the carving.

The pilgrims then got to their feet and went along a lane and across three fields to St Brigid's Well, saying the rosary as they went. On arrival they knelt and prayed for a few moments before walking three times around it. Water was drunk from the well and then they might either tie a piece of rag to a bush near it or deposit a religious medal, fragment of rosary, a small cross or other little token in the well or place them on the ground beside it. After doing this a third rosary was recited before the pilgrim returned home. It is intriguing that this particular well was described by John O'Donovan, an Ordnance Survey surveyor in 1836, as being within a few yards of the old graveyard and it is shown in this location on the earliest Ordinance Survey map of the area.

However, at some time after this survey an 'enemy' blocked up the well which later miraculously appeared in its present position, a not unusual occurrence with holy wells. When another well about 228m (748ft) from the graveyard was cleaned out in 1900 many religious objects were found in the mud at the bottom, so it is probable that veneration was initially switched to this well on the closure of the original one and this was used for the St Brigid Day *pattern* until the present well 'appeared'. The stone head of 'St Brigid' was cleaned of moss and lichen in the 1970s which revealed it was a corbel of a bearded man!

Changing the date of the pattern

In a number of cases the *pattern* held in honour of St Brigid was changed from 1 February to a more suitable date for an open-air pilgrimage, i.e. when the weather was warmer. For example, a pilgrimage to a St Brigid's well in Co Westmeath used to be held on 1 February, but in the 1950s or '60s Father Andrew Shaw, Curate of Mullingar Parish, made some changes and improvements to the area surrounding the well. This is now covered with a half dome of stone and turf, with nearby a small replica of an ancient Irish church, except its south side is open to accommodate a stone altar where Mass is celebrated in front of the pilgrims. There are also fourteen praying stations (*leachta*), a larger number than usual, but not unique and it is traditional after saying a prayer at each station to complete the pilgrimage by taking a drink from the well. Father Shaw also changed the pilgrimage day to the last Sunday in August when the number of people taking part increased considerably and 500 participants is not an unusual number.

Another pilgrimage that was dedicated to St Brigid and which was reorganised in the 1950s was that to the *Tobar Bhríde* (Brigid's Well) at Brideswell, Co Roscommon. This was moved from 1 February to the last Sunday in July, and a portable altar was set up near the well on the pilgrimage day where Mass was said by the priest or even occasionally the diocesan bishop. The well and its associated praying stations are on Brideswell Common and the well, surrounded by a stone wall with a gate and steps leading down to the water, which is deep and dark, rising from a spring in a limestone outcrop. The water from the well flows through a stone-lined channel and through an arch in the west wall of a square stone building, about 6m (20ft) square, called the Water House, and then out again through an arch in the east wall where it flows to a *turlough* (winter flood land) some distance off.

A stone slab on the west wall of the Water House bears the coat of arms of Randall MacDonnell, and the inscription: 'Built by the Right Honourable Sir Randall MacDonnell first Earl of Antrim, 1625.' He did this in gratitude to St Brigid as he and his wife were childless until they made a pilgrimage to *Tobar Bhríde* to pray for children. Following their visit his wife did conceive a son. It is probable that prayers were said for children by childless couples at many wells, but this aspect was not openly discussed and only a few wells are specifically associated with this 'problem'.

Pilgrims to the *Tobar Bhríde* began their pilgrimage at a ring of stones which they walked round saying five *Our Father's* and five *Hail Mary's*. Then they walked to an ash tree on a mound repeating the formula, before walking to a large stone in the middle of the common. From there pilgrims made their way through a nearby school yard, whose two gates were kept unlocked, to a modern statue of St Brigid. In earlier years the journey from the stone to the statue was done on their knees. The pilgrims then prayed and walked around the statue before returning to the Water House, which is entered by a door in the north wall where they knelt on a rocky outcrop to say the rosary, completing the pilgrimage by taking a drink of the water.

One of the most notable St Brigid wells is that at Liscannor, Co Clare, which is still regarded as a healing well and resorted to by many people seeking help. This is another whose pilgrimage date was moved from 1 February to the first Sunday in August and more recently the date was again changed to the 15 August. Some people used to spend the night by the well before carrying out the rituals. The well is covered by an elongated Well House built by Cornelius O'Brien in the 1830s, that contains a large collection of votive offerings, such as pictures, statues, rosaries, medallions, Brigid's crosses, writings and other such items, along with some medical items such as neck braces.

Local tradition says that the Well House was built in gratitude for a cure he was granted by taking the water of this well. It is also said to be a 'jumping well' because of a belief that the well moved from one side of the road to the other after a local pagan woman attempted to prevent St Brigid blessing it as a Christian holy well. Another version of the tale, however, explains that its 'relocation' was because it was insulted by a woman washing potatoes in it! At one time the well seems to have had eels in it and it may have been the site of pagan rituals as it is the point where two small rivers come together.

The site is surrounded by a wall with a gate through which pilgrims entered the well area to kneel before a statue of St Brigid to say five *Our Father's*, five *Hail Mary's* and five *Glorias*. They then walked around the statue saying prayers before climbing some steps to the graveyard and repeating five *Our Fathers*, *Hail Mary's* and *Glorias*. The pilgrims then moved to an old stone cross up a slight hill and repeated the prayers while walking round it. After this they returned to the well where they took a drink of water in the 'Name of the Father, the Son and the Holy Ghost' and, it is said, that requests made in their prayers would be granted immediately. At one time there was a poem at the well based on one published in *The Treasury of Irish Saints* by John Irvine:

> Sweet St Brigid of the kine,
> Bless these little fields of mine,
> The pasture and the shady trees,
> Bless the butter and the cheese.
> Bless the cows with coats of silk,
> And the brimming pails of milk.
> Bless the hedgerows and I pray,
> Bless the seed beneath the clay.
> Bless the hay and bless the grass,
> Bless the seasons as they pass,
> And Heaven's blessings will prevail,
> Brigid, Mary of the Gael.

Brigid's cures

Eye problems

The majority of holy wells are connected with healing, either for general cures or for specific complaints and one of the most common cures was for eye problems. Eye complaints were frequent in the past, due to many people living in smoke filled houses, sleeping on straw filled mattresses and doing jobs where motes easily got into the eyes. There were limited herbal medicines available to treat eye problems and getting things out of eyes took a skilled practitioner and the correct invocations. Very few people wore glasses so myopia was probably also a problem, all of which explains the large number of wells reputed to cure eye troubles.

St Brigid is associated with eye cures and complaints due to the story that she plucked out her eye (or eyes according to some versions) on one occasion to discourage a suitor, but later successfully replaced it. She was not the only saint to resort to this ploy! At least three St Brigid wells are claimed as being the actual one that she used to restore her sight, one lies in Killinagh Parish, Co Cavan, one at Dunleer in Co Louth and the third is the famous well at Faughart, also in Co Louth.

Whooping cough

In the past one of the most common childhood ailments was whooping cough and a number of wells were reputed to be able to cure this. One of these was the *Tobar Bhríde* (Brigid's Well) in Darragh Parish, Co Limerick. The sick child was taken to the well to drink the water, which was regarded as being at its most effective on 1 February, but could be visited to treat a child at any time. As part of the continuing treatment moss growing near the well was gathered and taken home where it was boiled in milk for the child to drink. If St Brigid's Well at Ardagh, Co Longford, was found to be dry then the pilgrims would gather moss from around the well and boil this in water to bring about a cure.

Toothache

Toothache must have been another common problem in the past when dentists were few and costs expensive, so its not surprising that many wells were said to be able to cure this. One of these was St Brigid's Well at Greaghnafarna townland in Co Leitrim, but to get an effective and long lasting cure the sufferer had to perform a complex and curious ritual. The sufferer went to the well where they said five Aves and a Creed before making a promise to St Brigid that they would not shave (if a man) or polish their shoes on a Sunday. They then blessed themselves with the water and also applied some to the painful tooth. Finally they had to add a stone to the pile left by other supplicants beside the well and deposit a token offering such as a coin or pin. However, if the promise made to the saint was broken then the toothache would return.

Headaches

Many wells had a reputation for curing headaches, but to obtain a cure a special ritual had to be carried out at many of them, which usually involved bathing the head before drinking from the water and saying prayers. A famous 'headache well' was St Brigid's Well, Kilranelagh Parish, Co Carlow, which is located in an old graveyard. A stream runs from the well into a stone trough in a nearby field and it is the water in the trough that is used to cure headaches.

For animals

While dipping or washing animals in many wells was regarded as an insult that might nullify the powers of the water, there were some wells that were known to be suitable for horses and other domestic animals. The most important were wells that were good for cows, by increasing their milk yield, improving their fertility or generally keeping them healthy.

In the case of these large animals it was usual to swim them in holy lakes, rivers or the sea where a holy spring flowed into it, a practice that is of great antiquity. One story, first recorded in the 1840s from the Parish of Burrishoole, Co Mayo, relates how St Brigid quarrelled with the local saint, a man called Marcan. Farmers used to visit Lough Marcan and swim their cattle in it to be cured of their diseases by St Marcan, but on driving the animals home they took great care to avoid Kilbride or any other place associated with St Brigid. This was because they believed she still remembered the quarrel they had had and would be displeased if St Marcan was credited with any miracles concerning cattle, an animal usually associated with St Brigid! While there was a very widespread belief that wells had the power to heal, things did not always turn out as expected, as Lady Gregory records in another story about St Brigid's Well at Burren, Co Galway:

> There was a beggar boy used to be in Burren, that was very simple like and had no health and if he would walk as much as a few perches it is likely he would fall on the road. And he dreamed twice that he went to Saint Brigit's blessed well upon the cliffs and that he found his health there. So he set out to go to the well and when he came to it he fell in and he was drowned. Very simple he was and innocent and without sin. It is likely it is in heaven he is at this time.

Votive offerings

Votive offerings were, and are, often left at holy wells (and other sites) and in Ireland consist of a wide range of items, the most common being small denomination coins, rosary beads, devotional cards, sacred pictures, small pieces of jewellery and trinkets, hair pins, nails, fish hooks, small statuettes, pieces of clothing, strips of cloth and flowers. In the past tobacco pipes, buttons and pins were deposited. More recently a variety of other offerings have been reported. For example, on a whitethorn bush (hawthorn) near a well at Seir Ciaran was seen a clothes peg, a metal corkscrew/bottle opener, a length of elastic and a baby's dummy, in addition to the more usual rosary beads, religious medals, small crosses and so on.

This is a custom that far pre-dates the introduction of Christianity, and was originally an offering to the gods and goddesses when appealing for their help or in thanksgiving. There was a widespread custom among the Celts of depositing valuable items such as swords, shields, jewellery and many other objects in water as an offering. This practice continued as rich pilgrims, especially in the medieval period, would make valuable offerings and donate quite large sums of money, so that many important shrines at the sites of wells grew rich on such gifts. There seems to have been a number of reasons for leaving tokens at holy wells. By leaving an item of a definitely pious nature it was a form of continuous prayer, and by leaving a gift after you had said it you would leave your sins behind and the saint would bestow their blessing on you and, hopefully, grant your wish. Many items, particularly rosaries and small items of jewellery, are left in the hope that the sufferer will, in effect, leave their ailment behind and so be cured. These objects are invariably left untouched as its 'unlucky' to remove them and there is still a widespread belief that if anyone takes them they will get the ailment for which the pilgrim was seeking a cure.

At some healing wells crutches, walking sticks, bandages and other medical accessories were left by pilgrims who no longer needed them after their visit, having been cured of their affliction while at the well. Many such cures were often psychosomatic and short lived, particularly in the case of *patterns* that involved the consumption of large amounts of alcohol! Such items acted as visible proof to others that the well really did have curative properties. In many places strips of cloth were tied to a tree or bush in the vicinity of the well. This too is a very old custom and still prevalent today. In the case of healing wells it is believed that the supplicant's disease would disappear as the cloth rotted away. Sometimes, before tying the cloth to the tree, it was dipped into the well water and rubbed onto the affected part in an attempt to transfer the disease to the strip of cloth as a form of sympathetic magic, a practice that was common at many wells including that at Bridewell, Co Dublin. In other cases the supplicant would spit on the cloth before tying it to a tree.

Today the rags can be of any colour, but old accounts suggest that traditionally they were either white, multicoloured or red, a colour that has all sorts of magical associations as it was believed to resist the power of evil spirits and also represents blood – the life force. In a few places pilgrims would take a cup or glass to the well with which to drink the water and would then leave the drinking vessel there as a votive offering. This was a tradition at St Brigid's Well, Castlemager, Co Cork, which was so famous as a healing well that people travelled great distances to drink from it.

There is a St Brigid well not far from a monastery at Cliffony, Co Sligo. It is said locally that the saint used to pray there while immersed in the water, no matter how poor the weather or cold the water, spending all night there weeping while praying. However, God felt her mortification was too severe, and one day when she arrived at the well she found it dry so she could not continue with this penance. Once she discontinued the practice the well once again filled with water. There used to be a *pattern* to this well on 1 February, but this no longer seems to occur. It is said that three phantom nuns, accompanied by a big spotted cow 'like an Ayrshire' used to be seen walking from St Brigid's Well at Ardagh, Co Longford towards Slieve Golry. Could this be a modified story originally concerning the Goddess Brighid in her aspect of three Brighids?

Brigid's wells in Scotland

The Scottish church actively discouraged the veneration of wells and other traditional customs, and in 1638 the General Assembly enacted penalties against those making pilgrimages to holy wells. This meant that many of the saint's dedications attached to wells were lost and Brigid/Bride wells today are very few. She was so venerated in the past, however, that there were almost certainly more at one time.

There was a charm stone, used for curing diseases, that was closely associated with a Brigid's Well. This was the Keppoch Stone, a rock crystal mounted in silver, which was said to have been taken to Australia by a branch of the McDonnell's of Keppoch in the mid-nineteenth century. Up to then it was the custom to dip this stone in the water of the *Tobar Bhríde* (St Bride's Well) and to call on the charm stone to heal all the ills of 'every suffering creature with this water that was kept pure by St Brigid, in the name of the Holy Apostles, Mary the Virgin of Virtues, the High Trinity and all the radiant angels.' Unfortunately, the location of the well today is no longer remembered.

Brigid: Goddess, Druidess, Saint

56 St Brigid's Well, Liscannor, Co Clare, Ireland. The well is at the end of a tunnel lined with a great variety of offerings left by pilgrims, including a medical neck brace.

57 Collecting holy water from St Brigid's Well, Kilcullen, Co Kildare, Ireland. The sculpture, by Henry Flanagan (1977), depicts St Brigid feeding the poor.

Brigid's wells in Wales

As in other parts of the Celtic world, wells (*ffynnon* – singular, *ffynhonnau* – plural) are often regarded as holy or sacred, as are springs (*tarddiannau*) and even water spouts (*pistylloedd*). It is interesting to note that the practice of tying rags to trees or bushes near the well is very rare in Wales. Of 1,179 wells surveyed by Francis Jones (*The Holy Wells of Wales*, 1954) only ten are known to have had rags placed nearby, of which seven are in Glamorgan. This was either not a widespread practice or it was a custom that generally died out a long time ago. Of the 437 Welsh wells that bear the names of saints, only eight are associated with St Ffraid/Brigid/Bride.

Brigid's wells in England

While there are many thousands of holy wells in England, very few are connected to St Brigid. However, Bridewell Streets and Lanes can be found in a number of towns and villages, although in many cases the well that was presumably there has now disappeared. One of the most famous wells dedicated to her, although it too has disappeared, was the St Bride's Well in Fleet Street, London. Excavation showed the area of the well has been occupied since Roman times, and the well must have been in use from that time, and gave its name to the nearby Bridewell Palace. Its water was regarded as special, but in this case its water seems to have been depleted because of a royal coronation as is clear from an account by William Hone in his *Everyday Book* published in 1831:

> The last public use of the waters of St Bride's Well drained it so much that the inhabitants of the parish could not get their usual supply. The exhaustion was caused by the sudden demand on the occasion of King George IV being crowned in Westminster in July, 1821. Mr Walker, of the hotel at number ten, Bridge Street, Blackfriars, engaged a number of men in filling thousands of bottles with the sanctified fluid from the pump.

Presumably the bottled water was sold either simply as a refreshing drink or, more likely, as a 'special water' in some way connected with the crowning of the monarch. After this the well dried up, having given at least 1,800 years of service to the inhabitants of that area. In 1850 the well was made deeper at great expense, but the water found was unfit for use. Mr Mason, the churchwarden of St Bride's, was paid to tap a fresh source nearby to provide an alternative usable supply for the local people and this yielded 300 gallons (1,364 litres) a day, but high salinity made it unsuitable for domestic purposes.

Twelve

Saint Brigid and the natural world

While St Brigid was well known for her generosity and acts of charity many stories about her from the seventh century onwards mention animals and birds that were connected with her or which she encountered. However, while a variety of animals appear in the stories concerning her, it does not seem possible to identify any, other than cattle, that might have been particularly associated with the Goddess Brighid in pre-Christian folk tales that may have later been 'transferred' to St Brigid.

Animals

Saint Brigid's cattle

The animals most closely associated with St Brigid are cattle, which feature in many of the early, i.e. pre-tenth-century *Lives*, and in later miracle stories about her. Some of the customs carried out on her Day were specifically designed to protect and bless cattle and she is also associated with the curing of many cattle diseases, while dairy products form an essential part of the special meal prepared to celebrate her special Day or its Eve.

In Scotland many general invocations or those specifically for the protection of cattle appealed to St Brigid and crooning songs (*cronan bleoghain*), sung to cows while milking them, also feature the saint. St Brigid is sometimes shown in Christian art accompanied by a cow, with a pair of milk pails or standing by a cow house. It is not surprising that she was long regarded as the patron saint of milkmaids. One of her titles was 'Christ's Milkmaid', although a number of modern illustrations show her with sheep rather than cows, but these animals appear in very few early stories about St Brigid compared to tales featuring cows. During the medieval period it seems she was so closely associated with the blessing of cows that some people started to object to this, as is clear from a verse in a sixteenth century poem:

> Priestes, pray no more
> To Sanct Anthone to save your sow,
> Nor to Sanct Bride to keipe your cow,
> That grieves God right sore.

Saint Brigid and the natural world

Cattle played a major part in Ireland's economy for thousands of years, and continued to be an important part of personal and national economy until quite recently. It is not surprising, therefore, that there are many stories and beliefs concerning them. Cattle were used to provide milk, butter, cheese, meat and leather, while oxen were used in ploughing and drawing wagons. Their bones were used to make sewing tools and other items, while horns provided opaque panels for lanterns and even windows. Possession of cattle was an indicator of their owner's wealth.

This strong connection between St Brigid and cattle is almost certainly a characteristic that has been 'transferred' to her from the pre-Christian period when the animals were sacrificed to the Celtic deities. The blessing of cattle was carried out by Druids who used to drive cattle between two fires, while chanting incantations, at the festival of *Beltane* (1 May) to protect them from diseases.

It is probable that the importance of cattle in ceremonies held at the temple of Brighid would have continued once the temple changed to a church and its Druidesses and Druids into nuns and monks, not in being sacrificed to the goddess but used in a way more appropriate to Christianity. Brigid, as Abbess and former Chief Druidess, would have continued to bless cattle and their produce, and to perform ceremonies to cure or prevent illnesses in the animals. This has come down to modern times in customs associated with the Brigid's crosses (see page 96), the left over material from making them (see page 105), and in the taking of cattle to some holy wells and springs, particularly those dedicated to St Brigid.

The Goddess Brighid also had a connection with cattle because these animals played such an important part in the Celtic economy as well as in religious beliefs and mythology. Despite their importance in the Celtic world as a whole, there is no trace of a cow goddess in Celtic literature, although several Irish goddesses, such as Bóann, Flidais and Brigid's mother, Mór-Ríoghain, possessed supernatural cows.

One of the earliest recorded Irish customs, noted in 1171 by Benedictus Abbas, concerns the 'baptising' of children with milk. This was inspired by one of the traditional stories about St Brigid, that is itself probably based on a story originally concerning the Goddess Brighid (see page 92). There are images of cattle used decoratively on Celtic objects and art works, but in many cases it is only the head that is depicted so it is uncertain whether the animal shown is a bull, representing strength, fierceness and virility, or an ox, a more sedate animal but one which was vital in agriculture for pulling ploughs and wagons, as well as supplying meat and hides.

St Brigid was said to have a special white cow that accompanied her (or her spirit) on her travels on the eve of her Day bestowing her blessing, and it was said her cow's milk never ran dry. Some Irish folk traditions identify her cow with the ancient legendary cow known as *Glas Ghoibhneann* (the Grey Cow of Goibhniú) which was supposed to have emerged from the sea (often seen as a way of entering/leaving the Otherworld) and got her power of inexhaustible milk from licking an object on or near the sea shore. This cow's supply of milk only dried up when when she was mischievously milked into a sieve. These elements seem to come together in a traditional Irish story:

> There was a cow at Foughart which was called St Brigid's Cow, but it was forbidden for anyone to put a small vessel under this cow to milk her in. This cow used to walk each day from Foughart to Narrow Water [near Meath, Co Armagh] and lick a barrel that was there in the water before returning to Foughart. This was the very barrel in which St Patrick arrived in Ireland. However, one day a poor woman at Foughart had no milk for her tea and so she went to St Brigid's cow with a pint tin to get a drop of milk, which she did. Either that day or the following day the cow took her calf and journeyed to Narrow Water, but this time they jumped

into the water beside the barrel, and went down Carlingford Lough and returned to Foughart no more. Today there are two big rocks out in the sea at Whitestown which are called the Cow and the Calf (*Bó* and *Lóig*).

Brigid's cow miracles from the early lives
Of the Cow Milked Three Times in One Day, as recorded by Cogitosus, *c.* 650 AD

Here is a deed which, among the rest of Saint Brigid's miracles, seems amazing and worthy of admiration. As bishops had arrived and were staying with her and since she had no food to give them, God's manifold power came to her aid, abundantly as usual, according as her needs demanded. So she milked one and the same cow three times in one day, contrary to normal practice. And what it normally took three of the best cows to produce, she, by an extraordinary turn of events, obtained from her one cow.

Of the Gift of the Best Cow and the Best Calf, as recorded by Cogitosus, *c.* 650 AD

As poor people and pilgrims were flocking to her from all sides attracted by the enormous renown of her miracles and of her lavish generosity, an unpleasant leper among them came up and demanded to be given the best cow of the herd together with the best calf of all the calves.

Nor did she put him off when she heard his request. But thereupon she freely gave the sick man who requested them the cow she had come to know as the best of the lot, and the sleek and prime calf of another cow. And in her kindness of heart she sent her own chariot with him for the long journey over the very broad plain and gave orders that the calf be placed in the chariot behind him, lest the sick man should be wearied by the long journey and endure hardship in driving the cow.

And so, the cow licking it with her tongue and loving it as if it were her own, without compulsion from anyone, followed behind to the place of destination.

Of the Cattle Stolen and Returned, as recorded by Cogitosus, *c.* 650 AD

And after some time had elapsed, some very wicked thieves, who had no respect for either God or man, came from another province to carry out a raid and, emerging from a large river which was easily forded on foot, stole Saint Brigid's cattle.

But, as they were going back the same way, the force of the huge river, resulting from sudden flooding, overwhelmed them. Indeed, the river rose up like a wall and did not allow this most criminal theft of blessed Brigid's cattle to pass through; but swamping the thieves it swept them away and the cattle, freed from their grasp, returned to their own flock and booley with the thongs hanging from their horns.

Of the Cow and the Calf, from the *Bethu Brigte*, *c.* 800 AD

A pious woman from Fid Eoin gave Saint Brigid a cow on Easter day. Two of them were driving the cow, namely the woman and her daughter, however, they were not strong enough to drive their cow. As they were coming through the wood they besought Brigid. That entreaty availed, it was their cow which comes before them on the way to the place where Brigid was.

'Verily this is good', said Brigid to her maidens 'for this is the first offering made to us since occupying this hermitage. Let it be taken to the bishop who blessed the veil on our head.' 'Of little use is that to him', said the maidens, 'the cow without the calf.' 'That is of no account', said

Brigid, 'the calf will come to meet its mother so that it will be together they will reach the stead'. It was done as she said.

The Deceit About a Cow, from the *Bethu Brigte*, *c.* 800 AD

On the Tuesday following there was a good man nearby who was related to Saint Brigid. He had been a full year ailing. 'Take ye for me', he said, 'the best cow in my yard to Brigid this day, and let her pray to God for me; perhaps I may get better.' The cow was brought and Brigid said to those who brought it: 'Bring it immediately to Mel.'

However, they brought it back to its house and changed it [for a lesser cow] unknown to the sick man. This was related to Brigid. Brigid was enraged at the deceit practised on her. 'Between a short time from now and the morning', said Brigid, 'wolves shall eat the good cow which was given into my possession and which was not brought to you', she told Mel, 'and they shall eat seven calves in addition to it.'

This was related to the sick man. 'Go ye', said he, 'take to her seven calves of the choice of the yard.' It was so done. 'Thanks be to God', said Brigid, 'let them be taken to Mel to his church; he has been preaching and saying Mass for us these seven days between the two Easters [between Easter Sunday and Low Sunday]; a cow each day for his labour, it is not greater than what he has given, and take ye a blessing with all eight, a blessing on him who sent them', said Brigid. No sooner did she say this than he was healed at once.

Irish cattle customs

As St Brigid's Day marked the end of winter and the beginning of spring, so it was expected that milk production would increase as the grass once again began to grow, and St Brigid was invoked to ensure all went well. There was an old saying that 'the cows milk went up into their horns from Christmas until after St Brigid's Day' – i.e. there was scarcity of milk during this time. In January, if anyone complained about the scarcity of milk, it would be pointed out that 'It won't be scarce very long now as Brigid and her white cow will be coming round soon'. This was a reference to the belief St Brigid went round the district accompanied by her cow, and so strong was this tradition that food was not only left out for the saint but feed for her cow (see page 130).

Cows only calve once a year, and so care was taken to see that the birth would coincide with the increase in grass growth, which in turn encouraged the production of milk so the calf would have abundant nourishment in the first few weeks of life. As the calf became independent, the amount of milk available for human use increased. There was a tradition of taking a blessed candle from the previous Candlemas Day (2 February) into the cow's stall on St Brigid's Eve and singeing the long hair on the upper part of the cow's udder to bring a blessing on its milk. It was also common to put cows under the protection of St Brigid, and candles blessed on Candlemas were used to bless the cow before it was milked for the first time after calving.

Another way of placing the newly calved cow under St Brigid's protection before it was put out to grass again was for the man and woman of the house to stand on each side of the cow and pass a lump of hot coal, held in tongs, over its back and under its udder three times, while repeating a prayer to St Brigid. Then the coal was quenched by throwing it into the drain of the cow house before a red rag, containing a cinder and a grain of salt, was tied to the cow's tail. A drop of holy water was then sprinkled over it before it was taken out to join the other cows in the field. The lead and the tongs were thrown after it, and the way the tongs fell was used to predict how lucky the milker would be, before being returned to the fireplace.

If a cow was unable to give milk then St Brigid's help was invoked in a charm called *Buarach Tháil* (Charm of the tying up). This involved making a *buarach* (lead or halter) from rushes cut from a single clump, preferably growing in or on the bank of a river forming the boundary between two townlands. The *buarach* was tied around the cows hind legs and the *Burach Tháil* was said, which seemed to be any 'cow prayer' that mentioned St Brigid, rather than a specific one for this problem. This would break the 'spell' from which such cows were believed to be afflicted.

There was a tradition, maintained by a number of Hebridean milkmaids, that each of the cows teats had different qualities. It was said that one contained more butter, another more casein, a third more sugar and the fourth had more fat in the milk. The four teats were often dedicated to four saints, one of whom (the butter teat), not surprisingly, was called the 'teat of Brigid.' Other ways of bring the blessing or protection of the saint on cows or curing problems was by the use of the *Brat Bhrigid*, Brigid's Cloak (see page 124). The blessing of the saint was also often invoked as soon as a calf was born to give the animal the protection of St Brigid. One of the ways of doing this was to sprinkle the calf with holy water while reciting a special prayer three times:

> God's blessing on thee, O cow!
> Twice blest be thou, O calf!
> May the Three who are in heaven bless you:
> The Father and the Son and the Holy Spirit!
> Come Mary, and sit down;
> Come Brigid, and start milking;
> Come, Blessed Michael, the Archangel, and bless the beef.
> In the name of the Father, the Son and the Holy Spirit,
> Amen, O God.

The church accepted that it was St Brigid who had to be appealed to in the case of cattle diseases and there was even a Latin prayer to be said by the priest in the case of lung disease in cattle or sheep called *Carmen pro lonsoucht* (*A Chant to Guard Against Infected Lungs*):

> In the name of the Father, Son and Holy Ghost. May God bless these herds/flocks; and by the power of these words no harm shall befall from the threat of lung disease or the danger of any other ailment: In and through the virtue of Holy Brigid, for God gave her authority to bless all creatures upon the earth. Amen.

Saint Brigid and the Wild Boar

According to a miracle story recorded by Cogitosus *c.* 650 AD in his *Life of Saint Brigid* :

> Once, a lone savage wild boar, fleeing in terror, arrived at the most blessed Brigid's herd of swine, rushing in headlong flight and, as she chanced to see it among her pigs, she blessed it. Thereupon, it remained with her flock of swine unafraid and tame.

This inspired a poem by the seventh-century St Broccán:

> A wild boar frequented her herd,
> To the North he hunted, the wild pig;

Brigid blessed him with her staff,
And he took up his stay with her swine.

The pigs driven by wolves
Another story told by Cogitosus concerns a gift of pigs to the saint:

> One day a man came from a distant province some three or four days away and offered Saint Brigid a gift of fat pigs, but asked that others be sent with him back to his farm, which was a long way off, to collect them. Accordingly the saint sent help with him to bring back the pigs, and after they had spent a day travelling back towards his farm they saw, in the distance, a herd of pigs coming towards them being guided along the road by a pack of wolves.
>
> As they came nearer the farmer realised that these were his own animals that he had offered to the saint. The wild wolves, out of respect for Brigid, had come from the great forests of Mag Fea and were working hard at the task of rounding up and driving the pigs, not harming them and acting with great intelligence contrary to their natural instincts. When they came up to the farmer and his helpers the wolves left the pigs and returned to their forests, and the next day the farmer and the helpers arrived back with the animals telling everyone of this wonderful occurrence.

The story of the tame and untamed Fox
Cogitosus also tells a story about two foxes:

> One day an uneducated man saw a fox walking in the king's palace and thought it was a wild animal, not being aware that it was tame and a pet of the king's court. This fox was very intelligent and had learnt various tricks and often entertained the king and the court. However, being unaware of this the man killed it in front of members of the court, all of whom denounced him, whereupon he was taken and put in shackles and brought before the king.
>
> The king was furious when he learnt what the man had done and ordered him to be executed and his wife and sons reduced to slavery, and all his possessions seized unless the man gave him a fox that matched his own animal in all the tricks it had done. When Saint Brigid learnt of this she prayed to God on behalf of the man who she felt had been unjustly sentenced. She ordered her horses to be put into the harness of her chariot and drove to the king's palace.
>
> The Lord had heard her prayers, and as she travelled a wild fox came running up to her chariot and nimbly jumped up into it and sheltered quietly under her cloak. So, when she came before the king she pleaded for the prisoner, saying he was ignorant of the true nature of the king's pet and should be released.
>
> However, the king refused to listen to her entreaties and said that he would not let the man go unless he was given back a fox as tame and clever as his own. Brigid then produced her fox for all to see and, in the presence of the king and the whole crowd, it showed all the traits and the tameness of the other fox and performed various tricks in the same way as the previous animal had done. When the king saw this, he was placated and, with all the courtiers applauding the tricks of the fox, he had the man unshackled and freed him.
>
> However, after the man had left and Brigid was on her way home the cunning fox slipped away out of the palace and, despite the horsemen and dogs sent out to recapture him, returned to the wild and his den. All admired what had taken place and venerated Saint Brigid who performed such feats because of her sanctity and many virtues.

Birds

Saint Brigid and the wild ducks
Cogitosus also tells the story of St Brigid and the ducks:

> Another day, when Saint Brigid saw ducks swimming in the water according to their natural instinct and occasionally flying through the air, she bade them come to her. In winged flight and with remarkable zeal for obedience, they began to fly to her in flocks without any fear, familiar with her calls as though domesticated. She touched them with her hand and took them in her arms and, after doing this for some time, she let them go back flying into the air on their own wings.
>
> And, through creatures visible, she praised the invisible creator of all things to whom all animate things are subject and for whom all things live, as one says in the recitation of the office. And from all this, it can be clearly understood that the whole of nature, beasts, cattle and birds, was subject to her power.

Saint Brigid and the oyster catcher
There was a tradition, found in both Scotland and Ireland, of associating the oyster catcher (*Haematopus ostralegus*), a bird found widely distributed on the coasts of Europe, with this generous saint. In Scotland it was referred to as *Giolla Bhríde* – 'the servant of Bride', the same name being used in Ireland where it was called *Gille Bhríde* , while on the Scottish islands of North and South Uist it was simply called *Bhridein* – 'bird of Bride'. In Ireland there is a traditional story concerning the oyster catcher which tells how it was as generous as St Brigid, foolishly as it turns out, lending its webbed feet to the seagull who never returned them. This explains the traditional Irish saying *Iasacht an roilligh don bhfaolóig* – 'A loan given by the oyster catcher to the seagull', in other words a loan that will never be returned.

In the Hebridean Islands another story is told about the oyster catcher. There a traditional tale said that St Brigid was the Foster Mother of Christ (*Muime Chriosda*). On one occasion when she had charge of him the young Jesus was chased by his enemies and ran to the seashore, where he sank down exhausted. A pair of oyster catchers laid him between two rocks and covered him with seaweed. When his enemies reached the shore the two oyster catchers flew up making a great noise, so the enemies of Christ thought they must have been the first there to disturb the birds and so Christ could not have passed that way. Because of this oyster catchers have always carried a white cross that can be seen when they fly up, and were regarded as St Brigid's faithful servants.

There was also a tradition in South Uist in the Outer Hebrides, that St Brigid had come ashore there at a place called Kilbride, with an oyster catcher perched on each wrist like a hunter carries a hawk. A description that sounds more like that of a goddess than a saint. The reappearance of oyster catchers on St Brigid's Day was regarded as heralding the beginning of spring. Another story says that when St Brigid went to Britain to convert the pagans there she carried an oyster catcher in her pocket. When she arrived safely she released it, and it flew back to her monastery at Kildare to bring news of her safe arrival.

Saint Brigid and the linnet
The linnet (*Carduelis cannabina*), in some parts of Scotland, was called *Bigein Bhríde*, 'Bride's little bird', and its whistling in the spring was said to be welcoming her return. It was also

believed that if the linnet was heard singing on St Brigid's Day it was a good omen for spring. In the summer it develops red feathers on its breast and head which was said to be scorch marks from going too close to St Brigid's sacred fire. In some parts of Ireland the linnet is called *Gille Bhríde* – 'the page or servant of Bride' (as, confusingly, is the oyster catcher) and is referred to by this name in the verse:

> Poor page of Bride,
> What cheeping ails thee?

The falcon of Kildare

Gerald of Wales relates a story that he presumably heard from locals at Kildare in the twelfth century about a falcon:

> From the time of Brigid a noble falcon was accustomed to frequent the place, and to perch on the top of the tower of a church. Accordingly it was called by the people 'Brigid's bird' and was held in great respect by all. This bird used to do the bidding of the townspeople and the soldiers of the castle, just as if it were tamed and trained in chasing and, because of its own speed, forcing duck and other birds of the land and rivers of the plain of Kildare from the air down to the ground to the great delight of the onlookers.
>
> For what place was left to the poor little birds, when men held the land and the waters and a hostile and terrible tyrant of a bird endangered the air? A remarkable thing about this bird was that it did not allow any mate into the precincts of the church where it used to live. When the season of mating came, it went far away from its accustomed haunts and, finding a mate in the usual manner in the mountains near Glendalough, indulged its natural instincts there. When that was finished it returned alone to the church.
>
> In this it showed a good example of honour to church men, especially when they are entrusted with divine office within the precincts of the church. Exactly at the time of the first departure of Lord John from Ireland, that bird, which had lived for so many generations, and had so agreeably added interest to the shrine of Brigid, having occupied itself without sufficient caution with the prey which it had caught and having too little feared the approaches of men, was killed by a rustic with a staff which he had in his hand. From which it is clear that one must ever fear in prosperity a turn of the tide and that one should have little confidence in a daily life that is delightful and well loved.

Insects

Saint Brigid and the bees

St Brigid was well known for her charity and was upset when she was unable to give as much as she wished. According to many of the traditional stories about her a miracle usually occurred that enabled her to give as generously as she wished. This is exemplified in this story:

> One day a man went to Brigid to ask for a *sextarius* [one and a half pints or 0.85l] of honey, but unfortunately she had none in stock and felt really bad about being unable to help. Suddenly the hum of bees was heard coming from beneath the floor of her house and when the spot where the humming came from was examined a sufficient quantity of honey to fulfil the man's requirements was found, and so the man received all he asked for and joyfully returned to his home.

Fish

Saint Brigid and the fishing season
As in Scotland, Irish fishermen regarded the start of the fishing season as beginning on *Fhéile Bhríde* (St Brigid's Day). As a group they tended to be particularly superstitious, understandable in view of the dangers of their job. They firmly believed it was very unlucky and almost a waste of time putting to sea if, on their way to their boat, they saw a women with red hair or barefoot, a priest, a hare, or a fox. If you wanted to ensure that a fisherman would have no luck at all that season you only had to say:

> May there be a fox on your fishing hook,
> And a hare on your bait,
> And may you kill no fish until St Brigid's Day!

Molluscs

The placing of the shellfish
It was believed in Ireland that the tide nearest to St Brigid's Day was the greatest spring tide of the year and was called the *Rabharta na Fhéile Bhríde*. People on the coast would gather seaweed to put on the fields as fertiliser and gather shellfish for consumption. However, in some villages around Galway Bay there was a tradition of placing a live shellfish such as a limpet (*bairneach*) or periwinkle (*faechóg*) at each corner of the house to bring good luck with the fishing in the coming year and ensure a good harvest from shore gathering.

Plants

The dandelion
A plant sacred to St Brigid in Ireland, the Highlands of Scotland and the Hebrides was the Dandelion (*Taraxacum sp.*), called *Caiserarbhan* in Irish and *Bearnan-Bhríde* in Scottish, which means 'the notched plant of Bride', a reference to its serrated leaves. This is one of the few plants that appear around St Brigid's Day and may also be associated with her because of its milk-like sap, which was believed to provide nourishment for early lambs, although the plant has many other uses.

St Brigid was often likened to the sun and there are indications that the Goddess Brighid may have had an indirect sun connection. Since the Dandelion resembles a sun and appears at start of the Celtic spring, it is an appropriate plant to represent both the goddess and the saint, so may, perhaps, have originally been sacred to the Goddess Brighid.

The picking of the Torranan
In the Hebridean islands St Brigid was invoked when picking the *Torranan*, a member of the Figwort family (probably *Gratiola officinalis*), a plant which has a white flower resembling a woman's breast. The cup of the flower is said to fill with dew as the tide comes in and dries as the tide falls. The plant was used in a ritual to increase the quality of the milk, so improving butter or other products made from it. The plant was picked when the tide was coming in, taken to the dairy and passed three times over the milk pail or churn while saying an invocation to St Brigid

to increase the yield and provide a blessing, before placing it under the milk pail or churn. This example, recorded by Carmichael, was recited by Isabel MacEachainn of Bunessan, Mull:

> I will cull my gracious root
> As Brigit culled it with her one hand,
> To put essence in breast and gland of milk,
> To put substance in udder and in kidney,
> Butter and curd, fat and cheese,
> Like stream pouring from breast of fortune,
> Like honey distilling from the love on high.
>
> Thou only anointed white One of the God of grace,
> Keep Thou for me mine own,
> Keep Thou to me the share of grace,
> Keep Thou from me the goods of foes,
> Keep Thou from me the folk of lies,
> Keep Thou from me the [word forgotten] of death,
> Keep Thou from me the [word forgotten] of harm,
> Keep Thou from me the repairing,
> Keep Thou from me the false calf. [i.e. aborted]
> The plant of heathgrass be theirs,
> The plant of substance be mine,
> [Line forgotten].
>
> There is blessing in the crop,
> There is fruitage in the kine,
> There is honour in the pairing,
> Thou only anointed White One of the God of grace.
> In dependence on the God of life,
> In dependence on the power of health,
> Keep Thou for me my share desired,
> Keep Thou for me my share beloved,
> Keep Thou from me the [word forgotten] of bane,
> Keep Thou from me the howling of foes.
>
> Brigit came homeward
> With butter, curd and cheese,
> And she laid under the nine firm locks
> The nine stocks securely—
> The stock of the God of life,
> The stock of the Christ of love,
> The stock of the Spirit Holy,
> The Triune of grace.

Bay laurel

Another plant associated with the saint in Scotland is the Bay Laurel (*Laurus noblis*), known as the 'Shrub of St Bride' and although it does not flower on her day, its yellow flowers do appear in spring to be followed by black berries on female plants. Its leaves are used to flavour many dishes.

Betony
In Wales the Betony (*Stachys officinalis*), a plant whose purple flowers appear between June and October, is called *cribau Sant Ffraid* (St Ffraid's comb), perhaps because its serrated leaves resemble an ancient comb.

Watercress
In one of the early Irish *Lives* of the saint is related a story concerning watercress (*Rorippa nasturtium aquaticum*):

> Saint Brigid was at a certain place where there were many rivulets, but these did not contain the water-herbs [watercress] that usually grow in a natural state on streams supplied by springs. While there a band of holy virgins, belonging to the place, came to visit and ask her a question: 'Why, O mother, do not the water-herbs on which holy men are accustomed to live, grow in these waters?' Brigid, knowing that they desired a growth of such herbs there, spent the following night in vigil and prayer. On rising the next morning those religious found the rivulets filled with such herbs, while others grew for a considerable distance around, where they had not hitherto been seen.

Saint Brigid and the carrots
In the Hebrides the wild carrot (*Daucus carota*) was once of great significance to the people as it symbolised fertility and children. The carrot was given by a woman to a man, but rarely from a man to a woman. It was the girls and women who gathered the wild carrots, and when a woman found a bifurcated carrot she cried out excitedly an invocation to saints Michael, Brigid and Mary:

> Little cleft one! Little cleft one!
> Joy of carrot surpassing to me!
> Little cleft one! Little cleft one!
> Fruitage of carrot surpassing to me!
>
> Michael militant will give me seed and fruit,
> Calm Brigid will give me passion,
> Fite Fith will give me wine and milk,
> And Mary mild will give me aid.

Trees
In Ireland there were many individual trees that were regarded as sacred (*bile*), and were venerated, or at least regarded as of great significance, into early Christian times, and some till quite recently. These trees were usually replaced if lost due to old age, disease or being cut down by an enemy, a not uncommon occurrence in the early days. Some of these trees were associated with saints and one, now vanished, was apparently connected with St Brigid. The townland name of Bridetree in Co Dublin almost certainly commemorates a tree, which has now disappeared, that locals regarded as sacred to St Brigid and this is backed up by the presence of a holy well there dedicated to her.

Mythological animals

The dragon who knew of Saint Brigid

There is a curious story concerning St Brendon, known as 'The Navigator' (*c*.486–575 AD) a contemporary of St Brigid and, according to tradition, a great friend of hers, who composed a hymn in her honour beginning 'Brigid a woman ever good'.

> One day Brendon came to visit Brigid and told her of a strange encounter with some sea dragons during one of his voyages, and asked her what might be the meaning of it. Brendon was standing on a lofty crag and saw below him two sea dragons trying to drown each other, when suddenly one of the monsters cried out to the other in a human voice 'I beseech you in the name of St Brigid to let me be'. On hearing this the other dragon at once withdrew to the depths of the sea. What puzzled Brendon was why the dragon had referred to St Brigid despite the fact that he himself was present. Brigid then said that each of them should make a statement of belief that might provide an answer to this puzzle.
>
> Brendon said 'I declare I have never crossed seven waves without turning my mind to God'. Then it was Brigid's turn: 'I confess that since I first fixed my mind on God, I have never taken it off and never will till doomsday.' Brigid then explained that because Brendon was constantly exposing himself to danger during his voyages it was natural that he had to concentrate on his own safety, and it was not because he forgot God that he thought of him only at every seventh wave. So Brigid had to kindly and modestly explain that the dragon had called on her as it felt she was the holier of the two!

Rocks

In Britain and Ireland there are many prehistoric standing stones and naturally formed rocks that have a distinctive shape that makes them stand out, and often these have names attached to them. Sometimes the names reflect their apparent shape, e.g the Devil's Arrows, North Yorkshire (prehistoric standing stones), or their supposed builder e.g. the Giant's Causeway, Co Antrim, Ireland (a natural rock formation).

There is a stone known as St Brigid's Chair at Lough Derg, Ireland, and there was an old stone with several crosses incised on it at Foughart Old Church in Co Louth that was called, in the nineteenth century, St Brigid's Stone. Near the ancient church of Killinagh, near the village of Blacklion, Co Fermanagh, Ireland, is a *Bullaun* rock known as St Brigid's Stone. This is nearly on the shore of Lough Macnean and consists of a boulder of red sandstone measuring 1.74m x 1.57m (5ft. 9in x 5ft 2in) and on its table-like surface are nine round depressions around a larger central one. This was described by W.F. Wakeman in his article *On Certain Markings on Rocks, Pillar-Stones, and Other Monuments, Observed Chiefly in the County Fermanagh* published in the *Journal of the Royal Society of Antiquaries of Ireland*, 1879–82.

At that time each depression contained a smooth, generally oval, stone which nearly filled it, not all the stones being of the same type. Wakeman was told by a lady whose family had lived in the neighbourhood for a long time that, when she was a young girl, she was often taken to the stone by an 'aged nurse' who performed some kind of devotions there and distinctly remembered her calling it 'the cursing stone'. The name for these 'monuments' comes from the Latin *bulla* – 'a bowl'. They are found in great numbers all over Ireland, often near old churches and monastic sites. No written evidence for how they were actually used survives.

Brigid: Goddess, Druidess, Saint

58 St Brigid's stone, near the ancient church of Killinagh, Blacklion, Co. Fermanagh, Ireland. This red sandstone boulder with its nine depressions was known as 'the cursing stone' in the ninteenth century.

Not far from this rock was a well dedicated to St Brigid which was covered with a Well House in the nineteenth century and was greatly venerated at that time, while a nearby megalithic tomb was known locally as 'St Brigid's House'.

The St Brigid well at Kildare with its prayer stones is much visited today and beautifully maintained. The water from the well flows under a stone arch and through two adjacent stones which are called St Brigid's Wooden Shoes *(Bróga cailteach Bhríde)* or St Brigid's clogs as they resemble these items of footwear. An isolated hill made of different material from the surrounding red sandstone rock at Kingenny, Angus, Scotland, is known as St Bride's Ring, while a St Bride's Rock is found at Tomantoul, Aberdeenshire.

It is rare for stones to be associated with the names of saints in England, but an exception may be found in the case of St Brigid whose name is attached to two groups of locally notable rocks. Although it should be borne in mind that the English examples may have a different origin for their name. One is a group of prehistoric stones at the junction of three valleys at Calderdale, Yorkshire, known as the Bride or Bridie stones. The other is the Bridestones, the remains of a prehistoric burial chamber on the Cheshire/Staffordshire border, which may be named after the saint, although a local story says they mark the spot where a Viking nobleman and his Anglo-Saxon wife are buried. At Locoal in the Peninsula of Plec, France, there is a chapel dedicated to St Brigid and nearby is a prehistoric standing stone nearly 2.75m (9ft) high which is called *la Quenouille de Ste. Brigitte* (St Brigid's fish/meat roll), while not far from this is another that is called *Brigitte Fuseau* Brigid's Spindle.

The Goddess Brighid and snowdrops

The snowdrop (*Galanthus nivalis*) is associated with the modern veneration of the Goddess Brighid, but has long been connected with the Virgin Mary. In the medieval period the image of Mary was removed from its niche on Candlemas Eve (1 February) and strewn with snowdrops. The statue was replaced the next day, Candlemas or the Feast of the Purification of the Virgin Mary. In the Hebrides the *Bhrideag* (image of St Bride) was decorated with snowdrops (see page 135). The snowdrop features on many modern *Imbolc* greetings cards and illustrations of the Goddess Brighid.

Thirteen

The relics of Saint Brigid

The veneration of the remains of important people occurs in many world religions, and the earliest cult of Christian relics is found in the Mediterranean area, particularly the Holy Land. The remains of early Christians buried in the catacombs of Rome also quickly became objects of veneration, and services were held at their tombs. It was not long, however, before their bones were disinterred, divided up, placed in protective casings or portable shrines called reliquaries and taken all over the Christian world. The cult of relics rapidly spread from its area of origin to parts of Europe where Christianity had taken hold. In 396 AD Victricius, Bishop of Rouen, travelled to Britain with relics he had obtained in Northern Italy, regarding himself to be under their protection and offering their protective power to the whole Christian congregation in Britain.

In Ireland this 'fashion' led to the veneration and enshrining of relics (*mind*) of local saints in reliquaries (*cumdach*) and as remains of the individuals were naturally limited and demand so great, then objects connected with these people such as bells, books, croziers etc. also became venerated as relics. Indeed, relics were so much in demand that even a piece of cloth that had touched the remains of the saint (*brandea*) came to be regarded as sacred and were also enshrined.

In Ireland relics date to the earliest days of Christianity. Even Palladius, the first known Christian missionary to arrive there in 431 AD brought with him relics of St Peter, St Paul and other saints. Less than two centuries later Ireland had an extensive cult of relics and they were being sold and even stolen. Many Irish relics of the saints were lost in Viking raids from the eighth century onwards, but the most popular time for relics was in the eleventh and twelfth centuries when many saints' bodies were disinterred in both Ireland and Europe and their remains divided up and placed in reliquaries. Many other saint's remains were relocated from their original resting place to a new position of greater honour, a process known as 'translation', from the Latin *translatio*.

For a church to possess a relic of its founding saint added greatly to its status, which was further enhanced if it had relics of a number of saints. This coincided with an increase in the writing of *Lives* of the saints which gave even greater credibility to the claims of the church for its relics. The Anglo-Normans, who invaded Ireland in the twelfth century, recognised the importance of relics and the part they could play in maintaining control over the areas they had colonised.

Brigid: Goddess, Druidess, Saint

In places under direct English control the Reformation saw the destruction of many relics and reliquaries in the sixteenth century. Despite this a number did survive, mostly in 'less English' areas and relics continued to be made down to the eighteenth century. Earlier ones were sometimes repaired and refurbished for centuries as they were often carried in processions on the saints' day, appeared at fairs and other gatherings, and were even hired out for use in the swearing of oaths and to effect cures. A relic associated with St Brigid was used for cures as early as the mid-seventh century at Kildare Cathedral.

The earliest reference to the relics of St Brigid is found in the description by Cogitosus, *c.*650 AD. He describes the tombs of St Brigid and St Conlaed which were situated on each side of the high altar in Kildare Cathedral, the greatest place of honour. The tomb shrines (*scrín*), adorned with gold, silver and precious stones and gems, were probably made of wood covered in metal plates to which the jewels were fixed (see page 53).

Hanging above them were crowns of gold and silver and Kildare is the only Irish church known to have used such votive crowns, which is a practice found in Europe. There it seems to have originated in the late 320s or 330s AD when the Roman Emperor Constantine donated a crown to St Peter's basilica where it was placed in front of the shrine of St Peter. He later gave four crowns to the basilica of St John Lateran and by the seventh century this was a custom found in churches with royal connections in France and Italy. It therefore seems likely that these crowns were royal gifts to the cathedral, although in the case of St Brigid it may also have been seen as an allusion to the Crown of Virginity. In one of the *Annals*, under the year 800 AD, it says that the relics of St Conlaed, first Bishop of Kildare, were enshrined, suggesting that fragments of his skeleton were taken from his shrine at Kildare at this date, put in reliquaries and distributed to other churches or monasteries.

However, there were other items associated with St Brigid at Kildare in the seventh century that were regarded as relics, such as the greenwood altar in the church (see page 227), and the ancient door on the north side of the cathedral which had come from the first church at Kildare and through which the saint herself used to enter the building (see page 55). A millstone placed by the entrance to the church enclosure was believed to have healing powers (see page 57). At Kildare pilgrims would also take away pieces of an ancient oak tree as relics, a sacred symbol that pre-dated the change of the site from a pagan temple to a Christian church (see page 51). The most important relic of St Brigid was her body, which was placed in a shrine in the cathedral on her death around 525 AD to which pilgrim's came from far and wide to pray and show their veneration to the saint. However, with the increase in Viking raids which began in 795 AD and which became more frequent and sustained from 830 AD, often penetrating far inland, there was great concern for the safety of her tomb. About 835 AD, or possibly a few years earlier, the remains of St Brigid were removed to *Dun* (later *Dun Pádraig*, now Downpatrick) for safety.

There they were deposited with the remains of St Patrick (*Pádraig*) who died *c.*464 AD, although the *Annals of Ulster* give his date of death as 493 AD, which is regarded as more accurate by some historians. Later the remains of St Columba (Columcille) were also re-interred alongside the other two saints. Columba died *c.*597 AD and was originally buried on the Scottish island of Iona, but his remains were taken to Ireland in 875 AD when Vikings started plundering that part of the Scottish coast. *Dun* also came under attack by the Vikings and it seems the remains of the saints were concealed, with their exact location being a closely guarded secret known only to a few people. However, over a period of time the place where the remains had been hidden seems to have been forgotten, although people continued to be aware that the relics were hidden somewhere there.

The Anglo-Norman invasion of Ireland began in 1169 during the reign of Henry II. One of the nobles who arrived the following year with 200 knights and 2000 men was Richard de Clare, Earl of Pembroke, nicknamed 'the Strongbow'. During his campaigning he was responsible for slaughtering many people, destroying towns and farmsteads and damaging Irish churches and other religious sites as well as enforcing various reforms of Pope Adrian. He was deeply resented by the Irish who regarded him as 'the greatest destroyer of the clergy and laity that had come to Ireland since the time of the Norse tyrant Tugesius.' However, according to tradition his sacrilege did not go unpunished as a fatal ulcer broke out on his foot 'through the miracles of St Brigid, St Columba and the other saints, whose sanctuaries he had desecrated.' Nor did he apparently die in peace as on his deathbed he had a vision in which he saw St Brigid in the act of killing him in punishment for the desecration of her holy shrines in Leinster! This was not the only instance of a ghostly appearance by St Brigid. At the battle of Almhuin in 720 AD she was 'seen over the men of Leinster', which assured their victory, despite their army of 9,000 being pitted against a force of 21,000 men led by Fearghal, High King of Ireland, who was killed.

In 1185 Malachy III was Bishop of Down and often prayed that the relics of Ireland's most important saints might once again be revealed and one night this did indeed apparently happen. Malachy was praying in the cathedral church when he saw a light like a heavenly sunbeam extending through the church until it stopped over the hidden graves of the saints. He was overjoyed at this vision and fervently prayed that the ray of light might remain until he uncovered the relics. He then went and quickly got some tools and began to dig until he came across the bones of the three saints. The grave of St Patrick was in a central compartment, with the bodies of saints Brigid and Columba on each side. He took up the remains and placed them in three coffins which he reburied in the same spot, carefully noting their exact location.

Malachy then went to John de Courcey who, the bishop knew, was sympathetic to the cult of relics as he had demonstrated some years before. John de Courcey, during his conquest of Ulster had captured a collection of holy relics and their shrines at the battle of Downpatrick in 1178 when the Irish tried to retake the city. They had taken the relics into battle as aids to success, but shortly after their capture de Courcey arranged for a number of them to be returned, although the main reason for this was probably to gain the support of the Irish. John de Courcey agreed with the bishop that news of the discovery should be sent to Pope Urban III in Rome with a request that the relics be translated to a place of honour. Pope Urban agreed to this and immediately dispatched Vivian, the Cardinal Priest of St Stephen, to Ireland to act as his representative.

An account of the discovery was also recorded in *Topographia Hiberniae* (*The Geography of Ireland*) written *c.*1187 by Gerald of Wales:

> Saint Columba and Saint Brigid were contemporaries of Saint Patrick. Their three bodies were buried in Ulster in the same city, namely, Down. They were found there in our times, in the year, that is, that Lord John first came to Ireland [1186], in a cave that had three sections. Patrick was lying in the middle, and the others were lying one on either side. John de Courcie, who was in command there, took charge when these noble treasures were, through divine revelation, found and translated.

The public translation of the remains of the three saints took place at Down Cathedral on 9 June 1186, St Columba's Day. The remains were taken up from their place of discovery with due

Brigid: Goddess, Druidess, Saint

reverence and solemnly carried to their new shrines located in a prominent position within the cathedral. In attendance were fifteen bishops, many abbots, provosts, deans, archdeacons, priors and other clergy, as well as important members of the laity. *The English Martyrology* records an inscription on the new tomb of the three saints which, interestingly, put St Brigid first:

In burgo Duno, tumulo tumulantur in uno,
Brigida, Patricius, atque Columba pius.

In the town of Down, entombed in a single burial mound,
are the Holy Brigid, Patrick and Columba.

There was another troubled period in Irish history in the thirteenth century, and the monastery and cathedral at Down was damaged or destroyed a number of times. At one point the monks offered the bodies of the three saints to King Henry III (1216–72) in return for letting them settle in some safe place of sanctuary in England, but he did not take up their offer. It seems that Irish saints, even of the importance of Patrick, Brigid and Columba were of little interest to the English.

The story of the rediscovery of the remains of the three saints was not quite settled. Despite their re-internment in Down in 1186, Nicholas mac Mael Íosa, Archbishop of Armagh, also claimed to have found the remains of Saints Patrick, Brigid and Columba at Sabhall in 1293!

59 Downpatrick Cathedral, Co Down, Ireland. The remains of St Brigid were moved from Kildare Cathedral to Downpatrick Cathedral about 835 AD because of increasing Viking attacks. They remained there until 1538 when the cathedral was destroyed and they were 'carried off'.

(Sabhall could be either Saul, Co Down, to the north east of Downpatrick or another Saul near Armagh). According to *The Annals of the Kingdom of Ireland by the Four Masters*, 'they were taken up after him and great miracles were afterwards wrought by means of them; and after being honourably covered they were deposited in a decorative shrine', although it is unclear in the *Annals* exactly where this was.

In 1315 occurs a reference to the 'See of Down where Patrick, Brigid and Columba lie in one tomb'. Shortly after this Edward Bruce, a Scot who had successfully fought the English in Scotland and had been invited to Ireland to drive out the English there, destroyed the cathedral at Down. However, the relics were obviously moved to a place of safety before this as there is a further reference to them in a document dated 1512 written by Tiberius, Bishop of Down and Connor. In 1538 Leonard Gray Governor of Ireland under Henry VIII, destroyed the rebuilt cathedral and, according to the *Annals of Ulster*, carried off the relics of the three saints but does not say where they were taken, and there is reference to statues of the three being destroyed. Another later writer, Friar Edmund McCana, stated that his grandfather saw Edward, Lord Cromwell, burn the remains of the three saints in 1607.

The cathedral was a place of pilgrimage as it housed the bodies of these three saints. However, after the destruction of the cathedral in 1538 Down gradually lost its prestige, and when Richard Stanihurst (*c.*1545–1618) described it he said it had 'lost much of its former splendidness and had contracted from a town to a village'. Although the cathedral remained a ruin for 250 years there was a strong local tradition that the graves of the three saints were still there, but concealed somewhere on the south side of the building, and there is a reference to a chapel in that area being dedicated to them. During the eighteenth century there was a local belief that the graves of the saints lay outside the cathedral walls, and this inspired John Irvine to write in his *Treasury of Irish Saints*:

> In Downpatrick on the hill,
> St Patrick, Bride and Colmkille,
> Underneath a leafy shade,
> In one grave at last were laid.

Another eighteenth century tradition said that when the cathedral was undergoing renovation three stone coffins were found under the high alter. It goes on to claim that they were reburied outside and the market cross erected to mark the spot, although this is rather unlikely. However, today most people still believe that the remains of the three saints are buried somewhere beneath the cathedral. A boulder of granite from the Mountains of Mourne, inscribed with the name Pádraig (Patrick) and a Celtic cross, was placed in the graveyard in 1901 as a memorial to the saint, not an indicator of the exact site of the burial. While originally there were two churches dedicated to St Patrick in Downpatrick, more recently two others have been built, one dedicated to St Brigid and the other to St Columcille (Columba).

However, not all of St Brigid's bones have been lost as the whereabouts of her skull is still known. Tradition says that after the re-internment ceremony in 1186 some 'Irish knights' took the head of St Brigid to the Duke of Vienna, Heinrich Jasonmirgett, who was a great supporter of the Irish monks on the continent, and even built a monastery for them, *Heilige Maria zu den Schotten* (St Mary of the Scots). The skull was initially placed in the town of Wiener Neustadt about 1192. Later it was taken to Prague where it was kept in the private chapel of the Emperor. The Holy Roman Emperor, Rudolf II, gave the skull to Juan Borgia in 1587 and it was brought, with other relics to Lisbon, Portugal where it remains today.

Brigid: Goddess, Druidess, Saint

A stone on the exterior of the parish church of St John the Baptist at Lumiar on the outskirts of Lisbon reads:

> Here in these tombs lie the three Irish Knights who brought the Head of the Blessed Saint Brigid, a native of Ireland, which relic is preserved in this Chapel; in memory of which the officers of the confraternity of the same saint caused this to be made in January 1283.

In 1884 Cardinal Moran was appointed Archbishop of Sydney, Australia and went to Cologne in Germany to obtain a relic of St Brigid for the Brigidine nuns of Sydney. In 1903 a fragment of St Brigid's skull from Lisbon was given to the church of St Brigid at Kilcurry, Co Louth, thanks to the efforts of Sister Mary Agnes of the Dundalk Convent of Mercy. In 1934 Cardinal McRory instigated a National Pilgrimage which is still held annually on the second Sunday in July at Foughart near Dundalk, Co Louth, traditionally the place where St Brigid was born. The holy relic of the saint is carried in procession from St Brigid's Church, Kilcurry to her shrine at Foughart, in the ornate reliquary which is normally kept beneath the high alter of Kilcurry Church. St Brigid's Church in Ardagh village, Co Meath, was completed in 1881 and also contains a skull fragment relic of St Brigid, received as a gift in 1903.

Other Saint Brigid relics in Ireland

Saint Brigid's Shroud (Recholl Bhríde)
The church of Scrín, Co Meath once housed a shroud in which St Brigid was supposed to have been wrapped on her death. This relic was listed in St Adamnan's catalogue of religious treasures compiled in the seventh century AD. In early Ireland people were often buried in their cloaks which acted as a shroud.

The relics of Saint Brigid at Armagh
There is a reference to 'relics of Saint Brigid' being preserved at Armagh church which is dedicated to St Brigid and said to have been founded by her. However, they were destroyed during an Anglo-Norman raid on the city in 1179 and two sculptured crosses associated with the church of St Brigid have also now disappeared. In more recent times the saint was commemorated by having a street named after her as well as a ward of the city.

Saint Brigid's Bell (Clog Bhríde)
St Brigid's Bell was once carried in procession in Ireland until 'banished' by Henry V (1413–22). This was claimed to be the very bell sent to St Brigid by St Gildas from his monastery in Brittany. It is said in some sources that Gildas visited Ireland and became great friends with St Brigid, and remained in contact with her when he left the country, an event that the *Welsh Annals* record as occurring in the year 484 AD. However, this is not the impression given in the *Life of Saint Gildas*, written *c.*884 AD:

> Now Saint Brigid, an illustrious virgin, who flowered at that time in the Isle of Hibernia, and was an Abbess over a community, on hearing the renown of Saint Gildas sent a messenger to him, saying with entreating words 'Rejoice, Holy Father and be always strong in the the Lord! I beseech thee to deem it worthy to send me some token of thy holiness, that the memory of thee may be, without ceasing, ever held in honour among us'.

Thus Saint Gildas, having heard the Holy Virgin's ambassador wrought with his own hands, a mould and according to her petition, constructed a bell, and dispatched it to her by means of the messenger whom she had sent. She joyfully took it, and gladly received it, as a heavenly gift sent to her from him.

The Shrine of Saint Brigid's Shoe (Scrín Bhróg Bhríde)

An unusual reliquary exists to contain a shoe that was reputed to have belonged to St Brigid. However, its original relic, the shoe, no longer exists, and it is not known when this was lost. The reliquary was preserved until the nineteenth century in a convent at Loughrea, Co Galway, and later came into the possession of Major Henry Sirr, and remained in his collection until it was purchased by the Museum of Antiquities of the Royal Irish Academy, now the National Museum of Ireland, where it is on display.

Made of copper alloy with the sole and heel soldered to the upper part and about 23 cms. (9in) long, it has on the front a head that an inscription identifies as John the Baptist. Below this are the clasps for a large gem, now missing, that was probably a rock crystal cut cabochon (rounded with a flat back), with a radiate design engraved around its edge. Below this is the figure of Christ crucified, which is above the remains of a clasp for another smaller, missing, squared jewel that also has a radiate design engraved around its edge.

Further engraved designs cover the sides and 'heel' of the reliquary. A Latin inscription indicates it was used for swearing oaths while another inscription bears the date 1410, although the museum regard this as a late example of a reliquary, perhaps dating to the sixteenth century. So common was the swearing of oaths on relics that the Irish word for oath, *mionn* in modern Irish, is derived from the word for relic, *mind*.

The Brigid Bell of Saint Patrick (Clog Bhríde ar leith Naomh Pádraig)

Bells are associated with many saints in the Celtic world and were used to perform miracles by their owners, but were also often given as gifts which were later venerated as relics. One of the most famous is St Patrick's Bell in the National Museum of Ireland, Dublin, but in the past there seem to have been a number of relic bells which were claimed to have once belonged to St Patrick, and he is supposed to have given a bell to St Brigid as a gift.

A bell reputedly associated with the two most venerated saints of Ireland would have been a very important relic, although it has now been lost. This bell is mentioned in the *Tripartite Life of Saint Patrick* written at the end of the ninth century. The saint spent forty days and forty nights on Cruachan Aigle where he was attacked by flocks of black demon birds:

> He sang maledictive psalms at them. They left him not because of this. Then his anger grew against them. He strikes his bell at them, so that the men of Ireland heard its voice, and he flung it at them, so that its gap [clapper] broke out of it, and that bell is 'Brigid's Gapling'.

Saint Brigid's Birth Stone (Bhríde Breithe Cloch)

There was a long standing tradition that St Brigid was born in a house at Foughart, near Dundalk, Co Louth where a church, now a ruin, is said to have been built on the site of this building. During the nineteenth century, and probably earlier, locals pointed out to visitors a stone at the back of the ruins that they said was the stone on which she was placed immediately following her birth. This was regarded as a very sacred relic and many miracles were said to have been accomplished though its 'instrumentality.' The graveyard also has a holy well dedicated to her.

Relics of Saint Brigid Outside Ireland

The Mantle of Brigid (Brat Bhríde)

In the cathedral of St Saviour in Bruge, Belgium, built about 1250, is *La Manteline de St Brigide d'Ireland* (The Mantle of St Brigid of Ireland). This is a rectangular piece of woollen cloth measuring 54.5 cm x 64 cm (21 x 25in), dark crimson in colour with the outer surface covered all over with tufts of curly wool so that it resembles a sheep's fleece. The relic was given to the church of St Donation at Bruge, along with jewellery and other gifts, by Princess Gunhild, the sister of King Harold II of England, who had taken refuge in Flanders following the death of her brother at the Battle of Hastings in 1066. It remained there till her death in 1087. In 1559 the church was elevated to the status of a cathedral. Gunhild got the relic from one of her brothers who had both been to Ireland. In 1051 her father, Godwine, Earl of Wessex, was exiled from England by King Edward the Confessor with whom he had quarrelled and took refuge in Flanders. His two sons, Harold and Leofwine, Gunhid's brothers went to Ireland where they stayed with Dermot, King of Leinster, while they raised an army and ships. This show of force reconciled Godwine and Edward, and the Earl and his sons returned from exile in 1052.

It is quite likely that these two young nobles would have visited the famous cathedral of St Brigid at Kildare, which lay in the kingdom of Leinster, as it was still a centre of pilgrimage although St Brigid's remains had been relocated to Down about 170 years earlier. At some point during their stay they acquired this relic, which they would have seen as conferring a blessing on their task of getting a pardon for their father so they could all return to England from exile. In addition, it is well documented that relics were carried on the battlefield to ensure victory, even those of female saints. For example when St Moninna was dying she addressed her nuns and bequeathed them various personal items, telling them 'through which, do not doubt, you will, by the grace of God, have victory if they are borne against any of your provincial enemies who come to devastate you'.

Gunhild chose to go to Flanders after the death of King Harold as there would have been friends there from the time of her fathers exile and took the relic with her. A cloak belonging to a saint would have been divided into sections to provide a number of relics. The first written record of the Brigid's cloak relic dates to 1347 when it appears in a list of property belonging to the Church of St Donation, and remained there until the eighteenth century. It was displayed to the people every 1 February, St Brigid's Day, to allow its veneration. The cathedral was destroyed at the time of the French Revolution, but the relic had already been moved to a place of safety and was later presented to the church of St Salvator, when it was adopted as Bruges's cathedral in place of the one which had been destroyed. Until the 1860s it looked like a mantle or shoulder cape made of yellow silk interwoven with gold thread and decorated with gold lace.

However, in 1866 it was closely examined and the 'cape' was found to be a decorative covering made around 1400. Inside this was found the actual relic which consisted of a piece of shaggy woven cloth with a lining of blue and green linen which showed signs of wear. At that time the relic was removed from its silk covering and a light lining of red silk taffeta replaced the blue and green linen. A further detailed examination of the relic took place in 1935. The woollen fabric's outer surface is covered by a curly fleece that gives it the appearance of astrakhan which comes from the young Karakul lambs of central Asia, while the inner side is perfectly smooth. The tufts of wool seem to have been made by incorporating unspun fleece at intervals with spun wool which, with beating and shrinking, eventually gave the appearance of a shaggy woollen cloth, a method quite unlike any carried out in Western Europe in modern times. This technique is an ancient one and similar material has been found in Bronze Age graves in Denmark.

Cloaks of this appearance were certainly used in Ireland. *Táin Bó Cuailgne* (The Cattle Raid of Cooley), gives descriptions of such items of clothing. This story was written down in the twelfth century although the language used show parts date to the eighth century, and some of the verses may even be two centuries older than this. In this the leaders of Ulster are described:

> Muremar son of Gerrcend who wore a dun coloured cloak of curly wool about him. Fergna, son of Findchoem who wore a cloak of red curly wool about him and the three sons of Conchobar who wore three curly red kirtles with brooches of silvered bronze…

This type of cloth was produced extensively in Ireland and several English writers in the sixteenth-century speak of Irish 'shag – rugs' or 'shag rug mantles'. So it seems this piece of woollen cloth, dyed crimson with madder, a plant-based dye, is ancient. While it cannot be proved it came from Ireland its early history suggests this is more than likely. So it may be a 'true' relic of St Brigid probably obtained from Kildare in the eleventh century, but whether it really did once belong to St Brigid is impossible to determine, and it may just have been a local tradition about it at the time. St Brigid's cloak appears in a number of stories about the saint, and this was probably the type of material that would have been used to make a cloak for someone of the rank of Abbess in the fifth and sixth century. In 1976 a further examination was undertaken by investigators in Brussels who concluded that this was an authentic sacred relic.

Saint Brigid at Abernethy

The people of the Hebrides had an ancient tradition that the remains of St Brigid were deposited not in Ireland but at Abernethy near Perth in Scotland. This was the former capital of the Picts, whose church was founded by King Nechtán Morbet, in thanks for the help of the saint who had enabled him to regain his kingdom when he prayed to her. When the Scots annexed the Pictish lands they are said to have treated the remains of the saint with great veneration. There was very close contact between Ireland and Scotland, particularly with the Pictish kingdom, from early times, with a daughter house of the Kildare monastery established there by 625 AD, so its quite probable that relics of St Brigid were indeed kept there, possibly even some of her bones.

For a long time Abernethy was the seat of the Metropolitan See and was a famous ecclesiastical centre from which the cult of St Brigid spread over Scotland. The seal of the church bears the image of St Brigid with a pastoral staff and her emblem, the cow. Some old writers actually claimed she was born in Scotland, but this was probably due to a misunderstanding. One of the early names for Ireland was Scotia Major, while Scotland was called Albanian Scotia or simply Albian, being grouped together with England, also called Albian.

The relics at Fulda

Candidus, a ninth century monk, gave a description of relics of St Brigid kept of Fulda in Germany and mentions that a chapel there was dedicated to her before 818 AD.

Brigid: Goddess, Druidess, Saint

60 St Brigid on a stained-glass window in St John Baptist church, Glastonbury, Somerset. She holds the relic bell from Beckery Chapel that was rediscovered in 1913 (see page 176). Above her head is a flame, while her blue robe and cloak symbolise her connection with the Virgin Mary. The wolf at her side relates to a miracle story (page 211).

The green wood altar
It was said that that a relic, a fragment of the wood that became green when touched by St Brigid (see page 56), was preserved at *Candida Casa* (the stone house), a chapel built by St Ninian on Whitethorn Island off the eastern coast of Galloway, Scotland. This is said to be the oldest Christian site in Scotland and housed the relic until the Reformation in the sixteenth century when it was lost.

The relics at Athol
A church of St Brigid at Athol in Scotland was once famous for miracles that occurred because of the relics of the saint that were housed there.

The Glastonbury relics
A bell, bag, necklace or collar and weaving tools, all said to have belonged to St Brigid, were kept at St Brigid's Chapel, Beckery, Somerset till the sixteenth century. The saint herself was to have left them there following her visit in 488 AD (see pages 166–67). One of St Brigid's ribs was among the many relics kept at Glastonbury Abbey.

Saint Brigid on Irish reliquaries

The Silver Sunday Shrine (Scrín Dhomhnach Airdid)
A reliquary in the form of a rectangular box made of wood, gilt copper alloy, silver, enamel and rock crystal contained fragments of a gospel book dating to the eighth or ninth century, the date of the oldest part of the reliquary. It was remodelled in the fourteenth century by John O'Cabry, Abbot of Clones. Flanking a large central crucifixion scene, is an elongated cross dividing the front into quarters, in which are the raised figures of saints, two in one of the quarters and three in each of the others. One of the saints is Brigid wearing an ankle length tunic with rounded neck, a hooded cloak (*cochal brat*) that falls almost to her ankles and pointed shoes. She holds the crosier of a bishop in one hand with a book in the other.

The Shrine of Saint Patrick's Tooth (Scrín Fiacal Naomh Pádraig)
This reliquary was made to contain the tooth of St Patrick which is recorded as falling out on the doorstep of the church of St Brone in Carboy, Co Sligo. It is shield shaped 30cm (12in) high by 24.1cm (9½in) wide and 3.8cm (1½in) thick and made of silver. Both the front and back are decorated with figures in relief, filigree work, and crystals, amber and glass. On the front is a crossbar with a crystal set in the centre, with above a Crucifixion scene and below the figure of St Patrick. Each quarter contains the figures of two saints, among which was St Brigid. Unfortunately her particular image is missing, but is known to have been there as her name is still inscribed below where it once was.

This is the only saint missing from the reliquary, and often pieces were removed from these for use in effecting cures by touching the part or being put into water which was drunk. So it may not be a coincidence that St Brigid, a saint particularly connected with healing, has been removed. The reliquary has a Latin inscription that shows that it was redecorated before 1376, retaining some portions of the original. This is now in the National Museum of Ireland, Dublin.

Fourteen

Stories of the Goddess-Saint

Tales of Brighid the Goddess

Brighid in pre-Christian times was widely venerated in Ireland as a goddess, and there must have been stories about her origins, deeds and interactions with other figures whether deities or humans or both. These would have originally been told by the *Bhaírds,* the Druids whose role was to tell the stories of the Irish deities and brave warriors, and recite the history of the people. However, these pagan stories were generally lost with the increasing Christianisation of the country, but a few of these very ancient stories were recorded by monks although given a Christian veneer.

Some of the deeds of the Goddess Brighid may have survived in the form of a few of the miracles attributed to St Brigid particularly in the earliest *Lives* written in the eighth and ninth century. The 'pre-Christian miracles' are likely to be those concerned with the multiplication of food and drink, those connected with livestock, particularly cattle, and those which show her ability to control the weather – all attributes Brighid got from her pagan deity parents (see page 29).

One of the pre-Christian stories concerning the Goddess Brighid was recorded in the twelfth-century *Lebor Gahbála Eireann* (*The Book of Invasions of Ireland*). This seems to be based on a ninth century work that itself recorded stories that had survived from earlier times in an oral tradition. This tells of successive invasions of Ireland and while it cannot be taken as an accurate account of the prehistoric and later periods in Ireland, it must be borne in mind that accounts of major events and upheavals can be remembered orally passed down the generations, although with embellishments and bits left out in successive recountings. These accounts and stories have to be read with both a modern understanding of history and archaeology, and making allowance for inaccuracies in copying. The 'recording' of stories and events orally was particularly strong in Ireland where the *Bhaírds* and later the *Senchaid*, the traditional reciters of stories in Christian times, kept alive the ancient tales.

There were many Irish gods and goddesses, their chief one being the Daghdha. These seem to have been a mixture of gods and goddesses that originated in Ireland, plus some who were imported into the country by various peoples who invaded or migrated to Ireland over the centuries. A number of deities were associated with one craft, effectively being its patron deity.

Stories of the Goddess-Saint

One aspect of early Irish Celtic belief that is unusual is that they regarded their gods and goddesses as their ancestors, not beings who had created the people or the world, which was a common belief among many other peoples. A number of Irish kings and important families claimed direct descent from one deity or another.

The medieval compilers of the *Lebor Gahbála Eireann* and other early books had to try and make sense of the pagan stories and a muddled genealogy of the deities. To add further confusion, several versions of the stories and genealogies would have survived in both oral and written forms. They not only had to make sense of the old traditions, but also had to present them in a way that suited the beliefs of Christians. To make the pagan gods more acceptable, they were sometimes turned into 'historical' characters from ancient times, but this gave rise to even more confusion and contradictions.

The early books were copied out by hand, and this too led to minor variations between the same story as given in different manuscripts. Translation from Irish into Latin (or vice versa) also gave rise to different versions. This explains why there are slight differences between the very few pre-Christian stories that have survived about the Goddess Brighid. For example the description of Brighid from three manuscripts of the *Lebor Gahbála Eireann* reads:

1. Brighid the poetess, daughter of the Daghdha, she had Fea and Feimhin, the two oxen of Dil, from whom are named Magh Fea and Magh Feimhin. With them was Triath, king of the swine, from whom is Treithirne. Among them were heard three demon voices in Ireland after plunder, to wit, whistling and outcry and groaning.
2. Brighid the poetess, daughter of the Daghdha, she who had Fe and Men, the two Royal oxen, from whom Femen is named. She had Triath, king of her boars, from whom Treithirne is named. With them were, and were heard, the three demoniac shouts after rapine in Ireland, whistling and weeping and lamentation.
3. Brighid the *banfhile*, daughter of the Daghdha, she who had Fe and Menn, the two Royal oxen, from whom Femen is named. She had Torc Triath, King of the boars of Ireland, from whom Mag Triathairne is named. With them were, and were heard, the three demoniac shouts after ravaging – battle uproar and wailing and outcry.

Brighid was the only daughter of the chief Irish god according to the *Lebor Gabála Érenn*:

The Daghdha had four children, Oengus mac Oc [or Oengus/Angus], Ord [or Aed], Cermat Coem, and a daughter called Brighid.

This genealogy is also reflected in a Scottish legend that says that Brighid spent the winter imprisoned within Ben Nevis by the *Cailteach* (Hag) until she was rescued by Angus (Oengus mac Oc), a young god who was her brother (see page 87).

The Daghdha was a member of the *Tuatha dé Danann* (the tribe of the Goddess Danu). These came to Ireland, according to a twelfth-century account and dispossessed some earlier inhabitants called the *Fir Bolg* after a battle at *Mag Tuireadh* (or *Moytura*), but they allowed the *Fir Bolg* to retain the Province of Connacht, while the *Tuatha dé Danann* held the rest of the country, building their capital at Tara.

During this battle the king of the *Tuatha dé Danann*, Nuada, lost his right hand. All early Irish kings had to be without any physical blemish, so he was forced to abdicate, and Breas was elected in his place. Breas's father belonged to another group of ancient Irish inhabitants, the *Fomhóire* (Fomorians), while his mother was of the *Tuatha dé Danann*. To further strengthen the alliance between the two groups, a dynastic marriage was arranged between Breas and Brighid, the daughter of the Daghdha. However, his reign was not successful as he was not a good king and eventually he too lost his throne, which led to a further battle. Known as the second battle of *Mag Tuireadh*, this battle was between the *Fomhóire* and the *Tuatha dé Danann* who were victorious.

The son of Brighid and Breas, Ruadhan, was taught to make weapons by the *Tuatha dé Danann* god of smithing, Goibhniú. He was famed for his magical weapons although another account says that Ruadhhan spied on him to learn his secrets. However, Ruadhán later seized a spear and plunged it into Goibhniú, but the smith had enough strength left before he died to kill Ruadhán. Brighid was heartbroken and keened for her dead son. 'At first she shrieked, and then she wept, then for the first time weeping and shrieking were heard in Ireland.'

While somewhat muddled, these few ancient stories about the goddess have come down to us from antiquity and probably date back as far as the 'conception' of the goddess in the first century AD. Cattle and boars were both of great significance in Celtic religion, the boar having many stories associated with it in Ireland, often involving shape shifting. It was a symbol of bravery, battle and warriors, but was also a symbol of hospitality. Pork was not only a favourite dish among the early Irish but also regarded as an Otherworld food.

Oxen [bulls] were not only of great agricultural significance but also represented power and, above all, fertility. They were regarded generally as sacred animals and used as sacrifices in some ceremonies. Often deities were associated with places in the landscape, and those mentioned in the account of Brighid's animals can still be identified today. Mag Fea (the Plain of the Barrow) is in Co Carlow, while Mag Femen is a plain in the south east part of Co Tipperary and Treithirne was the name of a plain in the west part of Co Tipperary. All these were locations within the kingdom of the Brigantes who had settled in Ireland, and a further link with this group and the Goddess Brighid.

These sacred animals cried out when Ireland was plundered or attacked suggesting some type of protective role, which may well have applied to Brighid since she had responsibility for these sacred animals. The mention of a great boar (*torc*) may be indicative of a warrior aspect, of which little is generally made, since the boar was symbolic of war. It was also representative of hospitality, which the Celts saw as almost a sacred ritual, using the feast to seal agreements between people or tribes in dispute as well as for general celebration and enjoyment. In the Celtic world, the boar is the most common animal depicted, being worn as a crest on the helmets of warriors, and appears on jewellery, in carvings and as bronze models, as well as making an appearance in many stories. Brighid's 'High' status is also indicated by the fact that her animals are described as 'Royal' or the 'King' of their kind.

The Goddess Brighid, a member of the *Tuatha dé Danann*, was married to King Breas so strengthening his links with the *Tuatha dé Danann* rather than the *Fomhóire*. This presumably conferred the blessing of the gods and particularly the chief of the gods, the Daghdha, on Breas since it was his daughter that he married. This may also represent the unification role that Brighid had when first 'conceived' by the Druids (see chapter two).

It is difficult to say exactly what is the meaning behind the treachery of Brighid's son. However, as Brighid was a patron goddess of smiths it may be a way of accounting for her having this attribute as she took on this role when her son killed the old god of smithing. Her son could not do this as he was also killed. Traditionally smiths were seen as intercessors with the gods as well as having almost magical abilities by turning ore into weapons and other useful items, initially of bronze and later of iron. So it would have been a powerful role to attribute to this goddess.

In Ireland smiths were specially honoured, and were classed among the *áes dana*, a privileged group which included judges, doctors and poets. They were granted certain privileges and also had the right to move freely between different tribes. In addition to making weapons and armour, Goibhniú was the host of the *An Saol Eile Fleadh Goibnenn*, the Otherworld Feast of Goibhniú. The participants of which drank of a special ale that he brewed which gave them immunity from disease and death, so Goibnhniu was also regarded as a god of healing (along with Dian Cécht). Yet another attribute the Goddess Brighid seems to have taken over on Goibnhniu's death.

The story also indicates that the tradition of *caoine*, anglicised to keening, a formal wailing of lamentations by women when a person dies, was said to have been invented by the Goddess Brighid. Keening was a common practice at Irish funerals until the nineteenth century and women were sometimes employed to do it. In *The White Goddess* Robert Graves looks at myth tales concerning the Goddess Brighid of varying dates that claim she was either the mother of the Daghdha, his mate or his three daughters (as well as his single daughter).

Another story says that three of the Daghdha's descendants, Brian, Luchair and Lucharba were the sons of Brighid (by her husband Tuireann). O'Rahilly in *Early Irish History and Mythology* discusses his interpretation of a ninth-century text that three gods, called Brian, Luchair and Uar were the sons of the *banfhili Brighid* (Brigid the woman poet). In one story her sons killed the god Cian, father of Lugh Lamhfhada when he was in the form of a pig. This could suggest that this too might be an old tradition, although these three figures in the *Lebor Gabala Érenn* are described as the sons of Mor-Rioghan, who was Brighids mother. This is a good example of the problems of trying to sort out the relationships between Celtic deities when they have come down to us in a confused form!

One story, purporting to be an incident from the life of St Brigid, sounds more as if it should apply to the Chief Druidess of Brighid or even the goddess herself. In Ireland many pagan stories feature magical or divine bulls, along with supernatural or Otherworld cows and, as in Scotland and Wales, they are often described as white with red ears. White indicating their sacredness and the red ears showing they were Otherworld animals. One ancient story says that St Brigid was only able to take nourishment from an Otherworld cow as a child, a story that appears in a number of *Lives* of the saint, such as this example from the *Old Irish Life of Saint Brigid*:

> When it was time to take her from the breast, the Druid became anxious about her; whatever thing he gave her, she vomited it at once, but her appearance was none the worse. 'I know', said the Druid 'what is the matter with the maiden, it is because I am impure'. Thereafter a white red-eared cow was set aside for her sustenance and she became well as a result.

The miracles of Saint Brigid

All saints are associated with miracles, and this is a qualification for sainthood in the Roman church. However, simply founding an abbey or building a church, plus leading a 'good life', was enough to gain a sainthood in the early Celtic church. In some cases miracles are associated with the saint's body after their death, but many were said to have performed miraculous acts during their lifetimes although the accounts of such events were often recorded long after their demise.

St Brigid, as a very popular saint with a widespread following, has lots of miracles attributed to her, many apparently carried out during her life, and the earliest account of her miracles

Brigid: Goddess, Druidess, Saint

61 St Brigid preaches to the people of Ireland. A carving in St Brigid's parish church, Kildare, Ireland. This panel was in the church before it was rebuilt in 1833.

62 A traditional story tells how when Brigid's father was in the King's castle, trying to sell her as a servant to him, he left Brigid in his chariot outside. A poor man asked her for alms and, having no money, she gave him her father's sword, a valuable item worth the equivalent of ten cows. This confirmed her father's reason for selling her as he was afraid he would be ruined because she gave so much away to the poor! (Illustration from *The Red Book of Saints' Stories*, 1952)

is found in Cogitosus's seventh-century *Life of Brigid*. These early *Lives* were not meant to be an accurate biography of the person but an affirmation of the saint's holiness, their faith and a confirmation of the power of God working through them. This is certainly the case with the miracles detailed by Cogitosus, who also uses them to emphasise St Brigid's many acts of charity. So the accounts of these miracles are not just a series of folktales gathered together, although there is an element of this. They are used to point out religious lessons of primary importance, and even compare them with events in the Bible to underline the fact that they are evidence of God's power. An example of this is the eighth miracle recorded by Cogitosus:

Stories of the Goddess-Saint

63 A stained-glass window in St Brigid's parish church, Kildare, Ireland, made by Patrick Pye, illustrating the story of St Brigid expelling ungrateful nuns from a dining hall. They had refused the hospitality of a kind host because they were making their journey during Lent.

Another wonderful occurrence was this: lepers asked the venerable Brigid for ale. Since she had none, seeing some water prepared for the baths and blessing it with the power of faith, she changed it into excellent ale and drew it in abundance for the thirsty men. For He who changed water into wine at Canna in Galilee also changed water into ale through the faith of this most blessed woman.

Using Cogitosus's writings, St Brigid's miracles, of which he relates thirty-two, can be divided into several categories and some examples have been given elsewhere in this work where appropriate. For a full account see *Cogitosus's Life of St Brigit* by S. Connolly & J. M. Picard. The miracles described by Cogitosus concern the following subjects:

Fire
Cows/Dairy Products/Butter
Her Virginity/Holiness
Charitable Acts
Harvest
Changes – water into ale and stone into salt
Sheep
Fertility/pregnancy
Cures
Those concerning Wild Animals/Birds
Miscellaneous Miracles
Post-mortem miracles

There are a number of pre-eleventh-century manuscript copies of St Brigid's *Life* still surviving attributed to the works of various writers, and much scholarly research has gone in to identifying both the authors of these works and their dates. Among the contenders for producing early accounts of St Brigid's life between the seventh and tenth century, according to tradition, oblique references or later surviving manuscript copies are St Fiech, St Ninnidh, St Ultan, St Aleran, St Coelan, St Kilian of Insis-Keltra and Animosus, whose Irish name was Anmchadh, a tenth-century bishop of Kildare. As time went on more 'histories' of St Brigid came to be written. By the medieval period she was credited with a great many miracles that do not appear in the earlier works, some of which were attributed to her although they originated with the acts of other saints.

Some of these later stories do not show St Brigid in quite the same all forgiving, charitable way as the earlier writers. For example one story tells how two lepers went to one of her sacred springs seeking healing. St Brigid appeared and told one man to wash the other and no sooner was this accomplished than the skin of the washed man was cleansed of disease. She then told the cured man to wash the other, but he was so disgusted by his companions condition that he refused to touch him. Brigid washed the second man herself and he was healed. However, Brigid was upset by the lack of compassion shown by the first man so she once again afflicted him with leprosy before he could leave.

Lives of St Brigid proliferated in Ireland, Britain, Europe and elsewhere. If all the different miracles now credited to her written from the seventh to the twentieth century are added up, it would come to around 200! A number of early hymns were written praising St Brigid and have the same problem of attributing authorship. Among the earliest writers of these hymns are said to be St Ultan and St Brendon, Bishop of Clonfort (both sixth century), and St Brogan Cloen of Rostuirk (seventh century). St Ultan wrote (in Latin):

> In our island of Hibernia,
> Christ was made known to man,
> By the very great miracles which he performed,
> Through the happy Virgin of Celestial Life
> Famous for her merits through the whole world.

After this hymns, songs and poems about her become numerous and are still composed today. As St Brigid's popularity spread, stories arose that have no basis in religious beliefs directly, although they concern miraculous, or perhaps more accurately, magical activities. For example, one folk tradition claims that St Brigid was the first person to practice weaving in Ireland, and into this first piece of cloth she incorporated healing threads. This in turn lead to, or was used to explain, the custom of putting out a piece of ribbon on St Brigid's Eve, which was then used to cure illnesses (see page 126). Another story claims that St Brigid wove the shroud that St Patrick's body was wrapped in after his death, something he had specifically requested.

In one story her early goodness is shown by her being joined by an angel when, as a small child, she tried to set up a little alter. Such stories became more and more fanciful, crediting her with such impossible acts as being the midwife present at the birth of Jesus, and coming up with explanations about how she got to the Holy Land, although faith in such stories glossed over such problems as the discrepancy in dates! (See pages 91 and 140.) Stories also evolved linking her directly with St Patrick. However, with events at this early period hard to date accurately, its difficult to know if St Brigid and St Patrick were contemporaries and, if so, whether St Brigid actually met him.

A story, in the ninth century *Tripartite Life of Saint Patrick*, says that St Brigid was among a congregation listening to him preach. She fell asleep, however, and had a vision in which she saw both the present state of the church in Ireland and all the problems that would confront it. St Patrick knew she had had a vision and when she awoke urged Brigid to describe what she had

Stories of the Goddess-Saint

seen. This she did, and said at first she had seen a herd of white oxen amidst white crops, then spotted ones of various colours, after which appeared black and dark-coloured oxen. These were succeeded by sheep and swine, then wolves and dogs fighting with each other. A ninth century poem entitled *Hail Brigid* is preserved in the *Book of Leinster*, and shows the saint triumphant while many ancient kings have passed away, and even the great fortress of Alenn lies in ruins:

> Hail Brigid
> Brigid, sit thou enthroned in triumph
> upon the sweet plain of Liffey,
> as far as the strand of the ebbing sea!
> Thou art sovereign queen with gathered hosts over all the sons of Cathair the Great.
>
> Though the glittering Liffey is yours today,
> this land fell to many before.
> When I gaze over the fair Curragh,
> the fate of those fine kings fills me with awe.
>
> Loegaire governed even to the sea,
> and Ailill Ane's was a mighty name.
> The light-filled Curragh still stretches away,
> but not one of those king's remains.
>
> Fine Labraid Longsech is no more,
> having trod his thirty years.
> Since the days when in Dinn Rig,
> he dealt doom to Cobthach the Slim.
>
> Lorc's grandson, Oengus of Roirin,
> seized rule over all of Erin.
> Maistiu of the freckled neck, son of Mug Airt,
> threw princes down in their graves.
>
> Far-famed Alenn! Hill of delight!
> Many a prince lies under your borne.
> Greater you grew than ever was dreamt,
> when Crimthan the Victorious held throne.
>
> Hear the victory shouts after each triumph,
> rise from a shock of swords, mettlesome mass;
> See the strength of your warrior-bands,
> in the dark-blue battle array;
>
> Hear the note of your horns above hundreds of heads.
> Hear the tuneful ring of your black bent anvils,
> the sound of songs on the tongues of bards;
> See the ardour of your men in the mighty fray;
> the beauty of your women at the banqueting.

There was drinking of mead there in every homestead;
There were noble steeds, numberless tribes;
The jingle of chains unto kings of men,
under bloody blades of five-edged spears.

Hear the sweet strains there at every hour;
See the wine-barque upon the purple flood;
The shower of silver of great splendour;
The torcs of gold from Gaulish lands.

Even to the sea of Britain,
the renown of your kings sped like a star.
Lovely Alenn, mighty and strong,
made sport of every law.

Bresal Brec, king over Erin,
Fiachra Fobrec with his fierce bands,
Fergus of the Sea, Finn son of Roth;
These all loved to dwell in high Alenn.

But your spells and auguries came to naught,
all your omens betokening death;
Seen clearly now, such things seem vain,
for Alenn is but a deserted dun.

Cathair the Great, the finest of men,
ruled over Erin of many hues;
He will not come forth from his rath again,
for all his glory has passed and gone.

Fiachna of Fomuin, glorious Bresal,
governed the sea with showers of spears;
Thirty kings ruled to the very ocean,
conquering lands around Tara of Bregia.

Handsome Feradach in his finery,
surrounded by crested bands of men;
With his blue-speckled helmet, his shining cloak,
Many a king he overthrew.

Dunlang of Fornochta, a generous prince,
he routed the sons of Niall.
Yea, though we sing these tales over and over,
It is no longer the world that was.

Enna's grandson, Illann, with his tribe,
thirty times battled against king and king;

He was a rock against fear;
Not a host rode to him but with royal banners.

Ailill was one who dealt in favours,
against him fierce and blood-dark hosts:
Cormac, Carbre, Colmán the Great, Brandub,
Ships bearing whole armies of men.

Faelan the Fair was a model of princedom,
and Fianamail with him, too;
Bran, son of Conall, with his grand deeds,
Was a wave beating on the cliff.

O Brigid, whose land I behold now,
Where all these have ruled in turn,
Thy fame has outshone them far,
Thou art indeed over them all!

Thou rulest evermore with the King,
beyond the land where your body lies.
Grandchild of Bresal, son of Dian,
Brigid, sit thou enthrowned in triumph!

In complete contrast, but following the ancient tradition of writing poems about St Brigid, is *The Givaway* from *The Love Letters of Phyllis McGinley* (Viking Press, New York, 1957). This tells something of her propensity to give items to the needy, even if they were not hers, and the problems she must have caused her father. This is inspired by stories that appear in the various *Lives* of the saint such as that preserved in *The Book of Lismore,* which tells how she even gave away her fathers sword to a leper, much to his horror as it was worth the equivalent of ten cows!

The Givaway
Saint Brigid was
A problem child.
Although a lass
Demure and mild,
And one who strove
To please her dad,
Saint Brigid drove
The family mad.
For here's the fault in Brigid lay:
She WOULD give everything away.

To any soul
Whose luck was out
She'd give her bowl
Of stirabout;
She'd give her shawl,

Brigid: Goddess, Druidess, Saint

Divide her purse
With one or all.
And what was worse,
When she ran out of things to give
She'd borrow from a relative.

Her father's gold,
Her grandsire's dinner,
She'd hand to cold and hungry sinner;
Give wine, give meat,
No matter whose;
Take from her feet
The very shoes,
And when her shoes had gone to others,
Fetch forth her sister's and her mother's.

She could not quit.
She had to share;
Gave bit by bit
The silverware,
The barnyard geese,
The parlour rug,
Her little niece's christening mug,
Even her bed to those in want,
And then the mattress of her aunt.

An easy touch
For poor and lowly,
She gave so much
And grew so holy
That when she died
Of years and fame,
The countryside
Put on her name,
And still the Isles of Erin fidget
With generous girls named Bride or Brigid.

Well, one must love her.
Nonetheless,
In thinking of her givingness,
There's no denial
She must have been
A sort of trial unto her kin.
The moral, too, seems rather quaint.
WHO had the patience of a saint,
From evidence presented here?
Saint Brigid?
Or her near and dear?

Fifteen

Venerating the Goddess-Saint

The order of Saint Brigid

St Brigid founded a double monastery (*monasteria duplicia*) at Kildare around 470 AD. Initially with members drawn from the community of *ban Druid* (Druidesses) and possibly some of the male members of the Druid Order with Brigid, their Chief Druidess, becoming their first Abbess and guiding their conversion to Christianity. They became the first Brigidine nuns and monks and lived by some form of Rule, probably not too different from the regulations and constraints they had lived by before their conversion. At this early period there is no evidence as to what guidelines they followed in their day to day lives.

As the fame of the new Christian establishment spread, aided by the proselytising work of St Brigid herself who continued the work of St Patrick in Christianising the Irish, others would have joined the monastery. They lived in a community typical of the early Celtic church, and Cogitosus, the writer of the first *Life of St Brigid* in the seventh century, claimed that her monastery at Kildare was 'the head of almost all the Irish churches with supremacy over all the monasteries of the Irish…'. While Cogitosus, almost certainly a monk of the Kildare community, may have been exaggerating, there seems little doubt that this Brigidine monastery was one of the most important, and probably the foremost Christian establishment at that time in Ireland.

Over the centuries changes occurred by both natural development and later increasingly under the influence of the Roman church. Even after the Anglo-Norman invasion of Ireland in the twelfth century the Kildare female community remained under Irish control. Later the old Celtic monastery came under the control of the Canons Regular of St Augustine, with a Dean and Chapter to direct them. The power and influence of Kildare eventually declined, and in the 1540s the Order was disbanded as part of Henry VIII's Reformation of the church in Ireland. The flame that had been maintained at Kildare from the time of St Brigid and probably from pre-Christian times, was finally extinguished, and Brigidine nuns and monks, were to disappear from ecclesiastical history for over 250 years.

However, the Rt Revd Daniel Delany, Bishop of Kildare and Leighlin, recreated the Order of St Brigid on 1 February 1807. He made it clear that he was 'restoring the ancient Order of St Brigid' not founding a new congregation, and give them the motto *Fortiter et Suavitar*

(Strength and Gentleness). This first modern Brigidine convent was founded at Tullow, Co Carlow, and initially consisted of just six nuns under the tutelage of Judith Wogan-Browne chosen by the bishop as the mentor of the new community. To emphasise the continuity between the original order and its re-foundation, Bishop Delany brought an oak sapling from Kildare which he planted in the grounds of the convent at Tullow, where it remains to this day. This was followed by the founding of a second convent in Mountrath, Co Laois, on 18 April 1809. Initially, on 1 February every year, St Brigid's Day, each nun, holding lighted tapers in their hands, made a solemn affirmation before their sisters and re-dedicated themselves to serve her and continue the work on which they were engaged in her honour. However, from 1812, this annual re-dedication was replaced by the taking of a perpetual vow on being admitted to the Order.

All the pre-reformation records of the Brigidine Order had been lost. Not even the old Rules of the Order survived and although there are ancient references to the *Regula Sanctae Brigidae* (Rules of St Brigid), no surviving source said what they were. Therefore, the new Brigidine nuns adopted a Constitution and Rule that was devised by Bishop Delany that aimed to promote charity, piety and religious education in every parish where one of their establishments was located. As an educational order they had, by the middle of the nineteenth century, established three day schools, a poor school, a benefit school and a boarding school in Ireland. The teaching of poor children and adults in sessions held on Sundays and the holy days in the parish church to which each convent belonged was especially important to the Order.

The convent at Tullow sent out two affiliations, one to Abbeyleix in Co Laois in 1842 and one to Goresbridge Co Kilkenny in 1858. Another was founded towards the end of the nineteenth century at Paulstown also in Co Kilkenny. In 1883 Brigidine nuns from the convent at Mountrath set up a convent at Coonamble, New South Wales, Australia and this was followed by the establishment of a number of other houses in Australia and New Zealand where they continued their teaching. Their work was so appreciated that the Irish convents had difficulty in supplying enough nuns to satisfy the requests made by Australian and New Zealand bishops.

The order is still a teaching one but now work in day schools instead of maintaining boarding schools, and tend to live in ordinary houses so as to be closer to the communities which they serve. The Noviciate is situated in Dublin where there are several Brigidine houses where the sisters carry out social work. There are about 800 Brigidine nuns today based in Ireland, Australia, Kenya, Mexico, New Zealand, Papua New Guinea, the United Kingdom and the USA. In 1992 two sisters of the Order of St Brigid, Mary Minehan and Philomena O'Shea opened a small centre for Celtic Spirituality at Kildare known as the *Solas Bhríde* Centre (Brigid's Light), whose aims are to reconnect with the spirit of St Brigid and her aims in a way appropriate for the modern world.

This centre is a place for prayer and exploring Celtic spirituality, revering the sacredness of all creation and celebrating the Celtic feasts, especially *Fhéile Bhríde* (St Brigid's Day). They organise pilgrimages to Kildare Cathedral, St Brigid's Well and other sites in the area, and hold services at the 'Fire House' at the cathedral (see page 63). A large part of the work of the sisters is in promoting action for justice, peace and reconciliation. There is an outreach community of women and men called *Cáirde Bhríde* (Friends of St Brigid), who have been inspired by the values that she represents to promote peace, justice and reconciliation.

In 1993 Sister Mary Teresa Cullen, the then head of the Brigidine sisters, re-lit Brigid's Fire, the perpetual flame once maintained by St Brigid (see page 78). This took place in the Market Square, Kildare at the opening of a Conference called 'Brigid: Prophetess, Earthwoman and

Venerating the Goddess–Saint

Peacemaker'. This was organised by Action From Ireland (AFRI) to celebrate the tenth anniversary of its project entitled the 'St Brigid's Peace Cross'. From that time the flame has been maintained at the *Solas Bhríde* Centre by the sisters, and flames taken from this to other centres of Brigid veneration in various parts of the world.

On 1 February 2006 the President of the Republic of Ireland, Mary McAleese, lit a flame in Kildare Market Square with fire taken from the flame maintained by the Brigidine sisters. The artist Alex Pentek was commissioned to create a sculpture to house the fire, which consists of a tall twisted bronze column surmounted by oak leaves and an acorn cup that holds the flame, symbolising the Druidism that preceded St Brigid's Christian belief, while the flame acts as a beacon of light, hope, justice and peace for the world. This is lit on 1 February each year by various invited people and remains alight for a week.

An annual event arose from the 1993 conference which starts on *La Fhéile Bhríde* and lasts for five days, organised by the Brigidine sisters, *Cáirde Bhríde* and Action From Ireland. There is a peace and justice conference, along with a pilgrimage and the celebration of the Eucharist at local churches and St Brigid's Well. However, there are many other events such as the weaving of Brigid's crosses and Irish dancing, as well as plays by local school children based on the stories about St Brigid, music, poetry and art exhibitions. Out of this conference came the idea of using Brigid's crosses as a symbol of a movement to promote justice, peace and human rights, which has now helped thousands of young people throughout Ireland. These are referred to as Brigid's Peace Cross and is made of twenty-one pieces of straw to represent peace in the twenty-first century. Today, St Brigid of Kildare retains her popularity all over western Europe, with increasing numbers of people making a pilgrimage to her cathedral.

64 A bronze sculpture by Alex Pentek incorporating the perpetual flame of Brigid. It was lit by Mary McAleese, President of the Republic of Ireland on 1 February 2006 from a flame maintained by Brigadine sisters since 1993. The flame, which is also associated with the Goddess Brighid, is housed in an acorn cup, the oak leaves symbolising both the Christian beliefs of St Brigid and her Druidic associations, as well as representing Kildare (Cill Dara), which means 'Church of the Oak'.

International veneration of the Goddess-Saint

The veneration of St Brigid in Europe was once widespread, due to many of these countries being Catholic. The spread of her fame was also due to the extensive travels of Irish monks, and Irish people settling in various parts of Europe and America. Churches, chapels and other institutions dedicated to her were or are found in Belgium, Brittany, France, Germany, Italy, Portugal and Spain. Dedications to St Brigid were widespread in the USA due to Irish migrations during the nineteenth century. There are at least sixty-eight churches dedicated to her and there were once many orphanages, schools and asylums under her patronage, while a town in Kansas is named St Brigid's. In Canada at least twelve churches dedicated to her, and other institutions there took her as their patron.

Brigid, as both goddess and saint, has always had a wide appeal that continues to the present day. With the introduction of easy world-wide communication, particularly the use of the Internet, it has been possible to unite people from many continents in prayer, meditation and even practical work inspired by or under the patronage of Brigid. Exploration of such organisations on the internet will provide details of such bodies as 'The Order of St Brigid and St Columcille' a religious order within the Independent Celtic Church International, the Canadian 'Daughters of the Flame' and others. Details of one group that follows a long standing Brigid tradition is given below.

The Ord Brighideach

The *Ord Brighideach* is an Order of the Keepers of Brigid's Flame. An international group established by Kim Diane in 1998 whose aim is to maintain a perpetual flame in honour of Brigid. The Flamekeepers are non-denominational and include both Christians and pagans, and may visualise her either as a saint or a goddess depending on their personal philosophy, but are united in being engaged in devotional work and prayer inspired by Brigid. Flamekeepers may belong to a small group or, increasingly, tend a flame as an individual. There are over 600 Flame Keepers in such countries as Brazil, Canada, England, Finland, France, Germany, Greece, the Irish Republic, Italy, Netherlands, New Zealand, Northern Ireland, Scotland, South Africa, the USA and Wales.

Each Flamekeeper is assigned a shift to tend their flame on a twenty-day cycle. There are only nineteen shifts as the twentieth is regarded as being tended by the spirit of Brigid herself, following the tradition of the ancient perpetual fire at Kildare which was tended by nineteen nuns (see page 78). The Flamekeeper lights their candle or oil lamp at sunset on the day of their shift, and tends the flame till sunset on the next day following the old Celtic method of calculating the day (see page 94). Devotions, meditation, prayer or ritual is carried out as desired.

Brighid of the witches

Witches venerate a Goddess and, to a lesser extent, her consort a male God, who goes by a variety of names, that in British Celtic being Cernunnos. The Goddess is also known by many names in various parts of the world, and these may be regarded as different aspects of the same deity, all of whom were originally derived from a Mother Goddess. Brighid, in her manifestation as Goddess is venerated by many modern witches, and has a particular appeal as she is a western goddess and is visualised as either one or as three, not an unusual attribute among Celtic deities. She represents many of the aspects of life that are important to witches.

Witchcraft is a nature based belief which celebrates the various seasons of the year, and often the witches Goddess is also visualised as being three: maiden, mother, and crone, representing

Venerating the Goddess–Saint

the passing of the seasons. The maiden represents spring, the mother summer and the crone autumn and winter. These three aspects of the Goddess are also seen as representing the path of life: childhood/youth, adulthood and old age. Brighid's three attributes embody aspects that witches, the majority of whom are female, still value today: wisdom, inspiration and poetry (spells); healing and midwifery, and fire which represents both the home (hearth) and industry, and which plays an important part in ritual.

The Friends of Bride's Mound

In 1995 the Friends of Bride's Mound was formed. Their aim is to ensure that this site in Somerset (see chapter ten), important to local archaeology, local history, traditional folklore and as a sacred site of national importance is protected, managed and interpreted to preserve it for future generations, with access available to all. The Brigid 'connection' means the site is visited by both local and international pilgrims, and the Friends have support from both the local community and people and bodies drawn from further afield. Eventually, the Friends want to establish a 'perpetual flame' at the site in honour of the goddess-saint.

As a local voluntary group they bring a high degree of understanding of the requirements of the site and a love for the land. Since their formation they have researched the history of Bride's Mound, commissioned archaeological surveys, put on exhibitions and arranged instructive walks and group gatherings there. In Spring 2005 the Friends purchased twenty-four acres of land adjoining Bride's Mound that included the western end of Beckery Island. Members of the Friends come from all parts of the world, offering support in protecting this important site. To join the Friends of Bride's Mound or obtain further details contact:

The Friends of Bride's Mound,
c/o Glastonbury Opportunities,
7 Abbey Mews,
56-58 High Street,
Glastonbury,
Somerset,
BA6 9DY.
Website: www.friendsofbridesmound.com

65 Procession from Glastonbury to Bride's Mound, Beckery, Glastonbury, Somerset. Participants carry Brigid's crosses and a ceremony to honour the Goddess Brighid is held on reaching the mound.

Further Reading

Amber, K. and Azrael Arynn, K., *Candlemas: Feast of Flames* (Living Light Essences, 2001)
Andrews, E., *Rush and Straw Crosses; Ancient Emblems of Sun Worship* (Man 22 No. 34, 1922) 49-52
Baring Gold, Sabine, *Lives of the British Saints* (A Llanerch Facsimile reprint of the 1907-13 four vol. edition, 2000) ISBN 1 86143 096 5
Beith, Mary *The Healing Threads* (Polygon, 1995)
Benham, Patrick, *The Avalonians* (Gothic Image Publications, 1993)
Benedictine Monks of St Augustine's *The Book of Saints: A dictionary of Abbey, Ramsgate, Kent. Servants of God canonised by the Catholic Church – extracted from the Roman and other martyrologies* (Adam and Charles Black, 1947, 4th edition)
Betz, Evak, *Saint Brigid and the Cows* (St Anthony Guild Press, 1964)
Berresford Ellis, Peter, *The Druids* (Constable, 1994) ISBN 0 09 474470 X
— *Celtic Women* (Robinson, 1999) ISBN 0 8028 3808 1
Bladey, Conrad, *Brigid of the Gael: a complete collection of primary resources* (Hutman Publications, 2000)
— *Brigid of the Gael, a guidebook for the study of St Brigid of Kildare* (Hutman Publications, Maryland, 2000)
Bord, J. and Bord, C., *Sacred Water, Holy Wells and Water Lore in Britain and Ireland* (Grenada Publishing, 1985)
Buchanan, R.H., 'Calendar Customs' (Ulster Folklife No. VIII 1962) pp15-34 and No. IX, 1963, pp61-79
Buck, Percy, *A Carol of Brigit* (A two-part song, words by Warner, S.T.) Dunhill Singing Class Music No. 107, 1918. Also produced as a record.
Buckton, Alice M., *The Coming of Bride: A Pageant Play.* (Elliot Stock, London, 1914)
Carey, F.P., *Faughat of Saint Brigid* (Irish Messenger Publications, 1950)
Carley, James, *An Identification of John of Glastonbury and a New dating of his Chronicle* (Medieval Studies Vol. 40, 1978, 478–83)
Carley, James (ed.), *The Chronicle of Glastonbury Abbey: John of Glastonbury's Cronica Sive Antiquitates Glastoniensis Ecclesie* (Boydell Press, 1985)

Carley, James, *Glastonbury Abbey: The Holy House at the Head of the Moors Adventurous* (Gothic Image Publications, 1996)
Carmichael, Alexander, *Carmina Gedelica Vol I* (Floris Press, Edinburgh, 1928)
— *Carmina Gedelica Vol II* (Floris Press, Edinburgh, 1928)
— *Carmina Gedelica Vol III* (Floris Press, Edinburgh, 1940)
— *Carmina Gedelica Vol IV* (Floris Press, Edinburgh, 1941)
— *Carmina Gedelica Vol V* (Floris Press, Edinburgh, 1954)
— *Carmina Gedelica Vol VI* (Floris Press, Edinburgh, 1971)
Coles, W.D., *New Light on St Brides in Argyll and the Isles* (Diocesan News, Advent, 1981)
Colgan, J., *Triadis Thaumaturgae* (E. de Witte, 1647)
Condren, M., *The Serpent and the Goddess: Women, Religion and Power in Celtic Ireland* (Harper and Row, 1998)
Connolly, S., *The Authorship and Manuscript Tradition of Vita 1 S. Brigitae* (Manuscripta No. XVI, 1972, 67–82)
— *Cogitosus's Life of St Brigit: Content and Value* (J. Royal Soc. Antiquaries of Irl. Vol. 117 (1987) 5–27)
Connolly, S. and Picard, J.M., *Cogitosus: Life of Saint Brigit* (J. Royal Soc. Antiquaries of Irl. Vol. 117, 1987, pp11-27)
Crawford, H.S., *A Descriptive List of Irish Shrines and reliquaries* (J. Royal Soc. Antiquaries of Irl. Vol. 53, (1923) 74–93 and 151–76)
Curtayne, Alice, *St Brigid of Ireland* (Browne and Nolan Ltd, 1934)
Cutting, Tracy, *Beneath the Silent Tor: The Life and Work of Alice Buckton, 1867–1944* (St Andrew's Press, 2004)
Danacher, Kevin, *The Year in Ireland: Irish Calendar Customs* (Mercier Press, 1972)
de Blacam, Hugh, *The Saints of Ireland: The Life Stories of SS Brigid and Columcille* (Bruce Milwaukee, 1942)
de Paor, Liam, *St Patrick's World, the Christian Culture of Ireland's Apostolic Age* (Blackrock, 1993)
de Paor, Maire and de Paor, Liam, *Early Christian Ireland* (Thames and Hudson, 1958)
Delaney, Frank, *The Celts* (BBC Publications, 1986)
Donnelly, Maureen, *Patrick, Brigid and Columcille* (Donard Publishing Co., 1977)
Dorman, Sean, *Brigid and the Mountain* (Fiction, Ruffeen Press, 1995)
Egan, Ann, *Brigit of Kildare* (Fiction, Kildare County Library Service, 2001) ISBN 0 520013 6 5
English, M.S., *A Hymn to Saint Brigid* (Music by M.S. English, translated into Irish by Dr D. Hyde) (Cary and Co., 1911)
Esposito, Mario, *On the Early Lives of St Brigid of Kildare* (Hermathena No.24 (1935) 120–65)
— , *On the Earliest Latin Life of St Brigid of Kildare* (RIA Proc. Vol. XXC (1912–13))
Ferrar, Janet and Ferrar, Stewart, *The Witches Goddess: The feminine principle of divinity* (Robert Hale, 1987)
Flanagan, Laurence, *Ancient Ireland: Life before the Celts* (Gill and Macmillan, 1998)
Friends of Bride's Mound, *Songs of Bride* (Produced as a CD)
Gaffney, J., *The Life of St Brigid* (Dublin, 1931)
Gailey, A. and Ó hÓgáin, D. (eds) *Gold Under the Furz: Studies in Folk Tradition presented to Coimhin O Danachar* (Glendale Press, n.d.)
Gillespie, Raymond (Ed.), *St Brigid's Cathedral, Kildare: A History* (Kildare Arch. Soc., 2001)
Graves, Robert, *The White Goddess* (Farra, Straus and Gioux, 1948. Reprinted 1970)
Green, Marian, *A Calendar of Festivals: Traditional Celebrations, Songs, seasonal Recipes and Things to Make* (Element Publishing, 1991)

— *The Gods of the Celts* (Alan Sutton, 1993) ISBN 0 86299 292 3

Gregory, Lady, *The Blessed Trinity of Ireland: Stories of St Brigid, St Columcille and St Patrick* (Colin Smyth, 1932. Reprinted 1985)

— *Lady Gregory's Complete Irish Mythology* (Two volumes, 1902 and 1904. Reprint by Chancellor Press, London, 1994)

Grian, Sinead Sula, *Brighde, Goddess of Fire* (Brighde's Fire, 1985)

Gwynn, A. and Hadcock, R.N., *Medieval Religious Houses: Ireland* (Longman Group Ltd., 1970)

Harbison, Peter, *Pre-Christian Ireland: From the first settlers to the early Celts* (Thames and Hudson, 1988)

Hartwell Jones, G. Celtic Britain and the Pilgrim Movement (London, 1912)

Hilliard, Richard, *Biddies and Strawboys* (Ulster Folklife Vol. VIII (1962) 100–02)

Hope, Robert, Charles *The Legendary Lore of the Holy Wells of England* (Elliot Stock, London, 1893)

Horstman, Carl, '*Nova Legenda Anglie': As collected by John of Tynmouth, John Copgrave and others, and first printed, with New Lives, by Wynkyn de Worde a.d. MDXVI.' Now re-edited with fresh material from ms and printed sources* (Clarendon Press, 1901)

Hyde, D., 'Hymn to St Brigid'. Music by English, M.S. (Carey and Co., 1911)

Ingram, J.A., *The Cure of Souls: A History of Saint Brigid's Church of Ireland, Stillorgan* (Dublin, 1997)

Iona, *The Story of Saint Brigid, in the form of a dialogue* (Talbert Press, 1929) John of Glastonbury, *Cronica Sive Antiquitates Glastoniensis Ecclesie – The Chronicle of Glastonbury Abbey* (Carley, J.P. (ed.) (Boydell Press, 1985)

John of Glastonbury, *Cronica Sive Antiquitates Glastoniensis Ecclesie – The Chronicle of Glastonbury Abbey*. Carley, J.P. (ed.). (Boydell Press, 1985) ISBN 0 85115 409 3

Jones, Kathy, *The Goddess in Glastonbury* (Ariadne Publications, 1990)

— *The Ancient British Goddess* (Ariadne Publications, 2001)

Jones, Francis, *The Holy Wells of Wales* (Univ. of Wales Press, 1954. Reprinted 2003)

Keating, Reg, *St Brigid's Cloak* (Tarantula Books, Dublin, 1997) (A children's book)

K, Amber and K, Arynn *Candlemas: Feast of the Flames* (Llewellyn Publications, 2001) ISBN 0 7387 007

Keating, Reg, *St Brigid's Cloak* (Tarantula Books, Dublin, 1997) ISBN 1 85534 755 5 (Children's book)

Keating, Geoffrey, *General History of Ireland c.1630* (trans. Daivid Comyn, Irish Texts Society, published in 4 Vols. 1902, 1905, 1908, 1913, reprinted 1987)

Kenny, James F., *The Sources for the Early History of Ireland – Ecclesiastical* (Pádraic Ó Táilliúir, Dublin, 1979)

Knowles, J.A., *St Brigid, Patroness of Ireland* (Brown and Nolan, Dublin, 1907)

Lawton, Liam, *Light the Fire* (Veritas, Dublin, 1996)

Leatham, Diana, *The Story of St Brigid of Ireland* (The Faith Press, 1955)

Lewis, Lionel Smithett, *Glastonbury – her saints AD 37–1539* (Mowbray and Co., 1925 (second edition 1927)

Llwyd, Edward *Archaeologia Britannica: An Account of the Languages, Histories and customs of Great Britain, from Travels Through Wales, Cornwall, Bretagne, Ireland and Scotland* (1707)

Livingstone, Sheila, *Scottish Customs* (Birlinn Ltd., Edinburgh, 1996)

Logan, Patrick, *The Holy Wells of Ireland* (Colin Smythe, Dublin, 1980)

Low, Mary, *Celtic Christianity and Nature: Early Irish and Hebridean Traditions* (Blackstaff Press, 1996)

Lucas, A.T., *The Sacred trees of Ireland* (Cork Hist. and Arch. Soc. J. Vol. XVIII (1963) 32–34)

— *The Social Role of Relics and Reliquaries in Ancient Ireland* (J. Royal Soc. Antiquaries of Irl. Vol.116 (1986) 5–37)

Mac Neill, Maire, *The Festival of Lughnasa: a study of the survival of the Celtic festival of the beginning of harvest* (Oxford University Press, 1962)

MacAlister, R.A. (ed.), *Lebor Gahbála Eireann: The Book of the Taking of Ireland. Part IV* (Irish Texts Soc.Vol XLI., 1941)

MacCana, Proinsias, *Celtic Mythology* (Hamlyn, London, 1970)

MacCloud Banks, M. (ed.), *British Calendar Customs: Scotland Vol. II: Fixed Festivals (Jan–May)* (Folklore Soc. Publication)

MacDonald, Ian, *Saint Bride* (Floris Books, Edinburgh, 1992)

MacDonald, James *Religion and Myth* (David Nutt, London, 1893)

MacMahon, B., *The Hill of the Goddess: Brigid of Cill Dara* (Millwork Press Ltd., 1999)

Mac Suibhne, Peered, *Saint Brigid and the Shrines of Kildare* (Leicester Leader Ltd., 1973)

Mackillop, James *Dictionary of Celtic Mythology* (Oxford University Press, 1998)

McCrickard, Janet, *Brighde: her folklore and mythology* (Fielder Arts & Design, 1987)

McGarry, Gina, *Brighid Healing* (Green Magic Publishers, 2005)

Malmsbury, William, *The Early History of Glastonbury* (Boydell, 1981)

Mason, T.H., *St Brigid's Crosses* (J. Royal Soc. Antiquaries of Irl.Vol. 75 (1945) 160–66)

Meehan, Gary, *Sacred Ireland – The Travellers Guide* (Gothic Image Publications, 2002)

Mulligan, Bryce, *Brigid's Cloak: An Ancient Irish Story* (Erdmans Books, 2002)

Minehan, Rita, *Rekindling the Flame: A Pilgrimage in the Footsteps of Brigid of Kildare* (Solas Bhríde Community, 1999)

Monmouth, Geoffrey, *A History of the Kings of England. c.1135* (Penguin, London, 1966)

Moorage, Dewi, *Phoenix of Fleet Street* (Charles Knight and Co., 1972)

Moorland, John, *St Bridget's Chapel, Becky* (Papers and lectures, Glastonbury Antiquarian Soc., 2nd Series (1891) 65–70)

Mould, D.D.C.P., *Saint Brigid* (Columnar and Reynolds, 1964)

Ni Dhonnchadha, Cait, *An Bhrighdeog, An Claidheamh Soluis*, 6 Aug. 1910, pp3-4

O Briain, Felim *Brigantia*. Z.Celt.Philol. 6 (1977) pp112-37

O' Bigein, M.O., *The Old Irish Life of St Brigid* (Irish Historical Studies Vol. 1 (1938) 121–34)

O' Cathain, Seamas, *Hearth Prayers and Other Traditions of Brigit: Celtic Goddess and Holy Woman* (J. Royal Soc. Antiquaries of Irl. 122 (1992) 12–34)

— *The Festival of Brigit* (DAB Publications, 1995)

O'Cathasaigh, Donal, *The Cult of Brigid: A study of Pagan-Christian Syncretism in Ireland*, in *Mother Worship-Theme & Variations*, Preston, J.J. (ed.) (Univ. of North Carolina Press, 1982)

Ó Danachair, Caoimhín, *February Festivals* (Biatas (March, 1958) 565–68)

— *St Brigid's Cross* (Ireland of the Welcomes, Vol. 2 No. 5 (1973), 15–18)

— *The Quarter Days in Irish Tradition* (Arv. Vol. 15 (1969), 47–55)

O'Hanlon, John, *Lives of the Irish Saints*, Vol.II (James Duffy and Sons, 1875)

Ó hAodha, Donncha, *The Early Lives of St Brigit* (J. Kildare Arch. Soc.Vol.15 (1971–76) 397–405)

— *Bethu Brigte* (Dublin Institute for Advanced Studies, 1978)

Ó hÓgáin, Daithi *Myth, Legend and romance: An Encyclopaedia of the Irish Folk Tradition* (Prentice Hall Press, 1991) ISBN 0 13 275959 4

— *The Sacred Isle: Belief and Religion in Pre-Christian Ireland* (Collins Press, Cork, 1999)

O'Meara, John (ed.), *Gerald of Wales: The History and Topography of Ireland* (Penguin Books, 1982)

O'Meara, John, *Giraldus Camrensis, the history and topography of Ireland* (Mountrath, 1982)
O'Rahilly, T.F., *Early Irish History and Mythology* (Dublin Institute for Advanced Studies, Dublin 1964)
O Suilleabhain, Sean and Mason, T.H. *St Brigid's Crosses* (J. Royal Soc. Antiquaries of Irl. Vol. 75 (1945) p. 164)
Ó Súilleabháin, Sean, *La Fheile Bride* (Clódhanna Teoranta, Dublin, 1977)
— *Irish Folk Custom and belief* (Mercier Press, 1967. (Second edition 1977))
O'Sullivan, John, *StBrigid's Crosses* (Folk Life, No. 11 (1973) 60–81)
Paterson, T.G.F., *Brigid's Crosses in County Armagh* (Ulster J. Arch. Vol. VIII (1945) 43–8)
Paton, C.I. (ed.), *Manx Calendar Customs* (Folklore Soc. Publication, 1939)
Pollard, Mary E., *In Search of St Brigid of Kildare* (Kildare, 1987)
Radford, C.A.R., *The Earliest Irish Churches* (Ulster J. Arch. Vol. XL (1977) 1–11)
Rahtz, P. and Hirst, S. *Becky Chapel Glastonbury, 1967–8* (Glastonbury Antiquarian Soc., 1974)
Rahtz, P., *The English Heritage Book of Glastonbury* (Batsford, 1993) ISBN 0 7134 6866 1
Robinson, J. Armitage, *William of Malmesbury: On the Antiquity of Glastonbury* (Somerset History Essays (1921) 1–25)
Robinson, J.L., *Saint Brigid and Glastonbury* (J. Royal Soc. Antiquaries of Irl. Vol. 83 (1953), 97–99)
Rodenburg, J.A., *Pilgrimage Through Ireland* (Trans. Wraxell, Sir Lascelles. London, 1860)
Ross, Anne, *The Folklore of the Scottish Highlands* (Batsford, London, 1976)
— *Druids* (Tempus, 1999)
Ryan, John, *Saint Brigid of Cill Dara* (Irish Messenger Publications, 1978)
Scott, John, *The Early History of Glastonbury: an edition, translation and study of William of Malmesbury's De Antiquitate Glastoniensis Ecclesia* (Boydell Press, 1981)
Sellner, E., *Brigid of Kildare: A Study in the Liminality of Women's Spiritual Power* (Cross Currents, Winter 1989–90)
Sharkey, P.A., *The Lily of Erin: Saint Brigid* (Printed privately, 1921)
Sharpe, Richard, 'Vitae S. Brigitae: The oldest texts'. Peritia VOL 1 (1982) pp 81-106.
Sherlock, W., *Some Account of St Brigid and the See of Kildare with its bishops, and the cathedral now restored* (Hodges Figgis, Dublin, 1896)
Smith, Jill, *Mother of the Isles* (A Meyn Mamvro Publications, 2003)
Stella, Maris Sister, *Frost for Saint Brigid and Other* (Sheed and Ward Inc., 1949)
Stokes, Whitely (ed.) The Rennes Dindshenchas Rev. Celt. Vol. XVI (1895)
— *Cormac mac Cuilleanáin's Glossery* (publisher unknown, 1868)
Sucking, N., *She: The Book of the Goddess* (Lakeside Publishers, 1998)
Thomas, Charles, *The Early Christian Archaeology of North Britain* (Oxford University Press, 1971) ISBN 0 19 214102 3
Thorpe, Lewis (ed.) *Gregory of Tours: History of the Franks II, 1431* (Harmondsworth, 1974)
Wakeman, W.F., *On Certain Wells Situate in the North West of Ireland etc* (J. Royal Soc. Antiquaries of Irl. Vol. 15 (1879–82) 365–84)
Walters, Kathie, *Celtic Flames: Patrick, Berndon, Brigid, Cuthbert, Columba, Kieran and Comquall* (Good News Ministries, 1999)
Watson, J.W., *The History of the Celtic Place Names of Scotland* (Birlinn, 2004)
Webster, Graham, *The British Celts and their Gods Under Rome* (Batsford, 1988)
Wilson, Stephen (ed.), *Saints and Their Cults: Studies in Religious Sociology, Folklore and History* (Cambridge University Press, 1993)
Zaczek, Ian, *Ireland, Land of the Celts* (Collins and Brown, 2000)

Index

Abernethy 225
Ale 126, 155, 129, 195, 231, 233
Alter 55, 154, 156, 225
Angels 34, 58, 151, 201, 234
Anglesey 11, 16, 13, 21, 48, 152-3
Animosus 49, 119, 126, 234
Anu (Danu) 22, 25, 229
Ashes 79, 86, 76, 134-5
Athena 18, 64

Badhbh 25, 26
Baptism 17, 33, 37, 38, 40, 125, 179, 188
Bards 12, 23, 27, 37, 38, 228
Bath, Somerset 18
Battle relics 97, 224
Bay Laurel 213
Beckery, Somerset 35, 120, 159, 164-80, 227
Bed, Brigid's 100, 130, 133, 134, 158, 162
Bees 167, 211
Bell 174-7, 222, 223, 226
Beltane 79, 86, 157, 159, 205
Beorhtwald, Abbot of Glastonbury Abbey 166, 169
Betony 214
Biddie Boys 110
Biddy Bed 157-9
Biddy's Eyes 158, 159
Birds 210-11
Birgitta, saint 73
Birthplace, of Brigid 69-71, 136

Blessing of Brigid 87, 111, 119, 123, 125, 128, 139, 142, 155, 158, 172, 196, 204, 208
Boann, goddess 205
Bondmaid, status 68
Breas, King 229, 230
Breeshey, saint 162-4
Bregans, god 16
Brendon, saint 215, 234
Bride's Island (Eilean Bhride) 132
Bride's Mound, Glastonbury 159, 169, 243
Bride's Stone, Glastonbury 177-8
Bride's Well, Glastonbury 177-80
Bridewell Prison 186
Bridget of Sweden, saint 66, 73-4
Briga, saint 72
Brigidine monks 57, 58, 154, 239
Brigidine nuns 56, 76, 166, 170, 174, 175, 189, 222, 239, 241
Brigandu 9, 18, 64
Brigantes 9-20, 22, 64, 230
Brigantia, goddess 7, 9-20, 22, 23, 64, 159
Brighid, goddess, her family 21, 25, 26, 229, 230, 231
Brigid as Abbess of Kildare 36, 40, 41, 75, 87, 173, 174, 188, 239
Brigid as Bishop 57, 81, 156, 176
Brigid of Cille Muine, saint 74, 150-1, 154
Brigid, date of birth 71
Brigid, date of death 71
Brigid, her family 42, 68, 71, 173, 174

Brigid as Foster-mother to Christ 210
Brigid as Midwife to the Virgin Mary 88, 89, 137, 138
Brigid, name variations 63-4
Brigid and the Natural World 204-16
Brigid of Opaco, saint 72
Brigid, place of birth 69
Brigid of Picardy, saint 72
Brigindo 9
Brignat, saint 150-1, 153, 180
Britta, saint 72
Broicsech, St Brigid's mother 37, 38, 42, 68, 81, 83
Bru na Boinne 25, 28, 193
Boar 50, 208, 229, 230
Book of Kells 47, 53, 58
Book of Kildare 58
Boudicca, queen 10, 14
Buckton, Alice 176-7, 180
Burial Place, of Brigid 51, 218-21

Cailleach/Cailteach 85, 136, 229
Cake, Bride's 129
Cake, wedding 131
Camma, Druidess 46
Candlemas 91, 94, 105, 132, 133, 137, 154, 155, 163, 207, 216
Candlemas story 83
Cannons Regular of St Augustine 239
Caractacus, British freedom fighter 13-14
Carmichael, Alexander 132, 135, 136
Carmina Gadelica 132
Carrots 214
Cartimandua, Queen 13-15
Cathedral, St Brigid's, Kildare 47, 50-61, 119, 218, 239
Cathedral, Downpatrick 218-221
Cathedral, St Sauveur, Bruges 35
Cattle 25, 27, 40, 204-8, 229, 230
Cattle customs 207-8
Cattle Raid of Cooley 35, 48, 225
Celtic day 92
Cenwealh, King of the West Saxons 166, 167, 169
Chalice Well, Glastonbury 159-160, 176
Changlings 124, 137
Charcoal blessings 125

Charm for the Evil Eye 144
Charm for the Eye 144, 145
Charm of Protection 148
Charm of Encompassing 140
Charm of the Sprain 144
Charm of the Thread 146
Charm of Transmutation 147
Charm of Tying up 208
Chickens 135
Church: St Bride's, London 127, 131
Church: St Bride's, Bridgwater, Somerset 166, 181
Church: St. Bridget's, Brean, Somerset 166, 181
Church: St Brigid's, Chelvey, Somerset 166, 181
Churching of women 90-1
Churning Invocation 143
Cloak, Brigid's 56, 102, 122-4, 224-5
Coal 82, 188, 207
Cogitosus 42, 50, 59, 68, 80, 232, 239
Columba (Columcille), saint 34, 48, 77, 93, 137, 193, 218, 219, 220
Conlaed, Archbishop 51, 57, 58, 218
Cormac mac Cuilleanain, Bishop 30, 64, 75, 92, 165
Cross, St Ffraids 152, 153
Crosses, Brigid's:
    As relics 97
    Brigid's Eve supper 101, 102, 130
    Bringing in the material 101-2, 129
    Blessing 102, 103, 104-5, 130
    Collecting the material 101, 114
    Cross, Shield & Veil Custom 108-9
    Designs 95, 100, 107, 172
    Fixing 99, 105
    How to make 107-9
    Making them 103-4, 123, 130
    Reasons for making 95, 96, 98, 99, 100
    Replacing 99, 103, 105, 107
    Stories to account for them 96-7
    Used for other purposes 99-104, 106-7
Cullen, Mary Teresa. OSB 240
Curragh, of Kildare 44, 235

Daghdha, goddess Brighid's father 21, 23, 25, 26, 67, 193, 228, 229, 230

Dandelion 134
Darerca, saint 150-1, 180
Darlughdacha, second Abbess of Kildare 41
Delany, Daniel. Bishop of Kildare & Leighlin 239
Dew on Brigid's Day 160
Dian Cecht, Irish god of medicine 23, 231
Douglas family, of Scotland 148
Downpatrick 93, 165, 218
Dragons 215
Druids 10-22, 24, 28, 32, 33, 34, 36, 38, 39, 42, 81, 83, 89, 92, 97, 134, 147, 196, 205, 228, 231, 239
Druidesses 19, 23, 29, 34, 36, 37, 46, 48, 62, 64, 76, 205, 237
Dubhthach, St Brigid's father 37, 42, 68, 70, 71, 96, 145, 155, 195
Ducks 210
Dunstan, saint 166

Eithne ingen Ui Suairt, Abbess of Kildare 174
Eponina, Druidess 46

Fairies 26, 64, 124, 137, 140
Fairs, Brigid's 132, 152, 162
Falcon 211
Fertility 83-91, 99, 111, 123, 141
Ffraid, saint 13, 131, 150-7
Fidelma, Druidess 35, 48
Filidh 24, 27
Fir Bolg 229
Fire House, Kildare 49, 53, 61-2
Fire Temple, location 62
Fish 136, 152, 172, 179, 193-5, 212
Flidais, goddess 205
Folklore Commission 95, 100
Food, for Brigid's Day 84, 88, 101, 113, 126-31, 133, 163, 204
Fomhoire (Formorians) 26, 229
Fothairt sept 42, 68, 71, 173, 174
Foughart, Co Louth 70, 205
Foxes 209
Frost gathering 126

Genealogy, of St Brigid 68, 139
Gerald of Wales 43, 45, 58, 59, 76, 152, 219

Ghost, of St Brigid 219
Gildas, saint 222
Girdle, Brigid's 117-22
Glastonbury Abbey 159, 165, 167, 168, 169, 173, 183, 184
Goddess Temple, Glastonbury 160, 181
Goibhniu, god of smithing 23, 205, 230, 231
Gwynn ap Nudd, god 20

Head, cult of 19
Herding Blessings 143
Henry of Dublin, Bishop 76
Holy Island 152, 153
Hymns & poems 82, 143, 146, 234, 235, 238

Image of Brigid (Bhrideog, dealbh Bhride) 102, 109-17, 128, 133, 134
Invasion of Ireland, by the Anglo-Normans 219, 239
Imbolc 28, 79, 83, 85, 87, 92, 95, 126, 157, 159-60, 189, 216
Irish Pilgrims Route 165-6
Island, Brides (Eilean Bhride) 132
Isle of Man 153, 158, 161-3

John of Glastonbury 165, 166, 167, 171, 173, 174, 175, 180

Keening 230, 231

Lambay Island, Co Dublin 14, 15-16
Lepers 206, 232, 233, 234, 237
Linnet 210
Love potion 41
Lugh 23, 41, 42
Luigne, territory of 70
Lughnasadh 79, 99

Macha 25
Magic 29, 32, 33
Mary Magdalene, Chapel of 165, 173
Malachy III, Bishop of Down 93, 219
Mathghean, Druid 37
May Day 157, 159, 186
Marriage 105, 125
McAleese, Mary. President of the Republic of Ireland 241

Mel, Bishop 52, 57, 81, 82, 207
Medb, Queen of Connacht 48
Milk, baptising in 205
Milk, washing in 90
Milking Croons 142
Milkmaids 204, 208
Minehan, Mary. O.S.B. 240
Minerva, goddess 16, 17, 18-20, 64, 159
Miracles 32, 33, 41, 83, 126, 228, 231-7
Miracle stories
    Brigid and the bees 211
    Brigid and the burning coals 82
    Brigid and the watercress 214
    Brigid and the wild boar 50, 208
    Brigid and the wild ducks 210
    Brigid's girdle 119
    Brigid plucks out her eye 145, 189, 199

    Changing bark into lard 126
    Changing nettles into butter 126
    Changing water into ale 126, 195, 233
    The best cow and the best calf 206
    The cattle stolen and returned 206
    The churned butter given to the poor 146
    The cow and the calf 206
    The cow that was milked three times a day 151, 206
    The deceit About a cow 207
    The door that grew 53
    The expanding cloak 56
    The fiery column 81
    The garments hung on a sunbeam 80
    The greenwood altar 225
    The healing well 194-5
    The loom that was burnt and restored 80
    The mill stone 55
    The phantom fire 81
    The pigs driven by wolves 209
    The tamed and untamed fox 209
    The tub of water that was not broken 191
Mor-Rioghain, Goddess Brighid's mother 21, 23, 25, 26, 33, 40, 67, 205, 231
Morrigan 25, 67
Morrigu 25
Mountrath, Co Laois 240

Names of the goddess-saint, variations of 63-4
National Pilgrimage, Ireland 7, 157, 189, 222
Neamhain 25
Nechtan, King of the Picts 166, 225
New Grange 25, 28
Ninnidh, saint 71
Nuada, king 229

O'Shea, Philomena. O.S.B. 240
Oengus mac Oc, son of the Daghdha 229
Ogham script 28
Oimelc 92
Ord Brighideach 242
Order of St. Brigid 239-40
Otherworld, the 11, 16, 20, 25, 187, 205, 231
Oyster Catcher 210

Palladius, missionary 30, 217
Patrick, saint 36, 41, 48, 93, 97, 125, 132, 137, 165, 177, 189, 193, 205, 218, 220, 219, 221, 227, 234, 239
Perpetual Fire 32, 43, 45, 61, 76, 240, 241
Pilgrims route 181
Pilgrims, safeguarding 141
Plants 212-14
Purification of the Virgin Mary 83, 90-1, 216

Reaping Blessing 141
Relics 35, 174-7, 217-27
Ribbon, Brigid's 124-5
Rocks 206, 215-16, 223
Rod, of Bride 134
Ruadhan, son of Brighid 230
Rule, of St. Brigid 58, 240
Rune of the Well 140

Samhain 79, 85, 126
Seasons and Brigid 85, 136, 139
Seaweed 143
Serpents 135, 136
Servants, hiring of 126
Shellfish 212
Shoes 189, 190, 216, 222
Shroud 222, 234
Sleep Consecration 137
Smooring the fire 77

Smooring Blessings 78-9
Snowdrops 134, 216
Solas Bhride Centre 240, 241
Somerset, Brigid in 157-61, 164-86
Sparling 151, 152
Spring, the start of 7, 85, 86, 92, 210
Star, of little Bride (rionnag Bhride) 114, 133
Sulis, goddess 18, 20
Summer, St. Brigid's 160
Swans 151, 180

Temple of Brighid, Glastonbury 160, 181-2
Titles, of St Brigid 65
Tomb, of St Brigid 51, 218-21
Torranan 212
Translation, of St Brigid's body 93, 219
Transmutation Charm 147
Trees, sacred 32, 44, 48, 49, 55, 192, 214
Tuatha de Danann 229, 230
Tullow, Co Carlow 240

Umeras, Co Kildare 70

Vates (Velitas) 12, 23, 37, 38
Veleda, Druidess 48
Vikings 217, 218

Virgin Mary 83, 88, 91, 121, 137, 138, 159, 163, 168 216

Warping Chant 144
Water Cress 214
Weather sayings 85, 139, 161, 163
Weaving 144, 234
Wells and Springs of St Brigid
  in England 168, 177, 181, 203
  in Ireland 188-201
  in Scotland 201
  in Wales 203
  Animals and wells 200
  Appearing 188
  Church wells 191-2
  Fish in wells 172, 193-5
  Healing wells 189-91, 199
  Offerings at wells 192-3, 196, 197, 198, 199, 200-1, 202
  Pilgrimages to wells 191, 195-7
  Stones at wells 192-3, 196
  Trees at wells 177, 192, 196
White Cow, Brigid's 109, 128, 205, 207, 231
William of Malmesbury 164, 167, 171, 172, 173, 180
Witches 90, 142

Visit our website and discover thousands of other History Press books.
www.thehistorypress.co.uk